Clinical Toxicological Analysis

Edited by
Wolf-Rüdiger Külpmann

Related Titles

Maurer, H. H., Pfleger, K., Weber, A. A.

Mass Spectral and GC Data of Drugs, Poisons, Pesticides, Pollutants and Their Metabolites

2007
ISBN: 978-3-527-31538-3

Bertholf, R., Winecker, R. (eds.)

Chromatographic Methods in Clinical Chemistry and Toxicology

2007
ISBN: 978-0-470-02309-9

Rösner, P., Junge, T., Westphal, F., Fritschi, G.

Mass Spectra of Designer Drugs

Including Drugs, Chemical Warfare Agents, and Precursors

2007
ISBN: 978-3-527-30798-2

Crocker, J., Burnett, D. (eds.)

The Science of Laboratory Diagnosis

2005
ISBN: 978-0-470-85912-4

Hodgson, E. (ed.)

A Textbook of Modern Toxicology

2004
ISBN: 978-0-471-26508-5

Clinical Toxicological Analysis

Procedures, Results, Interpretation
Volume 2

Edited by
Wolf-Rüdiger Külpmann

WILEY-VCH Verlag GmbH & Co. KGaA

The Editor

Prof. Dr. Wolf-Rüdiger Külpmann
Medizinische Hochschule Hannover
Klinische Chemie
Carl-Neuberg-Str. 1
30625 Hannover

Library of Congress Card No.: applied for

British Library Cataloguing-in-Publication Data
A catalogue record for this book is available from the British Library.

Bibliographic information published by the Deutsche Nationalbibliothek
The Deutsche Nationalbibliothek lists this publication in the Deutsche Nationalbibliografie; detailed bibliographic data are available on the Internet at http://dnb.d-nb.de.

Typesetting Thomson Digital, Noida, India
Printing betz-druck GmbH, Darmstadt
Binding Litges & Dopf GmbH, Heppenheim
Cover Design Adam-Design, Bernd Adam, Weinheim

Printed in the Federal Republic of Germany
Printed on acid-free paper

ISBN: 978-3-527-31890-2

Foreword

It is with great pleasure that I write these introductory sentences for the new *Handbook of Clinical Toxicological Analysis*. This book follows up on a German language textbook on the same topic and largely written by the same authors. The very good reception of the book in German-speaking countries stimulated the editor and the publisher to launch an updated and considerably enlarged English edition to make it more easily accessible to the international scientific community, rather than a second German version. The German Society of Clinical Chemistry and Laboratory Medicine (DGKL) has supported this very important comprehensive collection of analytical procedures and their use in medicine from the beginning. In fact, many of the authors have been or are members of the "Clinical Toxicological Analysis" working group of the DGKL. Thus, as current president of the DGKL, it is with pride to see how well the working group and its main project thrived over the years under the continuous guidance of W.R. Külpmann. He succeeded to gather a group of highly competent experts including several from the Society of Toxicological and Forensic Chemistry (GTFCh), who completed this English textbook, which will aid in the dissemination, the understanding, and the future standardization of the complex methodology needed for toxicological analyses. Another important aspect of this book is the description of the clinical relevance of toxicological findings, which makes it also an interesting reference for health care professionals beyond the analytical laboratory. In summary, I am confident that the up-to-date information included in this textbook will be of great value for the correct and timely analysis of intoxications and their treatment.

Karl J. Lackner, MD
President of the German Society of Clinical Chemistry
and Laboratory Medicine

Clinical Toxicological Analysis: Procedures, Results, Interpretation. Edited by Wolf-Rüdiger Külpmann

ISBN: 978-3-527-31890-2

Contents

Foreword *V*
Preface *XVII*
List of Contributors *XIX*
Disclaimer *XXIII*

Volume 1

1 Introduction *1*

2 Requirements for Toxicological Analyses *5*

3 Materials for Investigation *11*

4 Methods for Clinical Toxicological Analysis *25*

5 Practicability of Clinical Toxicological Analyses *59*

6 Quality Assurance *63*

7 Assessment of Analytical Results *77*

8 The Analytical Toxicological Report *81*

9 Medical Interpretation *89*

10 Forensic Aspects *93*

11 Strategy of Clinical Toxicological Investigations *95*

12 Screening Procedures for "General Unknown" Analysis *107*

13 Nonopioid Analgesics and Antirheumatics *189*

Clinical Toxicological Analysis: Procedures, Results, Interpretation. Edited by Wolf-Rüdiger Külpmann

ISBN: 978-3-527-31890-2

14 **Analgesics: Opiates and Opioids** *215*

15 **Antidysrhythmic Agents** *271*

16 **Anticonvulsants** *287*

17 **Anticoagulants** *301*

18 **Bronchodilators** *313*

19 **Calcium Channel Blockers** *317*

20 **Cardiac Glycosides** *327*

21 **Hypnotics: Barbiturates** *339*

22 **Hypnotics and Sedatives: Benzodiazepines** *351*

23 **Hypnotics and Sedatives (Except for Barbiturates and Benzodiazepines)** *367*

24 **Neuroleptic Drugs and Antidepressants** *393*

Volume 2

25 **β-Receptor Blocking Drugs** *455*
L. von Meyer and W.R. Külpmann
25.1 Immunoassay *455*
25.2 High-Performance Liquid Chromatography *455*
25.3 Gas Chromatography *457*
25.4 Medical Assessment and Clinical Interpretation *457*
References *462*

26 **Drugs of Abuse** *463*
H. Käferstein
26.1 Amphetamines *463*
G. Sticht and H. Käferstein
26.1.1 Immunoassay *466*
26.1.2 High-Performance Liquid Chromatography *467*
26.1.3 Gas Chromatography – Mass Spectrometry *467*
26.1.4 Medical Assessment and Clinical Interpretation *469*
26.2 Cannabinoids *470*
L. von Meyer and H. Käferstein
26.2.1 Immunoassay *470*
26.2.2 Thin-Layer Chromatography *471*

26.2.3 High-Performance Liquid Chromatography *475*
26.2.4 Gas Chromatography – Mass Spectrometry *475*
J. Hallbach and L. von Meyer
26.2.5 Medical Assessment and Clinical Interpretation *478*
L. von Meyer, J. Hallbach, and H. König
26.2.6 Addendum *479*
26.2.6.1 Spice *479*
26.3 Cocaine *480*
H. Käferstein and G. Sticht
26.3.1 Immunoassay *481*
26.3.2 High-Performance Liquid Chromatography *482*
26.3.3 Gas Chromatography–Mass Spectrometry *483*
26.3.4 Medical Assessment and Clinical Interpretation *486*
26.4 γ-Hydroxybutyrate *488*
C. Merckel, V. Auwärter, D. Simmert, and F. Pragst
26.4.1 Immunoassay *488*
26.4.2 Gas Chromatography – Mass Spectrometry *489*
26.4.3 Medical Assessment and Clinical Interpretation *491*
26.5 Ketamine *492*
H. König
26.5.1 Immunoassay *492*
26.5.2 High-Performance Liquid Chromatography *492*
26.5.3 Gas Chromatography – Mass Spectrometry *492*
26.5.4 Medical Assessment and Clinical Interpretation *493*
26.6 Lysergic Acid Diethylamide *494*
L. von Meyer and W.R. Külpmann
26.6.1 Immunoassay *494*
26.6.2 High-Performance Liquid Chromatography *495*
26.6.3 Gas Chromatography *498*
26.6.4 Medical Assessment and Clinical Interpretation *498*
26.7 Mescaline *498*
W.R. Külpmann
26.8 Phencyclidine *499*
H. Käferstein and G. Sticht
26.8.1 Immunoassay *499*
26.8.2 High-Performance Liquid Chromatography *500*
26.8.3 Gas Chromatography–Mass Spectrometry *500*
26.8.4 Medical Assessment and Clinical Interpretation *502*
26.9 Psilocybin/Psilocin *503*
G. Sticht and H. Käferstein
26.9.1 Immunoassay *503*
26.9.2 High-Performance Liquid Chromatography *503*
26.9.3 Gas Chromatography–Mass Spectrometry *504*
26.9.4 Medical Assessment and Clinical Interpretation *506*
References *507*

27 **Solvents and Inhalants** *511*
27.1 Highly Volatile Alcohols and Ketones *511*
27.1.1 Ethanol *511*
H.J. Gibitz
27.1.1.1 Enzymatic Determination *512*
27.1.1.2 Headspace Gas Chromatography *514*
27.1.1.3 Medical Assessment and Clinical Interpretation *514*
27.1.2 Other Highly Volatile Alcohols and Ketones *517*
F. Degel and H. Desel
27.1.2.1 Introduction *517*
27.1.2.2 Headspace Gas Chromatography *517*
27.1.2.3 Medical Assessment and Clinical Interpretation *520*
27.2 Aromatics (BTEX) *523*
F. Degel
27.2.1 Gas Chromatography *523*
27.2.2 Medical Assessment and Clinical Interpretation *527*
27.3 Glycols *531*
F. Degel and H. Desel
27.3.1 Gas Chromatography *532*
27.3.2 Medical Assessment and Clinical Interpretation *535*
27.4 Volatile Halogenated Hydrocarbons *539*
F. Degel, M. Geldmacher-von Mallinckrodt, and C. Köppel
27.4.1 Color Test *540*
27.4.2 Gas Chromatography *544*
27.4.3 Medical Assessment and Clinical Interpretation *548*
27.5 Inhalant Abuse *553*
H.J. Gibitz
27.5.1 Gas Chromatography *555*
27.5.2 Medical Assessment and Clinical Interpretation *555*
References *555*

28 **Pesticides** *559*
28.1 Introduction *559*
M. Geldmacher-von Mallinckrodt
28.1.1 Definition *559*
28.1.2 Classification *559*
28.1.3 Toxicity in Man *561*
28.1.4 Guidelines and Exposure Limits *561*
28.1.5 Frequency of Poisoning from Pesticides *563*
28.2 Carbamates *564*
M. Geldmacher-von Mallinckrodt and F. Degel
28.2.1 Screening Procedure (General Unknown) *565*
28.2.2 Thin-Layer Chromatography (Procedure 1) *565*
28.2.3 Thin-Layer Chromatography (Procedure 2) *569*

28.2.4 Medical Assessment and Clinical Interpretation *573*
28.3 Chlorinated Hydrocarbons *576*
M. Geldmacher-von Mallinckrodt
28.3.1 Introduction *576*
28.3.2 International Agreements on the Use of Chlorinated Hydrocarbons *577*
28.3.3 Screening Procedure (General Unknown) *579*
28.3.4 Thin-Layer Chromatography *579*
28.3.5 Medical Assessment and Clinical Interpretation *580*
28.4 Paraquat *582*
T. Daldrup and C. Köppel
28.4.1 Screening Procedure (General Unknown) *583*
28.4.2 Color Test *583*
28.4.3 Spectrophotometry *585*
28.4.4 Medical Assessment and Clinical Interpretation *590*
28.5 Organophosphorus Compounds *591*
M. Geldmacher-von Mallinckrodt
28.5.1 Screening Procedure (General Unknown) *593*
28.5.2 Thin-Layer Chromatography *593*
28.5.3 Spectrophotometry *595*
28.5.4 Medical Assessment and Clinical Interpretation *595*
28.6 Pyrethroids *599*
M. Geldmacher-von Mallinckrodt
28.6.1 Introduction *599*
28.6.2 Gas Chromatography – Mass Spectrometry *599*
28.6.3 Medical Assessment and Clinical Interpretation *601*
References *604*

29 Antidiabetics: Proinsulin, Insulin, C-Peptide, and Oral Antidiabetics *613*
29.1 Insulin, Proinsulin, and C-Peptide *613*
J. Hallbach
29.2 Oral Antidiabetics: Sulfonylureas *615*
K. Rentsch
29.3 Medical Assessment and Clinical Interpretation *619*
J. Hallbach and K. Rentsch
References *621*

30 Dyshemoglobins *623*
H.J. Gibitz
30.1 Carboxyhemoglobin *623*
30.1.1 Introduction *623*
30.1.2 Oximetry *624*
30.1.3 Spectrophotometry *625*

30.1.4 Medical Assessment and Clinical Interpretation *626*
30.2 Methemoglobin *628*
30.2.1 Oximetry *628*
30.2.2 Spectrophotometry *628*
30.2.3 Medical Assessment and Clinical Interpretation *630*
30.3 Sulfhemoglobin *632*
30.3.1 Introduction *632*
30.3.2 Spectrophotometry *632*
References *633*

31 Various Drugs and Toxic Agents *635*
31.1 Chloroquine *633*
M. Geldmacher-von Mallinckrodt and H. Käferstein
31.1.1 Chromatographic Procedures: HPLC and GC *635*
31.1.2 Thin-Layer Chromatography *635*
31.1.3 Medical Assessment and Clinical Interpretation *638*
31.2 Nicotine *639*
H. König
31.2.1 Immunoassay *640*
31.2.2 High-Performance Liquid Chromatography *640*
31.2.3 Gas Chromatography – Mass Spectrometry *640*
31.2.4 Medical Assessment and Clinical Interpretation *641*
31.3 Strychnine *642*
M. Geldmacher-von Mallinckrodt and H. Käferstein
31.3.1 Chromatographic Procedures: HPLC and GC *643*
31.3.2 Thin-Layer Chromatography *643*
31.3.3 Medical Assessment and Clinical Interpretation *645*
31.4 Cyanide *646*
31.4.1 Introduction *646*
M. Geldmacher-von Mallinckrodt
31.4.2 Indicator Tube Method *647*
M. von Clarmann and M. Geldmacher-von Mallinckrodt
31.4.3 Paper Strip *650*
A. Scholer
31.4.4 Absorption Spectrophotometry *652*
T. Daldrup
31.4.5 Potentiometry *656*
F. Pragst and M. Geldmacher-von Mallinckrodt
31.4.6 Medical Assessment and Clinical Interpretation *659*
M. Geldmacher-von Mallinckrodt
31.5 Fluoride *661*
F. Pragst
31.5.1 Outline *661*
31.5.2 Potentiometry *661*
31.5.3 Medical Assessment and Clinical Interpretation *663*

31.6 Thallium 664
F. Degel and H. J. Gibitz
31.6.1 Atomic Absorption Spectrometry 664
F. Degel
31.6.2 Inverse Voltammetry 665
F. Degel
31.6.3 Absorption Spectrophotometry 668
R. Aderjan, T. Daldrup and H. J. Gibitz
31.6.4 Medical Assessment and Clinical Interpretation 671
F. Degel and H.J. Gibitz
References 673

32 Chemical Warfare Agents 679
M. Koller and L. Szinicz
32.1 Overview 679
32.2 Nerve Agents 682
32.2.1 Free Nerve Agents 684
32.2.2 Nerve Agent Metabolites 686
32.2.2.1 Phosphonic Acid Esters 690
32.2.2.2 Other Nerve Agent Metabolites 699
32.3 Vesicants/Blister Agents 703
32.3.1 Sulfur Mustard 704
32.3.1.1 Sulfur Mustard in Blood 704
32.3.1.2 Sulfur Mustard in Urine 705
32.3.2 Sulfur Mustard Metabolites 706
32.3.2.1 Thiodiglycol (TDG) 707
32.3.2.2 Protein–Sulfur Mustard Adducts 718
32.3.2.3 Sulfur Mustard–DNA Adducts 722
32.3.3 Nitrogen Mustard 723
32.3.4 Nitrogen Mustard Metabolites 723
32.3.4.1 Nitrogen Mustard Hydrolysis Products 723
32.3.4.2 Nitrogen Mustard–DNA Adducts 725
32.3.5 Lewisite (2-Chlorovinylarsine Dichloride) 726
32.3.6 Lewisite Metabolites 727
32.3.6.1 2-Chlorovinylarsonous Acid 727
32.3.6.2 2-Chlorovinylarsonous Acid 728
32.3.6.3 Lewisite–Hemoglobin Adduct 729
32.4 Lung-Damaging Agents (Choking Agents) 729
32.4.1 Phosgene 730
32.4.2 Phosgene Metabolites 730
32.4.2.1 Phosgene–Globin Adducts 730
32.4.2.2 Phosgene–Albumin Adduct 731
32.4.3 Chloropicrin 732
32.5 Blood Agents 732
32.6 Psychotomimetic Agents 732

32.7 Eye Irritants *733*
32.8 Nose and Throat Irritants *734*
References *735*

33 Biochemical Investigations in Toxicology *745*
W.R. Külpmann
33.1 Basic Biochemical Investigations *745*
33.1.1 Introduction *745*
33.1.2 Relevant Biochemical Investigations *745*
33.2 Cholinesterase *755*
P. Eyer and F. Worek
33.2.1 Introduction *755*
33.2.2 Determination of Cholinesterase Activity *757*
33.2.3 Determination of Cholinesterase Activity in the Presence of Reversible Inhibitors *761*
33.2.4 Application of the Modified Ellman Method *766*
33.2.4.1 Exposure Monitoring *766*
33.2.4.2 Monitoring of Intoxicated Patients and Assessment of the Efficacy of Reactivators *767*
References *769*

34 Therapeutic Drug Monitoring *775*
W.R. Külpmann
34.1 Introduction *775*
34.2 Pharmacokinetics *776*
34.3 Interpretation *777*
34.4 Analytical Methods for Drug Monitoring *780*
References *783*

35 Poisonous Plants *785*
M. Geldmacher-von Mallinckrodt and L. von Meyer
35.1 General Aspects *785*
35.1.1 Frequency of Poisoning *785*
35.1.2 Symptoms from Poisoning by Plants *788*
35.1.3 Traditional Chinese Medicine *799*
35.1.4 Special Herbal Preparations *799*
35.1.5 Poisoning of Animals *800*
35.2 Suspected Plant Poisoning: First Diagnostic Steps *800*
35.3 Aids for Identification of Plants *801*
35.3.1 Handbooks *801*
35.3.2 Electronic Media *802*
35.4 Identification of Plants *802*
35.4.1 Macroscopic Identification *802*
35.4.2 Microscopic Identification *803*
35.5 Toxic Plants' Substances *803*

35.5.1 Toxic Compounds *803*
35.5.2 Methods for Detection *805*
35.6 Investigation of Biologic Materials *805*
References *806*

36 Poisonous Mushrooms *809*
F. Degel
36.1 Introduction *809*
36.2 Classification of Poisonous Mushrooms *810*
36.3 Frequency of Poisoning *810*
36.4 Identification of Mushrooms *815*
36.5 Detection of Mushroom Toxins *816*
36.5.1 Amatoxins *816*
36.5.1.1 Screening Procedure *816*
36.5.1.2 Immunoassay *817*
36.5.1.3 High-Performance Liquid Chromatography *819*
36.5.1.4 Liquid Chromatography–Mass Spectrometry *819*
36.5.2 Toxins from *Russula* Species *819*
36.6 Medical Assessment and Clinical Interpretation *819*
References *822*

37 Venomous and Poisonous Animals *825*
D. Mebs
37.1 Introduction *825*
37.2 Venomous and Poisonous Animals *826*
37.2.1 Marine Animals *826*
37.2.2 Terrestrial Animals *831*
37.3 Conclusion *834*
References *834*

Appendix A Abbreviations *835*

Appendix B Therapeutic and Toxic Concentrations of Drugs and Xenobiotics in Plasma or Serum *841*

Appendix C Biological Tolerance Values at the Workplace (BAT Values) *853*

Appendix D Antidotes *857*

Appendix E Poison Information Centers *861*

Appendix F List of Narcotic Drugs According to German Law *867*

Index *877*

Preface

This book presents a rather comprehensive collection of procedures for the detection and determination of drugs, such as analgesics, anticoagulants, antidiabetics, antidysrhythmics, cardiac glycosides, drugs of abuse, hypnotics, and neuroleptics as well as poisons, for example, solvents, pesticides, and chemical warfare agents. The procedures are described in detail to facilitate the setup in a laboratory. In so far it is structured similarly as "Basic Analytical Toxicology" published by the WHO (1995) but is focused on new and advanced techniques, for example, immunoassays, gas chromatography (including gas chromatography–mass spectrometry and headspace gas chromatography), high-performance liquid chromatography (including liquid chromatography–mass spectrometry), and ion-selective electrodes. The investigation of poisoning from plants and mushrooms as well as animals requires special procedures, which are described separately. Every chapter is completed by a summary on medical assessment and clinical interpretation to assist in the evaluation of the results.

In addition, the techniques used are described in detail, as well as quality control and quality assessment, strategy of clinical toxicological analysis, and screening procedures for "General Unknown" analysis. Relevant therapeutic drug monitoring and biochemical measurements, especially the determination of acetylcholinesterase activity, are described in separate chapters.

In the appendices, information on toxic concentrations of compounds is provided along with a summary of antidotes.

"Clinical Toxicological Analysis" is based on an international approach, which was organized and supported by the Deutsche Forschungsgemeinschaft (official research organization of Germany) under the leadership of Professor Dr. Dr. M. Geldmacher- von Mallinckrodt.

The book should assist all who are involved in clinical pathology, clinical toxicology, pharmacology, and legal medicine. Physicians of many disciplines, such as internal medicine, pediatrics, psychiatrics, and intensive care, will get an insight into the efficiency and limitations of actual toxicological analyses and the interpretation of their results.

Clinical Toxicological Analysis: Procedures, Results, Interpretation. Edited by Wolf-Rüdiger Külpmann

ISBN: 978-3-527-31890-2

"Clinical Toxicological Analysis" has been mainly supported by the "Deutsche Vereinte Gesellschaft für Klinische Chemie und Laboratoriumsmedizin" (DGKL; the German United Society for Clinical Chemistry and Laboratory Medicine). The DGKL assisted *inter alia* by establishing a working group on clinical toxicological analysis and by sponsoring the translation. But also the "Gesellschaft für Toxikologische und Forensische Chemie" (Society of Toxicological and Forensic Chemistry) has been involved with contributions from several members.

Translation of manuscripts was carried out, if necessary, by Mr R. Buyny and typewriting was performed by Mrs S. Lauer. Mrs. N. Denzau, Dr. Melanie Rohn, Dr. Susanne Viebahn, Mrs. S. Volk, Mrs. C. Zschernitz as well as Dr A. Sendtko and Dr F. Weinreich from the publisher Wiley-VCH committed themselves continuously to the completion of the book.

Authors and editor are indebted to all of them.

Hannover, January 2009 *W.R. Külpmann*

List of Contributors

Prof. Dr. Rolf Aderjan
Institut für Rechtsmedizin
Ruprecht-Karls-Universität
Voss-Str. 2
69115 Heidelberg
Germany

Dr. Volker Auwärter
Institut für Rechtsmedizin
Universitätsklinikum Freiburg
Albertstr. 9
79104 Freiburg
Germany

Dr. Torsten Binscheck
Berliner Betrieb für Zentrale
Gesundheitliche Aufgaben (BBGes)
Institut für Toxikologie
Klinische Toxikologie und Giftnotruf
Berlin
Oranienburger Str. 285
13 437 Berlin
Germany

Dr. Hans-Jürgen Birkhahn
Berliner Betrieb für Zentrale
Gesundheitliche Aufgaben (BBGes)
Institut für Toxikologie
Klinische Toxikologie und Giftnotruf
Berlin
Oranienburger Str. 285
13 437 Berlin
Germany

Prof. Dr. Max von Clarmann
Josef-Wiesberger-Str. 6a
85540 Haar
Germany

Prof. Dr. Thomas Daldrup
Institut für Rechtsmedizin
Heinrich-Heine-Universität
Moorenstraße 5
40225 Düsseldorf
Germany

Dr. Fritz Degel
Klinikum Nürnberg Nord
Institut für Klinische Chemie
und Laboratoriumsmedizin
Prof.-Ernst-Nathan-Str. 1
90419 Nürnberg
Germany

Clinical Toxicological Analysis: Procedures, Results, Interpretation. Edited by Wolf-Rüdiger Külpmann

ISBN: 978-3-527-31890-2

PD Dr. Ulrich Demme
Institut für Rechtsmedizin
Friedrich-Schiller-Universität
Fürstengraben 23
07743 Jena
Germany

Dr. Herbert Desel
Institut für Pharmakologie
und Toxikologie
Georg-August-Universität
Robert-Koch-Str. 40
37075 Göttingen
Germany

Prof. Dr. Peter Eyer
Walter-Straub-Institut für
Pharmakologie und Toxikologie
Ludwig-Maximilians-Universität
Goethestr.33
80336 München
Germany

Dr. Norbert Felgenhauer
Klinikum r. d. Isar
der Technischen Universität München
Medizinische Klinik II / Toxikologie
Ismaninger Str. 22
81675 München
Germany

Prof. Dr. Dr. Marika Geldmacher-von Mallinckrodt
Schlehenstraße 20
91056 Erlangen
Germany

Prof. Dr. Hans Jörg Gibitz
Elsenheim Str.13
5020 Salzburg
Austria

Dr. Thomas Grobosch
Berliner Betrieb für Zentrale
Gesundheitliche Aufgaben (BBGes)
Institut für Toxikologie
Klinische Toxikologie und Giftnotruf
Berlin
Oranienburger Str. 285
13 437 Berlin
Germany

Dr. Jürgen Hallbach
Klinikum Schwabing
des Städtischen Klinikum München
Department für Klinische Chemie
Kölner Platz 1
80804 München
Germany

Dr. Dieter Hannak
Klinikum Mannheim
der Universität Heidelberg
Institut für Klinische Chemie
Theodor-Kutzer-Ufer 1–3
68167 Mannheim
Germany

Prof. Dr. Herbert Käferstein
Institute of Legal Medicine
University Hospital of Cologne
Melatengürtel 60–62
50823 Köln
Germany

Dr. Marianne Koller
Institut für Pharmakologie
und Toxikologie der Bundeswehr
Neuherbergstr. 11
80937 München
Germany

Dr. Harald König
Helios Kliniken Schwerin
Institut für Laboratoriumsmedizin
Wismarsche Str. 397
19049 Schwerin
Germany

PD Dr. Claus Köppel
Wenckebach-Klinikum
Zentrum für Altersmedizin
Wenckebachstr. 21
12099 Berlin
Germany

Prof. Dr. Wolf-Rüdiger Külpmann
Medizinische Hochschule Hannover
Institut für Klinische Chemie
Carl-Neuberg-Str.1
30625 Hannover
Germany

PD Dr. Dagmar Lampe
Berliner Betrieb für Zentrale
Gesundheitliche Aufgaben (BBGes)
Institut für Toxikologie
Klinische Toxikologie und Giftnotruf
Berlin
Oranienburgerstr. 285
13437 Berlin
Germany

Dr. Marianne Lappenberg-Pelzer
Im Kieferngrund 8
14163 Berlin
Germany

Prof. Dr. h. c. Hans H. Maurer
Abteilung Experimentelle
und Klinische Toxikologie
Universität des Saarlandes
66421 Homburg/Saar
Germany

Prof. Dr. Dietrich Mebs
Zentrum der Rechtsmedizin
Institut f. Forensische Toxikologie
Kennedyallee 104
60596 Frankfurt/Main
Germany

Dr. Careen Merckel
Institut für Rechtsmedizin
Charité – Universitätsmedizin Berlin
Hittorfstr. 18
14195 Berlin
Germany

Prof. Dr. Ludwig von Meyer
Institut für Rechtsmedizin
Universität München
Nußbaumstr. 26
80336 München
Germany

Prof. Dr. Fritz Pragst
Institut für Rechtsmedizin
Abt. Toxikologische Chemie
Charité – Universitätsmedizin Berlin
Hittorfstr. 18
14195 Berlin
Germany

Dr. Dr. Frank Pluisch
Institute of Legal Medicine
University Hospital of Cologne
Melatengürtel 60–62
50823 Köln
Germany

PD Dr. Katharina Rentsch
Institut für Klinische Chemie
Universitäts-Spital Zürich
Rämistr.100
8091 Zürich
Switzerland

Prof. Dr. Achim Schmoldt
Institut für Rechtsmedizin
Universität Hamburg
Butenfeld 34
22529 Hamburg
Germany

Dr. André Scholer
Kantonsspital Basel
Universitätskliniken
Dept. Zentrallabor
Petersgraben 4
4031 Basel
Switzerland

Dagmar Simmert
Institut für Rechtsmedizin
Charité – Universitätsmedizin Berlin
Hittorfstr. 18
14195 Berlin
Germany

Dr. Ulrich Staerk
Medizinische Hochschule Hannover
Institut für Klinische Chemie
Carl-Neuberg-Str.1
30625 Hannover
Germany

PD Dr. Werner Steimer
Klinikum r. d. Isar (MRI)
Technische Universität München
Institut für Klinische Chemie und
Pathobiochemie
Ismaninger Str. 22
81675 München
Germany

Dr. Guido Sticht
Institute of Legal Medicine
University Hospital of Cologne
Melatengürtel 60–62
50823 Köln
Germany

Prof. Dr. Ladislaus Szinicz
Institut für Pharmakologie und
Toxikologie der Bundeswehr
Neuherbergstr.11
80937 München
Germany

PD Dr. Franz Worek
Institut für Pharmakologie und
Toxikologie der Bundeswehr
Neuherbergstr. 11
80937 München
Germany

Disclaimer

The work has been very carefully elaborated. However, editor, authors and publisher accept no responsibility for the correctness of any information, data, reference or recommendation and possible misprints.

Details on manufacturers and purity of conventional chemicals and reagents may be omitted. It is assumed that p.a. grade will be appropriate, if not otherwise stated. Nevertheless exceptionally a product of a manufacturer may not meet the requirements. In general, necessary basic equipment, e.g. glassware, spectrophotometer, centrifuge, is not listed in detail and its availability is taken for granted.

Clinical Toxicological Analysis: Procedures, Results, Interpretation. Edited by Wolf-Rüdiger Külpmann

ISBN: 978-3-527-31890-2

25
β-Receptor Blocking Drugs

L. von Meyer and W.R. Külpmann

25.1
Immunoassay

At present, reagent kits for the determination of β-receptor blocking agents ("β-blockers") are not available.

25.2
High-Performance Liquid Chromatography

Introduction

The determination of β-blockers by means of high-performance liquid chromatography (HPLC) can be performed in generally applicable as well as dedicated systems. For urine in particular, the "closed" Remedi system (BioRad) is recommendable. Even at therapeutic doses, the concentrations are high enough to detect the intake without splitting of the conjugates. The Remedi is, however, not more commercially available.

In the following, a generally applicable procedure developed by von Meyer with sensitive and specific fluorescence detection for the quantitative determination in serum, plasma, and blood is described.

Outline

After alkalinization of the sample with 0.01 mol/l of sodium hydroxide solution, the β-blockers are extracted with ethyl acetate. After separation on a C_8 or C_{18} reverse-phase column, they are identified by means of fluorescence and retention time.

Specimens

Serum, plasma, blood, and urine: minimal volume 1 ml.

Equipment

HPLC device for isocratic elution
Fluorescence detector

Clinical Toxicological Analysis: Procedures, Results, Interpretation. Edited by Wolf-Rüdiger Külpmann

ISBN: 978-3-527-31890-2

HPLC column RP-8 or RP-18, length 250 mm, internal diameter of 4 mm (e.g., RP-18 LiChroCart or RP Select-B, Merck, Darmstadt).

Chemicals (p.a. grade)

Acetonitrile
Calf serum
Ethanol
Ethyl acetate
Ethylene glycol
Phosphoric acid (85%)
Potassium dihydrogen phosphate
Sodium hydroxide 1 mol/l.

Reagents

Phosphate buffer pH 4.5: 6.66 g of potassium dihydrogen phosphate is dissolved in aqua bidest, adjusted to pH 4.5 with phosphoric acid and made up with aqua bidest to 1000 ml.

Mobile phase (HPLC): 450 ml of acetonitrile is mixed with 550 ml of phosphate buffer.

Internal standard: The selection of the internal standard depends on the suspected substances and their particular fluorescence characteristics.

Stock solution of the β-receptor blockers: Each 1000 mg/1000 ml ethanol.

Working solution of the β-receptor blockers: 100 mg/1000 ml: The stock solution is diluted with ethanol 1 + 9 (v/v). Calibrators and control samples are produced with calf serum and are spiked appropriately. They are treated like the sample under investigation.

Sample preparation

In a 10 ml centrifuge tube, 20 μl of the working solution of the internal standard is added to 0.5 ml of each of the samples, the calibrators and the controls. For alkalinization, 0.5 ml of sodium hydroxide solution is added and extraction is performed by 5 ml of ethyl acetate. The extracts are evaporated and then dissolved with a mixture of 50 μl mobile phase and 50 μl ethylene glycol.

HPLC analysis

Twenty microliters of the solution is introduced into the HPLC column. The flow rate is 0.8 ml/min. Table 25.1 shows the conditions of excitation and emission for fluorescence detection, the retention time for RP Select B as well as the respective yield and detection limit. Diode array detection can be employed alternatively.

Analytical assessment

For the determination of therapeutic blood concentrations of the β-blockers mentioned below, the method with fluorescence detection can be regarded as sufficiently sensitive.

Table 25.1 β-Receptor blocking drugs: HPLC.

Compound	Fluorescence Ex/Em (nm)	RT (min)	Recovery (%)	Detection limit (μg/l)
Acebutolol	330/460	3.30	65	10
Alprenolol	270/300	5.70	80	10
Atenolol	270/300	2.60	75	10
Betaxolol	270/300	5.35	90	10
Bisoprolol	270/300	4.50	90	10
Bunitrolol	280/340	4.00	70	10
Carazolol	280/340	5.30	80	10
Celiprolol	330/460	3.70	55	50
Labetalol	300/470	4.10	90	50
Mepindolol	280/340	4.00	70	50
Metipranolol	270/300	5.20	85	50
Metoprolol	280/340	3.80	75	10
Nadolol	270/300	2.70	30	10
Oxprenolol	270/300	4.60	75	50
Penbutolol	270/300	9.70	50	10
Pindolol	280/340	3.60	70	10
Propranolol	280/340	5.40	80	10
Sotalol	270/300	2.80	80	50
Talinolol	245/330	n.d.	n.d.	n.d.
Timolol	300/410	3.50	80	50
Toliprolol	270/300	4.40	85	10

n.d.: no data; Em: emission; Ex: excitation; RT: retention time.

25.3 Gas Chromatography

The sample preparation with ethyl acetate as described above can also be used in the gas chromatographic and gas chromatographic–mass spectrometric determination of the β-receptor blockers in serum. Ethylene glycol, however, must not be added since the extract has to be evaporated to dryness. Especially in case of low concentrations, these polar substances may be lost considerably through adsorption onto the inner tube walls. Moreover, the substances may be partly destroyed when passing the hot injection port with formation of artifacts. Therefore, determination by HPLC is preferred.

25.4 Medical Assessment and Clinical Interpretation

β-Blockers inhibit the β-receptor in a competitive way. Numerous substances, each exhibiting special cardiac and extracardiac adverse effects, are available nowadays. First, there are selective β_1-receptor blockers such as acebutolol, atenolol, bisoprolol, celiprolol, esmolol, metoprolol, and talinolol. Secondly, there are substances working on both the β_1- and β_2-receptors. Among these are alprenolol, bupranolol, carazolol, carteolol, mepindolol, nadolol, oxprenolol, penbutolol, pindolol, propranolol, and sotalol. In addition, carvedilol has an α-blocking effect. Even though betaxolol and timolol are applied locally in the form of eye drops in the treatment of glaucoma, they can also exhibit systemic effects.

Atenolol [1–5] (Figure 25.1)

Bioavailability	60%
Volume of distribution	0.7 l/kg body mass
Plasma protein binding	3%
Plasma half-life	4–14 h
Elimination	90% unchanged in urine

Concentration in serum/plasma	
Therapeutic range	0.1–0.6 (–1.0) m/l
Toxicity	from 2 mg/l
Comatose/lethal	from 27 mg/l (case report)

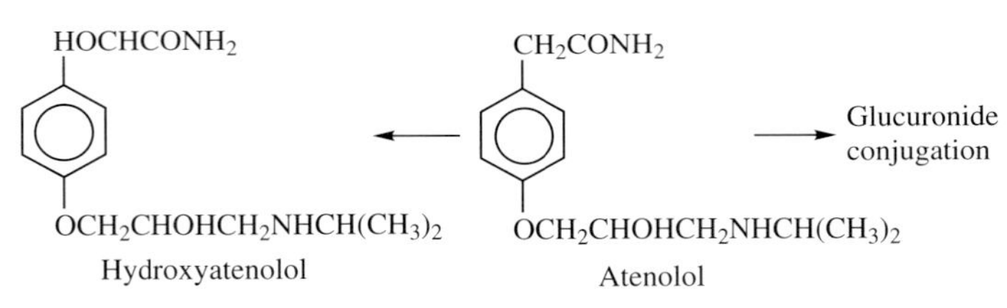

Figure 25.1 Atenolol metabolism.

Metoprolol [1–5] (Figure 25.2)

Bioavailability	50%
Volume of distribution	5.6 l/kg body mass
Plasma protein binding	10%
Plasma half-life	3–6 h
Elimination	<5% unchanged in urine metabolized in the liver with formation of an active metabolite

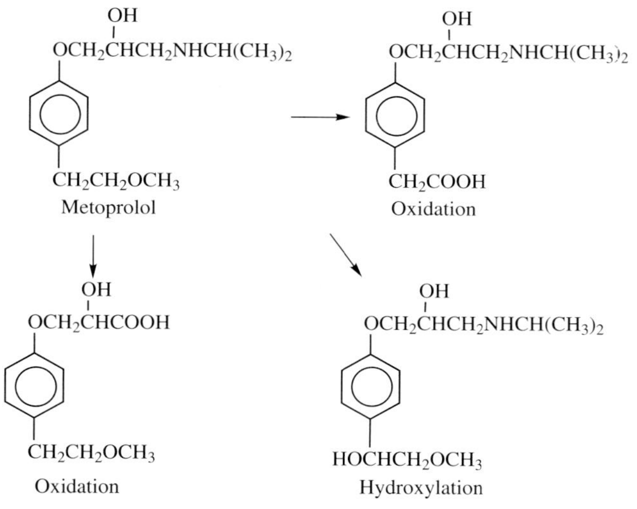

Figure 25.2 Metoprolol metabolism.

Concentration in serum/plasma	
Therapeutic range	0.1–0.6 mg/l 0.02–0.34 mg/l (trough)
Toxicity	0.65 mg/l (case report) from 1.0 mg/l
Comatose/lethal	4.7 mg/l (case report) from 12–18 mg/l

Oxprenolol [1–5] (Figure 25.3)

Bioavailability	40%
Volume of distribution	1.2 l/kg body mass
Plasma protein binding	75%
Plasma half-life	1–4 h
Elimination	<3% unchanged in urine

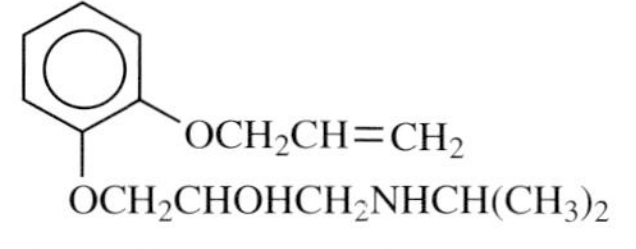

Figure 25.3 Oxprenolol.

Concentration in serum/plasma	
Therapeutic range	0.05–0.3 (–1.0) mg/l
Toxicity	from 2.0–3 mg/l
Comatose/lethal	from 10 mg/l

Propranolol [1–5] (Figure 25.4)

Bioavailability	40%
Volume of distribution	3.6 l/kg body mass
Plasma protein binding	93%
Plasma half-life	2–6 h
Elimination	<1% unchanged in urine metabolized in the liver with formation of an active metabolite (4-hydroxypropranolol, half-life <6 h)

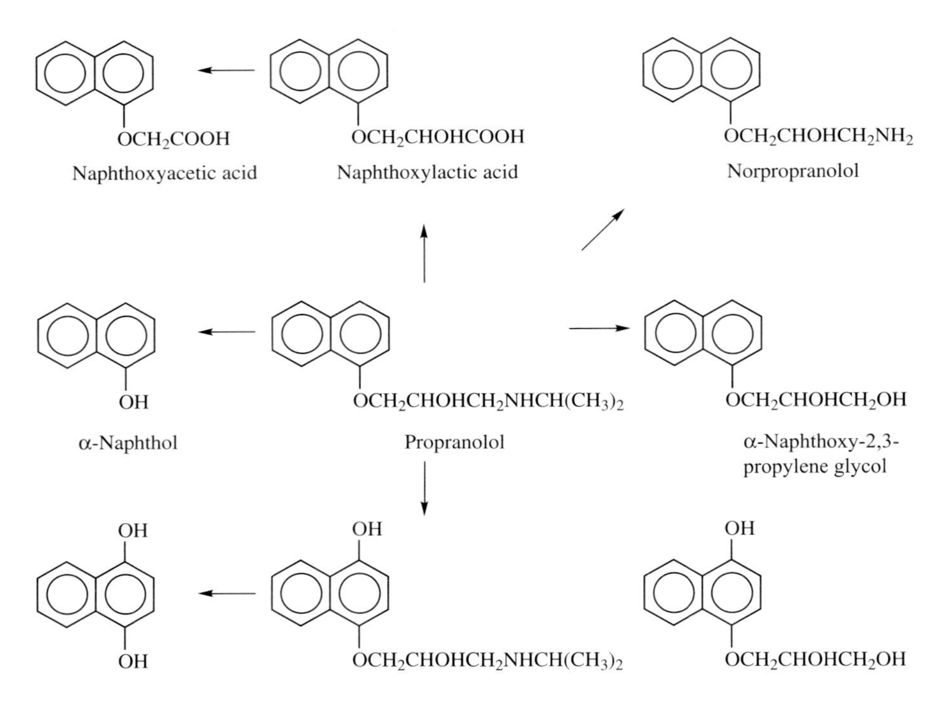

Figure 25.4 Propranolol metabolism.

Concentration in serum/plasma	
Therapeutic range	0.1–0.3 mg/l (peak)
	0.05–0.15 mg/l (trough)
Toxicity	from 1–2 mg/l
Comatose/lethal	from 4–10 mg/l

Sotalol [1–5] (Figure 25.5)

Bioavailability	100%
Volume of distribution	1.5 l/kg body mass
Plasma protein binding	<1%
Plasma half-life	5–13 (–17) h
Elimination	75% unchanged in urine

Concentration in serum/plasma	
Therapeutic range	0.5–3 (–5) mg/l
Toxicity	from 5–10 mg/l
Comatose/lethal	40 mg/l (case report)

Sotalol also exhibits properties of a class III antidysrhythmic drug.

$CH_3SO_2NH-C_6H_4-CH(OH)CH_2NHCH(CH_3)_2$

Figure 25.5 Sotalol.

Talinolol [1–5] (Figure 25.6)

Bioavailability	55%
Volume of distribution	3.3 l/kg body mass
Plasma protein binding	60%
Plasma half-life	10–14 h
Elimination	55% unchanged renally

Concentration in serum/plasma	
Therapeutic range	0.04–0.15 mg/l
Toxicity	No data
Comatose/lethal	5 mg/l (case report)

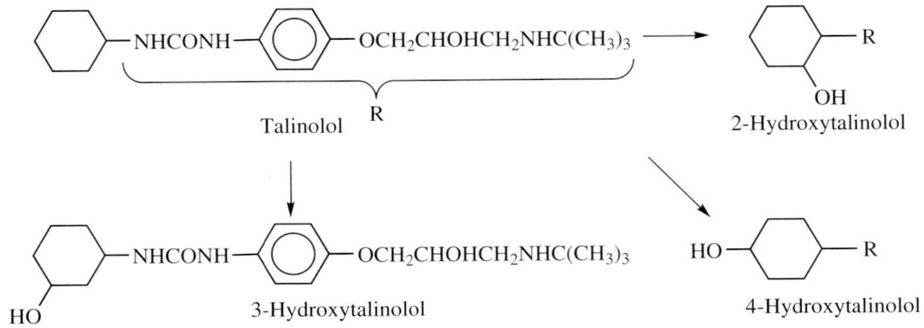

NHCONH — OCH$_2$CHOHCH$_2$NHC(CH$_3$)$_3$

R

Talinolol

R

OH

2-Hydroxytalinolol

NHCONH — OCH$_2$CHOHCH$_2$NHC(CH$_3$)$_3$

HO

3-Hydroxytalinolol

HO — R

4-Hydroxytalinolol

Figure 25.6 Talinolol metabolism.

Intoxication with β-blockers leads to a significantly reduced cardiac output per minute. Lipophilic compounds such as oxprenolol and propranolol have an adverse effect on the CNS and may cause convulsions. Enoximone in combination with glucagon and dopamine is used in the treatment of intoxication with β-receptor blockers. A thorough monitoring of the serum/plasma concentrations of the electrolytes (particularly potassium, calcium, and magnesium), glucose, and the cardiac markers (for differential diagnosis) is strongly indicated.

References

1. Bircher, J. and Sommer, W. (eds) (1999) *Klinisch-Pharmakologische Datensammlung*, 2nd edn, Wiss. Verlagsges, Stuttgart.
2. Baselt, R.C. (2004) *Disposition of Toxic Drugs and Chemicals in Man*, 7th edn, Biomedical Publications, Foster City, CA.
3. Moffat, A.C., Osselton, M.D. and Widdop, B. (2004) *Clarke's Analysis of Drugs and Poisons*, 3rd edn, Pharmaceutical Press, London.
4. Schulz, M. and Schmoldt, A. (2003) Therapeutic and toxic blood concentrations of more than 800 drugs and other xenobiotics. *Pharmazie*, **58**, 447–474.
5. TIAFT reference blood level list of therapeutic and toxic substances. Update 2005-03-03.

26
Drugs of Abuse

Introduction

H. Käferstein

This chapter deals with drugs mainly used by drug addicts apart from the opiates and opioides that are already discussed in Chapter 14 on analgesics.

In *Germany* (about 82 million inhabitants), 1835 persons died from drug abuse (including heroin) in the year 2001 and 1394 in 2007 (−24%). 22 551 persons became first-time users of illicit drugs in the year 2001 and 18 620 in the year 2007 (−17.4%).

Cannabis preparations, lysergic acid diethylamide (LSD), phencyclidine, and psilocybin are used as produced. On the other hand, cocaine and amphetamines, which are sold as powder, are often blended and the mass fraction of the drug may be below 10%. Caffeine, as well as lidocaine and tetracaine are used as adulterants. Lidocaine and tetracaine produce local anesthesia just as cocaine but without stimulating the central nervous system. Nevertheless, the substances used as adulterants may contribute to intoxication in addition to the illicit drug.

26.1
Amphetamines

G. Sticht and H. Käferstein

Introduction

The term "amphetamines," as used in this chapter, comprises amphetamine, methamphetamine as well as designer drugs (ecstasy) such as 3,4-methylenedioxyamphetamine (MDA), 3,4-methylenedioxymethamphetamine (MDMA) ("Adam") 3,4-methylenedioxyethylamphetamine (MDEA) ("Eve"), methylbenzodioxazolylbutanamine (MBDB) as well as benzodioxazolylbutanamine (BDB) (see Figure 26.1–26.4 and Appendix F).

Designer drugs are illicit drugs, which often contain a mixture of them. Opposite to amphetamine, designer drugs were used in Germany, but not before 1991. In the year 1997, 3799 persons became first-time designer drug abusers. In 2001,

Clinical Toxicological Analysis: Procedures, Results, Interpretation. Edited by Wolf-Rüdiger Külpmann

ISBN: 978-3-527-31890-2

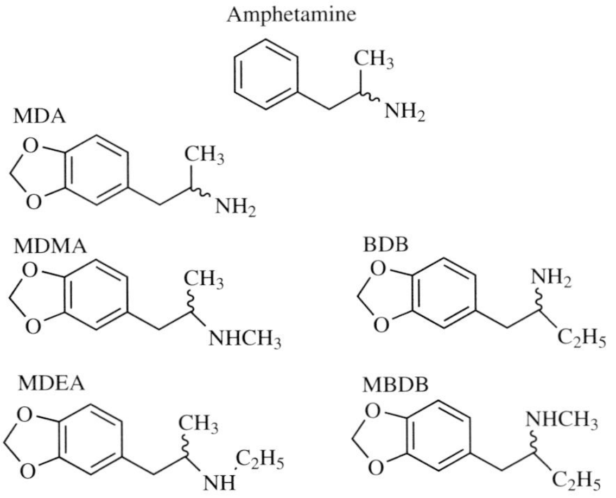

Figure 26.1 Structure of several amphetamines.

the number was 6097, in 2002 4737 and in 2007 the number decreased to 2038. In 2001, 4.5 million abuse units were confiscated, while in 2007, the number of confiscated units touched one million. The abuse of amphetamine, estimated from the number of first-time abusers, is still slightly increasing: 1997, 5535 persons; 2002,

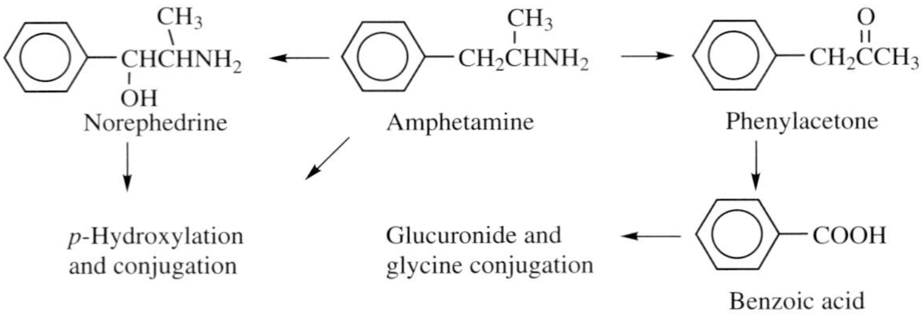

Figure 26.2 Amphetamine metabolism.

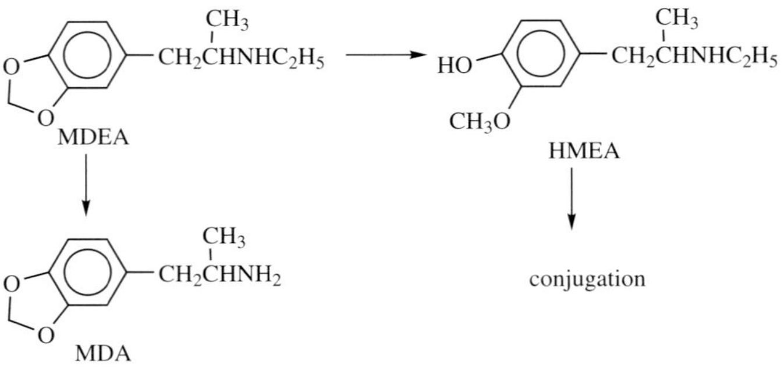

Figure 26.3 MDEA metabolism. HMEA: 4-Hydroxy-3-methoxyethylamphetamine.

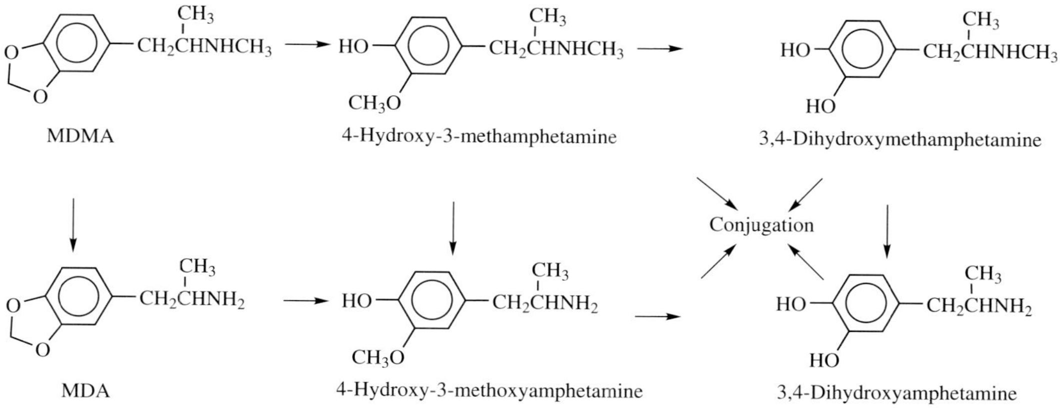

MDMA → 4-Hydroxy-3-methamphetamine → 3,4-Dihydroxymethamphetamine

MDA → 4-Hydroxy-3-methoxyamphetamine → 3,4-Dihydroxyamphetamine

Conjugation

Figure 26.4 MDMA metabolism.

6666 persons; 2007, 9949 persons [18]. 262 kg were confiscated in the year 2001 and 820 kg in 2007.

Usually, amphetamines are not sold as pure substances. Amphetamine is blended with caffeine, lidocaine, analgesics, or lactose. Designer drugs ("Ecstasy", "Adam", "Eve") may look like brand preparations, but indeed, they contain drug mixtures, that is, amphetamine and the substances used for adulterating amphetamine, which is not declared and is neither recognized nor anticipated by the consumers.

Considering the widespread abuse of amphetamines, for example, during Techno dancing events, routine drug screening for amphetamines is recommended in case of symptoms not attributable to organ failure, especially in case of concomitant hyperthermia.

Derivatives of amphetamine (e.g., aminorex, 4-methylaminorex, amfetaminil, benzphetamine, fenetylline, fenfluramine, dexfenfluramine, D-norpseudoephedrine, phendimetrazine, and phentermine) are used as slimming pills. These drugs are also used illicitly in sports for doping. They should not be used in combination with MAO inhibitors, which may increase the sympathomimetic effect in a life threatening way.

The attention deficit hyperactivity disorder (ADHD) is treated with the amphetamine derivate methylphenidate (chapter 24).

1-Benzylpiperazine (BZP) ("A2", "Frenzy", "Nemesis") was first used as an anthelminthic in the 1950s, but abandoned for adverse effects. In the 1970s it was investigated as a potential antidepressant, but rejected for amphetamine-like effects. BZP was used in a steadily increasing way since 1999 as a "pep pill", at first in New Zealand, and later in Canada, US and Europe in the party scene, as the drug was freely available. Nowadays, BZP is a Schedule I controlled substance in many countries (e.g. US), but may be still legal in others, e.g. in Canada and Ireland and is not controlled under any UN convention. It is rather easily accessible via the internet.

In general BZP will escape detection by the immunoassays for amphetamines. Most severe intoxications were observed in case of simultaneous abuse of BZP, MDMA and alcoholis beverages. The following symptoms were reported: seizures, tachycardia, hypertension, mydriasis, and hyperthermia.

Amphetamine derivates are present in some plants. *Ephedra* species such as *E. equiseta* and *E. sinica* contain L-ephedrine and ephedrine derivatives. They are used for the Chinese "Ma Huang." The Mediterranean ephedra species are less toxic, as they contain less ephedrine, but predominantly pseudoephedrine.

In African and Arab countries, khat obtained from the leaves of subbare khat or muktaree khat (*Catha edulis*) is chewed for hours. The drug contains cathinone, a compound with a structure similar to amphetamine.

26.1.1
Immunoassay

General aspects of immunoassays are described in Section 4.1. Grouptests for the detection of amphetamines are purchased by many manufacturers.

Some are produced for use on analyzers, for example,

CEDIA (Microgenics)
EMIT (Dade Behring)
FPIA (Abbott)
MTP (Mahsan)
ONLINE (Roche).

Immunoassays for point of care testing are also available:

ONTRAC (Roche)
TOXI-QUICK (Biomar)
TRIAGE 8 (Biosite).

Amphetamine or methamphetamine (which is seldom used in Germany) is the primary target of the antibody used in the assays. The other amphetamines are usually detected less sensitive as can be estimated from the cross reactivities (Table 26.1) that were stated by the manufacturers. Often assays based on a methamphetamine antibody detect methylenedioxyamphetamines as more sensitive

Table 26.1 Amphetamines (urine): cross-reactivity of several immunoassays.[a]

Assay	Amphetamine	MDA	MDMA	MDEA
EMIT (II) (Dade Behring)	100	33	17	n.d.
CEDIA (Microgenics)	100	2.2	70	40
CEDIA (Microgenics)[b]	100	112	220	206
TRIAGE 8 (Biosite)	100	50	50	n.d.
FPIA (Abbott)	100	153	98	46
MTP[c] (Mahsan)	100	162	0.07	n.d.

n.d.: no data.
[a]According to the manufacturer.
[b]Microgenics Amph/Ecstasy Assay. Cross-reactivity as compared to amphetamine (set to 100): MBDB, 123; BDB, 73.
[c]Data valid for blood.

than the amphetamine-based procedures. The cutoff concentrations ranges from 300 to 1000 μg/l. They are only valid for the respective calibrator (see Table 26.1). According to SAMHSA, [21] the cutoff concentration should be 500 μg/l, the EU Toxicology Experts Working Group [20] recommends 300 μg/l and the UKNEQAS uses 1000 μg/l in proficiency testing. Usually, the immunoassays are less sensitive to slimming drugs, methylphenidate, and ephedrine and its derivatives or 1-benzylpiperazine because of lower cross-reactivity with the antibody. These drugs will often escape detection, especially those assays dedicated specifically to the detection of designer drugs.

26.1.2 High-Performance Liquid Chromatography

The generally applicable HPLC procedure described in Section 12.1 can be used for the detection of amphetamines, but the Remedi HPLC is also applicable, especially for urine.

Outline

Remedi HPLC: The amphetamines are extracted by solid phase and their concentration is estimated by Remedi HPLC in a fully mechanized procedure. The system is not more available commercially, but is still used. It is considered as an example for other dedicated systems.

Specimens

Urine or serum, 1 ml.

For *equipment, chemicals, reagents, procedure,* and the calculation of results, refer Section 12.1. Amphetamine, methamphetamine, and MDA can be identified by their retention time and UV spectrum and their concentration can be estimated. MDMA and MDEA have nearly identical retention times and identical UV-spectra and cannot be detected separately, even though the spectra of the 3,4-methylenedioxyamphetamines are much more characteristic than those of amphetamine and methamphetamine.

26.1.3 Gas Chromatography – Mass Spectrometry

Outline

The amphetamines are extracted, converted to the perfluorobutyryl derivatives and injected into the gas chromatograph–mass spectrometer equipped with a capillary column. D_3-amphetamine, D_5-MDA, D_5-MDMA, and D_5-MDEA are used as internal standards.

Specimens

Serum, plasma, and urine: 1 ml.

Equipment

GC–MS with data system
Column: Ultra 1 (length 12 m, i.d. 0.2 mm, film thickness 0.33 μm) (Agilent)
Heating block 120 °C max, adjustable
Glass tubes 8 ml with standard ground joint.

Chemicals (p.a. grade)

1-Chlorobutane (for chromatography, gradient grade)
Ethyl acetate
Isohexane

Methanol
N-Methyl-bis-heptafluorobutyric acid (MBHFBA)
Sodium hydroxide 1 mol/l
Helium 6.0.

Reagents

Stock solution 1: Amphetamine, MDA, MDMA, MDEA, each 1.0 g/l methanol
Stock solution 2: D_3-amphetamine, D_5-MDA, D_5-MDMA, D_5-MDEA each 0.1 g/l methanol (Radian, Promochem)
Working solution 1: 10 ng/μl nondeuterated drugs; 100 μl stock solution 1 ad 10.0 ml with methanol
Working solution 2: 10 ng/μl deuterated drugs; 100 μl stock solution 2 ad 1.0 ml with methanol (volumetric flask).
Negative control sample: Drug-free serum
Positive control sample: 10 μl working solution 1 ad 1.0 ml with drug-free serum (volumetric flask).

Sample preparation
Ten microliters of working solution 2 is transferred to 1 ml sample (serum, urine) and to the negative and the positive control sample as well. To each tube, 0.5 ml sodium hydroxide is added. The samples are extracted with 5 ml isohexane/ethyl acetate 9 + 1 (v/v) and are centrifuged if necessary, to separate layers. Four milliliters of the organic (upper) layer are pipetted into a new vial and the solvent is evaporated (nitrogen or air). Immediately afterwards, the dry residue is dissolved with 70 μl MBHFBA and mixture is heated for 30 min at 120–130 °C in the heating block for derivatization. After cooling to room temperature, the solution is diluted with 630 μl 1-chlorobutane.

GC–MS analysis
One microliter of the solution is injected into the inlet port (split/splitless mode, 230 °C) of the GC–MS.

Table 26.2 Amphetamines: relevant ions (*m*/*z*) and retention indices (RI) (for details see the text).

Compound	Range of retention time 6–7 min *m*/*z*	8–9 min *m*/*z*	9–10 min *m*/*z*	RI
D_3-Amphetamine-PFB	121, 243			1322
Amphetamine-PFB	118, 240			1324
Methamphetamine-PFB	118, 254			1421
D_5-MDA-PFB		136, 380		1622
MDA-PFB		135, 375		1624
BDB-PFB			254, 389, (176)	1702
D_5-MDMA-PFB			258, 394	1731
MDMA-PFB			254, 389 (162)	1733
D_5-MDEA-PFB			273, 408, (162)	1772
MDEA-PFB			268, 403, (162)	1774
MBDB-PFB			268, 403, (176)	1789

PFB: perfluorobutyrate.

Column oven temperature program	
Initial temperature	60 °C (1 min)
Heating rate	15 °C/min
Final temperature	230 °C (1 min)
Carrier gas	Helium 6.0
Solvent delay	6 min

Evaluation: The amphetamines are identified by retention indices and characteristic *m*/*z* values (Table 26.2).

The *m*/*z* ratios of the deuterated and nondeuterated drugs in the positive control sample are used for the determination of the respective concentration by possibly subtracting their intensities in the negative control sample.

Analytical assessment

The procedure detects the different amphetamines, specifically at concentrations even below 10 µg/l. Apart from the substances mentioned before, ephedrine and norpseudoephedrine can also be detected along with other compounds, if they are extracted during sample preparation. The time required for analysis as well as the technician time is about 2 h for a sample.

26.1.4
Medical Assessment and Clinical Interpretation

During the therapy, 2.5–30 mg per day may be applied. In cases of abuse with ensuing tolerance, the daily dose may be 2 g or even 5 g. On an average, 50–100 mg of the methylenedioxy derivatives are daily taken by the addicts. Poisoning or even fatal outcome was already observed after the application of 2–2.5 tablets of ecstasy with unknown drug content or 160 mg amphetamine [2].

After oral administration of 10 mg amphetamine, the peak concentration in blood was 0.035 mg/l.

Amphetamine concentrations in the range 0.1–0.15 mg/l are observed during the treatment. During poisoning, concentrations exceed 0.2 mg/l. Coma or even death occurred at concentrations of 0.5 mg/l or higher [3, 4].

After 2 h of the intake of 50 mg MDMA, the peak concentration in plasma was 0.106 mg/l; 24 h after the intake, a peak concentration of 0.005 mg/l were measured [5]. Concentrations in serum between 0.1 and 0.35 mg/l are considered as harmless. In case of intoxications, serum concentrations ranged from 0.35 to 0.50 mg/l [4]. In fatalities, the concentration usually exceeded 1 mg/l and once it touched 45 mg/l blood [2, 6].

A considerable portion of the total amount of amphetamine is excreted unchanged in urine. The time span of excretion depends on the dose and the pH of urine. Only 2.4% of the dose was excreted by the kidneys at pH 8.0, while 60% of the same got excreted at pH 5.0. Minor portions of MDMA and MDEA are metabolized to produce MDA. Predominantly, the methylenedioxy part is split to yield the diphenol [1, 7]. In the late β-phase, only the hydroxy–methoxy metabolites of MDMA, MDEA, and MBDB could be detected in the urine. By FPIA (Abbott), a signal exceeding the cutoff concentration (300 μg/l) was obtained 40 h after the intake of 140 mg MDEA [8]. After the application of 100 mg MBDB-HCL, positive findings were observed by the FPIA during the 24 h, but only during the 4 h by EMIT. MBDB and MDB could be detected by chromatography up to 36 h [9].

Amphetamine is administered to stimulate the central nervous system. The methylenedioxy derivatives are abused for their additional disinhibiting effects with regard to feelings and communication reserve as well as for increasing self-esteem.

All amphetamines seem to have the following effects in common: increase in heart rate, blood pressure, and body temperature. After the intake, these ecstasies were observed: effects on the autonomous neural system, tachycardia, hypertension, nausea, hepatopathy (fatal in 5 of 28 patients), hyperthermia (fatal in 11 of 19 patients), blood coagulation disorders (fatal in 4 of 12 patients), nephropathy (fatal in 3 of 7 patients), and effects on the cardiovascular system (fatal in 7 of 12 patients). The dangerous hyperthermia is usually not survived, if the body temperature exceeds 42 °C [10].

26.2 Cannabinoids

L. von Meyer and H. Käferstein

26.2.1 Immunoassay

The general aspects of the immunochemical procedures are described in Section 4.1. Many manufacturers provide immunoassays for the diagnosis of cannabis abuse in urine. The primary target of the antibody is the most important metabolite of tetrahydrocannabinol 11-nor-Δ^9-THC-9-carboxylic acid. Other cannabinoids are also detected, but less sensitive.

Immunoassays are available to be run on mechanized analyzers from, for example, Abbott, BioRad, Dade Behring, Microgenics, and Roche. Also, assays for POCT are purchased for visual assessment by the investigator without any assistance by a measuring device (e.g., Biomar, Biosite, Mahsan, Roche).

Mechanized as well as POCTassays may be used for the diagnosis of cannabis abuse, even though the POCT procedures are less reliable in general. The cutoff concentrations that are recommended by the manufacturers, NIDA or UKNEQAS, range from 0.02 to 0.05 mg/l urine (SAMHSA, 0.05 mg/l; EU, 0.05 mg/l) [20, 21].

The use of the assays mentioned above is restricted to urine. For the examination of blood or serum special enzyme immunoassays on microtitre plates were developed by Mahsan and BioRad. Their cutoff concentration of 5 μg/l for cannabinoids may be often sufficiently sensitive for routine daily practice.

26.2.2 Thin-Layer Chromatography

Outline

In the following, a TLC procedure for urine is presented to confirm the intake of preparations of cannabis containing Δ^9-tetrahydrocannabinol (Δ^9-THC) (Figure 26.5). The assay is adequate to prove results by immunochemistry near to a decision limit of 20–25 μg/l. The procedure detects 11-nor-Δ^9-THC-9-carboxylic acid (THC-carboxylic acid), which is the main metabolite of Δ^9-THC.

Δ^9-THC is metabolized to the predominant metabolite Δ^9-THC-carboxylic acid and further to several conjugated and nonconjugated metabolites (Figure 26.6). In urine, THC-carboxylic acid is conjugated, mainly with glucuronic acid. Only after hydrolysis and extraction of free THC-carboxylic acid, the bigger part of the metabolite can be detected. After alkaline hydrolysis, the sample is extracted by C18-reversed-phase column. The eluate is applied to a thin-layer plate that is colored with Fast Blue BB salt after development. The phenolic OH-group of THC-carboxylic acid is involved in this reaction.

Specimens

Urine (spontaneous or collected), minimal volume 10 ml.

Storage: Stable for 4 weeks if kept at 4 °C and protected from light. For longer periods, storage at −20 °C is recommended.

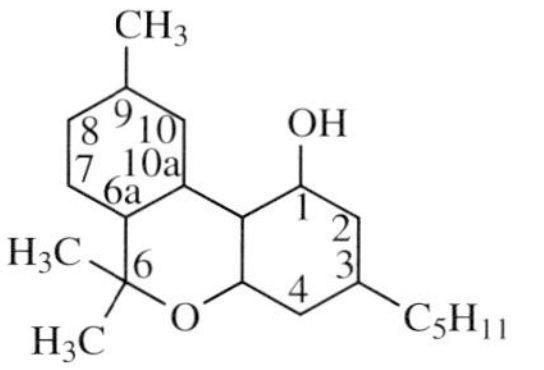

Figure 26.5 Cannabinoids: Dibenzofuran numbering.

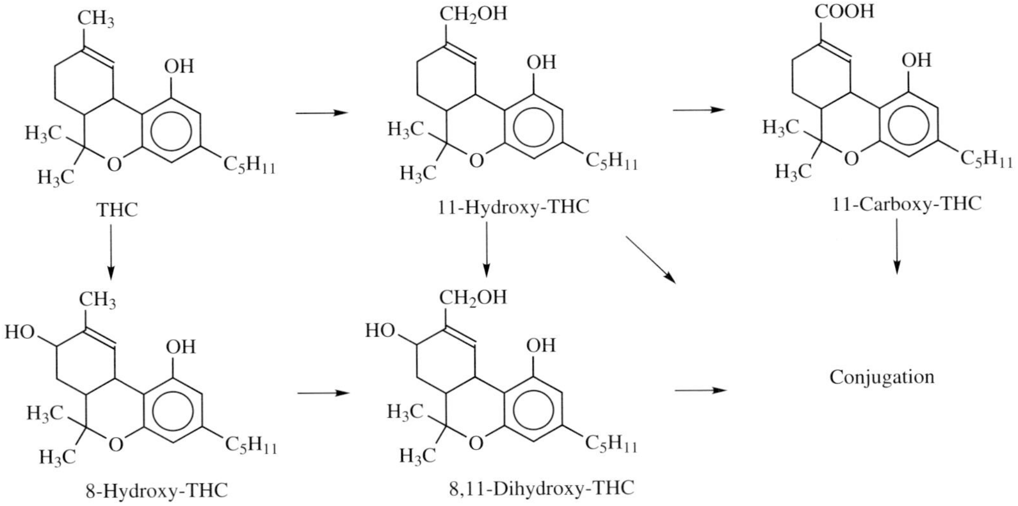

Figure 26.6 Δ^9-Tetrahydrocannabinol (THC) metabolism.

Equipment

Suction device for solid-phase extraction including tubes (e.g., Adsorbex SPU; Merck, Darmstadt)
Glass capillaries for application, 10 µl

DC plates ready for use:
(A) HPTLC plates with accumulation zone (e.g., Nano-SILGUR-20, 10 cm × 10 cm)
(B) Silica gel plates or foils, 10 cm × 10 cm.
Development tank
Dipping tank
C18 columns, 1 ml (e.g., Chromabond C18 ec, Macherey and Nagel, Düren)
Glass tubes with conical bottom and standard ground joint (10 and 20 ml)
Drying oven or other thermostatting device to keep 55 °C.

Chemicals

Acetic acid 96%
Acetone
Diethyl amine
Ethyl acetate
Fast Blue BB salt (e.g., Sigma)
n-Heptane
n-Hexane
Methanol
Sodium carbonate 0.1 mol/l

Sodium hydrogen carbonate
Sodium hydroxide 1 mol/l.

Reagents

Aqueous sodium hydrogen carbonate solution (5 g/100 ml).
Dipping solution: 100 mg Fast Blue BB salt is dissolved in 100 ml dichloromethane to yield an intensively yellow colored solution. The reagent is stable for several months, if kept in a refrigerator.

Solvent: *n*-Heptane/ethyl acetate/acetic acid (96%) 30 + 15 + 1 (v/v/v) to be used only during the day of preparation.
Negative control sample: Drug-free urine
Positive control sample: THC-carboxylic acid 10 µg/l
Drug-free urine is spiked appropriately with control urine: for example, Abbott TD_x Cannabinoid Controls (Abbott) or EMIT II Urine Cannabinoid Calibrator 100 µg/l (Dade Behring).

Sample preparation
Alkaline hydrolysis: 0.5 ml sodium hydroxide (1 mol/l) is added to 5 ml urine in a glass tube with conical bottom. The mixture is kept 15 min at 55 °C in a drying oven. Then, 2 ml acetic acid 96% is used to adjust to pH 4 (to be checked).

Extraction: C18 columns are put in the suction device and for conditioning, 5 ml methanol followed by 5 ml aqua bidest are rinsed through the column at reduced atmospheric pressure (1–2 inchHg vacuum), but not to dryness.

The hydrolyzed urine is applied and sucked at a reduced pressure (3–4 inchHg). The column is rinsed with

(1) 2.5 ml aqua bidest
(2) 2.5 ml sodium hydrogen carbonate solution
(3) 2.5 ml aqua bidest.

Five milliliters of methanol/aqua bidest (1 + 1; v/v) is applied to the column and sucked through at maximum vacuum to dryness. 0.5 ml *n*-hexane is pipetted on the column that is sucked at maximum vacuum to dryness. (Drops of water at the tip of the column are removed with a cellulose tissue.)

The relevant compounds are eluted with 1.5 ml acetone at a reduced pressure (1–2 inchHg). Acetone is evaporated to dryness at maximum 50 °C.

TLC analysis
The plates are not activated beforehand.

The extracts are dissolved with 50 µl methanol and the solutions are applied to the plates quantitatively with 10 µl glass capillaries (length of streak 7 mm, distance between streaks 5 mm). TLC is performed with ascending development in unsaturated tanks. The distance of the solvent front from the origin should be 80 mm.

Detection with fast blue BB: After drying with a hairdryer, the plate is dipped into a tank filled with the Fast Blue BB solution and the air is dried for 1 min. It is put into a TLC glass tank saturated with diethylamine vapor. (Glass tank filled with diethylamine up to 1 cm high. The plate is put onto a metal rack, so that it does not dip into the fluid. The tank is closed with a lid.)

Evaluation of the chromatogram: If THC carboxylic acid is present, a pink-colored zone on white background develops at a hR_f value of about 50. In case of more elevated concentrations of THC carboxylic acid, further pink zones at lower hR_f values are visible, indicating more polar metabolites.

Analytical assessment

Negative control sample: It shall not show a pink-colored zone at hR_f 50.

Positive control sample: A distinct pink zone shall be visible at hR_f 50.

Sensitivity: Maximum sensitivity (see Section 4.1) as estimated by control urine is 2 μg THC carboxylic acid/l urine and practical sensitivity (see Section 4.1) is estimated at 5 μg/l urine, if the HPTLC plates are used.

In case of conventional silica gel plates or foils, the maximum sensitivity is 5 μg/l urine and the practical sensitivity is 10 μg/l.

Specificity: Until now, no drugs became known, which show a pink zone at about hR_f 50, if the procedure is performed as described.

Practicability

The procedure can be used for a sensitive and reliable confirmation of results obtained by immunoassays. It can be easily performed, but not more than 2 h should elapse between the development and the detection. Otherwise the achievable intensity of the color of the zones decreases steadily and the sensitivity accordingly. Technicians familiar with the procedure will obtain the results after about 90 min.

Annotations

(1) Alkaline hydrolysis shall be performed at elevated temperature as described for a complete hydrolysis of the conjugates.
(2) The Fast Blue salts B, RR, and BB are very similar with regard to the color intensity of the zones. However, usually Fast Blue BB is preferred, because the other salts produce a yellow background on the TC plates [11]. Furthermore, Fast Blue B may be cancerigenic.
(3) The color can be preserved for about 2 months, if the plates are wrapped in paper and are kept at room temperature protected from light. Only the white background changes to a brownish color.
(4) A further TLC procedure is provided by ToxiLab (DRG Instruments). Its detection limit is 20 μg THC-carboxylic acid/l urine and is therefore appropriate at least for the confirmation of immunochemical procedures with a decision limit of 100 μg/l. Indeed, the mass fraction of THC carboxylic acid may be 0.14 of the total cannabinoids as estimated by immunoassays. In the mean, its fraction is 0.36, but may even exceed 1.0 and come to 1.11. Rarely, positive immunochemical results are falsely not confirmed by the TLC procedure. If the cannabinoid

concentration estimated by immunochemistry is only slightly exceeding the decision limit of 25 µg/l, the THC carboxylic acid concentration may be below 5 µg/l and escape detection by TLC.

26.2.3 High-Performance Liquid Chromatography

THC carboxylic acid is extracted (Sections 26.2.2 and 26.2.4), separated by HPLC on a reversed phase column (RP 8 or RP 18), and identified by a UV detector or preferably by a DAD detector [13, 14]. However, the sensitivity and specificity of the procedure are inferior to TLC (Section 26.2.2) or gas chromatography (Section 26.2.4). Therefore, HPLC is not recommended for the detection of cannabinoids in urine.

26.2.4 Gas Chromatography – Mass Spectrometry

J. Hallbach and L. von Meyer

Introduction

Immunochemical procedures are used for cannabinoid screening of urine, as they can be performed easily and rapidly. The following gas chromatographic–mass spectrometric procedure may be used for confirmation. It is focused on 11-nor-Δ^9-THC-9 carboxylic acid (THC carboxylic acid), the predominant metabolite of Δ^9-tetrahydrocannabinol. Besides THC carboxylic acid, several nonconjugated and conjugated metabolites are produced. As the time span between intake and urine sampling increases, the fraction of THC carboxylic acid of the excreted cannabinoids decreases steadily [12]. THC carboxylic acid is present in the urine mainly conjugated especially with glucuronic acid. For identification by gas chromatography, the conjugates must be hydrolyzed and the released compounds must be extracted.

Outline

Urine is hydrolyzed enzymatically followed by liquid extraction (ToxiLab A). The acetylated compounds are analyzed by the GC–MS and identified by the retention time and the characteristic mass fragments.

Specimen

Urine (spot or collected), minimal volume 10 ml

Storage: 4 °C protected from light. For longer periods −20 °C.

Equipment

Gas chromatograph with mass spectrometric detector (Quadrupol)
GC column: DB-5 ht (length 30 m, i.d. 0.25 mm, film thickness 0.1 µm; J & W)
Carrier gas: Helium 6.0
Heating block
Glass vials 10 ml with round bottom (e.g., Vacutainer, Becton, and Dickinson)
GC vials with conical hang-in cups.

Chemicals

Acetic anhydride
Ethyl acetate
Glucuronidase 300 000 units/g solid (e.g., G 0751, Sigma–Aldrich)
Phosphate buffer pH 5.5, 0.1 mol/l
Pyridine
ToxiLab A vials (Varian).

Reagents

Glucuronidase solution: Glucuronidase (100 000 units) is dissolved with 10 ml phosphate buffer. It is stored deep-frozen in appropriate portions; stable for at least 6 months.
Acetylation reagent: 60 μl acetic anhydride and 40 μl pyridine are mixed just before use.
Cutoff calibrator stock solution: THC carboxylic acid 120 μg/l.
Negative control sample: Drug-free urine.
Positive control sample: THC carboxylic acid 50 μg/l or commercially available control urine.

Sample preparation
Enzymatic hydrolysis: Five milliliters of urine is hydrolyzed with 500 μl glucuronidase solution in 10 ml glass vials at 50 °C in a heating block for 30 min.
Extraction: The hydrolyzed urine is distributed in two equal volumes and separately processed:

(A) 2.5 ml hydrolyzed urine
(B) 2.5 ml hydrolyzed urine + 0.5 ml calibrator stock solution (120 μg/l). Final concentration of THC carboxylic acid: 20 μg/l.

A and B are transferred to a ToxiLab A vial, vigorously shaken for 2 min and centrifuged (10 min, 2000 × g). 0.5 ml of the upper layer is pipetted into a 10 ml glass vial and the solvent is evaporated to dryness (water bath 35 °C, nitrogen flow).
Derivatization: The dry residues are dissolved each with 40 μl acetylating reagent and kept at 50 °C (heating block) in closed vials for 30 min. The solvent is evaporated with nitrogen flow. The residues are transferred completely with 50 μl ethyl acetate (two times) into the hang-in cups of GC vials.

GC–MS analysis
One microliter of the derivatized samples is injected splitless (30 psi for 1 min, column temperature at 70 °C). At constant carrier gas flow (1 ml/min), the following column oven temperature program is applied:

Initial temperature	70 °C; 2 min
Heating rate	20 °C/min
Final temperature	280 °C; 3 min

The retention time of THC carboxylic acid after acetylation is about 14.4 min. If not derivatized, the retention time is 13.6 min.

Settings for mass spectrometry

Transfer line 300 °C
Standard spectra autotune, 100 eV offset, 100 ms/ion.

The following ions (*m/z*) are recorded: 297.0, 314.0, and 271.0; 231.0, 299.0, and 243.0.

Evaluation of the chromatogram
Chem-Station-Integrator (Agilent) was used with the settings:

Initial area reject
Initial peak width 0.020
Shoulder detection off
Initial threshold 8.

For identification, the following criteria must be fulfilled:

(1) Sample A (not spiked with THC carboxylic acid): The typical fragments *m/z* in characteristic size relation are present at the respective retention time.
(2) The peak in A must fit into the corresponding peak in B, which results from spiking with THC carboxylic acid.
(3) The peak area of *m/z* 297 in A should be at least 50% of the relevant peak area in B to ensure that the concentration in A exceeds the cutoff concentration (20 μg/l).

Analytical assessment
Negative control sample: It shall not fulfill the criteria (see above)

Positive control sample: It shall meet the criteria 1 and 2 (see above).

In addition, a further positive control sample with a higher concentration of THC carboxylic acid may be used.

Specificity: Until now, no drugs were found that may interfere with the identification of THC carboxylic acid, if the respective criteria (retention time, mass relation of fragments) are observed.

Sensitivity: The detection limit is below a THC-carboxylic acid concentration of 5 μg/l as determined by a sequential dilution of control urine and depends on column performance and instrument settings. The given procedure with a cutoff concentration of 20 μg/l is appropriate for the confirmation of immunochemical assays with a cutoff concentration of 50 μg/l. The GC–MS procedure can be adjusted to realize the cutoff concentration 10 μg/l to function as a confirmation procedure for immunoassays with a cutoff concentration of 20 μg/l.

Practicability
The GC–MS confirmation procedure can be performed within 2 h.

26.2.5
Medical Assessment and Clinical Interpretation

L. von Meyer, J. Hallbach and H. König

The abuse of cannabinoids is wide spread. The amount of confiscated hashish and marijuana is huge and varying considerably from year to year (Germany [18]):

Year	2001	2002	2006	2007
Hashish (kg)	6863.1	5003.0	5606.1	3677.5
Marijuana (kg)	2078.7	6130.2	2954.1	3769.8

The material is detected often by chance and does not allow a reliable estimation of the actual abuse of cannabinoids.

The favorite method to prove a recent abuse of cannabis or a relapse during the withdrawal treatment is the determination of the actual serum concentration of Δ^9-tetrahydrocannabinol with GC–MS using the deuterated compound for calibration. Its concentration 6 h after smoking is below 1 µg/l in case of only incidental abuse. However, in case of regular abuse, Δ^9-THC may accumulate and concentrations exceeding 1 µg/l may be observed for 2 days.

As an alternative, a successful withdrawal treatment as well as a relapse can be proven by immunochemical estimation of the concentration of the cannabinoids every 4 days and by the concomitant determination of the creatinine excretion (Jaffe's method). The ratio of cannabinoids concentration to creatinine should steadily decrease, but in case of a relapse, a distinct increase is observed. The Jaffe method is preferred under these circumstances for the following reason: Creatine is a favorite adulterant to mask the factitious dilution of the urine sample. The Jaffe method determines creatinine specifically in the presence of nonphysiologically high concentrations of creatine, whereas, the results of enzymatic procedures may be false high under these circumstances.

Figure 26.7 shows a withdrawal without relapse and Figure 26.8 with a relapse.

Usually, a relapse takes place immediately after sampling. A relapse will not escape detection if drug screening takes place every 4 days.

Cannabinoids may be detectable by immunoassays up to 3 months in case of a previous intensive chronic abuse [15, 16].

Δ^9-Tetrahydrocannabinol is considered as the main pharmacologically active compound in hashish or marijuana. Usually, it is administered by smoking as hashish or marijuana and seldom orally with cakes or tea. Hashish oil does not play an important role for abuse. Repeatedly, the use of the oil from the seeds of hemp was cited as an explanation for positive findings. However, the concentration of cannabinoids in these oils is usually too low. Passive smoking of cannabis preparations may cause slightly elevated concentrations that exceed 20 µg/l THC carboxylic acid only under extreme (experimental) conditions (smoking in a motor-car) [17].

Typical clinical symptoms of THC abuse are tachycardia and (may be) reddish conjunctivae. Among the psychic effects are not only confined cognitive capabilities,

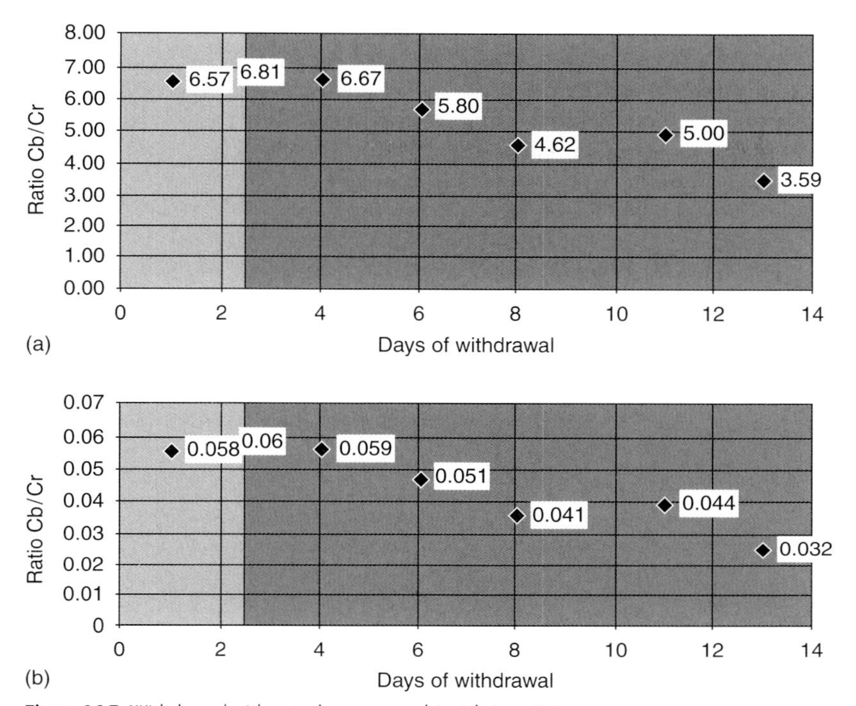

Figure 26.7 Withdrawal without relapse: cannabinoids/creatinine ratio in urine. (a) Ratio Cb/Cr: U-cannabinoids (μg/l)/U-creatinine (mmol/l). (b) Ratio Cb/Cr: U-cannabinoids (μg/l)/U-creatinine (mg/l). U: urine.

euphoria, excited mood, but also irritability, restlessness, and want of appetite. As these symptoms can also be connected with a psychosis, a positive or a negative finding for cannabinoids will assist in diagnosis.

Acute intoxication with cannabinoids, which needs medical treatment, is very rare.

26.2.6 Addendum

26.2.6.1 Spice

"Spice" is a street name for a mixture of herbs among which is **Leonorus sibiricus** (so-called "**Marihuanilla**") and which may also contain honey and vanilla. On the whole the herbs identified in different samples of Spice differ considerably. However, according to most recent investigations the effects of Spice are not evoked mainly by these constituents but by an added synthetic drug, which has been produced during drug research.

Note: Mixtures of herbs with effects similar to spice are **Chill-X**, **Insence**, **Sence**, **Smok(e)**, **Space** and **Yucatan Fire**.

The effects of Spice are similar to the effects of cannabinoids, even though the detected drug is chemically not related to the cannabinoids.

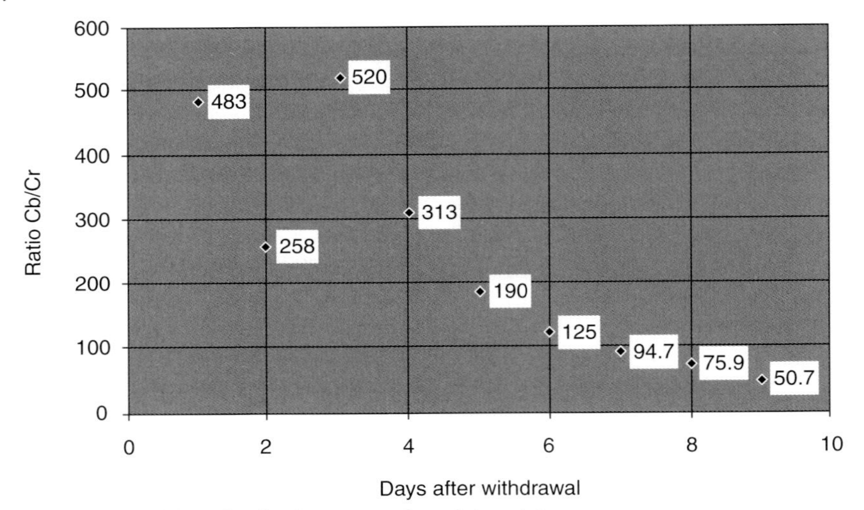

Figure 26.8 Withdrawal with relapse: cannabinoids/creatinine ratio in urine. Ratio Cb/Cr: U-cannabinoids (μg/l)/U-creatinine (mmol/l). U: Urine.

Actually, Spice is freely available in Germany and is smoked especially by teenagers as a substitute for cannabinoids. The detection of a cannabinoid like compound will probably pave the way for the prohibition of Spice, even though a purchase via internet may be still possible. Toxicological analyses of Spice and its constituents are not yet established and immunoassays for cannabinoids prove negative.

26.3 Cocaine

H. Käferstein and G. Sticht

Introduction

Cocaine and cocaine containing preparations may be prescribed in Germany, if the special requirements of the relevant law are fulfilled. However, the application of cocaine in medicine is considered as obsolete nowadays for most possible indications. In general, cocaine is used illicitly. According to an estimate by the Federal Criminal Bureau of Germany (Bundeskriminalamt) for 1992, there were 40 000–50 000 abusers who needed 3000–4000 kg of cocaine/year. The number of cocaine abusers registered for the first time by the authorities in Germany was 5691 in 1998, 4872 in 2001, 4933 in 2002, 4489 in 2005, 4225 in 2006, and 3812 in 2007 (82 million inhabitants).

Mass of cocaine detected in Germany: in the year 1998, 1133 kg; in the year 2001, 1288 kg, in the year 2002, 2136 kg, 1079 kg in 2005, 1716.6 kg in 2006 and 1877.5 kg in 2007 [18].

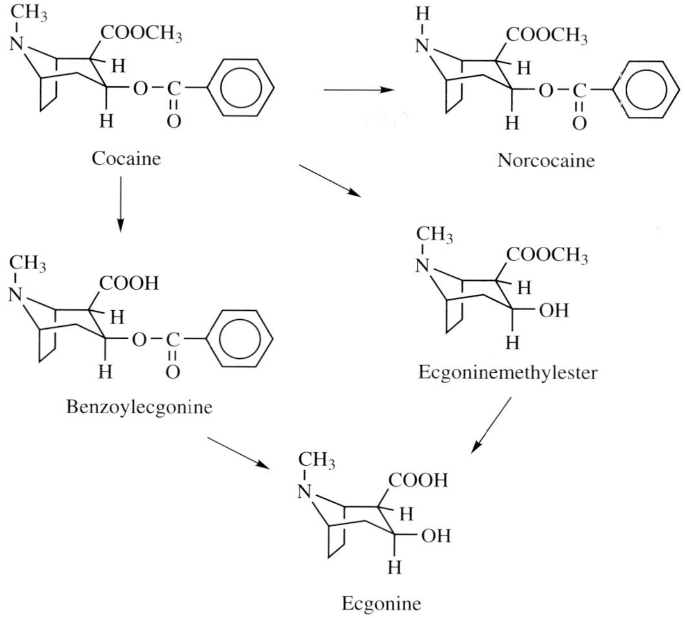

Cocaine

Norcocaine

Benzoylecgonine

Ecgoninemethylester

Ecgonine

Figure 26.9 Cocaine metabolism.

Cocaine is exported from countries such as Bolivia and Colombia as pure cocaine hydrochloride, but in the "scene," usually cocaine is adulterated not only with amphetamine, lidocaine, and caffeine, but also with pharmacologically ineffective substances such as lactose [19]. Street names are Coke, White Girl, and Crack (cocaine base). Because of the wide spread abuse and every case showing symptoms that are difficult to interpret, cocaine should be considered in differential diagnosis and not with regard to young patients only.

Cocaine is rather rapidly metabolized in humans to produce mainly benzoylecgonine, ecgonine methylester, norcocaine, and ecgonine (Figure 26.9). Benzoylecgonine is the most important metabolite for the analytical detection of cocaine abuse.

26.3.1 Immunoassay

For general information on immunoassays, see Section 4.1. Immunoassays for the detection of cocaine abuse are available from several manufacturers. Some of them are produced for use on analyzers of clinical chemistry, that is,

CEDIA (Microgenics)
EMIT (II) (Dade Bering)
FPIA (Abbott)
ONLINE (Roche).

Others can be performed in the field without the need of any further device, that is,

Ontrac (Roche)
Toxi-Quick (Biomar)
Triage 8 (Biosite).

The antibodies of the immunoassays are focused primarily on benzoylecgonine. Their cross-reactivity toward cocaine may be low or rather high. Usually, the manufacturers state a cutoff concentration of 300 μg/l for benzoylecgonine in urine, which is also recommended by the EU Toxicology Experts Working Group [20] and UKNEQAS, but not by SAMHSA (150 μg/l) [21].

The use of the immunoassays is authorized officially only for urine. Mahsan developed an immunoassay for use with blood or serum. Its cutoff concentration (10 μg/l serum) is surpassed for up to 48 h after abuse.

26.3.2 High-Performance Liquid Chromatography

Outline

After extraction (liquid/liquid or by solid phase), benzoylecgonine and cocaine are separated by a reversed phase column (mobile phase: acetonitrile/sulfuric acid (1 mmol/l)). The drugs are identified by a diode array detector.

Specimens

Serum or urine, minimal volume 3 ml

Equipment

HPLC instrument with thermostated column oven
Column: RP8, particle size 5 μm, 250 mm × 4 mm (e.g., VDS Optilab)
Glass tubes with standard ground joint, 8 ml.

Chemicals

Acetonitrile (gradient grade, for chromatography)
Aqua bidest (chromatography grade)
Chloroform
Disodium hydrogen phosphate
Propane-2-ol
Sodium hydroxide, 1 mol/l
Sulfuric acid, 1 mmol/l.

Reagents

Stock solutions 1 g/l
Cocaine 1 g/l acetonitrile (Radian, Promochem)
Benzoylecgonine 1 g/l methanol (Radian, Promochem).
Extraction mixture: chloroform/propan-2-ol 9 + 1 (v/v)
Mobile phase (HPLC): sulfuric acid/acetonitrile 60 + 40 (v/v).

Sample preparation

Sample (2 ml) and chloroform-propan-2-ol (5 ml) are pipetted into a glass tube with standard ground joint containing 0.1 g disodium hydrogen phosphate. The mixture is shaken vigorously for 1 min. After centrifugation (if necessary for separating of layers), the lower layer (chloroform layer) is filtrated through a filter paper into a glass vial (5 ml). The organic solvent is evaporated with a flow of nitrogen or air. The dry residue is dissolved with 50 μl acetonitrile and 10 μl is injected into the HPLC system.

HPLC analysis

Settings

Oven temperature 40 °C, flow rate 1.7 ml/min
Diode array detector: recording of the spectrum in the range 200–300 nm
UV detector: recording at 220 and 233 nm.

Evaluation For identification, the benzoylecgonine and cocaine standards are used.

(1) The retention times of sample and standards shall fit: Benzoylecgonine: about 2 min; cocaine: about 4 min.
(2) Peak height or peak area at wavelength 233 nm shall be about 50% bigger than those at 220 nm.
(3) The total spectrum of the sample extract as recorded by the diode array detector shall fit into the spectrum of the standard solutions, with a maximum at 233 nm and a minimum at 220 nm.

Analytical assessment

The UV spectra of benzoylecgonine and cocaine are rather characteristic. Benzoic acid shows a similar spectrum, but is separated during extraction. Therefore, the procedure is rather specific.

The recovery of benzoylecgonine after extraction exceeds 50%, while the same that of cocaine exceeds 90%.

Practicability

The time of analysis is about 90 min and is identical to the technician time.

26.3.3 Gas Chromatography–Mass Spectrometry

Outline

After solid-phase extraction, the dry residue is silylated and injected into the GC–MS system equipped with a capillary column. Deuterated benzoylecgonine and deuterated cocaine are used as internal standards.

Specimens

Serum, plasma: minimum volume 0.5 ml

Also, urine may be investigated.

Equipment

GC–MS system
Column: Ultra 1 (length 12 m, internal diameter 0.2 mm, film thickness 0.33 μm) (Agilent)
Heating block, adjustable up to 100 °C
Sucking device (e.g., Vac Master, IST)
High vacuum pump.

Chemicals (p.a. grade)

Acetic acid, 100%
Ammonia solution, minimum 25%
Disodium hydrogen phosphate dihydrate
Methanol
MSTFA (*N*-methyl-*N*-trimethylsilyl-trifluoroacetamide)
Potassium hydrogen phosphate
Potassium hydroxide
Helium 6.0
Nitrogen 5.0
Extraction columns: certify LRC 300 mg (Varian).

Reagents

Stock solutions 1.0 g/l
Cocaine 1 g/l acetonitrile (Radian, Promochem)
Benzoylecgonine 1 g/l acetonitrile (Radian, Promochem)
D_3-cocaine 1 g/l acetonitrile (Radian, Promochem)
D_3-benzoylecgonine 1 g/l methanol (Promochem).
Working solutions

(1) *D3-cocaine 10 mg/l, D3-benzoylecgonine 20 mg/l:* D_3-cocaine 10 μl and D_3-benzoylecgonine 20 μl are made up to 1 ml by methanol.
(2) *Phosphate buffer pH 8.0:* Potassium dihydrogen phosphate 13.61 g is dissolved with 900 ml aqua bidest in a volumetric flask, adjusted to pH 8.0 and made up to 1000 ml with aqua bidest.
(3) *Acetate buffer, pH 4.0:* Acetic acid, 5.7 ml is mixed with 800 ml aqua bidest in a volumetric flask. Potassium hydroxide (1 mol/l) 16 ml is added and the pH is adjusted to 4.0. The solution is made up to 1000 ml.
(4) *Aqueous methanol solution:* Methanol/aqua bidest 30 + 70 (v/v).
(5) *Eluent:* Methanol/ammonia solution 98 + 2 (v/v) to be prepared just before use.

Sample preparation

Extraction: Sample 0.5 ml is spiked with 5 μl working solution 1 (equivalent to 50 ng D_3-cocaine and 100 ng D_3-benzoylecgonine) and diluted with 5 ml phosphate buffer.

Preparation of extraction columns The following liquids are sucked through the columns at weak vacuum avoiding dryness:

(1) Methanol, 2 ml
(2) Phosphate buffer, 2 ml.

The prepared sample (see above) is applied to the column (flow rate 1–2 ml/min). The column is rinsed with 2 ml aqua bidest, 2 ml acetate buffer and 2 ml aqueous methanol. The column is connected with a high vacuum pump (pressure below 20 inchHg) for 2 min to remove the remaining liquid.

Elution: The drugs are eluted with the methanol–ammonia mixture in two steps of 2 and 1 ml at weak vacuum. The organic solvent of the eluate is evaporated at 40 °C with a nitrogen flow.
Derivatization: The residue is silylated with 50 µl MSTFA for 15 min at 80 °C (heating block) in a vial closed with Parafilm foil.

GC–MS analysis
Ten microliters of the derivatized extracts is injected into the GC–MS system (split 1 : 10).
Settings: inlet temperature 250 °C.

Column oven temperature program

Starting temperature	200 °C
Heating rate	5 °C/min
Final temperature	250 °C, 1 min

Carrier gas: Helium 6.0, flow 0.8 ml/min.
Solvent delay: 5 min.

Evaluation Cocaine and benzoylecgonine are identified by (see Table 26.3):

(1) Retention indices (determined by the authors and in agreement with Ref. [22])
(2) Fragments *m/z*.

Table 26.3 Cocaine and benzoylecgonine: relevant ions (*m/z*) and retention indices (RI).

	Range of retention time		
	5–6 min	**6–7 min**	
Compound	***m/z***	***m/z***	**RI**
Cocaine-d_3	185, **306**		2178
Cocaine	182, **303**		2180
Benzoylecgonine-d_3		243, **364**	2258
Benzoylecgonine		240, **361**	2260

Bold values: molecular ion.

The concentration of cocaine and benzoylecgonine can be determined with the characteristic m/z of the sample in comparison with the respective m/z of external standards. The deuterated compounds are used to correct the losses during sample preparation.

Analytical assessment
The procedure identifies very specifically cocaine and benzoylecgonine. It detects reliably the drugs, if present in concentrations that are higher than 20 µg/l. Good trueness is achieved by the use of the deuterated substances as internal standards.

Practicability
The procedure can be performed fairly easily, even though it is more demanding than the HPLC procedure (see Section 26.3.2). The time of analysis and technician time are identical and come up to 5 h.

Annotation
The procedure can be used for the detection of morphine and morphine derivatives after adequate adaptation (spiking with the relevant deuterated opiates).

26.3.4
Medical Assessment and Clinical Interpretation

The pharmacological activity of cocaine depends on the route of administration and the degree of tolerance. For nontolerant users, 16 mg cocaine is effective, if smoked or injected intravenously. Twenty milligrams is needed, if snuffed and 200 mg, if administered orally. 1.0–1.2 g for oral adminstration is considered as a lethal dosage. Maximum concentration of cocaine in plasma observed in "healthy" abusers with regard to dose and route of administration are shown in Table 26.4. Serum cocaine concentrations of 0.5 mg/l may be considered as toxic [3].

Toxicity is proven for serum concentrations exceeding 1 mg/l [19].

The degree of poisoning can be estimated by the concentration of cocaine in the serum. The concentration in urine is less meaningful. Cocaine is metabolized rapidly in the body (Figure 26.9). Independent of the route of administration, about 86% of the applied dose is excreted within 24 h, 96% within 48 h, and 98–99% within 72 h [23]. Only a minor part of the drug is excreted unchanged and is detectable in the urine during 8–12 h [24]. The renal elimination of the main metabolites benzoylecgonine and ecgonine methylester is nearly complete within 3 days after application. Benzoylecgonine in urine represents 36–39% of the dose and while ecgonine methylester accounts for 18–22% of the dose [23]. A nomogram was published [25] to estimate the applied cocaine dose from the benzoylecgonine concentration in urine taking into account the time of application and the sampling of urine. For example, 10 mg benzoylecgonine/l urine will be found 16 h after a dosage of 50 mg as well as 42 h after a dosage of 500 mg.

The effects that the drug exhibits in the central nervous system are the main reason for cocaine abuse. The intensity of these effects depends on the rapidity of transfer to

Table 26.4 Cocaine: peak plasma concentrations.

Application	Dose	Peak *C* (mg/l)	*t* (min)	Reference
i.n.[a]	1.5 mg/kg bm	0.12–0.47	60	[27]
i.v.[a]	16 mg	0.18	5	[28]
p.o.[a]	2 mg/kg bm	0.10–0.42	50–90	[29]
Inhalation[b]	50 mg	0.2	5	[26]
i.v.[a]	20.5 mg	0.2	—	[26]
i.n.[a]	94 mg	0.2	44 ± 7	[26]
i.v.[a]	23 mg	0.18	5	[23]
i.n.[a]	106 mg	0.22	30	[23]
Inhalation[b]	39 mg	0.20	5	[23]
i.n.[a]	64 mg	0.07	37 ± 7	[26]
i.n.[a]	96 mg	0.13	41 ± 5	[26]

i.n. = nasal, i.v. = intravenous, p.o. = oral, peak *C* = peak plasma concentration, bm = body mass, *t* = time after application.

[a]Cocaine hydrochloride

[b]Cocaine base

the CNS, which is influenced by the dose, the preparation, and the route of application and especially by the individual sensitivity to cocaine.

Three phases can be distinguished toxicodynamically:

(1) Euphoria
(2) Ecstasy
(3) Depression.

During the *first phase* (1), cocaine mediates an increase of physical strength and suppresses hunger, thirst, and feelings of exhaustion and fatigue. It stimulates fantasy, may launch a flood of ideas, alleviates communication, increases self-assurance, and stimulates sexually.

The transition to the *next phase* (2) advances slowly. Gradually, the cognitive facilities and the assimilation of stimuli change and produce feelings of stress, anxiety, and displeasure. The abuser relates objects or events in the surroundings to himself, even though they do not have any impact on him, indeed. Hallucinations relating to hearing or the sense of touch (a feeling as though being attacked by bugs) as well as persecution mania may occur.

The *following phase* (3) of depression is characterized by fatigue and exhaustion. The individual is downhearted and phlegmatic and this stimulates the repeated abuse. Therefore, the danger of dependency development for cocaine is rather great as compared to other drugs.

In case of intoxication, the following symptoms may be observed: dizziness, tremor, paleness, cold sweat, headache, vomitus, respiratory disturbances, frequent, and weak pulsation. Severe intoxications lead to seizures, coma, or even to death.

A toxicological analysis is always needed as the final proof of cocaine poisoning. For this purpose, urine should be examined rapidly by an appropriate immunoassay and the positive finding should be confirmed by the identification of benzoylecgonine with an adequate chromatographic procedure. As the time of application cannot be estimated from these results, the relation between the concentration of benzoylecgonine and cocaine should be considered. A ratio exceeding 100 is compatible with a drug administration before more than 10 h [25].

The detection of cocaine in urine must not be associated with distinct symptoms of cocaine poisoning. On the other hand, there may be a negative finding even though symptoms of cocaine intoxication, such as cardiac arrhythmia, coma, respiratory depression, or mydriasis, are present. Probably, the sampling of urine was made very early after cocaine abuse. A sample of urine taken 1 or 2 h later will produce a distinctly positive finding under these circumstances. Otherwise acute cocaine intoxication is unlikely, if the identity of the sample is assured and any analytical error can be excluded.

If body packing is suspected, X-ray of the abdomen should be performed, even though it yields sometimes a false negative finding [19].

General remarks on treatment: Treatment is focused on supporting vital functions. In case of hyperthermia, ice package or other cooling devices are applied [26]. Diazepam may be indicated to suppress seizures and for sedation [26]. Chlorpromazine may be helpful because of its antidopaminergic, antiadrenergic, antihyperthermic, and sedative effects. Arrhythmia may be treated with β-receptor blocking agents (e.g., propranolol) or preferably with α-β-blocking drugs (e.g., labetolol) [19]. One should bear in mind that acute cocaine poisoning can very rapidly and unexpectedly lead to death and therefore therapeutic measures should be applied as quickly as possible.

26.4 γ-Hydroxybutyrate

C. Merckel, V. Auwärter, D. Simmert, and F. Pragst

Introduction

γ-Hydroxybutyrate (GHB) (Figure 26.10) is used since a long time for anesthesia. Drug addicts became aware of the substances in 1998, which is increasingly used since then as "liquid ecstasy." Often GHB is obtained via the internet and applied in the party and techno scene. One should be aware that 1,4-butanediol and γ-butyrolactone are metabolized to GHB and therefore exhibit the same effects. An alternative to butyric acid derivatives are γ-hydroxyvaleriate and γ-valerolactone.

26.4.1 Immunoassay

Immunoassays for the detection of GHB or the related drugs are not available.

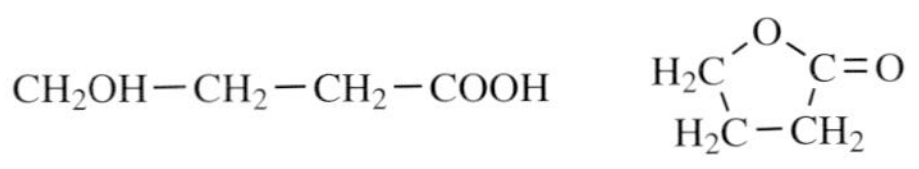

$CH_2OH-CH_2-CH_2-COOH$

$H_2C-O-C{=}O$ / H_2C-CH_2 (ring)

γ-Hydroxy butyrate (1) γ-Butyrolactone (2)

Figure 26.10 Structure of γ-Hydroxybutyric acid (1) and γ-butyrolactone (2).

26.4.2
Gas Chromatography – Mass Spectrometry

Outline

GHB is converted at acidic pH to γ-butyrolactone that is extracted by headspace – solid-phase microextraction (SPME). The charged fiber is introduced into the inlet of the gas chromatograph. The desorbed compounds are separated by gas chromatography and identified by mass spectrometry. Losses during extraction are taken into account by spiking the sample with GHB-d_6 as internal standard before analysis (Figures 26.11 and 26.12).

Specimens

Blood or serum: 1 ml
Urine: 0.05 ml.

Equipment

Gas chromatograph–mass spectrometer (e.g., GC 6890, MS 5873 (Agilent))
Sampling device (e.g., multipurpose sampler MPS 2 (Gerstel) or manual fiber holder)
SPME fibers (preferentially Carboxen – PDMS).

Chemicals

Sodium sulfate
Sulfuric acid (25%)
Helium 6.0 (carrier gas).

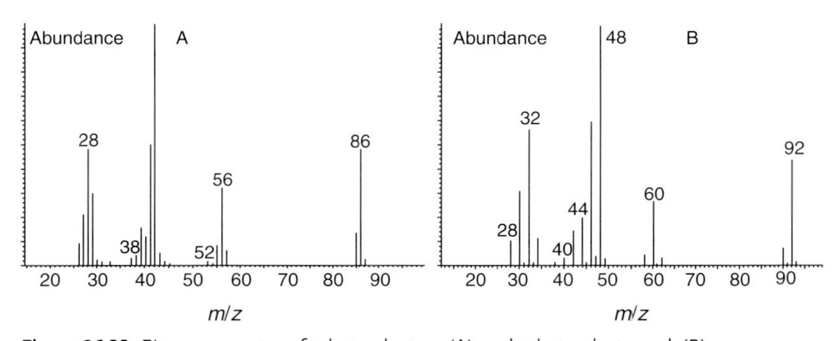

Figure 26.11 EI mass spectra of γ-butyrolactone(A) and γ-butyrolactone-d_6(B).

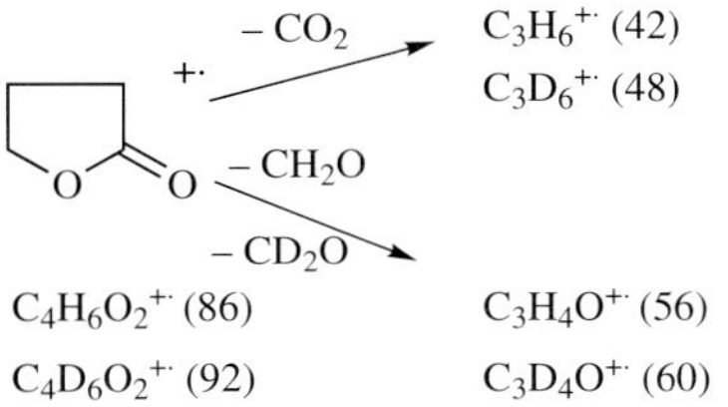

$-CO_2$ → $C_3H_6^{+\cdot}$ (42), $C_3D_6^{+\cdot}$ (48)

$-CH_2O$ / $-CD_2O$ → $C_3H_4O^{+\cdot}$ (56), $C_3D_4O^{+\cdot}$ (60)

$C_4H_6O_2^{+\cdot}$ (86)

$C_4D_6O_2^{+\cdot}$ (92)

Figure 26.12 Mass spectrometric fragmentation of γ-butyrolactone and γ-butyrolactone-d_6.

Reagents
Stock solutions

γ-Hydroxybutyrate 10 mg/ml methanol (e.g., Promochem)
γ-Hydroxybutyrate-d_6 10 mg/ml methanol (e.g., Promochem).

Methanol is evaporated from the 1 ml commercial standard solutions before use and the dry residues are dissolved with 1 ml aqua bidest as methanol disturbs the headspace extraction.

Calibrator solutions

Drug-free serum is spiked appropriately with GHB.
Final concentrations: 2, 5, 10, 20, 50, and 100 mg/l.
Drug-free urine is spiked appropriately with GHB.
Final concentrations: 10, 20, 50, 100, 200, and 500 mg/l.

Sample preparation
One milliliter of blood or serum (0.05 ml urine) and the calibrator solutions are spiked with 50 μl of the aqueous GHB-d_6 solution, and 0.5 g Na_2SO_4 and 1 ml sulfuric acid (25%) are added. The mixture is heated to 65 °C. The temperature is maintained for 15 min, while the headspace extraction of the mixture with a fiber is performed by shaking (250 rpm, 30 s on, 30 s off).

GC–MS analysis
The fiber is injected into the inlet of the gas chromatograph. It is desorbed and reconditioned for 4 min.

Settings

Injection splitless, split opened after 2 min
Inlet temperature: 260 °C
Column: HP-5MS; 25 m, carrier gas flow 1 ml/min.

Column oven temperature program	
Initial temperature	45 °C (2 min)
Heating rate	20 °C/min
Final temperature A	150 °C
Heating rate	30 °C/min
Final temperature B	280 °C (2 min)
Transfer line	280 °C
MS Quad	150 °C
MS Source	230 °C
Single ion monitoring	γ-Butyrolactone *m/z*: 42, 56, 86 Butyrolactone-d_6 *m/z*: 48, 60, 92
Solvent delay	≤ 3 min

Evaluation: The concentration is determined by the use of the calibration curve. It is constructed from the ratios of the peak areas for the fragment ions 86/92 versus the related concentrations.

In urgent need, the GHB concentration may be estimated:

Blood or serum	$C = 50 \times P_{86}/P_{92}$
Urine	$C = 1000 \times P_{86}/P_{92}$

C, concentration of GHB (mg/l); P_{86}, peak area related to m/z 86; P_{92}, peak area related to m/z 92.

The evaluation should be validated by the estimation of the concentration using peak areas m/z 42 and 48 as well as 56 and 60.

Analytical assessment

The procedure determines the concentration of γ-hydroxybutyrate plus γ-butyrolactone specifically. The detection limit was estimated for blood/serum: 0.6 mg/l and for urine 5 mg/l.

Annotations

(1) If GBL shall be determined separately, the analysis is performed without sulfuric acid. The concentration of GHB can be calculated as the difference of the result of the procedure with H_2SO_4 minus the result without H_2SO_4.
(2) Suspect materials or solutions are diluted with water (1 + 99; w/w). One milliliter is used for analysis (see blood samples).
(3) An artifact with the characteristic m/z values of GBL, but longer retention time was observed, when GHB-d_6 in methanol was used. Therefore, aqueous calibrator and internal standard solutions are recommended.
(4) Among the m/z from GHB-d_6 also 42 is present. Therefore, the corresponding peak area ratio is not suitable for quantitative determination.
(5) Other procedures for the determination of GHB are listed in the literature [30–39].

26.4.3
Medical Assessment and Clinical Interpretation

GHB is administered orally, but sometimes the spray is inhaled. In poisoning, the applied doses come up to 2 g or more. Usually, the patients cannot know them because the GHB content of the solvents stemming from many different sources differs considerably. GHB is also an endogenous compound, which may be present at a concentration of up to 4 mg/l in serum and in urine up to 10 mg/l in the absence of abuse.

The relation between GHB concentration in blood and the grade of consciousness is shown below:

State of consciousness	GHB (blood) (mg/l)
Awake	<50
Light sleep	50–150
Sound sleep	150–260
Deep sleep	>260

Plasma half-life ranges from 0.3 to 1.0 h. Five percent of the dose is excreted unchanged in urine. Twelve hours after application, the concentration in urine is normalized.

At low dosage GHB brings about euphoria and is stimulating. For these effects, GHB is named "liquid ecstasy." In case of overdosage, coma and respiratory suppression develop. The effects are fully present 15 min after the intake and may last up to 3 h. The symptoms may be modified or aggravated by alcohol that is drunk simultaneously during the respective parties and by the individual sensitivity. GHB is used as knockout drops that are poured secretly into drinks for date rape. Withdrawal symptoms may occur after ending chronic abuse.

26.5 Ketamine

H. König

26.5.1 Immunoassay

An immunoassay for the determination of ketamine is not available.

Immunoassays for the detection of phencyclidine may produce a false positive finding in the presence of ketamine. If phencyclidine can be excluded, a positive result may be an indication of the presence of ketamine.

26.5.2 High-Performance Liquid Chromatography

The retention time of ketamine related to MPPH is 0.159 and detection limit 0.10 mg/l in that procedure (see Table 13.2 and Section 13.1.1).

26.5.3 Gas Chromatography – Mass Spectrometry

See Section 13.1.2. The retention index is 1835 and the characteristic m/z 180, 209, and 152 (see Table 13.3). For further procedures see Refs [40–42].

26.5.4
Medical Assessment and Clinical Interpretation

Serum concentrations [3]	
Therapeutic concentration	1–6 mg/l
Toxicity	7 mg/l (abuse)
Comatose – lethal	7 mg/l (abuse)
Elimination half-life (serum/plasma)	1–3 h

Ketamine (Figure 26.13) is a racemic derivative of cyclohexanone and is closely related to phencyclidine. It exhibits a strong analgesic effect. It is used in medicine (also in veterinary medicine) as a narcotic for injection (strong analgesia combined with dissociated anesthesia). Often obtained illegally, it is used by drug addicts as a drug substitute to induce hallucinations.

After an intravenous administration, ketamine rapidly accumulates in the brain with maximum concentrations already after 1–2 min. The effect sets in after 15 s, and 15 s later the individual is comatose for 10–15 min. Hallucinations are present during 1 h after a dosage of 1–2 mg/kg body mass. After an intramuscular injection, ketamine is absorbed rather quickly and is analgesically effective: bioavailability 93%. Plasma protein binding is about 47%. Through the processes of *N*-demethylation (forming norketamine) and dehydrogenation, ketamine is rapidly and almost completely metabolized. The metabolites are excreted mainly in the urine [43].

Numerous adverse effects are described, depending on the applied dosage and the duration of the administration. Apart from nausea, vomiting, and hypersalivation, there are to be mentioned cardiac (increase in blood pressure and heart frequency, but a drop in blood pressure and possibly respiratory arrest in patients in a state of shock) and psychomotor effects of different degrees of severity (visual defects, agitation, and hallucinations). These adverse effects can be alleviated by administering benzodiazepines or neuroleptics.

In cases of severe overdosage or intoxication, convulsions, dysrhythmia, and respiratory arrest are observed, which can be treated by diazepam (in case of its ineffectiveness phenytoin or phenobarbital) supported by assisted respiration [43].

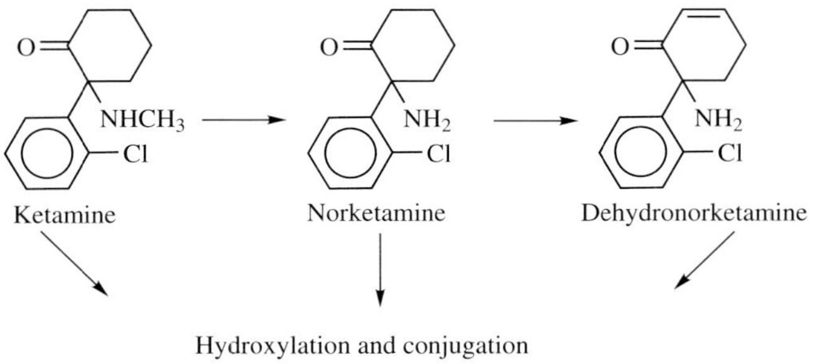

Figure 26.13 Ketamine metabolism.

26.6
Lysergic Acid Diethylamide

L. von Meyer and W.R. Külpmann

Introduction

For some years now, LSD (Figure 26.14) is being abused again, for example, at the disco parties. Rather often it is administered concomitantly with ecstasy. As compared to the seventies of the last century, the doses are lower and "horror trips" occur less often. The number of LSD abusers, who became known to the German authorities for the first time, decreased from 549 in the year 2001 to 229 in the year 2002 and to 145 in the year 2007. In the year 2001, 11 441 trips were detected 30 144 trips were reported in the year 2002, 12488 in 2006 and 10525 in 2007 [18].

26.6.1
Immunoassay

Immunoassays that are run on clinical chemical analyzers are provided by only a few manufacturers, that is,:

CEDIA (Microgenics)
EMIT II (Dade Behring)
MTP-Benzo-EIA (Mahsan)
Online (Roche).

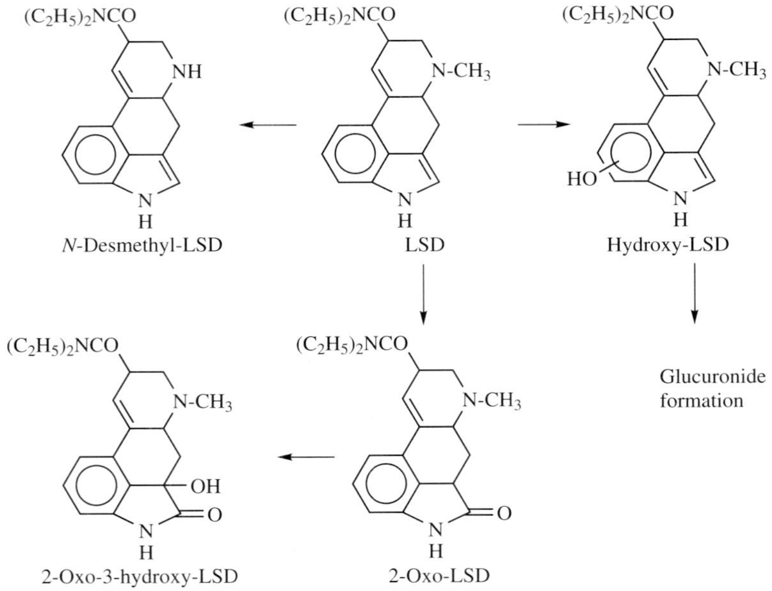

Figure 26.14 LSD metabolism.

For the CEDIA assay, the cutoff concentration is 0.5 μg/l urine. POCT assays that are sufficiently sensitive, are not available. The detection of LSD by immunoassays is hampered by the low concentration of the drug. At low concentrations, immunoassays are rather susceptible to cross-reactivity by other compounds and no commercial LSD assay should be considered as free of interference. The evaluation of the assays of Dade Behring, Mahsan, and Microgenics discovered false positive findings in the presence of several therapeutic drugs:

Often false positive findings were observed, if urine from patients, who were treated with psychotropic drugs, was examined with the Dade Behring assay. The same holds true for the Mahsan assay, although to a lesser extent. Ambroxol and bromhexine interfere with the LSD assay from Microgenics. There are hints that metabolites of fentanyl too may cause false positive findings.

Interferences are observed more often in urine than in blood or serum. The immunoassays from Mahsan and Microgenics may be applied not only to urine but also to blood and could be preferred for this reason.

Nevertheless, the probability of a false positive finding is always rather high and it is only a poor indication of LSD abuse, which should be confirmed by an appropriate method.

26.6.2 High-Performance Liquid Chromatography

Introduction

HPLC is preferred for the detection or determination of LSD. In gas chromatography, LSD may be adsorbed to the glass walls and therefore this method is less suitable. As a future alternative to HPLC, liquid chromatography/mass spectrometry should be considered. At the moment, appropriate systems are not yet commonly available and therefore, a procedure for this technique is not given.

Outline

Urine, plasma, or serum is applied to a column with a resin charged with immobilized antibodies to LSD. After elution with methanol, the dried extract is dissolved with ethylene glycol and the mobile phase of the HPLC and separated by a C_8 phase at pH 2.9. A fluorescence detector is used for detection and identification [44].

Specimens

Urine, serum, and plasma (minimal volume 2 ml).

Equipment

HPLC system with

Pump (e.g., Beckman M 114 (Beckman))
Sampling device (e.g., Merck-Hitachi L 7200 (Merck))
Fluorescence detector (e.g., Merck-Hitachi (Merck)).

Suction device for solid-phase extraction including vials for the eluate (e.g., Adsorbex SPU, Merck).

Centrifuge glass tubes, graduated with round bottom and standard ground joint (NS 12.5), 10 ml.

Centrifuge glass tubes with conical bottom and standard ground joint (NS 12.5), 10 ml.

Polypropylene tubes with round bottom and thread with screw cap, 10 ml.

Vials for sampling device: polypropylene micro vials 300 µl height 11 mm with crimp caps N11 TB/aA (butyl rubber red/PTFE 1.1 mm).

Chemicals

The following chemicals can be recommended, as they did not cause any interference during extraction, chromatography, and detection. If chemicals from other manufacturers will be used, they should be evaluated beforehand.

Acetonitrile (Far UV HPLC, A/0627, Fisher Scientific)
Dipotassium hydrogen phosphate trihydrate (p.a., VWR)
Ethylene glycol (p.a., VWR)
LSD (L7007, Sigma)
LSD ImmunElute (No. 18 68 756, Microgenics)*
Methanol (prepsolv, VWR)
Potassium dihydrogen phosphate (p.a., VWR)
o-Phosphoric acid 85% (p.a., VWR)
Sodium chloride (p.a., VWR)
Sodium hydroxide 1 mol/l (p.a., VWR)
Water (aqua ad injectabilia).

Stock solutions

Lysergic acid methylpropylamide (LAMPA) 100 µg/l acetonitrile (Radian)
LSD 25 µg/l methanol
LSD 25 µg/l acetonitrile (for direct injection into the HPLC system).

The solutions are stable for 6 months if kept in the refrigerator.

Reagents

Phosphate buffer 0.02 moll, pH 7.0: 6.8 g KH_2PO_4, 14.9 g $K_2HPO4 \times 3H_2O$, and 8.77 g NaCl are dissolved with 1000 ml aqua ad inject. The solution is adjusted to pH 7.0 with sodium hydroxide and filtrated through acid resistant filter paper. The buffer is stable for 6 months if kept in the refrigerator.

Calibrator solution: The solutions are prepared immediately before use by spiking drug-free urine or serum with the methanolic stock solution.

* LSD ImmunElute is at present not available. Liquid/Liquid extraction can be performed with 1-Chlorobutane.

Calibrator solution 1: 0.5 μg LSD/1000 ml
Calibrator solution 2: 3.0 μg LSD/1000 ml.

Sample preparation

Extraction

Urine: 1 ml urine and 5 ml phosphate buffer (0.02 mmol/l, pH 7.0) are pipetted into a polypropylene tube. Five microliters (0.5 ng LAMPA) of the internal standard LAMPA solution is added and the mixture is shaken mechanically. After centrifugation the supernatant is transferred concomitantly with 200 μl LSD-ImmunElute resin suspension to a Polyprep extraction column C (which is part of the LSD ImmunElute kit). After mixing, the suspension shall settle in 15 min. The column is rinsed once with the phosphate buffer (0.02 mmol/l, pH 7.0) and two times with 10 ml aqua bidest each. After sucking for 2 s only 1 ml methanol and 20 μl ethylene glycol are pipetted on to the column. It is closed with a stopper and left for 5 min. The stopper is removed and the eluate is collected in a centrifuge glass tube with a conical bottom. The elution procedure is repeated once more. Methanol of the combined eluates is evaporated at 50 °C by the use of a rotary evaporator. Seventy microliters of the mobile phase (see below) is added to the ethylene glycol residue. The mixture is shaken for a short while and transferred to a vial of the HPLC sampling device.

Blood, serum or plasma: 1 ml of blood, serum, or plasma and 4 ml of methanol are pipetted into a polypropylene tube. Five microliters (0.5 ng LAMPA) of the internal standard LAMPA solution is added and the mixture is shaken vigorously. After centrifugation, the supernatant is transferred to two test tubes and centrifuged again (12 000 rpm). The two supernatants are combined and evaporated (rotary evaporator) to a remaining volume of 0.5 ml. It is diluted with 5 ml phosphate buffer (0.02 mmol/l) and then processed as described for "urine." Calibrators are prepared in the same way for HPLC analysis.

HPLC analysis
Thirty microliters to the solution in the vial is applied to a RP-Select B column 250 mm × 4 mm (internal diameter) (Merck). The mobile phase consists of a mixture of acetonitrile and triethylammonium phosphate buffer pH 2.9, 1 mol/l 80 + 20 (v/v) that is rinsed with a flow rate of 1.3 ml/min at room temperature (23 °C). The fluorescence spectra are registered for an excitation wavelength of 310 nm and an emission wavelength of 420 nm.

The needle of the mechanized sampling device is cleaned after every injection with a mixture of methanol and water 1 + 1 (v/v).

Data acquisition and evaluation is assisted by HP Chemstation, software for LC, GC, A/D, and CE (revision A.04.01).

Evaluation: The data are evaluated by the use of the self-made library and the calibration program of the software.

Analytical assessment
The selective extraction by LSD antibodies yields chromatograms that can be easily evaluated. The detection limit of the procedure including fluorescence detection is 0.05 μg/l, which is far below the cutoff concentration of the immunoassay (CEDIA): 0.5 μg/l urine and 0.2 μg/l blood, serum, and plasma.

Therefore, the procedure is useful to confirm immunochemical findings.

26.6.3 Gas Chromatography

The sample preparation, as described before (Section 26.6.2) including immunoaffinity chromatography, can be used as well for corresponding gas chromatography or gas chromatographic–mass spectrometric methods. However, for these procedures, the extract must be evaporated to dryness. During this step, a considerable fraction of polar compounds like LSD, especially if present at low concentrations, may be lost by adsorption to the glass walls. Furthermore, the high temperature of the inlet may lead to losses by pyrolysis and artifact formation. The examination by HPLC should be preferred.

26.6.4 Medical Assessment and Clinical Interpretation

LSD is administered predominantly per os. The doses are rather low and ranges from 0.05 to 0.3 mg. LSD is absorbed rapidly and peak concentrations in plasma are reached within 30–60 min. The volume of distribution was estimated at 0.27 l/kg [45]. After hydroxylation, LSD is glucuronidated and excreted in the bile. Plasma half-life is 3 h. Only a small portion of a dose is excreted unchanged in the urine and is detectable within 1–5 days after the intake. Abusers become tolerant rather quick, but normal susceptibility is restored within 4–5 days after the withdrawal. Plasma concentrations exceeding 2 μg/l are considered as toxic. LSD is a hallucinogenic and a psychedelic drug as well. One hour after the intake, not only the optical and acoustic disturbances occur, but also a dissociation of the personality is observed.

Among the clinical symptoms, mydriasis, tachycardia, elevated body temperature, and hyperglycaemia (sympathomimetic effects) are predominant and may last up to 8 h. After the intake of high doses, salivation and lacrimation are increased, and paresthesia, hyperreflexia, ataxia, respiratory disturbances, and nausea leading to vomitus occur. Fatal outcome is seldom directly due to intoxication. More often LSD may lead the individual to suicide or may become the main cause for a deadly accident.

26.7 Mescaline

W.R. Külpmann

Mescaline (Figure 26.15) is a phenethylamine and therefore related to amphetamine. The hallucinogenic drug can be obtained from peyote cactus (*Lophophora williamsii*)

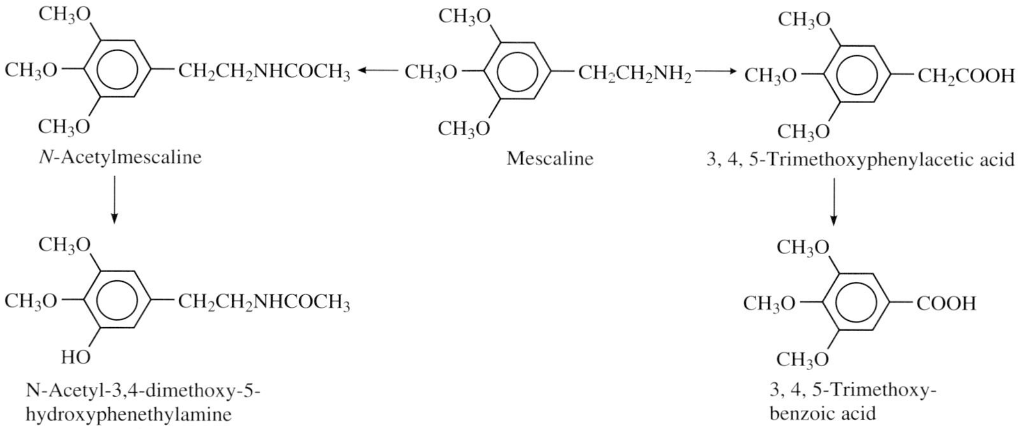

CH_3O / CH_3O / CH_3O — $CH_2CH_2NHCOCH_3$ ← CH_3O / CH_3O / CH_3O — $CH_2CH_2NH_2$ → CH_3O / CH_3O / CH_3O — CH_2COOH

N-Acetylmescaline   Mescaline   3, 4, 5-Trimethoxyphenylacetic acid

↓   ↓

CH_3O / CH_3O / HO — $CH_2CH_2NHCOCH_3$ CH_3O / CH_3O / CH_3O — COOH

N-Acetyl-3,4-dimethoxy-5-hydroxyphenethylamine   3, 4, 5-Trimethoxy-benzoic acid

Figure 26.15 Mescaline metabolism.

and has been used for centuries in Mexico and the southwest of USA to induce psychedelic effects.

It can be detected by GC–MS (see Chapter 12).

Serum concentrations were about 3.8 and 1.5 mg/l, respectively, 2 and 7 h after the intake of an oral hallucigenic dose of 500 mg. Plasma half-life was estimated as 6 h. A percentage of 87% of the dose is excreted within 24 h in the urine, about 60% unchanged.

26.8 Phencyclidine

H. Käferstein and G. Sticht

Phencyclidine (1-*p*henyl*c*yclohexyl*p*iperidine, PCP), Angel Dust, Peace Pill) is an anesthetic and opioid analgesic drug, that may not be prescribed as per German law. Abuse is by smoking. In the United States, phencyclidine is a fairly popular drug of abuse, whereas in Germany it is seldom used by addicts.

26.8.1 Immunoassay

General information on immunoassays is given in Section 4.1. Immunoassays for the detection of PCP are available from many manufacturers, because it belongs to the five drugs that are mentioned explicitly by the NIDA. Usually, the cutoff concentration of the assays is 25 µg/l urine, which is identical to the recommendation of SAMHSA [21].

26.8.2
High-Performance Liquid Chromatography

Outline
PCP can be detected in urine by the Remedi system (BioRad) (see Section 12.1). After solid-phase extraction, phencyclidine can be detected and its concentration can be estimated by an on-line procedure. The Remedi system is not sufficiently commercially available, but is still in use. It may be considered as an example for other dedicated systems.

Specimen
Urine 1 ml.

Regarding *equipment, chemicals, reagents,* and details of *procedure* see Section 12.1.

Evaluation
Phencyclidine and its metabolite 4-hydroxyphencyclidine (Figure 26.16) are registered in the library of spectra. The two compounds absorb only the UV light with short wavelength, but not all at 230 nm and longer. Among others, morphine, quinine, and quinidine as well as a metabolite of doxylamine interfere, as these substances are separated neither by sample preparation nor by chromatography.

26.8.3
Gas Chromatography–Mass Spectrometry

Outline
The sample is prepared by liquid/liquid extraction as described for amphetamines (see Section 26.1.1). After re-extraction the sample is investigated by capillary gas chromatography. Deuterated PCP (D_5 –PCP) is used as an internal standard.

Specimens
Serum, blood, and urine: minimal volume 2 ml.

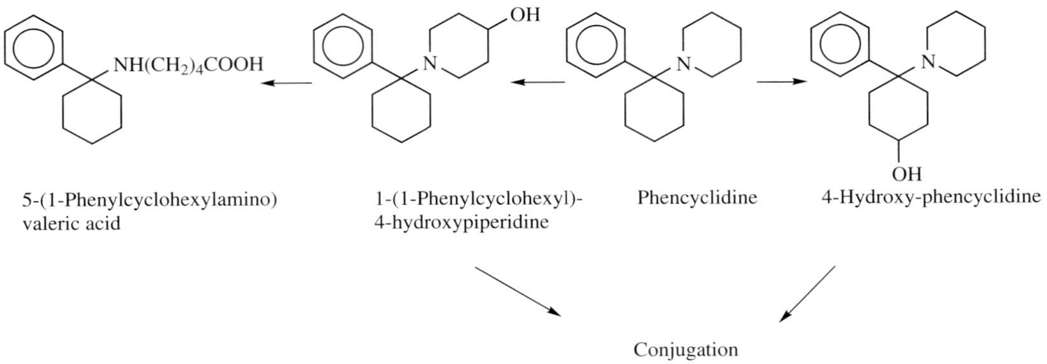

$NH(CH_2)_4COOH$
OH
N
N
N
OH
5-(1-Phenylcyclohexylamino) valeric acid
1-(1-Phenylcyclohexyl)-4-hydroxypiperidine
Phencyclidine
4-Hydroxy-phencyclidine
Conjugation

Figure 26.16 Phencyclidine metabolism.

Equipment

Gas chromatograph–mass spectrometer
GC column: Ultra 1 (length 12 m, internal diameter 0.2 mm, film thickness 0.33 μm) (Agilent)
Centrifuge tube with a conical bottom and a standard ground joint.

Chemicals

Ammonia solution 25%
1-Chlorobutane (for chromatography, gradient grade)
Chloroform
Isohexane
Methanol

Sodium hydroxide 1 mol/l
Helium 6.0.

Reagents
Stock solutions

(1) Phencyclidine 1 g/l methanol (Radian/Promochem)
(2) Phencyclidine-d_5 0.1 g/l methanol (Radian/Promochem).

Working solutions

Working solution 1: 100 μl stock solution 1 is made up with methanol ad 10.0 ml in a volumetric flask. PCP concentration: 10 ng/μl.
Working solution 2: 100 μl stock solution 2 is made up with methanol ad 1.0 ml in a volumetric flask. PCP-d_5 concentration: 10 ng/μl.
Negative control sample: Drug-free pooled serum.
Positive control sample: 1.0 ml drug-free pooled serum is spiked with 5 μl working solution 1.

Sample preparation
Extraction: 1.0 ml of the sample as well as 1.0 ml of the negative and positive control samples are spiked with 5 μl working solution 2 (deuterated PCP).

One milliliters sodium hydroxide and 5 ml iso-hexane are added and the mixture is well shaken. It is centrifuged, if necessary, for separation of the layers. Four milliliters of the organic layer is transferred to a centrifuge tube. One milliliter sulfuric acid (0.05 mol/l) is added for re-extraction and the organic layer is discarded. The aqueous layer is alkalinized with 0.5 ml ammonia solution and extracted with 1 ml chloroform. Chloroform is evaporated with a flow of nitrogen. The dry residue is dissolved with 50 μl 1-chlorobutane.

GC–MS analysis
Two microliters of the dissolved residue is injected in split/splitless mode (240 °C) into the gas chromatography–mass spectrometry system.

Column oven temperature program	
Initial temperature	60 °C (1 min)
Heating rate	15 °C/min
Final temperature	280 °C
Carrier gas	Helium 6.0

Evaluation: For the quantitative determination of the PCP concentration in the sample, the mass ratios of deuterated and nondeuterated PCP (after subtracting the related intensities of the negative control sample) are used. (By this way, it is taken into account that PCP-d_5 will contain a small fraction of nondeuterated PCP). Retention index for PCP is 1900. Typical m/z for PCP is 186, 200, and 243 and for PCP-d_5 is 191, 205, and 248. For quantitative determination, the ratios 243/248 and 200/205 are recommended.

Analytical assessment
The procedure detects specifically phencyclidine. The detection limit is below 5 μg/l. Several other drugs can be detected by this procedure (Section 26.1).

Practicability
The investigation of one sample takes 2 h. The time of analysis and technician time are identical.

26.8.4 Medical Assessment and Clinical Interpretation

One "cigarette" may contain 1–100 mg PCP. During chronic abuse, daily dose may go up to 1 g. Volume of distribution is rather high: 6.2 l/kg. In case of withdrawal after chronic abuse, the PCP concentration decreases more quickly during the first 9 days than it does afterward. Usually, positive findings are obtained at least for 2 weeks and negative findings at least after 30 days [47].

The phencyclidine concentration in blood is not closely related to clinical findings and does not give clear indications of the necessary therapeutic measures. Therapeutic concentrations are not relevant and not applicable. A maximum concentration of 3 μg/l blood was measured after oral administration of 1 mg. Concentrations exceeding 100 μg/l blood are considered as toxic and concentrations exceeding 300 μg/l as possibly fatal. Among traffic offenders, PCP concentration in blood ranged between 12 and 118 μg/l with a median of 44 μg/l [48]. In a case of extreme overdosage, serum concentration was found to be 1879 μg/l. Even though there was a hyperthermia of 41.3 °C, the patient survived.

Symptoms that require hospitalization are seizures, rhabdomyolysis, and hyperthermia. Coma may follow or precede delirium. Treatment is guided by symptoms. However, most fatalities are not caused directly by PCP intoxication, but occur as follow-up to insufficient muscle coordination leading to severe injuries and drowning.

26.9
Psilocybin/Psilocin

G. Sticht and H. Käferstein

Introduction

Psilocybin and psilocin were first isolated from the mushroom *Psilocybe mexicana*. Other psilocybin-containing species are *Panaeolus, Conocybe, Copelandia, Gymnopilus, Pluteus,* and *Stropharia*. The drugs are present in mushrooms, such as *Panaeolus subalteatus* and *Stropharia coronilla* in Europe. Usually, the content of psilocybin exceeds the content of psilocin that lacks the stabilizing phosphate group. Both, psilocybin and psilocin are illicit drugs that may not be prescribed. They are abused increasingly for their hallucinogenic effects by eating the relevant mushrooms (see above).

26.9.1
Immunoassay

Immunoassays for the detection of psilocybin/psilocin are not commercially available.

26.9.2
High-Performance Liquid Chromatography

Outline

The Remedi system (BioRad) (Section 12.1) can be used for the detection of the drugs. Hydrolysis with β-glucuronidase is followed by solid-phase extraction. Psilocin is separated on-line and identified by the HPLC system. The system is not sufficiently commercially available, but is still in use.

Specimen

Urine, 1 ml

For *equipment, chemicals, and reagents* see Section 12.1.

Additionally needed:

β-Glucuronidase from *E. coli* K12 (Roche)
Thermostat (e.g., water bath).

Sample preparation

One milliliter urine and 10 μl glucuronidase solution are mixed and maintained for 1 h at 45 °C. After cooling to room temperature, the sample is prepared as described (Section 12.1).

Evaluation

Psilocin presents a spectrum with maximum absorption at 224 nm and minimum absorption at 247 nm, which is typical for tryptamines. Metoprolol and moclobemide metabolites, haloperidol, and metipranolol and some other drugs also have a similar retention time as psilocin. They might interfere as they are not separated during sample preparation. Detection limit of psilocin is 100 μg/l urine.

26.9.3
Gas Chromatography–Mass Spectrometry

Outline
After solid-phase extraction, the sample is silylated and analyzed by capillary gas chromatography. Deuterated morphine is used as internal standard. If the sample is hydrolyzed enzymatically before extraction, psilocin including psilocin glucuronide is measured.

Specimens

Serum and plasma: minimal volume 2 ml
Urine: minimal volume 1 ml.

Equipment

Gas chromatograph–mass spectrometer
Column: Ultra 1 (length 12 m, i.d. 0.2 mm, film thickness 0.33 μm)
Heating block adjustable up to 100 °C
IST Vac Master (ict)
High vacuum pump.

Chemicals (p.a. grade)

Ammonia solution, 25%
Acetic acid, 100%
Disodium hydrogen phosphate dihydrate
Methanol
MSTFA (*N*-methyl-*N*-trimethylsilyl-trifluoroacetamide)
Potassium dihydrogen phosphate
Potassium hydroxide, 1 mol/l
Helium 6.0
Nitrogen 5.0
Certify LRC 300 mg (Varian).

Reagents
Stock solutions

(1) Psilocin 0.1 g/l methanol (Radian/Promochem)
(2) Morphine-d_3 1.0 g/l methanol (Radian/Promochem)
(3) Internal standard: 6.25 μg morphine-d_3/ml methanol: 25 μl stock solution 2 diluted with methanol ad 4.0 ml.

Working solutions

(1) *Internal standard:* 25 ng morphine-d_3/40 μl methanol. Stock solution 3 diluted with methanol 1 + 9 (v/v)

(2) *Phosphate buffer, pH 8:* Potassium dihydrogen phosphate 13.61 g is dissolved with about 900 ml aqua bidest in a volumetric flask. The pH is adjusted to 8.0 and the solution made up to 1000 ml with aqua bidest.
(3) *Acetate buffer, pH 4.0:* Acetic acid 5.7 ml is pipetted to about 800 ml aqua bidest in a volumetric flask. After addition of potassium hydroxide solution (16 ml), the pH is adjusted to pH 4.0 and the volume is made up to 1000 ml with aqua bidest.
(4) *Methanol–water mixture:* Methanol 30 ml + aqua bidest 70 ml.
(5) *Eluent:* Methanol/ammonia solution 98 + 2 (v/v), to be prepared just before use.

Sample preparation

One milliliter serum or 0.5 ml urine are spiked with 25 ng morphine-d_3 (40 µl working solution 1) and diluted with 5.0 ml phosphate buffer. *Extraction:* The Certify LRC columns are prepared by rinsing at slightly reduced atmospheric pressure with 2 ml methanol and 2 ml phosphate buffer, and by avoiding dryness.

The prepared sample is applied to the column adjusted to a flow rate of 1 to 2 ml/min. Then, 2 ml aqua bidest, 2 ml acetate buffer, and 2 ml methanol–water mixture are rinsed through the column. A high vacuum pump is attached to suck the column to dryness (vacuum below 20 inchHg). The drugs are eluted with the methanol/ammonia solution: First, 2 ml and then 1 ml is rinsed at slightly reduced pressure. The organic solvent of the combined eluates is evaporated in a heating block at 40 °C with nitrogen flow. The residue is dissolved with 50 µl MSTFA. The tube is closed with Parafilm foil and kept 15 min at 80 °C in a heating block for derivatization.

Determination of psilocin including psilocin glucuronide: The sample is mixed with 40 µl working solution 1 and 10 µl glucuronidase solution. The solution is maintained at 45 °C for 1 h in a water bath. It is diluted with 5 ml phosphate buffer and further processed as described.

GC–MS analysis

Ten microliters of the derivatized extract is injected into the GC–MS system.

Settings: Inlet temperature 250 °C.

Column oven temperature program	
Initial temperature	200 °C (5 min)
Heating rate	5 °C/min
Final temperature	250 °C (1 min)

Carrier gas, Helium	0.8 ml/min
Solvent delay	7 min

Evaluation Retentions indices and typical m/z are used for identification

	RI	m/z
Psilocin	2099	290, 291, 348
Morphine-d_3	2520	404, 417, 432, 433

RI, retention index.

External standards are used for quantitative determination. Morphine-d_3 is used to correct for losses during extraction.

Analytical assessment
The procedure allows identifying psilocin specifically at concentrations exceeding 10 μg/l serum. Morphine-d_3 is preferred to morphine as internal standard in case of concomitant abuse of morphine/diamorphine. An alternative to morphine-d_3 as an internal standard would be psilocin-d_{10} (Radian/Promochem). Psilocin is detected by the GC–MS procedures described for morphine (Chapter 14) and cocaine (Section 26.3).

Practicability
The procedure takes about 5 h. The time for analysis and the technician time are identical.

26.9.4
Medical Assessment and Clinical Interpretation

Psilocybin is dephosphorylated rapidly in the human organism to produce psilocin (Figure 26.17). Therefore, only psilocin can be detected in the biological fluids. However, about 65% of total psilocybin in serum and about 80% in urine are glucuronidated [49]. The conjugates will escape detection without the application of glucuronidase.

There is no information on toxic concentrations.

Magic mushrooms are eaten solely to produce hallucinogenic effects. A state of slight euphoria is achieved already after oral administration of 4 mg psilocybin, whereas 6–20 mg lead to distinct psychic effects. Temperature, heart rate, and blood pressure may be increased and be accompanied by mydriasis. Not only euphoria is observed, but also anxiety and panic attacks are experienced [50]. Treatment with sedative drugs will be necessary only for severe excitation. Usually, the effects disappear within some hours without any sequels. Withdrawal symptoms have not been observed [50].

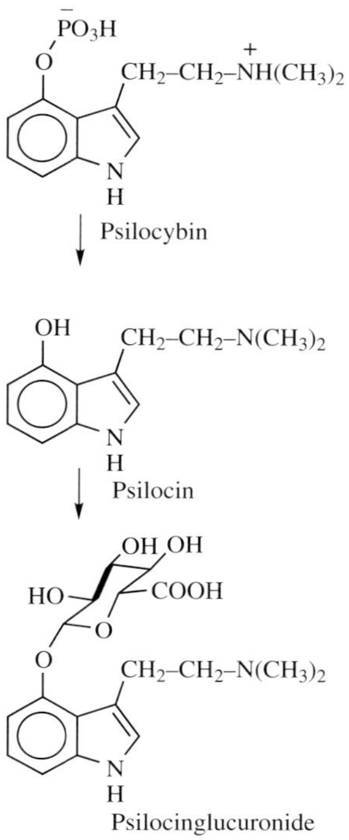

PO_3H ... $CH_2–CH_2–\overset{+}{N}H(CH_3)_2$ ... N H

Psilocybin

OH $CH_2–CH_2–N(CH_3)_2$ ... N H

Psilocin

OH OH, HO, COOH, O, O $CH_2–CH_2–N(CH_3)_2$ ... N H

Psilocinglucuronide

Figure 26.17 Psilocybin metabolism.

References

References for Section 26.1

1 Maurer, H.H. (1997) Designer drugs of the amphetamine type – metabolism and detection in urine by immunoassay and GC–MS. Proceedings to the Satellite Symposium "Designer-Drugs" of the Amphetamine Type – "Ecstasy," Mosbach, 17.04.1999, Abbott, Wiesbaden.

2 Cox, D.E. and Williams, K.R. (1996) "Adam" or "Eve"? – A toxicological conundrum. *Forensic Science International*, **77**, 101–108.

3 Schulz, M. and Schmoldt, A. (2003) Therapeutic and toxic blood concentrations of more than 800 drugs and other xenobiotics. *Pharmazie*, **58**, 447–474.

4 Uges, D. (1999) Toxic and therapeutic ranges. *Bulletin of The International Association of Forensic, Toxicologists* **29**, 15–16.

5 Verebey, K., Alrazi, J. and Jaffe, J.H. (1988) The complications of "ecstasy" (MDMA). *The Journal of the American Medical Association*, **959**, 1649–1650.

6 Fineschi, V. and Masti, A. (1996) Fatal poisoning by MDMA (ecstasy) and MDEA: a case report. *International Journal of Legal Medicine*, **108**, 272–275.

7 Kraemer, T. and Maurer, H.H. (1998) Determination of amphetamine, methamphetamine and amphetamine-derived designer drugs or medicaments in blood and urine. *Journal of Chromatography B*, **713**, 163–187.

8 Ennslin, H.K., Kovar, K.-A. and Maurer, H.H. (1996) Toxicological detection of the designer drug 3,4-methylenedioxyethyl-amphetamine (MDE, "Eve") and its metabolites in urine by gas chromatography–mass spectrometry and fluorescence polarization immunoassay. *Journal of Chromatography B*, **683**, 189–197.
9 Kintz, P. (1997) Excretion of MBDB in urine, saliva, and sweat following single oral administration. *Journal of Analytical Toxicology*, **21**, 570–575.
10 Tehan, B., Hardern, R. and Bodenham, A. (1993) Hyperthermia associated with 3,4-methylenedioxyamphetamine ("Eve"). *Anaesthesia*, **48**, 507–510.

References for Section 26.2

11 de Faubert-Maunder, M.J. (1974) Preservation of cannabis thin-layer chromatograms. *Journal of Chromatography*, **100**, 196–199.
12 Joern, W.A. (1987) Detection of past and recurrent marijuana use by a modified GC/MS procedure. *Journal of Analytical Toxicology*, **11**, 49–52.
13 ElSohly, M.A., ElSohly, H.N. and Jones, A.B. (1983) Analysis of the major metabolite of Δ-9-tetrahydrocannabinol in urine – II. A HPLC procedure. *Journal of Analytical Toxicology*, **7**, 262–264.
14 Karlsson, L. and Roos, C. (1984) Combination of liquid chromatography with ultraviolet detection and gas chromatography with electron capture detection for the determination of Δ-9-tetrahydrocannabinol-11-oic acid in urine. *Journal of Chromatography*, **306**, 183–189.
15 Huestis, M.A., Mitchell, J.M. and Cone, E.J. (1995) Detection times of marijuana metabolites in urine by immunoassay and GC–MS. *Journal of Analytical Toxicology*, **19**, 443–449.
16 Johansson, E. and Halldin, M.M. (1989) Urinary excretion half-life of delta 1-tetrahydrocannabinol-7-oic acid in heavy marijuana users after smoking. *Journal of Analytical Toxicology*, **13**, 218–223.
17 Mulé, S.J., Lomax, P. and Gross, S.J. (1988) Active and realistic passive marijuana exposure tested by three immunoassays and GC/MS in urine. *Journal of Analytical Toxicology*, **12**, 113–116.

References for Section 26.3

18 Drogen- und Suchtbericht (2008) der Bundesregierung Deutschland (Report on drugs and addiction (2008) of the Federal Government of Germany.
19 Hollander, J.E. (2004) Cocaine, in *Medical Toxicology*, 3rd edn (ed. R.C. Dart), LippCincott Williams and Wilkins, Philadelphia, PA.
20 de la Torre, R., Segura, J., de Zeeuw, R. and Williams, J. (1997) Recommendations for the reliable detection of illicit drugs in urine in the European Union, with special attention to the workplace. *Annals of Clinical Biochemistry*, **34**, 339–344.
21 Substance Abuse and Mental Health Services Administration (SAMHSA) (2004) *Federal Register.* **69** (71), 19673–19732.
22 de Zeeuw, R.A. (1992) *DFG-REPORT XVIII – Gas Chromatographic Retention Indices of Toxicologically Relevant Substances on Packed or Capillary Columns with Dimethylsilicone Stationary Phases*, 3rd revised and enlarged edn, Wiley-VCH Verlag GmbH, Weinheim.
23 Jeffcoat, A.R., Perez-Reyes, M., Hill, J.M., Sadler, B.M. and Cook, L.E. (1989) Cocaine disposition in humans after intravenous nasal insufflation (snorting) or smoking. *Drug Metabolism and Disposition*, **17**, 153–159.
24 Hamilton, H.E., Wallace, J.E., Shimek, E.L., Land, P., Harris, S.C. and Christenson, J.G. (1977) Cocaine and benzoylecgonine excretion in humans. *Journal of Forensic Sciences*, **22**, 697–707.

25 Ambre, J. (1985) The urinary excretion of cocaine and metabolites in humans: a kinetic analysis of published data. *Journal of Analytical Toxicology*, **9**, 242–245.

26 Caplan, Y.H. (1988) *Cocaine: Abuse Drugs Monograph Series*, Abbott Laboratories, Diagnostic Division, Irving, Texas.

27 van Dyke, C., Barash, P.G., Jatlow, P. and Byck, R. (1976) Cocaine: Plasma concentrations after intranasal application in man. *Science*, **191**, 859–861.

28 Javaid, J.J., Dekirmenjian, H., Davis, J.M. and Schuster, C.R. (1978) Determination of cocaine in human urine, plasma and red blood cells by gas–liquid chromatography. *Journal of Chromatography*, **152**, 105–113.

29 van Dyke, C., Jatlow, P., Ungerer, J., Barash, P.G. and Byck, R. (1978) Oral cocaine: plasma concentrations and central effects. *Science*, **200**, 211–213.

References for Section 26.4

30 Blair, S., Song, M., Hall, B. and Brodbelt, J. (2001) Determination of gamma-hydroxybutyrate in water and human urine by solid phase microextraction-gas chromatography/quadrupole ion trap spectrometry. *Journal of Forensic Sciences*, **46**, 688–693.

31 Frison, G., Tedeschi, L., Maietti, S. and Ferrara, S.D. (2000) Determination of gamma-hydroxybutyric acid (GHB) in plasma and urine by headspace solid-phase microextraction and gas chromatography/positive ion chemical ionization mass spectrometry. *Rapid Communications in Mass Spectrometry*, **14**, 2401–2407.

32 Bosman, I.J. and Lusthof, K.J. (2003) Forensic cases involving the use of GHB in The Netherlands. *Forensic Science International*, **133**, 17–21.

33 Elian, A.A. (2001) GC-MS determination of gamma-hydroxybutyric acid (GHB) in blood. *Forensic Science International*, **122**, 43–47.

34 LeBeau, M.A., Montgomery, M.A., Miller, M.L. and Burmeister, S.G. (2000) Analysis of biofluids for gamma-hydroxybutyrate (GHB) and gamma-butyrolactone (GBL) by headspace GC–FID and GC–MS. *Journal of Analytical Toxicology*, **24**, 421–428.

35 Villain, M., Cirimele, V., Ludes, B. and Kintz, P. (2003) Ultra-rapid procedure to test for gamma-hydroxybutyric acid in blood and urine by gas chromatography–mass spectrometry. *Journal of Chromatography B*, **792**, 83–87.

36 de Vriendt, C.A., van Sassenbroeck, D.K., Rosseel, M.T., van de Velde, E.J., Verstraete, A.G., Vander Heyden, Y. and Belpaire, F.M. (2001) Development and validation of a high-performance liquid chromatographic method for the determination of gamma-hydroxybutyric acid in rat plasma. *Journal of Chromatography B*, **752**, 85–90.

37 Baldacci, A., Theurillat, R., Caslavska, J., Pardubska, H., Brenneisen, R. and Thormann, W. (2003) Determination of gamma-hydroxybutyric acid in human urine by capillary electrophoresis with indirect UV detection and confirmation with electrospray ionization ion-trap mass spectrometry. *Journal of Chromatography A*, **990**, 99–110.

38 Duer, W.C., Byers, K.L. and Martin, J.V. (2001) Application of a convenient extraction procedure to analyze gamma-hydroxybutyric acid in fatalities involving gamma-hydroxybutyric acid, gamma-butyrolactone, and 1,4-butanediol. *Journal of Analytical Toxicology*, **25**, 576–582.

39 Ferrara, S.D., Tedeschi, L., Frison, G., Castagna, F., Gallimberti, L., Giorgetti, R., Gessa, G.L. and Palatini, P. (1993) Therapeutic gamma-hydroxybutyric acid monitoring in plasma and urine by gas chromatography–mass spectrometry. *Journal of Pharmaceutical and Biomedical Analysis*, **11**, 483–487.

References for Section 26.5

40 Wieber, J., Gugler, R., Hengstmann, J.H. and Dengler, H.J. (1975) Pharmacokinetics of ketamine in man. *Anaesthesia*, **24**, 260–263.

41 Stiller, R.L., Dayton, P.G., Perel, J.M. and Hug, C.C. (1982) Gas chromatographic analysis of ketamine and norketamine in plasma and urine: nitrogen-sensitive detection. *Journal of Chromatography*, **232**, 305–314.

42 Blednov, Y.A. and Simpson, V.J. (1999) Gas chromatography-mass spectrometry assay for determination of ketamine in brain. *Journal of Pharmacological and Toxicological Methods*, **41**, 91–95.

43 Caravati, E.M. (2004) Hallucinogenic drugs, in *Medical Toxicology*, 3rd edn (ed. R.C. Dart), Lippincott Williams and Wilkins, Philadelhia, PA.

References for Section 26.6

44 Bergemann, D., Geier, A. and von Meyer, L. (1999) Determination of lysergic acid diethylamide in body fluids by high-performance-liquid chromatography and fluorescence detection – a more sensitive method suitable for routine use. *Journal of Forensic Sciences*, **44**, 372–375.

45 Caravati, E.M. (2004) Hallucinogenic drugs, in *Medical Toxicology*, 3rd edn (ed. R.C. Dart), Lippincott Williams and Wilkins, Philadelphia, PA.

References for Section 26.7

46 Charalampous, K.D., Walker, K.E. and Kinross-Wright, J. (1966) Metabolic fate of mescaline in man. *Psychopharmacologia*, **9**, 48–63.

References for Section 26.8

47 Caravati, E.M. (2004) Hallucinogenic drugs, in *Medical Toxicology*, 3rd edn (ed. R.C. Dart), Lippincott Williams and Wilkins, Philadelphia, PA.

48 Kunsmann, G.W., Levine, B., Costantino, A. and Smith, M.L. (1997) Phencyclidine Blood concentrations in DRE cases. *Journal of Analytical Toxicology*, **21**, 498–502.

References for Section 26.9

49 Sticht, G. and Käferstein, H. (2000) Detection of psilocin in body fluids. *Forensic Science International*, **113**, 403–407.

50 Schonwald, S. (2004) Mushrooms, in *Medical Toxicology*, 3rd edn (ed. R.C. Dart), Lippincott Williams and Wilkins, Philadelphia, PA.

27
Solvents and Inhalants

27.1
Highly Volatile Alcohols and Ketones

27.1.1
Ethanol

H.J. Gibitz

Introduction

In Germany, an estimated 2.5 million people out of a total population of 82 million (including children) are considered as alcoholics. Ten percent of the alcoholics are young individuals. Every year ethanol causes death of 40 000 people, out of which 17 000 die of liver cirrhosis. In comparison to this, 1513 persons died of drug abuse in 2002. Alcohol plays a decisive role in every fourth act of violence with lethal outcome and in 17% of fatal traffic accidents. About 2200 children with disorders caused by alcohol are born every year. Up to the age of 60, the lethality of alcoholics is 22%, that of smokers is 18%, and that of smoking alcoholics is 31% [1].

In patients who had a traffic, industrial, or leisure accident or in cases of a suspected exogenous intoxication, it is therefore justified to always check the ethanol concentration in blood. In these cases, it is important to obtain quickly valid results that need not comply with all the requirements of forensic blood alcohol determination but should meet appropriately clinical needs. This aspect has to be taken into consideration in case of a possible later forensic evaluation of the measurements.

Ethanol can be produced either technically by synthesis or in a natural way by fermentation of natural materials containing sugar or starch with yeast. Ethanol is applied in various ways as a solvent in laboratories and in industry. Alcoholic beverages are characterized by very different ethanol concentrations. For example, beer comes to an ethanol concentration of 35–45 g/l, wine 75–90 g/l, brandy 300–320 g/l, and absinth even up to 600 g/l (see annotation 4).

In practice, the quantitative determination of ethanol is carried out via either headspace gas chromatography or enzymatic analysis. The enzymatic method can be carried out easily and quickly in plasma/serum and is therefore well suited for clinical toxicology. It can be applied in emergency laboratories, too.

Clinical Toxicological Analysis: Procedures, Results, Interpretation. Edited by Wolf-Rüdiger Külpmann

ISBN: 978-3-527-31890-2

27.1.1.1 Enzymatic Determination

Outline
In a catalytic reaction, ethanol is oxidized to acetaldehyde by alcohol dehydrogenase (ADH). As per the present amount of ethanol, NAD^+ is reduced to NAD-H_2, which is measured by its absorption at 334, 340, or 365 nm.

Specimens
Serum, plasma, or urine (0.5–1.0 ml).

A separate sample tube should be used for the determination of ethanol. If not available, the respective well-closed container is used for the determination of ethanol first, and later for other assays. Immediately after taking the analytical portion, the tube is closed gastight again.

Only alcohol-free disinfectants (e.g., 0.1% oxycyanate solution) shall be used for the disinfection of the puncture for blood sampling. It is strongly recommended to use closed sterile sampling systems with anticoagulative additives such as heparin or EDTA (ethylenediaminetetraacetic acid disodium salt). The additives do not interfere with the determination of ethanol, if present at concentrations typical for clinical chemistry.

Blood samples containing anticoagulative additives (e.g., 1.5 ml reaction vessels with plastic caps) can be centrifuged immediately. The use of a microcentrifuge (approximately 10 800 × g) allows spinning for only 1–2 min, thereby markedly reducing the time needed for the preparation of the samples, which is particularly important in cases of emergency.

Hemolytic, lipemic, or icteric serum or plasma can be deproteinized by 6% aqueous trichloroacetic acid (TCA) before analysis. A mixture containing one part of the sample and two parts of TCA is centrifuged. The clear supernatant is further analyzed like native serum, but the result has to be multiplied by the dilution factor of 3.

Samples for the determination of ethanol are to be stored at +4 °C tightly closed and labeled unambiguously until the need of an additional forensic examination can be eliminated.

Enzymatic analysis
The enzymatic determination of ethanol is carried out in hospital laboratories by means of partially or fully mechanized analyzers with dedicated reagents (e.g., Abbott, Roche). For information about the procedure, detailed instructions of the manufacturers offering both the equipment and the reagents are available. In the case of open analytical tubes, single measurements should be performed. During time-consuming long series, ethanol may evaporate and lead to erroneously low results.

Reagents for the manual determination of ethanol in serum, plasma, or urine are still available (e.g., Randox, Rolf Greiner). The reagents are delivered with detailed instructions that should be followed carefully. The analysis of a patient sample comprises the measurement of reagent blank solution, standard solution, patient sample, and control sample.

Table 27.1 Ethanol: analytical assessment of a manual procedure.

	Imprecision CV (%)	Bias (%)	Error of measurement (%)
Requirement[a]	< 3	< 3	< 9
Results, obtained[b]	2.16	2.27	6.4

[a]Valid for Germany [2] for concentrations > 1.0 g/l.
[b]Twenty single measurements.

Control samples for both the mechanized and the manual determination of ethanol are commercially available (e.g., Bio-Rad: Liquichek Ethanol/Ammonia control; Roche: Ammonia/Ethanol/CO_2 control; Medichem: Ethanol S plus).

Analytical assessment

Precision, linearity range, and detection limit: Table 27.1 shows the evaluation of measurements for control samples, which were used over 20 days for manual determination. Imprecision, bias, as well as error of measurement met the respective requirements [2].

In the manual analysis, the linearity range was 0–4 mg/l, the detection limit was 0.07 mg/l, and the recovery of spiked ethanol (0.5–3.0 mg/l) ranged from 93 to 105%.

Details on the analytical evaluation of the enzymatic determination of ethanol by mechanized analyzers are provided by the manufacturers. Imprecision, bias, and error of measurement meet the respective requirements (Table 27.1), which are related in Germany to clinical needs and the current state of the art.

Specificity: Alcohol dehydrogenases are not strictly specific to ethanol. The cross-reaction of related substances was measured at a concentration of 3 g/l in duplicates or triplicates (Table 27.2).

As for clinical toxicological cases, the specificity of the enzymatic method can be regarded as appropriate. However, the measurement of the ethanol concentration gives erroneously high results in serum and plasma samples containing high concentrations of LDH (>8000 U/l) and lactic acid [3].

For forensic investigations, an additional determination of the ethanol concentration by another method, for example, gas chromatography, for confirmation is mandatory.

Table 27.2 Ethanol by enzymatic determination: cross-reaction.

		Bias	
Substance	Concentration (g/l)	(g/l)	(%)
Methanol	3.0	0.0	0
1-Propanol	3.0	1.0	33
2-Propanol	3.0	0.1	3
1-Butanol	3.0	0.24	7
Acetone	3.0	0.0	0
Ethylene glycol	3.0	0.0	0

External quality assessment: In Germany, each laboratory has to take part four times a year in an external quality assessment schema. The ring trials are organized by approved reference institutions and comprise two samples each, the concentrations of which are unknown. The laboratories that meet the requirements receive a certificate valid for 6 months. The long-term results of these ring trials can be used for an objective evaluation of the accuracy of the respective procedure.

Practicability

The application of mechanized analyzers has the advantage of a quick availability of the results and low personnel costs. However, it also faces the disadvantage of storage costs for the ready-for-use but perishable reagents needed. Therefore, the analyzers should only be used, if an appropriate number of ethanol determinations are carried out regularly.

The technician time spent on the manual analysis of one patient sample (including reagent blank solution, standard solution, sample, control sample) is 8.5 min and that for a series of five patient samples is 14 min. If the incubation time of 15 min is added, the total time spent on the analysis from the beginning to the final result is 23.5 min per patient sample and 29 min for five samples in a series. Details are given in Section 5.2 and Table 5.2. For details on the direct personnel costs see Section 5.3.

27.1.1.2 Headspace Gas Chromatography

See Sections 4.6 and 27.2.1.

In clinical toxicological and, in particular, in forensic toxicological analysis, headspace gas chromatography is regarded as a method of reference for the determination of ethanol in serum. It is also suitable for the analysis of whole blood.

27.1.1.3 Medical Assessment and Clinical Interpretation

Medical assessment

In clinical toxicological and clinical chemical laboratories, ethanol is almost exclusively determined in serum or plasma but not in whole blood. As a number of assessment criteria usually refer to the concentration of g/kg (‰) in whole blood, a conversion of the results may be necessary. The following formula may be used for the calculation of the blood alcohol concentration (g ethanol/kg whole blood (=‰)) in forensic practice:

$$E(B) = E(S) \times D^{-1} \times W^{-1},$$
$$E(B) = E(S) \times 0.81,$$

where D is density (serum) (=1.026 g/kg), $E(B)$ is ethanol (blood) g/kg, $E(S)$ is ethanol (serum) g/l, and W is the serum versus blood partition coefficient for ethanol (=1.2).

The partition coefficient of 1.2 is only a mean value. It may fluctuate individually between 1.10 and 1.25. It is found in the different distribution of ethanol in plasma and blood cells and depends on hematocrit. Also, density is not a constant and varies individually. Nevertheless, in the conventional formula, the numbers are fixed as shown.

The –so-called endogenous alcohol can reach a level of 0.015 g/l of serum (0.012 g/kg whole blood). The maximal ethanol concentration, which can be attained through inhalation, is 0.24 g/l serum (0.2 g/kg whole blood).

Blood ethanol concentrations between 4.2 and 4.8 g/l of serum (3.5–4.0 g/kg whole blood) are regarded as potentially lethal in adults, according to the literature. There are reports, however, that people survived an alcohol level of 7.5 g/l serum (6.2 g/kg of whole blood) presumably due to tolerance from chronic abuse.

After oral intake and quick absorption in the gastrointestinal tract, ethanol is distributed in the whole body fluid. Only a small fraction (2–10%) is excreted unchanged via the lungs or the kidneys. Mainly in the liver, 90–98% of ethanol is oxidized enzymatically via acetaldehyde (catalyzed by ADH) to acetic acid (catalyzed by ALDH) over a wide range, at a constant rate practically independent of the concentration. In the case of blood alcohol concentrations >1.2 g/l serum (1.0 g/kg whole blood), the microsomal ethanol oxidizing system (MEOS) is also involved in the process of metabolic degradation [4].

The mean elimination rate per hour for ethanol is 0.18 g/l serum or 0.15 g/kg whole blood. Regular intake of alcohol results in tolerance and frequently also in an increased elimination rate.

Clinical interpretation

Acute intoxication: The effect of ethanol strongly depends on the dosage. In small doses, it has a stimulating effect but with increasing doses it begins to inhibit the central nervous system (CNS). In cases of acute intoxication, ethanol first stimulates the respiratory center, increases the pulse rate, the blood pressure, and the cardiac output per minute, and leads to vasodilation in the skin. During the further course of intoxication, attentiveness and reactions are slowed down, power of vision and hearing is reduced, and coordination disorders occur. This is later followed by an impaired sense of balance, slurred speech, loss of inhibitions, and an increased preparedness to take risks. At higher doses, even dimming of consciousness, coma, and death by respiratory paralysis can occur.

After the same dosage, the severity of symptoms may differ considerably. This may be due to tolerance, hypersensitivity, individual situation, enzyme deficiency (e.g., aldehyde dehydrogenase), and so on. The direct effects of ethanol may be accompanied by toxic effects of the metabolite acetaldehyde. Among these are tachycardia and flush. Further, a genetic variability of the enzymes, which metabolize ethanol, has to be considered. A race-dependent atypical aldehyde dehydrogenase (ALDH) can slow down the degradation of ethanol and can lead to distinct symptoms (e.g., in people from China or Japan) already after ingestion of small volumes.

Clinical chemical signs of serious ethanol intoxication are metabolic acidosis, distinct hypoglycemia, and acetonuria without glucosuria. The osmolal gap is increased; for example, 1 g ethanol/l extends the osmolal gap by 22 mmol/kg.

Interactions with drugs: When ethanol is taken simultaneously with drugs, their mode of action can be altered considerably, for example, by influencing absorption, distribution, and excretion of the drug. A similar inhibitory effect of the drugs on the CNS can quickly and suddenly cause the death of the patient.

Impaired pharmacokinetics of drugs can be due to a direct or indirect effect of metabolic interference from ethanol. The interference may lead to an inhibition or activation of the drug metabolism or to an impairment of the ethanol metabolism caused by drugs and other xenobiotics, which can lead to the so-called "antabus alcohol syndrome." It is observed after simultaneous ingestion of disulfiram (antabus) and related compounds from mushrooms (Chapter 36) or calcium carbide on the one hand and ethanol on the other.

The evaluation of the data of toxicological wards has clearly shown that rather often intoxication with drugs or poisons had occurred in conjunction with the ingestion of ethanol causing a considerable alteration and usual aggravation of the clinical picture.

Interpretation of findings: The fact that a state of inebriation has been discovered does not rule out the simultaneous presence of toxic substances or drugs (e.g., hypnotics) that can have a crucial influence on the clinical course of the intoxication.

Furthermore, one has to take into consideration that apart from the state of inebriation, there could be other causes responsible for the clinical condition of the patient (e.g., an undetected fractured skull).

For clinical purposes, the determination of ethanol in urine can be seen as an easily practicable way of controlling abstinence. It is not possible, however, to calculate reliably the blood ethanol concentration of a particular patient from his ethanol concentration in urine.

Annotations

(1) References to other analytical procedures for the determination and detection of ethanol in whole blood, urine, oral fluid, and other biological fluids can be found in Ref. [5].

(2) For details on the calculation of the blood alcohol concentration from the amount of ingested alcohol or the calculation of the blood alcohol concentration at a particular time in the past from the current concentration, see Ref. [6].

(3) Devices for the determination of ethanol in breath are very often used in police traffic checks. Details about the calculation of the blood alcohol concentration from ethanol concentration in breath are found in Refs [7, 8].

(4) Before 1991, the alcoholic beverage absinth was prohibited in many countries. But in 1991, the EU implemented new guidelines lifting the prohibition and since then the beverage has developed into a fashionable drink that is very popular on the scene. Basically, absinth is an extract from wormwood (*Artemisia absinthium*) containing the substance thujone, a bicyclic monoterpene in high-proof alcohol. Details on the history of absinth, its toxicity, and the detection of thujone (DC, GC–MS) are found in [9].

(5) The determination of ethanol in serum or urine reflects only the short period of the current alcohol consumption. The amount of alcohol consumed over the last

3 weeks can be estimated rather accurately from the carbohydrate-deficient transferrin (CDT) concentration in serum [10]. The determination of ethylglucuronide (ETG) in serum or urine reflects the alcohol consumption over the last 3 days [11], while the 5-hydroxytryptophol (5HTOL) level in urine gives an indication of the recent alcohol intake over the last 24 h [12]. Over the last years, CDT, ETG, and 5HTOL have become useful markers for the detection of alcohol abuse and the monitoring of alcohol abstinence. These markers have proved worthwhile in the selection of individuals for liver transplantation [13].

27.1.2 Other Highly Volatile Alcohols and Ketones

F. Degel and H. Desel

27.1.2.1 Introduction

Very often low molecular weight alcohols and ketones are essential parts of solvent mixtures, cleaning agents, antifreeze (windscreen cleaner with antifreeze), and disinfectants. They are the most frequently occurring toxic agents in acute solvent intoxication. Several years ago, methanol was added illegally to wine on a large scale. Furthermore, denatured industrial alcohols must be considered as potential sources of intoxication. The metabolism of low molecular weight alcohols is characterized by the occurrence of toxic, acidic metabolites causing metabolic acidosis of a greater or lesser degree of severity. Without appropriate therapy, methanol intoxication can lead to loss of sight or even death. In case of an early diagnosis, it is possible to alleviate the intoxication and prevent further damage to the organs by administering antidotes and/or through extracorporeal detoxification (in most of the cases hemodialysis). Therefore, a quick detection or exclusion of these compounds is mandatory. For that purpose, one can start with the screening procedures such as chromometric gas analysis, which is capable of detecting methanol by means of a sequential application of the gas detection tubes “alcohol” and “formaldehyde” (e.g., Dräger). Nowadays, however, the most reliable and the most frequently used method for the identification and quantitative determination of these highly volatile components is headspace gas chromatography (headspace GC). The applied procedure must be capable of separating methanol from other frequently occurring volatile compounds such as acetone that occurs physiologically and the principal and concomitant substances of alcoholic beverages (ethanol, acetaldehyde, ethyl acetate, etc.). The presented isothermal gas chromatographic method meets these requirements.

27.1.2.2 Headspace Gas Chromatography

Outline

See Section 4.6.

Specimens

Whole blood (anticoagulant EDTA or oxalate), serum, and urine.

See Sections 4.6 and 27.1.1 for general recommendations on sampling.

Equipment
Gas chromatograph with flame ionization detector (FID) and supply units for carrier gas and fuel gases. Not necessary but advantageous: mechanized headspace sample injector (e.g., HS 40, Perkin-Elmer).

Stationary phase: Fused silica quartz capillary column, length 30 m, internal diameter 0.53 mm, RTX BAC2 (basis Carbowax), coating 2 μm, (e.g., Restek, Bad Homburg), or any other comparable polar phase.

Chromatographic settings	
Carrier gas: helium; mean flow rate	25 cm/s
Column oven temperature program: isothermal	40 °C (5 min)
Injector temperature	150 °C
Detector temperature	280 °C

Settings of the autosampler system HS 40 (Perkin-Elmer): sample temperature, 60 °C; needle temperature, 70 °C; transfer temperature, 110 °C; thermostat time, 20 min; pressure build-up time, 2 min; injection time, 0.06 min; dwell time, 0.5 min; pressure HS 40, 70 kPa; GC column pressure, 40 kPa.

Further equipment and accessories needed: see Section 4.6.

Sampling
General information about sampling: see Section 4.6.

Chemicals (p.a. grade)

Reference substances for gas chromatography: acetone, ethanol, methanol, 2-propanol.
Anticoagulants: ethylenediaminetetraacetic acid disodium salt.
Internal standard: *tert*-butanol.

Reagents
Standard solutions for calibration: It is recommended to use commercially available standard mixtures containing acetone, ethanol, methanol, and 2-propanol for the analysis of highly volatile substances, for example, UTAK Volatiles Reagent A, UTAK (Germany: SL Marketing, Radolfzell).

This standard mixture can be used directly for calibration without prior dilution and contains the following substances: acetone (800 mg/l), ethanol (1500 mg/l), methanol (400 mg/l), and 2-propanol (800 mg/l).

Alternatively, the user can prepare an appropriate combined standard solution in aqueous solution by himself, if necessary, by adding other relevant compounds.

Control samples: For qualitative screening methods, the solutions mentioned above can also serve as positive control samples.

Whole blood, serum, or control material used in clinical chemistry can serve as negative control samples, if they do not contain the relevant substances.

It is recommended to use commercially available control material for quantitative measurements, for example, UTAK Volatiles Serum Toxicology Control, UTAK

(Germany: SL Marketing, Radolfzell) or Liquichek Serum Volatiles Control, Level 1 and 2, BioRad Laboratories, Munich.

Internal standard solution: tert-Butanol/aqua bidest 0.08 + 100 (v/v).

Sample preparation

For general information, see Section 4.6. In case the samples have not already been sent in headspace vials, a constant aliquot of the primary sample (range 0.5–2 ml) is transferred to the headspace vial with a pipette. Then, a constant volume of the internal standard solution is added to each headspace vial. The vial is tightly sealed by putting on both a butyl rubber septum and a star spring before it is finally closed gastight with an aluminum crimp cap. However, if the blood samples have already been sent in closed headspace vials, a constant volume of the internal standard solution is injected into the sample liquid through the septum using a gastight syringe. It is important that the total levels of the calibration standard solution and the samples are identical.

Headspace GC analysis

For general information, see Sections 4.6 and 12.4.

For headspace analysis, the samples are incubated in headspace vials at 60 °C for 20 min.

Evaluation: The first step is the identification of the volatile substances occurring in the samples by comparison with the standard mixture and/by the standard addition of a suspected volatile substance. Quantitative determination is carried out by the method of internal standard.

Analytical assessment

Repeated measurements of a serum-based standard solution have been carried out to scrutinize the imprecision of the method for various sampling procedures. Results are shown in Table 27.3. It is obvious that a satisfactory precision can be obtained even

Table 27.3 Headspace analysis: imprecision [24].

Substance	**CV%**		
	Headspace manual (20 µl) (*n* = 10)	**Headspace SPME (polyacrylate 85 µm) (*n* = 5)**	**Headspace mechanized HS 40, Perkin-Elmer (*n* = 10)**
Methanol	8.12	5.33	1.51
Ethanol	7.14	2.22	0.77
Acetone	4.17	1.38	0.55
Propan-2-ol	6.52	1.55	0.75
tert-Butanol	4.66	1.14	0.73

n: number of measurements; CV: coefficient of variation.
SPME: solid-phase microextraction.

by means of manual sampling. Better results can be achieved, however, from SPME sampling and the mechanized headspace sampling procedure, which is the most precise technique. The linearity of the method was evaluated by analyzing a serum sample spiked with ethanol or methanol. A linear calibration function in the range of 0–4000 mg/l was obtained. Samples containing higher concentrations have to be diluted with isotonic sodium chloride solution. Due to the similarity of their physical properties, a similarly high linearity range can be expected for other low molecular weight alcohols and ketones for measurements by a flame ionization detector.

The accuracy of the method has been evaluated for methanol, ethanol, acetone, and isopropanol by analysis of a commercially available quality control sample in each analytical series. An analytical bias below 10% can be easily achieved. The detection limit of the highly volatile compounds mentioned above was 0.1 mg/l in the mean.

External quality assessment schemes are available for external quality control (Germany: Arvecon, Walldorf).

Practicability

The method allows a reliable chromatographic separation and quick identification of this group of volatile substances. The technician time spent on this procedure (one sample, standard mixture, positive control, and negative control) is 15 min, whereas the complete analysis requires a total of 60 min.

27.1.2.3 Medical Assessment and Clinical Interpretation

Medical assessment

Some of the members of the group of highly volatile alcohols and ketones also occur as concomitant substances of alcoholic beverages. Reference ranges and the biological tolerance values for compounds used at the workplace (BAT) are shown in Table 27.4.

Acetone

Acetone is removed from the body through excretion via the lung and the kidneys. Acetone is oxidized probably to carbon dioxide via acetate and formate [15]. In healthy subjects, the endogenous blood acetone concentration is in the range of 1–6 mg/l [16]. Concentrations >50 mg/l are regarded as clinically relevant.

Table 27.4 Highly volatile alcohols and ketones: reference intervals and BAT values [15, 22, 25, 26]).

Substance	Compound	Reference interval	BAT value	Material
Acetone	Acetone	<5 mg/l	n.d.	Blood
	Acetone	<2.5 mg/l	80 mg/l	Urine
Ethanol	Ethanol	<15 mg/l	n.d.	Serum
Methanol	Methanol	<2.5 mg/l	30 mg/l	Urine
2-Propanol	2-Propanol	<5 mg/l	n.d.	Serum
	Acetone	<5 mg/l	50 mg/l	Blood
	Acetone	<2.5 mg/l	50 mg/l	Urine

BAT: biological tolerance value for exposure to the solvents at the workplace; n.d.: no data.

Concentrations of 200–300 mg/l are classified as potentially toxic. There have been reports of lethal concentrations of 550 mg/l [17]. On the other hand acetone concentrations exceeding 2000 mg/l following isopropanol ingestion have not been lethal [18, 19].

Ethanol
See Section 27.1.1.

Isopropanol
After fast absorption and distribution (in case of oral administration 80% in approximately 30 min; volume of distribution: 0.6–0.7 l/kg), isopropanol (propane-2-ol) is metabolized to acetone catalyzed by alcohol dehydrogenase. Therefore, acetone can be detected in serum following the ingestion of isopropanol already after approximately 30 min. However, the maximum concentration of acetone is reached after about 4 h. Three hours after isopropanol ingestion at the latest, the metabolite acetone is detectable in urine (by urine test strip for "ketone bodies"). Twenty to fifty percent of a dose is excreted unchanged through the kidneys rather quickly. The metabolite acetone is excreted through the lung and the kidneys more slowly.

In the blood of type I insulin-dependent, acetonemic diabetics, endogenously produced isopropanol is found in concentrations of up to 0.3 g/l [14]. In healthy adults, the blood isopropanol concentration is <5 mg/l.

Methanol
Ninety percent of ingested methanol is metabolized in the liver. First, it is degraded to formaldehyde by alcohol dehydrogenase and then oxidized to formic acid (formate) by aldehyde dehydrogenase. Finally, formate is metabolized to carbon dioxide by means of a tetrahydrofolate-dependent mechanism. Since the oxidation process of methanol is 10 times slower than that of ethanol, it has a noticeably longer elimination half-life. The affinity of ethanol to alcohol dehydrogenase is 10–20 times higher than that of methanol. Therefore, ethanol can be used as a competitive agent to slow down the conversion of methanol into its toxic metabolites formaldehyde and formic acid. It is important to monitor the dosage of the applied ethanol by determining the ethanol concentration (target value approximately 1 g/l) during treatment. The enzymatic determination of ethanol is not interfered because of the very low cross-reactivity of methanol (Table 27.2). According to some authors, the blood formate level correlates much better with the clinical picture of methanol intoxication (ocular toxicity) than the serum methanol concentration [20], which is contradicted by others [21]. The maximal formic acid concentration in blood and urine is reached only after 2–3 days. Small amounts of methanol are excreted unchanged in urine and expired air, whereas 10% of a dose is excreted as formic acid in urine. The endogenous blood methanol and urine formate concentrations are 1.5 and 12–17 mg/l, respectively [15]. The data on lethal doses for methanol vary considerably. However, 100–200 ml is reported as being lethal for adults, whereas others report potentially lethal doses of only 30 ml. Permanent loss of sight was observed already after the ingestion of only 10 ml. Anyhow, methanol concentrations

of >100 mg/l are commonly associated with toxic effects, and concentrations of >200 ml have to be classified as severe intoxication [15].

Methanol as marker of alcoholism: Slightly elevated blood methanol concentrations are considered by many authors as an indication of chronic alcoholism, because the level of methanol, as a concomitant substance of alcoholic beverages, only drops at blood ethanol concentrations of < 0.4 g/kg (‰). In case of chronic alcohol abuse, this can result in an accumulation of methanol. A methanol concentration of 8–10 mg/l has been proposed as the decision limit.

Clinical interpretation

All aliphatic alcohols exhibit narcotic effects, the intensity of which increases with the respective carbon mass number and the lipid solubility. Mother compound and/or metabolites may contribute more or less to toxicity depending on the individual substance. In the case of ethanol, the acute toxic effect is mainly caused by the substance itself, whereas in the case of the homologous alcohols methanol and isopropanol, the metabolites are more toxic. For higher molecular weight compounds, the mother compound is increasingly again the main toxic agent. Alcohols, ketones, and other water-soluble volatile substances may increase the osmolal gap considerably according to their amount of substance concentration. In primary alcohols, the corresponding aldehydes and carbonic acids are produced through oxidative metabolic degradation by means of alcohol dehydrogenase. The concentration of the aldehydes and carbonic acids is mainly decisive for the severity of the developing metabolic acidosis.

Acetone

Acetone exhibits a relatively low toxicity. In adults, doses of 15–20 g did not produce any toxic symptoms. The ingestion of 200–400 ml of pure acetone resulted in a depression of the CNS, which was, however, short-lived [23]. Even though the clinical symptoms are similar to those of ethanol intoxication, the anesthetic properties of acetone are more distinct. A moderate intoxication results in lethargy, headache, ataxia, vomiting, and slurred speech, while a severe intoxication can lead to sopor and coma [22].

Ethanol

See Section 27.1.1.

Isopropanol

The effect of isopropanol is similar to that of ethanol but its toxicity is twice as high. Serious central nervous depression may develop and even cardiovascular depression in case of high doses. The metabolite acetone can intensify and extend the central nervous effects. Through the metabolic degradation of acetone to acetic acid and formic acid, a mild acidosis may develop [22].

Methanol

Compared to ethanol, the narcotic effect of methanol is considerably milder, but the state of inebriation lasts longer due to its slower oxidation and excretion. According to recent studies, formic acid, instead of the relatively ephemeral formaldehyde, is mainly responsible for toxicity and metabolic acidosis as well. Apart from an often distinct acidosis with a blood pH value dropping below 7.0, another main characteristic symptom is the occurrence of visual defects that develop in two phases. In the first phase beginning with the third day, the developing visual defects are still reversible. If there is no therapy, the second phase of the intoxication beginning with days 5–6 is characterized by an irreversible degeneration of the optic nerve resulting in permanent loss of sight. Methanol increases the osmotic gap (1 g/l methanol increases the osmolality by 31 mmol/kg). In case of methanol intoxication, ethanol is a highly effective antidote that inhibits competitively the metabolic degradation to formaldehyde and formic acid. Since several years, fomepizole (4-methylpyrazol) is used increasingly as an effective inhibitor of alcohol dehydrogenase. More common therapeutic measures are compensation of the acidosis and hemodialysis. The effectiveness of hemodialysis for eliminating methanol is still discussed controversially, but it supports, indeed, reversal of acidosis.

27.2 Aromatics (BTEX)

F. Degel

Introduction

The aromatics (BTEX: benzene, toluene, ethyl benzene, xylenes) are in frequent use as components of fuels and solvent mixtures for industry and household, such as paint thinners, brush cleaners, and degreasing agents. Solvent sniffing is a frequently occurring way of drug abuse.

Acute intoxications with benzene or related aromatic hydrocarbons are regarded as particularly problematic. The chromometric gas analysis (Dräger tubes toluene and benzene), as described in Sections 4.6 and 12.4, can be applied for screening and fast identification of these substances in residues. For confirmation analysis and quantitative determination of individual substances, gas chromatographic headspace analysis is the method of choice at present.

27.2.1 Gas Chromatography

Outline

See Sections 4.6 and 12.4.

Specimens

Whole blood (anticoagulants: EDTA or oxalate).

General recommendations for sampling: see Section 4.6.

Equipment

Gas chromatograph with flame ionization detector and supply units for carrier gas and fuels.

Stationary phase: Fused silica quartz capillary column, internal diameter 0.53 mm, length 30 m; coating 2 μm RTX BAC2 (basis Carbowax) (Restek, Bad Soden), or comparable polar phase.

Chromatographic settings: Carrier gas, Helium; mean flow rate $u = 49$ cm/s.

Column oven temperature program	
Initial temperature	40 °C (5 min)
Heating rate	10 °C/min
Final temperature	190 °C (10 min)

Injector temperature: 150 °C; detector temperature: 280 °C.

Mechanized sampling system HS 40 (Perkin-Elmer): sample temperature, 60 °C; needle temperature, 70 °C; transfer temperature, 110 °C; thermostat time, 30 min; pressure build-up time, 2 min; injection time, 0.06 min; dwell time, 0.5 min; pressure HS 40, 70 kPA; column pressure on GC, 40 kPA.

Sample vials: 20 ml headspace vials with PTFE-coated butyl rubber septa, ferrules, and aluminum crimp caps (e.g., Perkin-Elmer, Rodgau-Jügesheim).

Headspace vials and the butyl rubber septa in particular have to be dried for 2 days in a drying oven at 110 °C prior to use for BTEX determination. They have to be stored in an airtight screw cap vial or a desiccator. On no account, the butyl rubber septa may be used twice.

Pipettes: For whole blood samples, 2 ml positive volume displacement pipettes are used (e.g., Gilson Microman from Abimed, Langenfeld, or Socorex, Rennes, Switzerland). Due to the volatility of the analytes, conventional pipettes with air displacement cannot be recommended. Positive volume displacement pipettes (e.g., Hamilton HPLC syringe, 100 μl, or similar) are also used for the standard stock solution.

Further devices and equipment needed: see Section 4.6.

Chemicals (p.a. grade)

Reference substances for gas chromatography: benzene, ethyl benzene, toluene, *o*-, *m*-, and *p*-xylene

Ethylenediaminetetraacetic acid, disodium salt

Propylene carbonate

Sheep blood (whole blood with EDTA) (e.g., Fiebig-Nährstofftechnik, Idstein).

Reagents

Due to the high volatility and low water solubility of the analytes, it is recommended to use commercially available ready-for-use standard mixtures instead of self-made standard solutions. They are offered by various manufacturers, for example, BTEX Standard Solution (100 μg/ml) ERB 039S (Cerilliant) (Promochem, Wesel).

The low water solubility of the pertinent substances necessitates the storage of the stock solutions in appropriate solvents (e.g., propylene carbonate). The calibration solutions are prepared freshly by mixing stock solution and sheep blood, anticoagulated by EDTA. The absence of interfering peaks in the blood within the time span of the BTEX standard substances has to be proven in a preceding analysis.

Preparation of the standard stock solution: By means of a 1000 μl HPLC syringe (or volume displacement pipette, for example, Gilson Microman, Langenfeld), 500 μl of the standard solution is quickly transferred from the ampoule (cooled down to −20 °C) to a 100 ml volumetric flask, which is already two-thirds filled with cooled propylene carbonate. Propylene carbonate is made up to slightly below the ring mark. The flask is closed, its content well mixed, and after reaching a temperature of +20 °C (water bath), it is filled up to the mark. The stock solution is stored at 4–8 °C in a refrigerator. It can be kept for 6 months, but if it is been opened frequently, it has to be replaced earlier. The concentration of the standard stock solution is 5000 μg analyte/l propylene carbonate.

Preparation of the standard working solution for calibration: A 1960 μl portion of analyte-free sheep blood is transferred to a headspace sample vial. A 40 μl aliquot of the standard stock solution is added by an HPLC syringe. The vial is closed gastight with a PTFE-coated septum and aluminum crimp cap, and the content is well mixed on a vortex mixer for 1 min. All steps should be carried out quickly to avoid loss of the volatile compounds. Table 27.5 shows the concentrations of the analytes in the standard working solution for calibration. It is stable for about 3 months, if kept in the closed headspace vial at −20 °C.

Control samples: Matrix-based positive control samples are prepared in the same way at two different concentrations. Sheep blood or proven analyte-free whole blood is used as negative control sample.

Sample preparation

In case the samples have not already been sent in headspace vials, a constant volume (2 ml) is transferred from the primary vial into the headspace vial with a volume displacement pipette. The vial is tightly closed by putting on a PTFE-coated septum and a star spring before it is closed gastight with an aluminum crimp cap using a hand crimper. If the blood sample is obtained already in a headspace vial, it can be used directly for the analysis. Volumes of calibration standard solutions and of samples in the headspace vials shall be identical.

Table 27.5 Aromatics (BTEX).

Substance	Ampoule concentration (mg/l)	Standard working solution concentration (μg/l)
Benzene	100	100
Toluene	100	100
Ethyl benzene	100	100
m-Xylene	100	100
p-Xylene	100	100
o-Xylene	100	100

Table 27.6 Aromatics (BTEX): headspace GC; imprecision and detection limit.

Substance	Imprecision (interassay) CV%	Imprecision (intraassay) CV%	Detection limit (μg/l)
Benzene	2.69	7.5	5.0
Toluene	3.11	5.8	5.0
Ethyl benzene	2.73	5.2	5.0
m-Xylene	2.88	6.1	5.0
p-Xylene			
o-Xylene	3.21	8.2	5.0

Imprecision: a standard working solution (100 μg/l) was analyzed repeatedly; detection limit: negative control sample (analyte-free blood) was analyzed.

Headspace GC analysis
See Section 4.6.

Evaluation: At first, the volatile substances of the samples have to be identified. The respective peaks are compared to the peaks of the standard working solution. In doubtful cases, the suspected aromatic is added to the prepared sample for confirmative analysis. Quantification is supported by chromatographic software using the method of external standard.

Analytical assessment
The imprecision of the method was determined by repeated measurements of a matrix-based standard working solution. The results are shown in Table 27.6.

The linearity of the procedure was evaluated by the analysis of blood samples spiked with increasing amounts of benzene and toluene. The linear calibration function ranges from 0 to 3000 μg/l and exceeds the concentrations usually observed in cases of acute intoxication. For an accurate determination of higher concentrations, the samples have to be diluted with analyte-free whole blood. The trueness of the procedure was estimated from recovery. Different amounts of the standard substances mentioned in Table 27.5 were added to an analyte-free whole blood sample. Recovery ranged from 96 to 106%.

For the determination of the lower limit of detection, the mean area value of the basic noise plus its threefold standard deviation of an analyte-free whole blood sample in the respective retention time range was measured. This value was related to the response factor (area/concentration) of the individual analyte. The results are shown in Table 27.6.

In Germany, ring trials for external quality assessment are offered by the German Society of Industrial and Environmental Medicine (project group quality assurance).

Practicability
Technician time for one analysis (one sample, standard mixture, positive control, and negative control) is about 15 min, whereas total analysis takes about 185 min. To rule out a solvent intoxication, it is recommended to restrict oneself to a qualitative analysis of the patient blood sample and one positive control. If no relevant peaks are observed in the patient sample, no further analysis is necessary. Otherwise, a

Table 27.7 Aromatics (BTEX): reference intervals and BAT values [32, 33].

Substance	Analyte	Reference interval	BAT value	Material
Benzene[a]	Benzene	< 0.2 μg/l (NS)	No data	Whole blood
	Benzene	< 0.5 μg/l (S)	No data	Whole blood
	Phenol	< 15 mg/l	No data	Urine
	t,t-Muconic acid	< 1 mg/l	No data	Urine
	S-Phenylmercapturic acid	< 5.0 μg/l	No data	Urine
Toluene[a]	Toluene	< 1.14 μg/l (NS)	1.0 mg/l	Whole blood
	Toluene	< 2.0 μg/l (S)	No data	Whole blood
	o-Cresol	< 0.5 mg/l	3.0 mg/l	Urine
Ethyl benzene[a]	Ethyl benzene	< 0.5 μg/l	No data	Whole blood
Xylene[a] (all isomers)	Xylene	< 3.0 μg/l	1.5 mg/l	Whole blood
	Methylhippuric acid	< 10 mg/l	2000 mg/l	Urine

BAT value: biological tolerance value for compounds at the workplace; NS: nonsmoker; S: smoker.
[a]Health hazard from percutaneous absorption.

comprehensive quantitative determination comprising all quality control samples is carried out. The qualitative analysis takes only 115 min.

27.2.2 Medical Assessment and Clinical Interpretation

Medical assessment

Data on aromatic compounds, such as reference values, biological tolerance values, and exposure equivalents for carcinogenic substances, are presented in Tables 27.7.

Benzene

In most cases, benzene is incorporated via inhalation. Resorption is fast, and the fraction of a benzene/air mixture that is retained from inhalation is 40–50%. After ingestion, it can be readily absorbed through the gastric mucosa. To a rather limited degree, it can enter the organism through skin absorption. Due to its good lipid solubility, benzene preferably accumulates in the fatty and nerve tissue; because of its high vapor pressure, 10–50% of a dose is exhaled unchanged through the lung (three phases with half-lives of 1, 3–4, and 20–30 h). Only small amounts of benzene are excreted unchanged in urine. Approximately 50% of a dose is metabolized. The degradation takes place in the liver through the mixed functional P450 oxidase system and results in epoxide formation. This is converted to phenylmercapturic acid through glutathione conjugation and through ring split to muconic acid, but most of it is metabolized to phenol (Figure 27.1). Further, phenolic components such as hydroquinone and catechol are produced as by-products. The phenols are excreted through the kidneys mainly as conjugates of glucuronic acid or sulfate.

Ethyl benzene

The skin resorption rate of ethyl benzene is 22–33 mg/cm^2 per hour. No detailed information is available about other routes (lung and gastrointestinal tract) and their respective absorption rates or the distribution of the substance in the organism. It can

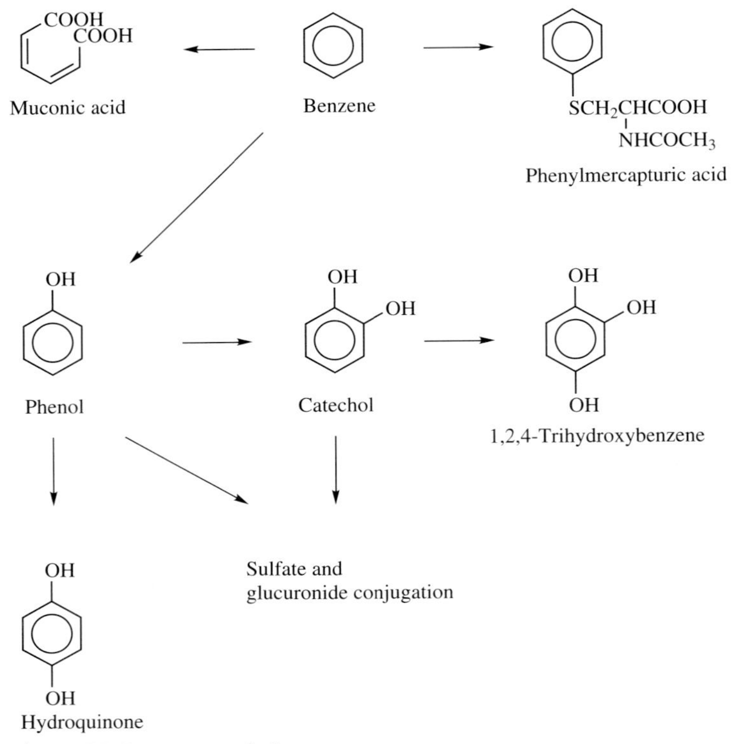

Figure 27.1 Benzene metabolism.

be assumed, however, that the data of ethyl benzene are similar to those of the other alkyl benzenes. The substance is eliminated mainly through the kidneys after metabolic degradation. Predominantly, the degradation concerns the side chain (Figure 27.2) whereas ring oxidation plays only a minor role (4% of total). The end products are mandelic acid and phenylglyoxylic acid. About 6% are found as glutathione conjugates and unchanged ethyl benzene.

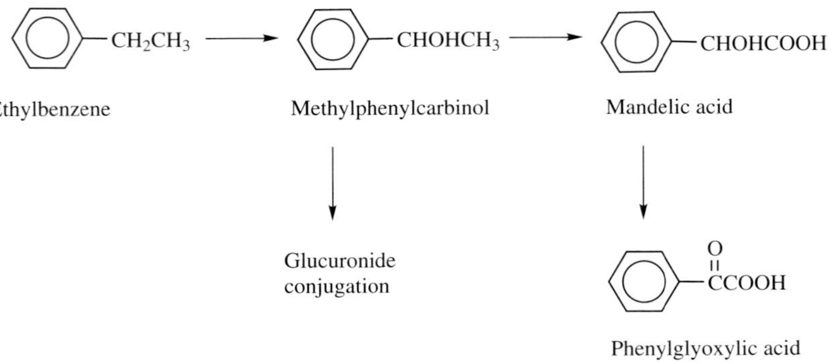

CH_2CH_3 → $CHOHCH_3$ → $CHOHCOOH$

Ethylbenzene   Methylphenylcarbinol   Mandelic acid

Glucuronide conjugation

O
CCOOH

Phenylglyoxylic acid

Figure 27.2 Ethyl benzene metabolism.

CH_3 → $COOH$ → Conjugation

Toluene Benzoic acid

Figure 27.3 Toluene metabolism.

Toluene

During inhalation, 40–60% of the inhaled substance is absorbed by the human organism. In case of ingestion, there is almost a complete resorption, even though it is much slower than in the case of inhalation, and peak blood concentrations are reached after about 2 h. Skin resorption is a slow process, which rarely leads to toxic manifestations. The resorption rate of liquid toluene is 14–23 mg/cm^2 of skin per hour. According to the literature, mild to moderate exposure resulted in blood concentrations of 0.3–0.6 mg/l [27]. In cases of lethal intoxication, the concentrations were $\geq$10 mg/l, but, in case of habituation, higher doses are tolerated with scarcely any symptoms. Similar to benzene, the distribution of toluene in the organism is determined by the lipid content of the tissues and organs. Approximately 20% of the absorbed toluene is exhaled unchanged through the lung, and 80% is excreted through the kidneys after mainly hepatic degradation. Oxidation of the methyl group and the ring is catalyzed by cytochrome P450-dependent monoxygenases (Figure 27.3). About 80% of the absorbed toluene is conjugated with glycine after oxidation of the methyl group, and finally excreted through the kidneys as hippuric acid. Only after saturation of this mechanism, the metabolite is conjugated with glucuronic acid. Further oxidation steps lead to benzaldehyde and benzoic acid. Only about 1% is hydroxylated at the ring to produce cresol. Toluene inhibits the metabolic degradation of benzene, styrene, xylene, and trichloroethane, whereas ethanol inhibits the degradation of toluene.

Xylene (all isomers)

The most common route of exposure for xylene is inhalation. Pulmonary absorption is fast, with a resorption fraction between 60 and 65%. An increased absorption is observed for physical exercise and the resultant increase of pulmonary ventilation. After oral intake, the resorption from the gastrointestinal tract is almost complete. The dermal resorption rate is between 0.7 and 4.3 μg/cm^2 per minute. Xylene shows a distribution in the organism very similar to that of the other alkyl benzenes, with the fatty tissue, spleen, and CNS exhibiting higher concentrations. In the liver, 95% of the absorbed xylene is rapidly metabolized to methylhippuric acid (Figure 27.4). This involves the mixed functional oxygenase system through oxidation of a methyl group and the ensuing conjugation with glycine. Ring hydroxylation (to dimethylphenols) and conjugation with glucuronic acid play a minor role (2% of total). Other metabolites to be mentioned here are methylbenzoic acid and xylenols (below 2% of the total). Xylenes are eliminated unchanged via the lung and after metabolic degradation via the kidneys. The elimination proceeds in two phases with half-lives of 1–2 and 20 h, respectively.

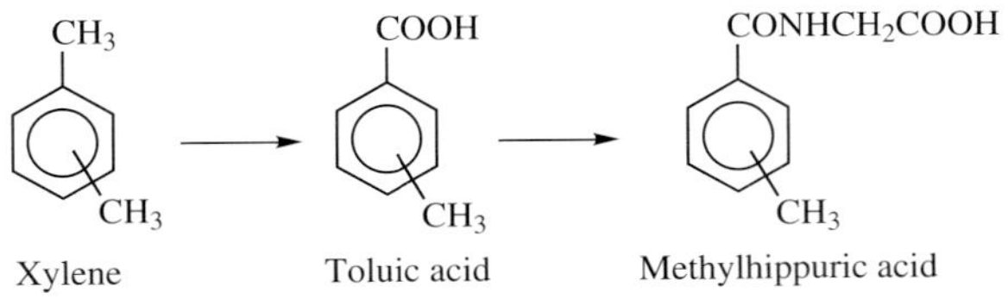

Figure 27.4 Xylene metabolism.

Clinical interpretation

Benzene

Benzene has CNS depressant effects and produces irritation of both the skin and the mucous membranes. Mild intoxication is characterized by inebriation with euphoria (benzene abuse), vertigo, headache, and vomiting. Higher doses result in convulsions, unconsciousness, arrhythmia, and finally death by central respiratory depression or circulatory collapse. Relatively high air concentrations are necessary to cause acute intoxication. Exposure to concentrations of 60 g/m^3 (number fraction 0.02/20 000 ppm) for only 5–10 min and 22.5 g/m^3 (number fraction 0.0075/7500 ppm) for 30 min is reported to be lethal. An air concentration of 2.3 g/m^3 (number fraction 0.0007/700 pm) is supposed to cause unconsciousness after 30–60 min. A fatal outcome was observed after the ingestion of 10 ml of benzene [27]. Blood concentrations of 1–20 mg benzene/l are reported in case of serious intoxications. If the patient survives the acute phase of the intoxication, usually he recovers fast. In poisoning, the blood count does not show any change at first and neither the liver nor the kidneys are damaged severely. However, benzene turns into a hemotoxin after repeated, extended, or single but massive exposure. Erythropoiesis, leukopoiesis, and thrombopoiesis are impaired. Frequent working in benzene-polluted air increases the risk of falling ill with acute myelocytic and monocytic leukemia [28]. Moreover, benzene or benzene derivatives are capable of causing genetic changes and a carcinogenic effect cannot be ruled out.

Ethyl benzene

Symptoms observed in acute poisoning are only irritations of the mucous membranes and serious systemic manifestations in humans have not been reported. In animal experiments with high doses (44 g/m^3 air), however, pulmonary edema and generalized visceral hyperemia developed [30].

Toluene

Irritation of the mucous membranes as well as narcotic and neurotoxic effects are the dominant symptoms of acute toluene poisoning. Medium to higher doses can cause cardiac and hepatic disorders and the bone marrow and the blood count can be affected negatively after extended exposure. As benzene is always present in toluene, it is difficult to assess its possible involvement. High doses can lead to CNS depression, convulsions, nausea, coma, and finally unconsciousness and death. Toluene has become widely known as an agent for sniffers. In case of this abuse, motorial weakness, intentional tremor, ataxia, and, rather seldom, cerebral atrophy [29] are observed.

Xylene (all isomers)
High doses produce prenarcotic and narcotic effects. Typical symptoms are fatigue, giddiness, headache, poor concentration, and vertigo. Chronic exposure results in irritation of respiratory tract and even in the formation of acute pulmonary edema. Further complications are hepatic, renal, and cardiac disorders. Ingestion results in severe gastrointestinal disorders, whereas inhalation causes aspiration pneumonia, pulmonary edema, and hemorrhage [31].

27.3 Glycols

F. Degel and H. Desel

Introduction

The toxicity of short chain glycols is known since many years. Over a hundred people died in the United States in 1937 after ingesting a sulfanilamide "elixir" containing the solvent *diethylene glycol* [34]. Since then, mass poisonings with diethylene glycol as solvent for drugs that are taken orally have occurred in five countries. A recent incidence occurred in Haiti in 1996 [35, 36]. In Germany, a dermally applied preparation for the pyrethrum treatment of pediculosis (mainly head lice infestation) containing diethylene glycol (40%) as solvent (Goldgeist forte) is frequently used and occasionally ingested by little children by mistake. In the "Glycol Scandal" of 1985, diethylene glycol was illegally added to wine to increase its sweetness.

Today, however, poisonings with *ethylene glycol* occur more frequently. It is an antifreeze in car radiator coolants and a solvent in many household products. It is also used as an industrial solvent (e.g., synthetic resins), as a plasticizer, and at low concentration in hydraulic and brake fluids. Poisonings can be either accidental, because of its pleasant taste or for attempting suicide. Also, alcoholics often take ethylene glycol as a substitute due to its inebriating effects.

At present, little is known about the acute toxicity of low molecular weight *glycol ethers* occurring in brake fluid and recently to an increasing extent in industrial and household products. Experiments indicate that, compared to free glycols, glycol ethers tend to show a higher toxicity in mammals. Glycol poisoning can be suspected after a diabetic ketoacidosis has been ruled out if the following signs are observable:

(1) state of inebriation but no fetor and no detectable ethanol,
(2) a severe metabolic acidosis that is not explicable otherwise, and/or
(3) oxaluria.

Oxalate crystals that preferably occur in ethylene glycol poisoning with high urine calcium and oxalate concentrations are dihydrates ("tent" or "envelope" form). If oxaluria occurs for other reasons than ethylene glycol poisoning, a calcium oxalate crystal modification called monohydrate (“coffin lid”) predominates.

27.3.1
Gas Chromatography

Outline
A direct gas chromatographic determination after deproteinization by means of acetone is described.

The procedure is capable of detecting not only ethylene glycol and other short chain-free glycols but also the corresponding glycol ether without prior separation and derivatization. The difficult extraction of these extremely polar substances from an aqueous medium can be bypassed through direct injection of the serum sample after deproteinization with acetone.

Specimens

Serum, plasma
Scene residue after dilution with water 1 + 1000 (v/v).

Equipment
Gas chromatograph with flame ionization detector and supply units for carrier gas and make-up gas.

Stationary phase: fused silica quartz capillary column, internal diameter 0.32 mm, stationary phase: Nukol (basis: polyethylene glycol/nitroterephthalic acid ester), length 30 m, coating 0.25 μm (Sigma/Supelco, Deisenhofen, Germany).

Chromatographic conditions: carrier gas, helium, 24 ml/min; make up-gas, nitrogen, 40 ml/min.

Column oven temperature program	
Initial temperature	60 °C (0.5 min)
Heating rate	45 °C/min
Final temperature A	90 °C (6 min)
Heating rate	20 °C/min
Final temperature B	180 °C (15 min)

Injector temperature: 200 °C; volume of injection: 1 μl; split-mode split ratio: 1 : 2; detector temperature: 250 °C.

Further devices and accessories needed: positive displacement pipette 200 μl (e.g., Gilson Microman).

Chemicals (p.a. grade)
Reference substances for gas chromatography: ethylene glycol, diethylene glycol, diethylene glycol monomethyl ether, diethylene glycol monoethyl ether, diethylene glycol monobutyl ether, 1,2-propylene glycol, triethylene glycol, tetraethylene glycol, and acetone.

Reagents

Calibration standard solutions:

Solution 1: 10 mg ethylene glycol in 20 ml acetone (final concentration: 500 mg/l).

Solution 2: 10 mg diethylene glycol in 20 ml acetone (final concentration: 500 mg/l).

Solution 3: 10 mg of each of the compounds: ethylene glycol, diethylene glycol, triethylene glycol, diethylene glycol monomethyl ether, diethylene glycol monoethyl ether, and diethylene glycol monobutyl ether in 20 ml acetone (final concentration: each 500 mg/l as a combined standard solution).

Internal standard solution (solution 4): 1,2-propylene glycol, 10 mg in 20 ml acetone (final concentration: 500 mg/l).

Control samples: analyte-free serum or analyte-free control material is used as the negative control sample.

Positive control samples can be prepared by spiking a negative control sample appropriately.

The following control sample is commercially available but contains only ethylene glycol: Liquichek Serum Volatiles Control, Level 1 and 2, BioRad Laboratories, Munich.

Sample preparation

The internal standard solution (solution 4) is added to the serum, to the control samples, and to the standard solution (see Table 27.8).

The carefully mixed sample is centrifuged at least at 8000 × g for 5 min using a microcentrifuge. A 1 µl of the supernatant is injected into the gas chromatograph.

GC analysis

To rule out poisoning, the blank value is injected first, followed by the deproteinized patient serum.

In case only the peak of the internal standard appears, the calibration standard solutions containing the suspected substances are analyzed to exclude malfunction.

Table 27.8 Glycols: sample preparation for GC analysis.

	Blank (µl)	Calibration sol.[a] (µl)	Sample (µl)	Control (µl)
Aqua bidest	200	—	—	—
Calibration standard solution(s): 1, 2, or 3	—	200	—	—
Sample	—	—	200	—
Control	—	—	—	200
Internal standard (solution 4)	200	200	200	200

For pipetting, a positive displacement pipette (e.g., Gilson Microman) is used. For further details see the text.

[a]Calibration standard solution.

Table 27.9 Glycols: gas chromatographic retention times.

Substance	Retention time (min)
Diethylene glycol monomethyl ether	9.017
1,2-Propylene glycol (IS)	9.184
Diethylene glycol monoethyl ether	9.363
Ethylene glycol	9.629
Diethylene glycol monobutyl ether	10.917
1,4-Butandiol[a]	11.940
Diethylene glycol	12.422
Triethylene glycol	14.544

IS: internal standard. For further details see the text.
[a]Precursor of γ-hydroxybutyrate.

If relevant peaks appear, the adequate calibration standard solutions are injected for calibration, followed by the sample extract and the relevant control samples.

Evaluation: In a first step, the peaks obtained in the sample extract have to be identified. This is done by direct comparison of the standard mixture and/or spiking the sample extract with the suspected component. The retention times of different glycols and glycol ethers under the given settings are shown in Table 27.9. The presence of the glycols and glycol ethers included in the calibration standard solution can be ruled out in concentrations exceeding the detection limit of 10 mg/l, if no peak is detected in the relevant retention time range. After identification, the determination is carried out using the method of internal standard.

Analytical assessment

The imprecision of the procedure has been evaluated by repeated measurements of a serum-based standard solution (Table 27.10).

The linearity of the method has been evaluated by adding increasing amounts of ethylene glycol to an analyte-free serum sample. A linear calibration function up to 5000 mg/l was observed. This range fits into the range of concentrations found in cases of acute poisoning. Calibrations and measurements at higher concentrations have not been carried out. A similarly high linearity range can be expected for other

Table 27.10 Glycols: imprecision of gas chromatographic procedure.

Substance[c] (concentration)	n	Coefficient of variation (CV) (%)		
		CV within series	CV between days[a]	CV between days[b]
Ethylene glycol (293 mg/l)	10	1.84	3.47	6.82
Ethylene glycol (98 mg/l)	10	2.72	6.37	9.35

n: number of measurements.
[a]Calibrated once.
[b]Calibrated daily.
[c]Sample: Serum Volatiles Control (BioRad).

glycols and glycol derivatives due to their comparable physical properties, if a flame ionization detector is used.

To estimate the detection limits, the mean area for noise of an analyte-free serum sample at the respective retention time range plus its threefold standard deviation was measured. The value was related to the response factors (area/concentration) of the individual analytes. For ethylene glycol and diethylene glycol, the resultant average detection limit was 10 mg/l. For ethylene glycol, the procedure has been checked in each analytical series by analyzing commercially available quality control samples. Deviations from the target value have always been <10%.

Practicability

The procedure allows a quick exclusion or a reliable chromatographic separation of this group of polar compounds. For excluding the presence of toxic ethylene glycol concentrations, technician time is approximately 15 min. The total analysis takes about 45 min. The direct personnel time for ruling out the presence of all the glycols and glycol esters mentioned (one sample, positive control) is approximately 15 min, and in this case the total analysis requires about 60 min. The quantification of ethylene glycol (one sample, standard mixture, quality control samples) requires a direct personnel time of 15 min and a total analysis time of about 75 min. The respective times for the quantification of all the glycols and esters described (one sample, standard mixture, quality control samples) are about 15 and 90 min.

Annotations

Free glycols as well as various glycol ethers can also be determined by the procedure described by Maurer and Kessler [37] after derivatization with pivalinic acid (see Chapter 4 for standard procedure GC/MS). For the sole determination of ethylene glycol, the procedure of Porter *et al.* for vicinal OH groups after derivatization by means of phenylboronic acid [38] can be applied using common, nonpolar stationary phases for chromatography. Moreover, it is possible to determine ethylene glycol after periodate splitting via the end-product methanol using a headspace gas chromatograph. Ethylene glycol and its metabolite glycolic acid can be determined simultaneously by GC–MS after protein precipitation, conversion into dimethylformamide, and derivatization [39]. Another reliable method for the determination of ethylene glycol is HPLC.

27.3.2 Medical Assessment and Clinical Interpretation

Medical assessment

Diethylene glycol

Current reports on the metabolic degradation of diethylene glycol (Figure 27.5) are contradictory and at least there have been no data as yet on increased concentration of oxalate in poisoning [45].

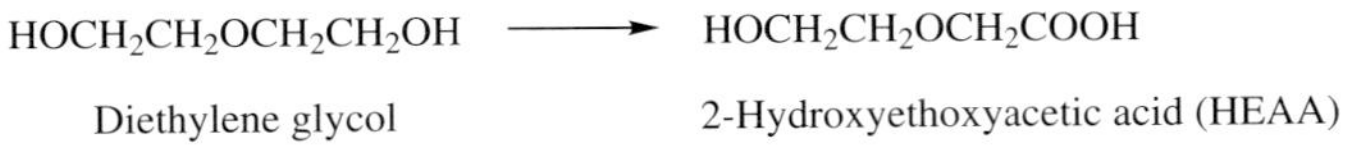

$HOCH_2CH_2OCH_2CH_2OH \longrightarrow HOCH_2CH_2OCH_2COOH$

Diethylene glycol → 2-Hydroxyethoxyacetic acid (HEAA)

Figure 27.5 Diethylene glycol metabolism.

In rats, the metabolites 2-hydroxyethoxyacetaldehyde and 2-hydroxyethoxyacetic acid have been detected. Metabolites that require the splitting of the ether bridge have not been found. For the most part, diethylene glycol is excreted unchanged through the kidneys [46].

Ethylene glycol

After oral ingestion, ethylene glycol is rapidly absorbed, and quickly distributed through the organism due to its good water solubility. The substance is glomerularly filtrated in the kidneys and passively reabsorbed to a large extent [40]. Almost 20% of a dose of 1 g/kg body mass is excreted unchanged [41]. Catalyzed by alcohol dehydrogenase, ethylene glycol is degraded in the liver to glycolaldehyde, which is converted to glycolic acid and glyoxylic acid. About 0.3–2.5% of the dose are metabolized to oxalic acid, the final product of that oxidative metabolic process [42] (Figure 27.6). The acidic metabolites dissociate at physiological pH, thereby releasing acidic equivalents (protons). Plasma half-life of ethylene glycol is 3–5 h (without therapy). The antidotes ethanol and fomepizole [43] exceed the affinity of ethylene glycol to ADH by far: ethanol 100 times [44] and fomepizole 8000 times. Therefore, they inhibit competitively the enzymatic degradation of ethylene glycol considerably and extend its half-life to 17 and 20 h, respectively. The production of the toxic metabolites is slowed down [41]. In case of high blood ethylene glycol concentrations (>0.5 g/l), it is not sufficient to inhibit ADH by means of ethanol alone, but additional hemodialysis is necessary to accelerate the elimination of the parent substance and its toxic, acidic metabolites. It has to be taken into consideration, however, that ethanol will be dialyzed simultaneously. In any case, the monitoring the ethanol concentration (target value approximately 1 g/l) is mandatory to adjust the dosage of ethanol appropriately. Toxicological data (biological tolerance values, BAT) are shown in Table 27.11.

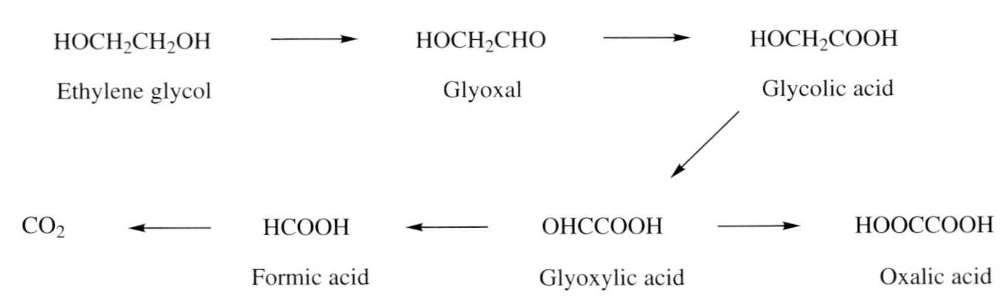

Figure 27.6 Ethylene glycol metabolism.

Table 27.11 Glycols: BAT values.[a]

Substance	Analyte	BAT value[a]	Material
Ethylene glycol		No data	—
Ethylene glycol monobutyl ether	Butoxyacetic acid	100 mg/l	Urine
Ethylene glycol monoethyl ether	Ethoxyacetic acid	50 mg/l	Urine

[a]Biological tolerance value at the workplace [51].

Glycol ether

The derivatives of ethylene glycol are rapidly and completely absorbed through the lung, the gastrointestinal tract, and the skin. The dermal absorption rate of ethylene glycol monomethyl ether is 10 times that of acetone or methanol, for example, and is also faster than the derivatives of diethylene glycol. Metabolism mainly takes place in the liver. In animal experiments (rat), glycols with a free hydroxyl function are metabolized to the corresponding acid and its glycine conjugates. For example, 50–60% of an ethylene glycol monomethyl ether dose (Figure 27.7) is metabolized to methoxyacetic acid and 18–25% to methoxyacetyl glycine. As in the case of other glycol derivatives, ethanol is capable of inhibiting that metabolic pathway competitively. In humans, the elimination half-life of 2-methoxyacetic acid is 77 h and that of 2-ethoxyacetic acid is 21–42 h. Toxicological data (biological tolerance values, BAT) are provided in Table 27.11.

Clinical interpretation

Short chain glycols and glycol ethers have narcotic effects (CNS depression, coma) in common. Other toxic effects are regarded as molecule specific and are assigned mainly to the products of oxidation. Both the intermediate and the final acidic products of metabolism are responsible for the development of a possibly serious metabolic acidosis. In the process of enzymatic oxidation, NAD is converted into NADH. The changed NAD/NADH ratio is responsible for shifting the redox system to the reduced side, thereby also contributing to the formation of metabolic acidosis via an increase of lactate and hydroxybutyrate [41].

$HO{-}CH_2{-}CH_2{-}O{-}CH_3 \longrightarrow HO{-}C({=}O){-}CH_2{-}O{-}CH_3$

Ethylene glycol monomethyl ether
(methylcellosolve)

Methoxyacetic acid

$\downarrow$

$H{-}C(COOH)(H){-}N(H){-}C({=}O){-}CH_2{-}O{-}CH_3$

Methoxyacetylglycine

Figure 27.7 Glycol monomethyl ether.

Diethylene glycol
The first symptoms of an acute intoxication include nausea and vomiting, followed by hematemesis, diarrhea and renal as well as abdominal pain and finally, damage of lung, liver, kidneys, and intestine [41]. Serious poisonings involve central nervous symptoms similar to those mentioned for ethylene glycol. The lethal dose mentioned in the literature is 1–2 g/kg body mass [41]. A secondary phenomenon following the organic disorders is the development of lactacidosis. As in the case of ethylene glycol, appropriate measures for treating serious symptoms in conjunction with high doses include the administration of sodium bicarbonate to compensate acidosis and hemodialysis for removal of toxins and pH adjustment.

Ethylene glycol
According to the time course and the effect of the metabolites, the following three stages can be distinguished clinically:

(1) Central nervous and metabolic symptoms
(2) Cardiopulmonary complications
(3) Renal manifestations.

First stage: After 30 min to 12 h of the ingestion, the first symptoms of poisoning are observed. The primary neurological symptoms are similar to those of ethanol intoxication. These can be followed at a later stage by convulsions and in case of high doses by somnolence and deep coma. Several hours after ingestion, a serious metabolic acidosis may develop from the formation and accumulation of organic acids (glycolic acid in particular) as described above. Anion and osmotic gap are enlarged. Progressive ethylene glycol metabolic degradation leads to a shrinking osmotic gap and a widening anion gap. The organic acids do not increase the osmotic gap but the anion gap by consumption of bicarbonate buffer. Therefore, the ratio of anion gap/osmotic gap can be regarded as an indicator of the degree of metabolic degradation [47, 48]. The citric acid cycle is inhibited by an increase in NADH and some metabolic products of ethylene glycol. The enhanced anaerobic energy production leads to an increase in the lactate concentration [41]. As only 0.3–2.5% of ethylene glycol is degraded to oxalate, this metabolite plays only a minor role in the development of acidosis.

Approximately 12 h after ingestion, the *second stage* of the intoxication sets in. It is characterized by cardiorespiratory complications that are supposed to be directly caused by the toxicity of ethylene glycol itself.

The *third stage* begins 24–72 h after ingestion. Although already present in the early stages, the impairment of renal function becomes more distinct, and oxalate crystals, protein, and blood can be observed in the urine. Renal pain and finally acute renal failure occur.

The transition between the stages is blurred and death can occur at any time. According to the literature, the lethal dose is 1.4–1.6 ml/kg body mass, which means an ingestion of about 100 ml for an adult of 70 kg [49]. However, much higher doses have been survived, whereas, on the other hand, there is a report on a fatal outcome after ingestion of only 30 ml [49]. In cases of lethal intoxication, the blood ethylene glycol concentrations ranged between 0.3 and 4.3 g/l and the concentrations in urine between 0.6 and 10.8 g/l [50]. There is a good chance of surviving even high doses if the therapy sets in at an early stage.

An effective antidote for counteracting ethylene glycol intoxication is ethanol, which competitively inhibits the formation of the toxic aldehydes and carbonic acids. Fomepizole (4-methylpyrazol) is another antidote that has become available only recently. It is also capable of inhibiting the alcohol dehydrogenase efficiently [43]. The administration of sodium bicarbonate to compensate acidosis and hemodialysis for removal of toxins and adjustment of pH are common therapeutic measures in case of serious symptoms and high doses.

Glycol ethers

The symptoms following the ingestion of high doses of glycol ethers resemble those of ethylene glycol and diethylene glycol. They include metabolic acidosis, renal disorders, and central nervous symptoms. An ingestion of 100 ml has caused agitation, confusion, nausea, cyanosis, hyperventilation, mild tachycardia, metabolic acidosis, and renal disorders. There has been a report on a lethal intoxication after ingesting a mixture of 400 ml of ethylene glycol monomethyl ether and an unknown quantity of ethanol. Among the findings of a postmortem examination was acute hemorrhagic gastritis, a fatty degeneration of the liver, and black colored kidneys showing degenerative damage of the tubules. Even without conclusive clinical data, a plausible therapeutic strategy may be the administration of ADH inhibitors such as ethanol as in ethylene glycol intoxication.

27.4 Volatile Halogenated Hydrocarbons

F. Degel, M. Geldmacher-von Mallinckrodt, and C. Köppel

Introduction

Volatile halogenated hydrocarbons (VHHC) are widely used as degreasing and cleaning agents as well as solvents in industrial products. These substances and carbon tetrachloride, in particular, are characterized by their high toxicity. Some of the compounds are potentially addictive, and often abused as inhalants (see Section 27.5). For quick screening, especially of scene residues, chromometric gas analysis with detection tubes for chloroform and carbon tetrachloride may be used. Chlorinated hydrocarbons have a higher density than water. Unlike their nonhalogenated analogues, they will sink to the bottom when transferred to a test tube (glass) filled with an aqueous solution. The Fujiwara reaction can be carried out in case of suspected poisoning with "trichloro" compounds. Headspace gas chromatography with electron capture detection (ECD) is the method of choice for a sensitive detection, identification, and quantification of VHHC. The procedure described here follows a DIN method (procedure DIN 38407-F4, ENISO 10301) for the determination of these substances in water.

R−CCl_3 → [pyridinium−C(Cl)(Cl)−R]$^{+}$ Cl^{-} —Alkali hydroxide solution→ HC(=N−C(Cl)(Cl)−R)−CH=CH−CH=CH−ONa → HC(=O)−CH=CH−CH=CH−ONa

	I	II	III
Wavelength_{max} (nm)	390–460	530	365

Figure 27.8 1,1,1-Trichlorohydrocarbons: Fujiwara reaction.

27.4.1 Color Test

Outline

According to Fujiwara, many chlorinated and (to a lesser extent) some brominated and iodinated hydrocarbons react with pyridine and alkali when heated to form usually red colored products [52]. The mechanism of the reaction is not fully understood yet (Figure 27.8).

Specimens

Urine, serum, stomach contents and gastric lavage (approximately 5 ml), and scene residues.

As plastic material is capable of absorbing chlorinated hydrocarbons, it is mandatory that the specimens are collected and stored only in glass containers with appropriate stoppers. It is necessary to keep the vials tightly closed to prevent evaporation of substances.

Equipment

Hood
Water bath
Test tubes made from glass.

Chemicals (p.a. grade)

Acetone
Potassium (or sodium) hydroxide (pellets)
Pyridine (chromatography grade) (*Note*: Immediately after use, the bottle must be tightly closed again. Otherwise, pyridine is deteriorated soon and too high reagent blank values are obtained.)
Trichloroacetic acid.

Reagents

Sodium (or potassium) hydroxide solution: 20 g sodium (or potassium) hydroxide pellets is dissolved in 100 ml of distilled water (*Caution*: generation of heat).

Trichloroacetic acid solution: Under vigorous stirring, 30 mg trichloroacetic acid is dissolved in 1000 ml of aqua bidest.

Negative control sample: The reagent blank solutions are used as negative controls (see "Examination procedure").

Positive control sample: Trichloroacetic acid solution serves as the positive control.

Sample preparation

Urine, serum: not applicable
Stomach contents or gastric lavage: filtration and centrifugation if turbid.

Examination procedure
See Table 27.12.

Under a hood, 2 ml of the alkali hydroxide solution and 5 ml of pyridine are heated in the water bath for 2 min at 100 °C, under permanent shaking.

Caution: Strong alkaline solution and danger of boil over. Safety goggles are mandatory.

After adding 1 ml of the specimen (e.g., urine), the mixture is heated and shaken again in the water bath for 2 min.

Color of the pyridine layer and its intensity are assessed immediately.

The intensity of colors steadily increases after the end of heating.

Analytical assessment
Negative control: The reagent blank values should not show any change in color.

Positive control: A distinct red coloring of the pyridine layer must be observed.

Sensitivity (see Section 4.1.4):

Maximal sensitivity: 3 mg/l urine
Practical sensitivity: 11 mg/l urine.

Specificity: Taking into consideration the absorption maximums shown in Figure 27.8, it becomes clear that different colors (from dark red to yellow) can develop depending on the chlorinated compound(s) present, the reaction tempera-

Table 27.12 Fujiwara reaction: pipetting scheme.[a]

	Sample (ml)	Positive control (ml)
Sodium hydroxide solution[b]	2	2
Pyridine	5	5
The mixture is heated in a water bath for 2 min at 100 °C and shaken permanently		
Sample[c]	1	—
Trichloroacetic acid	—	1
The mixture is heated in a water bath for 2 min at 100 °C and shaken permanently. Immediately afterwards, the color of the (upper) pyridine layer and its intensity are assessed		

[a]Use of hood is mandatory.
[b]Potassium hydroxide solution may be used as well.
[c]Urine, serum, or gastric lavage.

Table 27.13 Fujiwara reaction according to Table 27.12: detection limits.

Compound	($\varepsilon \times 10^3$)	Detection limit[a] (mg/l)
Trichloromethane (chloroform)	8.9	2
Trichloroacetic acid	11.4	3
Chloral hydrate	8.7	4
Tribromoethene	7.2	6
1,1,2,2-Tetrachloroethane	5.2	6
Trichloroethene	4.2	6
Tribromomethane (bromoform)	5.7	9
Tetrabromomethane (carbon tetrabromide)	3.2	18
α,α,α-Trichlorotoluene	0.8	45
1,1,1-Trichloroethane	0.5	45
α,α-Dichlorotoluene	0.4	90
Dichloroacetic acid	0.3	90
Tetrachloroethene	0.2	180
Pentachloroethane	0.15	180
Triiodomethane (iodoform)	0.1	720
Hexachloroethane	0.05	900

[a] Calculated from the molar extinction coefficient ε at the respective absorbance maximum (368, 470, or 535 nm) [56].

ture, and the time. The Fujiwara reaction also yielded positive results for some brominated and iodinated hydrocarbons (Table 27.13). In general, the reaction with the chlorinated compounds is more sensitive than with the respective bromide and iodine derivatives (see Table 27.14).

Chlorine water, bromine water, and iodine water give positive reactions as well [53].

Some halogenated hydrocarbons, however, do not show a positive reaction even under modified test conditions (Table 27.15).

Urine from patients after application of iodinated X-ray contrast media, for example, iohexol, iopromide, and urographine, proved negative and also the pure contrast media did not react.

With regard to other halogenated compounds described in the publications mentioned before, different observations can be due to either modified procedures or extremely different concentrations (see Ref. [53]). Moreover, the presence of solvents such as ethanol, acetone, acetonitrile, toluene, and so on may also exert a considerable influence on the reaction [54–56]. Stock solutions may have been produced with organic solvents because of the low water solubility of many halogenated hydrocarbons. Some of the authors may have regarded that fact as trivial and did not mention it in their publication.

Detectability of tetrachloromethane by means of the Fujiwara reaction: In the literature, a lot of contradictory opinions about the suitability of the Fujiwara reaction for the detection of tetrachloromethane are found. According to own experiments, a weak red coloring is obtained only for very high carbon tetrachloride concentrations under

Table 27.14 Fujiwara reaction: detectability of compounds.

	References
Intensive red color	
Chloral	[53]
Chloral hydrate	[54]
1,1,2,2-Tetrachloroethane	[53]
Tribromomethane (bromoform)	[53, 54]
Trichloroacetic acid	[53]
Trichloroethene ("Tri")	[53]
Trichloromethane (chloroform)	[53, 54]
Trichloro-*tert*-butanol	[53]
Triiodomethane (iodoform)	[53, 54]
Red color	
Dibromomethane	[53]
1,1,2,2-Tetrabromoethane	[53]
1,1,2,2-Tetrachloroethene (perchloroethylene)	[53]
Tribromoethene	[53]
Tribromomethane	[53]
1,1,1-Trichloroethane	[53]
Yellow color	
Trichloroethanol	[55]

the conditions described. However, after addition of 0.1 ml of acetone and another short heating period, a distinct red color was observed.

A positive result of the Fujiwara reaction is an important indication on the presence of one of the compounds mentioned in Table 27.14. Yellow coloring indicates the

Table 27.15 Fujiwara reaction: not detectable compounds.

Compound	References
Inorganic chlorides	[59]
Bromomethane	[53]
α-Bromotoluene	[53]
Chlorobenzene	[53, 59]
Chloroacetic acid	[53, 55]
Chloroethane	[55]
Chloroethanol	[55]
Chloroethene	[55]
Chloromethane	[53]
o-Chlorotoluene	[53]
DDT	[53, 55]
1,4-Dichlorobenzene	[53]
2,2′-Dichlorodiethyl ether	[55]
1,2-Dichloroethene	[55]
1,2,3,4,5,6-Hexachlorocyclohexane	[53]
Hexachloroethane	[53, 55]
Acid chlorides	[59]

excretion of trichloroethanol. Positive results must be confirmed by a gas chromatographic procedure. As the reaction varies considerably in color intensity, an estimation of the concentration is not possible.

It is a qualitative group test and one can only conclude from a positive finding that reacting halogenated hydrocarbons are present. The clinical interpretation of a positive result cannot be started unless the compounds are identified by, for example, gas chromatography. A negative Fujiwara result is not a proof of the absence of halogenated hydrocarbons for the following reasons:

- Some VHHC exhibit no or only weak reaction (Table 27.15).
- VHHC concentration in urine may be too low.
- VHHC compound does not form reacting metabolites.

All three items apply to, for example, 1,1,1-trichloroethane.

Annotations
It is important that the (upper) pyridine layer remains colorless (reagent blank value). Reasons for a change in color of the pyridine layer may be contaminated vials, contaminated pyridine (see above: chemicals), or vapors of chlorinated solvents in the air of the laboratory.

In case of interference from pyridine, the chemical must be distilled or replaced.

Practicability
The total analysis is carried out within 19 min.

27.4.2
Gas Chromatography

Outline
See Sections 4.6 and 12.4.

Specimens
Whole blood (anticoagulant EDTA). Use glass sampling system or transfer the sample immediately after sampling into prepared and closed glass headspace tubes with EDTA additive by injection through PTFE-coated septum by a syringe: sampling systems with gel are, in general, not suitable. For more information about sampling, see Section 4.6.

Equipment
Gas chromatograph with electron capture detector and supply units for carrier gas and make-up gas.

Stationary phase: Fused Silica quartz capillary column, internal diameter 0.53 mm, length 30 m, RTX 1701, coating 3 μm (Restek, Bad Soden), or comparable phase.

Chromatographic settings:

Carrier gas: helium, average flow rate 49 cm/s
ECD make-up gas: nitrogen.

Column oven temperature program	
Initial temperature	40 °C (6 min)
Heating rate	5 °C/min
Final temperature	110 °C (10 min)

Injector temperature: 150 °C; detector temperature: 300 °C
ECD amplifier: range: 1, attenuation: 1

Settings for the automated sampling system HS 40 (Perkin-Elmer): sample temperature, 60 °C; needle temperature, 90 °C; transfer temperature, 110 °C; thermostat time, 30 min; pressure build-up time, 2 min; injection time, 0.04 min; dwell time, 0.5 min; HS 40 pressure, 70 kPA; column pressure on GC, 39 kPA.

For clinical toxicological analyses of acute poisonings, the settings for the thermostat time and the pressure build-up time are appropriate for highly volatile halogenated hydrocarbons. Results for less volatile substances, in particular, may be less accurate, because the equilibrium has not been completely reached before sampling for GC. In those cases, it is recommended to extend the time settings as follows: thermostat time 90 min and pressure build-up time 6 min for less urgent analyses.

Sampling: Due to the volatility of the analytes, direct sampling into appropriately prepared headspace vials with EDTA is recommended (see Section 4.6). If not possible, the following procedure is applied:

Sample vials: 20 ml headspace vials with PTFE-coated butyl rubber septa with star springs and aluminum crimp caps (e.g., Perkin-Elmer, Shelton, USA; Germany: Rodgau-Jügesheim). Before use, the headspace vials, and the butyl rubber septa in particular, should be heated for 1 day in a drying oven at 110 °C. They have to be stored in an airtight screw cap container or a desiccator to avoid contamination by the environmental air. On no account, butyl rubber septa should be used twice.

Pipettes: Because of the volatility of the analytes, 2 ml positive displacement pipettes are used (e.g., Gilson Microman from Abimed, Langenfeld or Socorex, Ecublens, Switzerland) for whole blood samples. Positive displacement pipettes (e.g., Hamilton HPLC syringe, 100 µl, or similar) are also used for the standard stock solution.

Further devices and equipment: see Section 4.6.

Chemicals (p.a. grade)

Reference substances (GC grade)
Ethylenediaminetetraacetic acid, disodium salt
Propylene carbonate (4-methyl-1,3-dioxolane-2-on)
Sheep blood (whole blood with EDTA) (e.g., Fiebig-Nährstofftechnik Idstein).

Reagents

Standard mixtures are provided by various manufacturers, for example, standard solution volatile halogenated hydrocarbons in pentane, code W 1250 (Promochem, Wesel).

Due to the low water solubility of the substances, the standard mixtures contain appropriate solutizers (e.g., propylene carbonate). Because of the high volatility and low water solubility of the analytes, it is strongly recommended to use commercially available ready-for-use standard mixtures in ampoules instead of self-made standards.

VHHC stock solution: The standard mixture (ampoule) and the 500 μl HPLC syringe (Hamilton) are cooled down to −20 °C as well as a 15 ml propylene carbonate aliquot.

Five hundred microliters of the standard solution is quickly transferred with the HPLC syringe from the ampoule to a 10 ml volumetric flask, which is filled already two-thirds with cooled propylene carbonate. Cooled propylene carbonate is added up to slightly below the ring mark. The flask is closed, its content well mixed and after reaching a temperature of +20 °C (water bath), it is made up to the mark. The stock solution is stored at 4–8 °C in a refrigerator. It is stable for 6 months, but in case it is frequently opened, it has to be replaced earlier. Table 27.16 shows the concentrations of the analytes in the standard stock solution.

Standard calibration solution: The standard calibration solutions are prepared freshly by spiking blood (sheep blood or human blood, both with EDTA additive) with stock solution. A preliminary headspace analysis has to demonstrate the absence of interfering peaks within the time span of the volatile halogenated hydrocarbon standard substances in this blood.

Eighteen hundred microliters of cooled (4–8 °C) sheep blood or human blood is filled into a headspace vial. Using a cooled HPLC syringe, 200 μl of the standard stock solution is added. The tube is closed with a PTFE-coated septum and the aluminum crimp cap and the sample is mixed on a vortex mixer for 1 min. It is important to carry out the aforementioned steps quickly. Table 27.16 shows the concentrations of the analytes in the standard calibration solution. If kept in the closed headspace tube at −20 °C, it is stable for about 3 months.

Control samples: It is necessary to prepare whole blood-based control samples by oneself, because at present these are not commercially available.

The standard stock solution can also be used for the preparation of control samples at different concentrations. For this purpose, cooled sheep blood (4–8 °C) is filled into

Table 27.16 Volatile halogenated hydrocarbons: stock solution for headspace gas chromatography.

Substance	Concentration (ampoule) (mg/l)	Standard calibration solution (μg/l)
Dichloromethane	200	500.00
Trichloromethane	5.0	12.50
1,1,1-Trichloroethane	1.0	2.50
Tetrachloromethane	0.25	0.63
Trichloroethene	2.5	6.25
Bromodichloromethane	1.0	2.50
Tetrachloroethene	0.6	1.50
Dibromochloromethane	1.5	3.75
Tribromomethane	4.5	11.25

appropriate headspace vials to which an aliquot of the standard stock solution is added with a cooled syringe (total volume: 2 ml). The tube is quickly closed and mixed on a vortex mixer for 15 s. At −20 °C, the control samples are stable for 3 months.

Anticoagulated analyte-free sheep blood or human blood can serve as negative control samples.

Sample preparation

In case the samples are not obtained in headspace vials, aliquots of always the same volume (e.g., 2 ml) are pipetted from the primary vial into the headspace vial. The vial must be closed gastight by attaching a PTFE-coated butyl rubber septum with star spring and aluminum crimp cap using a hand crimper.

If the headspace vials are obtained already filled with the blood sample, they can be used directly for analysis. Due to the low distribution coefficient of VHHC, it is essential that the filling levels of the calibration standards and the samples are identical.

Headspace GC analysis

For general information see Section 4.6.

For headspace analysis, the samples are incubated at 60 °C for 30 min.

Evaluation: At first, the volatile substances present in the samples have to be identified. For computation, the chromatographic software is applied using the method of external standard. Since the ECD may be saturated already at rather low concentrations, it is especially important to pay attention to the limit of linearity.

Analytical assessment

The imprecision of the procedure was estimated by repeated measurements of a standard calibration solution based on sheep blood. The results are shown in Table 27.17.

For evaluation of linearity, increasing amounts of 1,1,1-trichloroethane, trichloroethene, and tetrachloroethene were added to samples of analyte-free blood. With optimized detection limit, a linear calibration function of 0–40 μg/l was obtained. Much higher concentrations can occur in acute poisoning. For an unbiased measurement, the samples have to be diluted with analyte-free sheep blood or human blood to the concentration range of the linear calibration function. The trueness of the procedure was estimated from recovery experiments. When blood was spiked with different amounts of the standard substances mentioned in Table 27.16, the recovery ranged from 97 to 112%.

Detection limits were calculated from mean area of noise and its threefold standard deviation of an analyte-free blood sample at the respective retention time range. This value was then related to the response factors (area/concentration) of the individual analytes. The results are shown in Table 27.17.

Ring trials for external quality assessment are organized, for example, by the German Society of Industrial and Environmental Medicine (project group quality assurance).

Table 27.17 Volatile halogenated hydrocarbons: headspace gas chromatography.

Substance	Concentration (μg/l)	Imprecision (interassay)[a] CV%	Imprecision (intraassay)[a] CV%	Detection limit[b] (μg/l)
Dichloromethane	500.00	2.0	5.2	6.0
Tetrachloro-methane+1,1,1-trichloroethane	12.50+2.50	5.5	7.4	0.8
Trichloromethane	0.63	3.2	7.3	0.2
Trichloroethene	6.25	3.0	5.4	0.5
Bromodichloromethane	2.50	4.4	6.5	0.2
Tetrachloroethene	1.50	4.1	7.8	0.2
Dibromochloromethane	3.75	5.0	8.1	0.2
Tribromomethane	11.25	4.5	8.7	0.5

CV%: coefficient of variation (%).
[a]Positive control (number of measurements: 10).
[b]Negative control (number of measurements: 10).

Practicability

Technician time for carrying out one VHHC analysis (one sample, standard calibration solution, positive control, and negative control) is about 15 min, whereas the total analysis takes about 135 min. If only poisoning with volatile halogenated hydrocarbons has to be ruled out, it is recommended to restrict oneself to the qualitative analysis of the patient blood sample and one positive control. If no relevant peaks are detectable in the patient sample, no further analysis is necessary. Otherwise, a comprehensive determination comprising all quality control samples has to be carried out. The shorter qualitative analysis requires only 75 min.

27.4.3 Medical Assessment and Clinical Interpretation

Medical assessment

Toxicological data about volatile halogenated hydrocarbons (reference values, biological tolerance values and exposure equivalents for carcinogenic substances) as well as the relevant materials and analytes are provided in Tables 27.18.

Apart from rare and mostly accidental cases of oral ingestion, volatile halogenated hydrocarbons are usually inhaled. As vapors of chlorinated hydrocarbons are considerably heavier than air, people laying tiles or parquet and little children, for example, inhaling runout stain remover are particularly at risk. It should be mentioned that skin absorption is also possible. After inhalation, peak blood concentrations are reached very soon, but in case of oral ingestion the same is reached only after 1–2 h. Preferred sites of accumulation of chlorinated hydrocarbons are fatty tissues and lipophilic organs such as the brain, which is due to the high lipid solubility of the substances. As they are capable of entering the placenta, the compounds are also detectable in the fetus. In an oxidative process, halogenated hydrocarbons are degraded in the liver by cytochrome P450. For some of the substances, the metabolism via glutathione-*S*-transferase is of toxicological importance. The substances are

Table 27.18 Volatile halogenated hydrocarbons: reference intervals.[a]

Substance	Analyte	Reference interval	BAT value	Material
Dichloromethane[c,d]	Dichloromethane	<1 μg/l	—	Blood
	CO-hemoglobin	<3%[b]	—	Blood
Tetrachloroethene[c,d] (perchloroethylene)	Tetrachloroethene	<1 μg/l	—	Blood
Tetrachloromethane (carbon tetrachloride)	Tetrachloromethane	<0.3 μg/l	3.5 μg/l	Blood
1,1,1-Trichloroethane (methylchloroform)	1,1,1-Trichloroethane	<1.3 μg/l	550 μg/l	Blood
Trichloroethene[c,d] (trichloroethylene)	Trichloroethene	<1 μg/l	—	Blood
	Trichloroethanol	<0.3 μg/l	—	Blood
	Trichloroacetic acid	<1.0 mg/l	—	Urine
Trichloromethane[d] (chloroform)	Trichloromethane	<2.0 μg/l	—	Blood

BAT: biological tolerance value at the workplace; EKA: exposure equivalents for carcinogenic substances.
[a]According to Refs [57, 61].
[b]Percentage of total hemoglobin.
[c]Carcinogenic solvent. EKA values: see Ref. [57].
[d]BAT value not established.

also partially excreted via the lung. In some compounds, the narcotic effects (e.g., chloroform, trichloroethene) are dominant, while others exhibit mainly neurotoxic effects (tetrachloroethene, PER) or hepatotoxic effects (carbon tetrachloride, tetrachloroethane) [58].

Dichloromethane (methylene chloride)
Dichloromethane is quickly resorbed through the lung. Its metabolic degradation to carbon monoxide is of toxicological relevance. If exposed to high concentrations, this oxidative route of degradation via cytochrome P450 becomes saturated. A glutathione-dependent alternative metabolism leads to the formation of CO_2 via formaldehyde. However, in an exposure range up to a number fraction of $200/10^6$ (200 ppm), a clear dose and time-dependent formation of CO-hemoglobin is observed. The workplace tolerance value is 3% CO-hemoglobin of total hemoglobin [60]. On an average, this percentage is reached from exposure to the "maximal threshold limit value for industrial environments" (MAK), which is set to a number fraction of about $75/10^6$ (75 ppm).

Trichloroethene (trichloroethylene, "Tri")
Most often trichloroethylene is absorbed by the lungs, but percutaneous resorption is also possible. Most of the resorbed substance is metabolized via oxidative decomposition. Trichloroacetaldehyde (chloral) is degraded to two metabolites: trichloroethanol, which is partly excreted as glucuronide in urine, and trichloroacetic acid (Figure 27.9). There are large individual differences among humans with regard to the proportion of trichloroethanol excretion to trichloroacetic acid excretion. In humans, the half-life of trichloroethanol is 10 h, that of the metabolite trichloroacetic acid about 100 h, because it is bound strongly to proteins. Among the other

Trichloroethylene

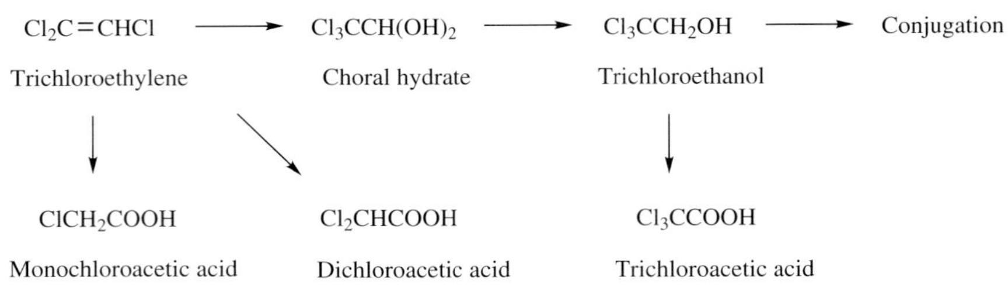

Figure 27.9 Trichloroethene metabolism.

metabolites, the neurotoxic dichloroacetylene deserves particular attention for its adverse effects on cerebral nerves. As elimination is predominantly renal, the Fujiwara reaction is appropriate even for monitoring industrial exposure.

Tetrachloroethene (tetrachloroethylene/perchloroethylene, "Per")

After inhalation of tetrachloroethene, most of the dose is exhaled unchanged again. The main process of metabolism is oxidative and the end product is trichloroacetic acid. However, there is great individual variation in the fraction of metabolized products. As described above, the half-life of trichloroacetic acid is very long (100 h) and is the reason for a long lasting excretion in urine. Only about 1% of the absorbed tetrachloroethene is excreted in urine as trichloroacetic acid within the first 70 h. The subordinate, glutathione-dependent metabolic pathway results in *S*-trichloroethenylmercapturic acid and other nephrotoxic metabolites. Intermediate metabolites of the oxidative pathway are probably responsible for the hepatotoxic effect of tetrachloroethene.

1,1,1-Trichloroethane (methylchloroform) and 1,1,2-trichloroethane

After inhalation, only about 2% of 1,1,1-trichloroethane is metabolized, whereas about 98% of a dose is exhaled unchanged. Therefore, only very high doses can lead to a positive Fujiwara reaction. The products occurring in urine are trichloroethanol and trichloroacetic acid (see above) (Figure 27.10). Due to its moderate toxicity, methylchloroform has a comparatively high threshold limit value for industrial environments (MAK value) with a number fraction of $200/10^6$ (200 ppm) [57]. The more toxic 1,1,2-trichloroethane is metabolized to a considerably higher extent. The main metabolites are chloroacetic acid and its derivatives *S*-carboxymethylcysteine and 2,2-thio-diacetic acid (thiodiglycolic acid), which are produced by conjugation with glutathione. The threshold limit value for industrial environments of 1,1,2-trichloroethane is set to a number fraction of $10/10^6$ (10 ppm).

Trichloroethane

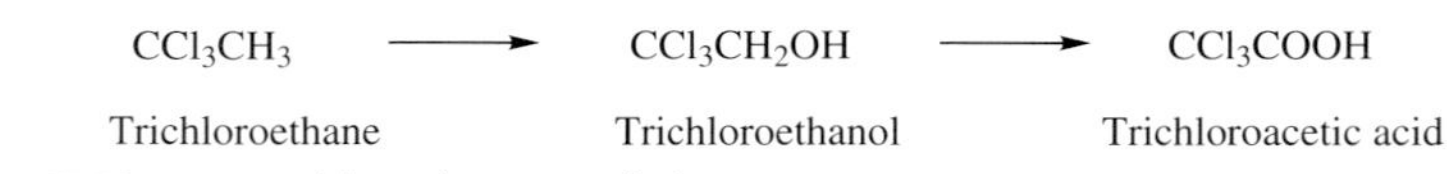

Figure 27.10 1,1,1-Trichloroethane metabolism.

Trichloromethane (chloroform)
Inhalation is the main route of entry into the organism for chloroform. In an oxidative process in the liver, it is metabolized to carbonylchloride (phosgene) by splitting HCL. Phosgene is a highly reactive cell poison, which may cause serious cell damage. In view of its nephrotoxic and hepatotoxic effects, it has a relatively low threshold limit value (MAK value) with a number fraction of $0.5/10^6$ (0.5 ppm). A biological tolerance value (BAT value) has not yet been established.

Tetrachloromethane (carbon tetrachloride, "Tetra")
Usually, tetrachloromethane is taken up by inhalation. First, an activating mechanism produces the trichloromethyl radical, which is more toxic than the mother substance. It may react with unsaturated fatty acids to form chloroform. The remaining free fatty acid radicals can further react through lipid peroxidation or diene conjugation. Then, either oxygen is attached to the fatty acid diene radical to form peroxides or residues of fatty acids combine with each other or it comes to a fragmentation of unsaturated fatty acids resulting in malondialdehyde, ethane and pentane. The latter two can be found in expired air. If "tetra" is inhaled stepwise, an otherwise acute lethal dose of the agent may be survived for the following reason: cytochrome P450 is mainly involved in the metabolism of tetrachloromethane. The first metabolic product, the trichloromethyl radical, can react irreversibly with the enzyme and destroy it. A marked decrease in the concentration of cytochrome P450 after the initial dose inhibits the conversion of carbon tetrachloride into the toxic trichloromethyl radical. By virtue of the strong hepatotoxic effect of the substance, there is a low threshold limit value with a number fraction of $0.5/10^6$ (0.5 ppm). The biological tolerance value for tetrachloromethane is 3.5 μg/l blood (after the end of exposure).

Clinical interpretation
At elevated concentrations, halogenated hydrocarbons mainly develop narcotic effects. Within a few minutes after inhalation, CNS depression is observable. High doses or a continuous exposure over many years can impair the central nervous system irreversibly. The substances not only have hepatotoxic as well as mild nephrotoxic effects but can also trigger tachyarrhythmia. Oral ingestion causes gastrointestinal irritation, and very high doses can even lead to hepatorenal failure and death. Tetrachloromethane, 1,2-dichloroethane, dichloromethane, tetrachloroethene, and trichloromethane are suspected of being carcinogenic [58]. Metabolites contribute considerably to serious toxic effects in VHHC poisoning. Shortly after ingestion, the halogenated compounds are visible in the X-ray of the stomach as a "contrast medium."

Dichloromethane (methylene chloride)
The CNS depressant effect of dichloromethane is less distinct than that of chloroform. Irritations of the respiratory tract up to toxic pulmonary edema have been observed. At present, carcinogenicity in humans is not assured.

Trichloroethene (trichloroethylene, "Tri")
Narcotic effects of trichloroethane appear after exposure to a number fraction of approximately $1000/10^6$ (1000 ppm). The metabolic product trichloroethanol has a distinct depressant effect on the CNS. It occurs also as a pharmacologically active metabolite of the hypnotic chloral hydrate. High doses and/or continuous exposure cause liver and kidney damage. In case of acute poisoning, renal damage or even renal failure have been described.

Like other halogenated hydrocarbons, trichloroethene sensitizes the myocardium to the effects of catecholamines. The metabolic product dichloroacetylene, which is produced in the presence of alkali, shows a distinct neurotoxicity and has, at least in animal experiments, a carcinogenic effect on the kidney.

Tetrachloroethene (tetrachloroethylene/perchloroethylene, "Per")
An exposure to tetrachloroethene at a number fraction of about $100/10^6$ (100 ppm) causes irritations of the mucous membranes and at a number fraction between $500/10^6$ and $1000/10^6$ (500–1000 ppm) of the respiratory system. Its narcotic effect is slightly stronger than that of chloroform. "Per" sensitizes the myocardium to catecholamines, and extrasystoles may occur. Hepatotoxicity and sometimes, at high concentrations, nephrotoxicity are observed. Among the described symptoms of acute poisonings are CNS depression, vomiting, headache, and loss of consciousness. There have been reports that continuous exposure or high doses can lead to a drop in cerebral performance from brain damage and to a change in personality. Teratogenic effects of the agent have not been verified yet. There is still considerable uncertainty with regard to the carcinogenicity of the substance. It has therefore been classified as potentially carcinogenic.

1,1,1-Trichloroethane (methylchloroform) and 1,1,2-trichloroethane
The depressant effect of 1,1,1-trichloroethane on the central nervous system is less severe than that of chloroform or trichloroethene. Higher concentrations cause an irritation of the mucous membranes of the respiratory tract. As compared to other chlorinated solvents and its isomer 1,1,2-trichloroethane, the toxicity of 1,1,1-trichloroethane is less marked. The hepatotoxicity and nephrotoxicity of 1,1,2-trichloroethane probably stem from the intermediate chloroacetic chloride. The lower toxicity of 1,1,1-trichloroethane may be explained by the absence of a comparable, toxic metabolite as well as by its low metabolic rate. There have been also no signs as yet of a potential mutagenicity or carcinogenicity of 1,1,1-trichloroethane.

Trichloromethane (chloroform)
Chloroform has been used as a narcotic and analgesic for a long time. However, because of its narrow therapeutic range, respiratory depression, and negative inotropic as well as hepatotoxic effects, its application was discontinued. In acute poisoning, chloroform causes centrilobular damage with necroses and fatty degeneration of the liver. A clinical characteristic of acute liver damage from chloroform narcosis is an increase of transaminases and other "hepatic" enzymes reaching their peak concentration on the third or fourth day. In most cases, these damages have

been reversible. After inhalation, the following symptoms and complications have occurred (depending on the concentration): deep breathing, increase in respiratory rate, hyperthermia, hypotension, depression of gastrointestinal mobility, respiratory acidosis, hyperglycemia, and a prolonged prothrombin time. Liver and kidneys may be damaged chronically. Experiments with rats and mice have clearly shown that chloroform damages the fetus. Moreover, chloroform is suspected of being potentially carcinogenic.

Tetrachloromethane (carbon tetrachloride)
In cases of chronic exposure as well as acute intoxication, the substance has shown high hepatotoxicity and nephrotoxicity. Like chloroform, carbon tetrachloride causes centrilobular liver necrosis. It can be assumed that a long-term inhalation of the poison at a number fraction between $10/10^6$ and $100/10^6$ (10 and 100 ppm) leads to the respective damages. Furthermore, the chronic liver damage is suspected of being connected with the development of liver tumors. There is no evidence yet of a teratogenicity of the substance. At the moment, it is not settled whether carbon tetrachloride may damage the fetus.

27.5 Inhalant Abuse

H.J. Gibitz

Introduction

Virtually, any volatile substance may be found in inhalant abuse, but most of them belong to one of the following categories: volatile solvents, aerosols, gases, and nitrites.
The following procedures are used for inhalant abuse:

Huffing: inhaling from liquid-soaked material.
Bagging: inhaling from a bag in which the material is sprayed or poured.
Sniffing or snorting: inhaling directly from containers, such as lighters or whipped-cream containers.

Inhalant abuse is popular mainly among children, young people, and young adults to reach a "high" and may lead to a special form of intoxication with organic solvent vapors [62]. The inhalants commonly occur in household, handicrafts, and industrial products. They are easy to obtain, or already present in many households, and usually cheap. A list of commonly abused inhalants is given in Table 27.19.

Table 27.20 provides some typical components of the inhaled products.

It must be taken into account that the commercial products rarely consist of a single substance but are composed from a mixture of 10, 20, or even more substances. Therefore, in case of an acute intoxication, analytical identification can be difficult and time consuming.

Table 27.19 Products for inhalant abuse.

Glue thinner
Glue: household/model airplane glue and so on
Bicycle tube/rubber glue
Paint and varnish thinner
Nitro thinner
Trichloroethylene
Nail polish and nail polish remover
Spot remover
Dry cleaning fluid
Typewriter correction fluid
Degreaser
Wax remover
Sealing compound for car radiator
Felt-tip pens
Butane lighters
Propane for (camping) stove
Sprays and aerosols such as hairspray, furniture polish, varnish/paint spray, cleaning sprays, deodorants, and so on
Chloroethyl surgical spray
"Poppers"
Gasoline

Table 27.20 Chemicals of inhalants.

Aliphatic hydrocarbons
For example, *n*-hexane
Cycloaliphatic hydrocarbons
For example, cyclohexane
Aromatic hydrocarbons
For example, benzene, toluene, xylene, and cumene
Chlorinated hydrocarbons
For example, methylene chloride, chloroform, carbon tertrachloride 1,1,1-trichloroethane, 1,1,2-trichloroethylene, 1,1,2,2-tetrachloroethylene, and chlorobenzene
Alcohols
For example, methanol, ethanol, hexanol, methylbenzyl alcohol, and cyclohexanol
Ketones
For example, acetone, cyclohexanone, and methylethylketone
Esters
For example, methyl acetate, ethyl acetate, and *n*-butyl acetate
Ether and glycol ether
For example, diethyl ether, tetrahydrofuran, ethylene glycol monomethyl ether, and ethylene glycol monoethyl ether
Organic nitrites
For example, ethyl nitrite, isoamyl nitrite, and isobutyl nitrite

27.5.1
Gas Chromatography

See Sections 4.6 and 12.4.

27.5.2
Medical Assessment and Clinical Interpretation

During the first stages of inhalation, the compounds may produce euphoria and increased excitement, followed by illusionary misjudgments, impaired color, and acoustic perception and hallucinatory experiences. People familiar with the procedure are able to find the appropriate dose and to reproduce the desired effects over a period of 10–12 h. Continuous inhalation can cause a serious solvent intoxication involving impaired consciousness and possibly coma.

Since chronic abuse results in irreparable brain damage including personality alterations and neurological disorders, it is paramount that the preventive measures are taken early, already at the stage of experimenting with inhalants.

References

References for Section 27.1.1

1 Batra, A. and Buchkremer, G. (2001) Beziehung von Alkoholismus, Drogen- und Tabakkonsum. *Deutsches Ärzteblatt*, **98**, C2070–C2073.

2 Richtlinie der Bundesärztekammer zur Qualitätssicherung quantitativer laboratoriumsmedizinischer Untersuchungen. *Deutsches Ärzteblatt*, **100**, A3335–A3338 (2003); (a) Richtlinie der Bundesärztekammer zur Qualitätssicherung laboratoriumsmedizinischer Untersuchungen. *Deutsches Ärzteblatt*, **105**, C301–C315(2008).

3 Nine, J.S., Moraca, M., Virji, M.A. and Rao, K.N. (1995) Serum ethanol determination: comparison of lactate and lactate dehydrogenase interference in three enzymatic assays. *Journal of Analytical Toxicology*, **19**, 192–196.

4 Lieber, C.S. (1983) Ethanol metabolism and toxicity. *Reviews in Biochemical Toxicology*, **5**, 267–311.

5 Gibitz, H.J. and Fenninger, H. (1981) Blood alcohol determination with the Automatic Clinical Analyzer (ACA DuPont) compared with headspace gas chromatography and a manual enzymatic technique. *Journal of Clinical Chemistry and Clinical Biochemistry*, **18**, 721–722.

6 Widmark, E.M.P. and Baselt, R.C. (eds) (1981) *Principles and Applications of Medicolegal Alcohol Determination*, Biomedical Publications, Davis, pp. 1–163.

7 Mason, M.F. and Dubrowsky, K.M. (1976) Breath-alcohol analysis: uses, methods and some forensic problems. *Journal of Forensic Sciences*, **21**, 9–41.

8 Sivilotte, M.L.A. (2004) Ethanol, isopropanol, and methanol, in *Medical Toxicology*, 3rd edn (ed. R.C. Dart), Lippincott Williams and Wilkins, Philadelphia, PA.

9 Caravati, E.M., McCowan, C.L. and Marshall, S.W. (2004) Plants, in *Medical Toxicology*, 3rd edn (ed. R.C. Dart), Lippincott Williams and Wilkins, Philadelphia, PA.

10 Arndt, T. (2001) Carbohydrate-deficient transferrin as a marker of chronic alcohol abuse. *Clinical Chemistry*, **47**, 13–27.
11 Schmitt, G., Aderjan, R., Keller, T. and Wu, M. (1995) Ethylglucuronide: a usual ethanol metabolite in humans. *Journal of Analytical Toxicology*, **19**, 91–94.
12 Beck, O. and Helander, A. (2003) 5-Hydroxytryptophol as a marker for recent alcohol intake. *Addiction*, **98/2**, 63–72.
13 Beresford, T.P. (2003) Predictive factors for alcoholic relapse in the selection of alcohol-dependent persons for hepatic transplant. *Liver Transplant*, **3**, 280–291.

References for Section 27.1.2

14 Bailey, D.N. (1990) Detection of isopropanol in acetonemic patients not exposed to isopropanol. *Clinical Toxicology*, **28**, 459–466.
15 Moffat, A.C., Osselton, M.D. and Widdop, B. (2004) *Clarke's Analysis of Drugs and Poisons*, 3rd edn, Pharmaceutical Press, London.
16 Levey, S., Balchun, O.J., Medrano, V. and Jung, R. (1964) Studies of metabolic products in expired air. II. Acetone. *The Journal of Laboratory and Clinical Medicine*, **63**, 574–584.
17 Stead, A.N. and Moffat, A.C. (1983) A collection of therapeutic, toxic, and fatal blood drug concentrations in man. *Human Toxicology*, **3**, 437–464.
18 Jones, A.W. (1992) Driving under the influence of isopropanol. *Clinical Toxicology*, **30**, 153–155.
19 Kelner, M. and Beuley, D.N. (1983) Isopropanol ingestion: interpretation of blood concentrations and clinical findings. *Journal of Toxicology – Clinical Toxicology*, **20**, 497–507.
20 Fraser, A.D. and MacNeil, W. (1989) Gas chromatographic analysis of methyl formate and application in methanol poisoning cases. *Journal of Analytical Toxicology*, **13**, 73–76.
21 D'Alessandro, A., Osterloh, J., Chumers, P., Quinlan, P., Kell, T. and Becker, C. (1993) Formate in serum and urine following controlled methanol exposure at the threshold limit value. *Veterinary and Human Toxicology*, **35**, 358.
22 Sivilotti, M.L.A. (2004) Ethanol, isopropanol, and methanol, in *Medical Toxicology*, 3rd edn (ed. R.C. Dart), Lippincott Williams and Wilkins, Philadelphia, PA.
23 Gildson, S., Werczberger, A. and Herman, J.B. (1966) Coma hyperglycemia following drinking of acetone. *Diabetes*, **15**, 810–811.
24 Degel, F. (1996) Comparison of new solid phase extraction methods for chromatographic identification of drugs in clinical toxicological analysis. *Clinical Biochemistry*, **29**, 1–12.
25 Schaller, K.H. and Triebig, G. (1996) Biologische Arbeitsplatz-Toleranzwerte. Teil IX: Arbeitsmedizinische Bewertung von organischen Lösungsmittelgemischen. *Arbeitsmedizin Sozialmedizin Umweltmedizin*, **31** (12), 504–512.
26 Deutsche Forschungsgemeinschaft (DFG) (2007) *List of MAK and BAT-Values 2007*. Commission for the investigation of Health Hazards of Chemical Compounds. Report No. 43. Wiley-VCH Verlag GmbH, Weinheim.

References for Section 27.2

27 Moffat, A.C., Osselton, M.D. and Widdop, B. (eds) (2004) *Clarke's Analysis of Drugs and Poisons*, 3rd edn, Pharmaceutical Press, London.
28 Infante, P.F., Wagoner, J.K., Rinsky, R.A. *et al.* (1977) Leukemia in benzene workers. *Lancet*, **2**, 76–78.
29 Fishbein, L. (1985) An overview of environmental and toxicological aspects of aromatic hydrocarbons. II. Toluene. *The Science of the Total Environment*, **42**, 267–288.
30 Fishbein, L. (1985) An overview of environmental and toxicological aspects of aromatic hydrocarbons. IV. Ethylbenzene.

The Science of the Total Environment, **44**, 269–287.

31 Fishbein, L. (1985) An overview of environmental and toxicological aspects of aromatic hydrocarbons. III. Xylene. *The Science of the Total Environment*, **43**, 165–183.

32 Schaller, K.H. and Triebig, G. (1996) Biologische Arbeitsplatz-Toleranzwerte. Teil IX: Arbeitsmedizinische Bewertung von organischen Lösungsmittelgemischen. *Arbeitsmedizin Sozialmedizin Umweltmedizin*, **31**, 504–512.

33 Deutsche Forschungsgemeinschaft (DFG) (2007) *List of MAK and BAT Values 2007. Commission for the Investigation of Health Hazards of Chemical Compounds. Report No. 43*. Wiley-VCH Verlag GmbH, Weinheim.

References for Section 27.3

34 Geiling, E.M.K. and Cannon, P.R. (1938) Pathologic effects of elixir of sulfanilamide (diethylene glycol) poisoning. *Journal of the American Medical Association*, **111**, 919–926.

35 Scalzo, A.J. (1996) Diethylene glycol toxicity revisited: the 1996 Haitian epidemic. *Journal of Toxicology – Clinical Toxicology*, **34**, 513–516.

36 Wax, P.M. (1996) It's happening again – another diethylene glycol mass poisoning. *Journal of Toxicology – Clinical Toxicology*, **34**, 517–520.

37 Maurer, H.H. and Kessler, C. (1988) Identification quantification of ethylene glycol and diethylene glycol in plasma using gas chromatography–mass spectrometry. *Archives of Toxicology*, **62**, 66–69.

38 Porter, W.H. and Auansakul, A. (1982) Gas chromatographic determination of ethylene glycol in serum. *Clinical Chemistry*, **28**, 75–78.

39 Porter, W.H. Rutter, W.R. and Yao, H.H. (1999) Simultaneous determination of ethylene glycol and glycolic acid in serum by gas chromatography–mass spectrometry. *Journal of Analytical Toxicology*, **23**, 591–597.

40 Beasly, V.R. and Buck, W.B. (1980) Acute ethylene glycol toxicosis: a review. *Veterinary and Human Toxicology*, **22**, 255–263.

41 Joliff, H.A. and Sivilotti, M.L.A. (2004) Ethylene glycol, in *Medical Toxicology*, 3rd edn (ed. R.C. Dart), Lippincott Williams and Wilkins, Philadelphia, PA.

42 Baselt, R.C. (ed.) (2004) *Disposition of Toxic Drugs and Chemicals in Man*, 7th edn Biomed Public, Foster City, CA.

43 Barceloux, D.G., Krenzelok, E.P., Olson, K. and Watson, W. (1999) American Academy of Clinical Toxicology: Practice Guidelines on the treatment of ethylene glycol poisoning. Ad-hoc Committee. *Journal of Toxicology: Clinical Toxicology*, **37**, 537–560.

44 Peterson, D.C., Collins, A.J. and Himes, J.M. (1981) Ethylene glycol poisoning: pharmacokinetics during therapy with ethanol and hemodialysis. *The New England Journal of Medicine*, **304**, 21–23.

45 Bowie, M.D. and Mackenzie, D. (1972) Diethylene glycol poisoning in children. *South African Medical Journal*, **46**, 931–936.

46 Heilmair, R., Lenk, W. and Luhr, D. (1993) Toxicokinetics of diethylene glycol (DEG) in the rat. *Archives of Toxicology*, **67**, 655–666.

47 Thomas, L. (1998) Chloride, in *Clinical Laboratory Diagnostics* (ed. L. Thomas), TH Books, Frankfurt/M.

48 Ten Bokkel Huinink, D., De Meijer, P.H.E.M. and Meinders, A.E. (1995) Osmol and anion gaps in the diagnosis of poisoning. *Netherlands Journal of Medicine*, **46**, 57–61.

49 Eder, A.F., McGrath, C.M., Dowdy, Y.G., Tomaszewski, J.E., Rosenberg, F.M., Wilson, R.B., Wolf, B.A. and Shaw, L.M. (1998) Ethylene glycol poisoning: toxicokinetic and analytical factors affecting laboratory diagnosis. *Clinical Chemistry*, **44**, 168–177.

50 Moffat, A.C., Osselton, M.D. and Widdop, B. (2004) *Clarke's Analysis of Drugs and Poisons*, 3rd edn, Pharmaceutical Press, London.

51 Deutsche Forschungsgemeinschaft (DFG) (2007) *List of MAK and BAT Values 2007*. Commission for the Investigation of Health Hazards of Chemical Compounds. Report 43. Wiley-VCH Verlag GmbH, Weinheim.

References for Section 27.4

52 Fujiwara, K. (1914) Über eine neue sehr empfindliche Reaktion zum Chloroformnachweis. *Sitzungberichte und Abhandlungen der Naturforschenden Gesellschaft zu Rostock*, **6**, 33–43.

53 Truhaut, R. (1949) La réaction colorée alcalino-pyridinique de Fujiwara et les réactions du même type. Leurs applications à la caractérisation et au dosage des substances toxiques et médicamenteuses. *Bulletin de la Fédération Internationale Pharmaceutique*, **23**, 432–451.

54 Webb, F.J., Kay, K.K. and Nichol, W.E. (1945) Observations on the Fujiwara reaction as a test for chlorinated hydrocarbons. *Journal of Industrial Hygiene & Toxicology*, **27**, 249–255.

55 Bonnichsen, R. and Maehly, A.C. (1966) Poisoning by volatile compounds. II. Chlorinated aliphatic hydrocarbons. *Journal of Forensic Sciences*, **11**, 414–427.

56 Lugg, G.A. (1966) Fujiwara reaction and determination of carbon tetrachloride, chloroform, tetrachloroethane, and trichloroethylene in air. *Analytical Chemistry*, **38**, 1532–1536.

57 Deutsche Forschungsgemeinschaft (DFG) (2007) *List of MAK and BAT Values 2007*. Commission for the Investigation of Health Hazards of Chemical Compounds. Report No. 43. Wiley-VCH Verlag GmbH, Weinheim.

58 Wang, R.Y. (2004) Hydrocarbon products, in *Medical Toxicology*, 3rd edn (ed. R.C. Dart), Lippincott Williams and Wilkins, Philadelphia, PA.

59 Ross, J.H. (1923) A color test for chloroform and chloral hydrate. *The Journal of Biological Chemistry*, **58**, 641–642.

60 World Health Organization (WHO) . (2000) Dichloromethane, in *Air Quality Guidelines for Europe*, 2nd edn, *European Series No. 91*, WHO Regional Publications, pp. 83–86.

61 Schaller, K.H. and Triebig, G. (1996) Biologische Arbeitsplatz-Toleranzwerte. Teil IX: Arbeitsmedizinische Bewertung von organischen Lösungsmittelgemischen. *Arbeitsmedizin Sozialmedizin Umweltmedizin*, **31**, 504–512.

References for Section 27.5

62 Flanagan, R.J. (2004) Volatile substances, in *Clarke's Analysis of Drugs and Poisons*, 3rd edn (eds A.C. Moffat, M.D. Osselton and B. Widdop), Pharmaceutical Press, London.

28
Pesticides

28.1
Introduction

M. Geldmacher-von Mallinckrodt

28.1.1
Definition

Pesticides in the broadest sense are substances that are applied against organisms that are harmful to men or in conflict with their interests. Insecticides, which represent a class of pesticides, are used against insects. These organisms constitute 75% of the 800 000 species living on earth, which gives an idea of the importance and the magnitude of the use of insecticides and henceforth pesticides [1].

28.1.2
Classification

There are different systems for classification of insecticides [2].

Classification according to main use

In most cases, only a single use is given. This is only for identification purposes and does not exclude other possible uses.

Examples:

Acaricides	Ixodicides
Aphicides	Larvicides
Bacteriostats (soil)	Molluscicides
Fumigants	Miticides
Fungicides (treatment)	Nematocides
Fungicides (for seed)	Plant growth regulators
Herbicides	Repellants
Insecticides	Rodenticides
Insect growth regulators	

Clinical Toxicological Analysis: Procedures, Results, Interpretation. Edited by Wolf-Rüdiger Külpmann

ISBN: 978-3-527-31890-2

Classification according to chemical structure

This classification is preferred for grouping substances according to their mechanism of action and metabolism in men or the respective organisms. It should be mentioned that some pesticides may belong to more than one class.

Examples [2]:

Arsenic compounds	Organotin compounds
Carbamates	Phenoxyacetic acid derivatives
Coumarin derivatives	Pyrethroids
Chloronitrophenol derivatives	Pyridyl derivatives
Organic mercury derivatives	Thiocarbamates
Organophosphorus compounds (OPCs)	Triazine derivatives

WHO recommended classification of pesticides by hazard

This classification was approved in 1975 and has since gained wide acceptance. The hazard referred to in this recommendation is the acute risk to health (i.e., the risk of single or multiple exposures over a relatively short period of time) that might be encountered accidentally by any person handling the product in accordance with the directions by the manufacturer or in accordance with the rules laid down for storage and transportation by competent international bodies. The classification distinguishes between the more and the less hazardous forms of each pesticide as it is based on the toxicity of the technical compound and its formulations. The classification is based primarily on the acute oral and dermal toxicity to rat, measured as the LD_{50}, since these determinations are standard procedures in toxicology [2].

(The LD_{50} value is defined as a statistical estimate of the number of milligrams of toxicant per kilogram of body weight required to kill 50% of a large population of test animals.)

The WHO recommended classification of pesticides by hazard with regard to toxicity [2] is provided in Table 28.1.

Table 28.1 WHO recommended classification of pesticides by hazard [2].

	LD_{50} for the rat (mg/kg body)			
	Oral		Dermal	
Class	Solids[a]	Liquids[a]	Solids[a]	Liquids[a]
Ia. Extremely hazardous	5 or less	20 or less	10 or less	40 or less
Ib. Highly hazardous	5–50	20–200	10–100	40–400
II. Moderately hazardous	50–500	200–2000	100–1000	400–4000
III. Slightly hazardous	>500	>2000	>1000	>4000

[a] The terms "solids" and "liquids" refer to the physical state of the active ingredient at room temperature.

It is assumed that substances that may be toxic only at doses distinctly exceeding LD_{50} in class III will not be acutely hazardous in normal use, that is, an oral dose of a solid of more than 2 g/kg body weight and 3 g/kg body weight of a liquid. Substances that are volatile or gaseous at room temperature are outside the scope of this classification. For these substances, WHO recommends to estimate the hazard from data of industrial medicine (e.g., MAC value, see Section 28.1.4). WHO points out explicitly that the solvents or vehicles of products may be more hazardous than the pesticides they contain. Therefore, some products have to be classified as more dangerous than the pesticide itself.

28.1.3 Toxicity in Man

It is difficult to provide very reliable data on the acute toxic dose of a pesticide for men. The dose depends on many influences, for example, route of application (oral, intravenous, inhalation, percutaneous), composition of industrial product (e.g., solvents, emulsifiers, adsorption to vehicles), health status and age of patient, and genetic factors. It is still more difficult to contribute reliable data on doses that lead to chronic toxicity. However, several publications, for example, on exposure limits will guide how to avoid hazards (see Section 28.1.4).

28.1.4 Guidelines and Exposure Limits

All over the world, many investigations were performed, data were collected, and then safety regulations were issued to protect men and environment from the toxic effects of pesticides.

Drinking water quality

Since 1983, WHO has been publishing "Guidelines for Drinking Water Quality," in which pesticides are also discussed. The latest guidelines, published in 2006 [3, 4], supersede those published in previous editions and addenda, as well as previous international standards. Information on many chemicals, including pesticides, has been revised. This includes information on chemicals that were not considered previously, revisions to take into account new scientific information, and, in some cases, lesser coverage where new information suggests a lesser priority.

Acceptable daily intake

The acceptable daily intake (ADI) for humans is an estimate of JECFA (Joint FAO/WHO Expert Committee on Food Additives) of the amount of a food additive and contaminants, expressed on a body weight basis, that can be ingested daily over a lifetime without health risk (standard man = 60 kg) [5–6].

The ADI value is derived from measurements or estimations of the highest dose in daily food that does not cause a significant alteration to the investigated measurands.

For OPCs or insecticide carbamates, the activity of the erythrocyte cholinesterase (ACHE; EC 3.1.1.7) or the serum cholinesterase (CHE; EC 3.1.1.8) can be used for this purpose.

Threshold limit values

The threshold limit values (TLVs) are provided by the American Conference of Governmental Industrial Hygienists (ACGIH) of the United States [7]. These values refer to airborne concentrations of substances and represent conditions under which it is believed that nearly all workers may be repeatedly exposed day after day without adverse effects. Three categories of TLVs are specified (for details see Ref. [8]):

1. threshold limit value – time-weighted average (TLV-TWA) is the time-weighted average concentration for a normal 8 h working day and a 40 h working week, to which nearly all workers may be repeatedly exposed, day after day, without any adverse effect.
2. Threshold limit value – short-term exposure limit (TLV-STEL) is the concentration to which the worker can be exposed continuously for a short period of time without suffering from (i) irritation; (ii) chronic or irreversible tissue damage; or (iii) narcosis of sufficient degree to increase the likelihood of accidental injury, impair self-rescue, or materially reduce work efficiency, and provided that the daily TLV-TWA does not exceed.
3. Threshold limit value – ceiling (TLV-C) is the concentration that should not exceed during any part of the working exposure.

Maximum allowable concentration

The term MAC (maximum allowable concentration) is widely used, for example, in the Netherlands and Germany as well as in the former Soviet Union and Central and East European countries [9, 10]. In Germany, the MAK (Maximale Arbeitsplatz-Konzentration) is defined as the maximum concentration of a substance (in the form of gas, vapor, or particulate matter) in the workplace air that generally neither does have adverse effect on the health of an employee nor does cause unreasonable annoyance (e.g., by nauseous odor), even when the person is repeatedly exposed for long periods, usually for 8 h daily, assuming on average a 40 h working week. The justifications are available in English in Ref. [11].

Another term for occupational exposure is MCP (maximum permissible concentration) used in Argentina, Finland, and Poland, among others. This term is also used in relation to a chemical's concentration in drinking water in several countries, including Japan, Germany, and the United States. The term PEL (permissible exposure limit) is used in the United Kingdom and by the US Occupational Safety and Health Administration.

Biological monitoring and biological limits

Biological monitoring of chemical exposure aims to measure either the amount of a chemical a worker has absorbed or the effect the absorbed chemical has on the worker. Biological monitoring involves taking a sample of body fluid (usually blood or urine) and measuring the level of the chemical or its metabolite. Alternatively, an effect of that chemical on the body may be determined by measuring the level of an enzyme or other chemical in the blood or urine. For parathion, this could be the concentration of *p*-nitrophenol in urine and the reduction of the activity of acetylcholine esterase in the erythrocytes to 70% of the reference value. Scientifically justified threshold limit values in biological material are being compiled and published at present by two institutions [12].

Biological tolerance values

One of the institutions to define biological tolerance is the German Senate Commission for the Investigation of Health Hazards of Chemical Compounds in the Work Area of the German Research Foundation. The BAT values are especially drawn up by the working group on "setting of threshold limit values in biological material." The justifications for the BAT values are available in English in Ref. [13].

Biological exposure indices

The biological exposure indices (BEI values) are developed by the BEI Committee working in parallel with the Threshold Limit Values Committee of the ACGIH. Descriptions of the procedures used by the ACGIH in the evaluation of the exposure limits can be found in the appropriate sections of the "Documentation of the TLV and BEI" (see Ref. [8]).

28.1.5
Frequency of Poisoning from Pesticides

In 2002, the Swiss Toxicological Information Centre received 24 772 calls concerning humans [14], of which acute accidental exposure represented the largest group. Of the total complaints, 765 persons (3.1%) suffered from intoxication by "agricultural and horticultural products" (including pesticides), 348 were adults and 411 children. The toxicological potential of agricultural products varies greatly from only slightly toxic (e.g., liquid fertilizers) to some highly toxic pesticides. Although 54% of cases reported involved children, none of them led to severe poisonings. In adults, four cases of severe poisonings were seen; among these, two cases were of self-poisoning with the insecticide carbofuran and one case was a suicide attempt with rat poison.

In the 2001 Annual Report of the American Association of Poison Control Centers (AAPCC), the Toxic Exposure Surveillance System (TESS) data were compiled by the AAPCC in cooperation with the majority of US poison centers [15]. Among the substances most frequently involved in human exposures, pesticides come to 4.0% (90 010 cases). Pesticides belong to the category with largest number of deaths (17 cases or 0.019%). To these fatal intoxications belonged one caused by a fumigant, one by chlorothalonil (fungicide), four by paraquat, three by insecticides (acephate, carbaryl, and endosulfan), and two by rodenticides (brodifacoum and zinc phosphate).

28.2
Carbamates

M. Geldmacher-von Mallinckrodt and F. Degel

Introduction

Usually, insecticide carbamates are esters of monomethylcarbamic acid:

$$CH_3 - NH - COO - R,$$

where R symbolizes most often an aryl derivative (Figure 28.1). Like organophosphates, the carbamates are toxic for humans by inhibiting cholinesterases. Symptoms of intoxication by carbamates are similar to those of organophosphorus poisoning (Section 28.5), but the inhibition of cholinesterases and clinical symptoms are observed for a shorter period of time. Therefore, measurement of cholinesterase activity in serum and erythrocytes to detect carbamate poisoning is only indicated shortly after intake (Section 28.5.2). Identification of the pesticide

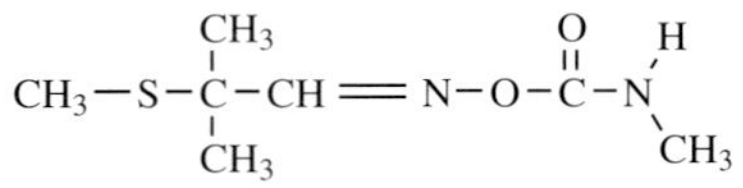

Aldicarb

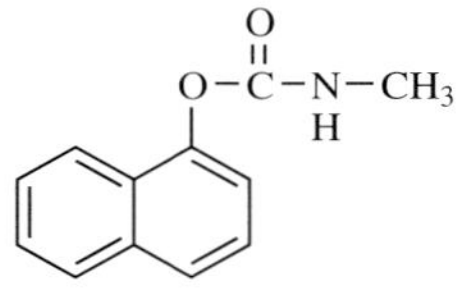

Carbaryl

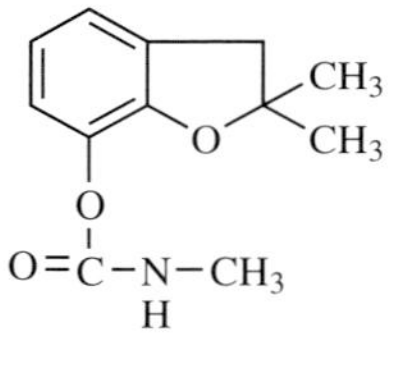

Carbofuran

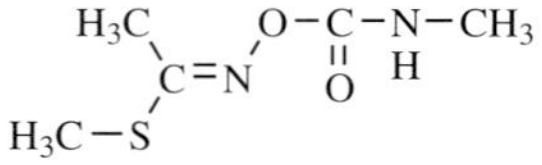

Methomyl

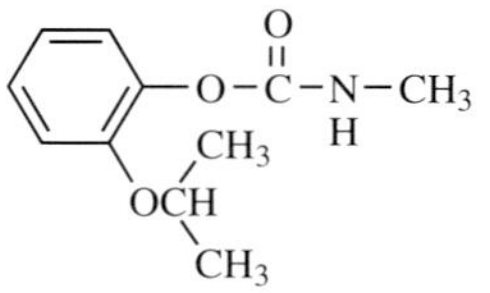

Propoxur

Figure 28.1 Insecticide carbamates.

involved in poisoning is important for treatment. The application of oximes (e. g., obidoxime) is not recommended in addition to atropine in carbamate poisoning [16].

28.2.1 Screening Procedure (General Unknown)

Many insecticide carbamates can be detected by HPLC and GC–MS general screening procedures as described in Chapter 12. It should be noticed that carbamates can be degraded to artifacts during GC–MS analysis.

28.2.2 Thin-Layer Chromatography (Procedure 1)

Outline

After liquid–liquid extraction, the sample is analyzed by thin-layer chromatography (TLC) with two different standardized mobile phases. Cholinesterase inhibiting carbamates are enzymatically detected very sensitive insecticide organophosphorus compounds (Section 28.5) [17, 18]. After the TLC development, the plate is sprayed with bovine liver preparation, containing liver esterase, incubated at 37 °C, and sprayed again with the appropriate substrate (2-naphthylacetate).

The following reaction is used: By the catalytic activity of cholinesterase, 2-naphthylacetate is degraded to 2-naphthol, which forms a pink-violet complex with Fast Blue Salt B. In the presence of insecticide carbamates, cholinesterase activity is inhibited and it cannot hydrolyze 2-naphthylacetate and white spots appear on the pink-violet TLC plate.

Simultaneously, two additionally developed TLC plates are screened by exposing them to UV light. They are sprayed with palladium chloride solution. Some insecticide carbamates are detectable by yellow spots further enhancing the specificity of the whole procedure.

Specimens

Stomach contents including material from gastric lavage
Vomitus
Suspect materials, for example, residues in drinking glass, vials.

Storage: 4 °C protected from light; long-term storage at −20 °C.
Transport: separated from blood and urine (to avoid contamination).

Equipment

Homogenizer
Drying oven (37, 110 °C)
Rotary evaporator or other device for evaporation with nitrogen flow
Device for spraying
UV lamp (254 nm).

Chemicals (p.a. grade)

Dichloromethane
Ethanol
Fast Blue Salt B (Sigma–Aldrich)
n-Hexane
Hydrochloric acid 32%
2-Naphthylacetate (Sigma–Aldrich)
Palladium chloride (Sigma–Aldrich)
Pentane
Phosphate buffer 0.02 mol/l, pH 7.0
Sodium sulfate, anhydrous
Toluene.

Bovine liver, approximately 20 g (no loss of cholinesterase activity for 1 year if stored at −20 °C)

Reference substances (Riedel de Haën, Promochem):

Triazophos	Carbofuran
Parathion	Azinphos-methyl
Pirimiphos-methyl	Methidathion
Quintozene	Parathion-ethyl

(Some organophosphorus compounds are added as reference substances because they can be investigated simultaneously by the procedure (Section 28.5).)

Reagents

Solvent mixtures used for TLC development:

S1: *n*-hexane/acetone 80 + 20 (v/v)
S2: toluene/acetone 95 + 5 (v/v).

Substrate solution (for enzymatic reaction):

(a) 2-Naphthylacetate (20 mg) is dissolved in 8 ml ethanol
(b) Fast Blue Salt B (50 mg) is dissolved in 32 ml aqua bidest.

The two solutions are mixed just before use. The mixture should be transparent, not cloudy.

Enzyme stock solution (E): Bovine liver (10 mg) is suspended in 90 ml phosphate buffer (0.02 mol/l, pH 7.0) and homogenized. After centrifugation (5 min, 3000 × g), portions of the supernatant are stored at −20 °C. Cholinesterase activity will be preserved for 1 year.

Enzyme working solution: Shortly before spraying, the enzyme stock solution is diluted with phosphate buffer (P) (E + P, 1 + 9 (v/v)).

Stock solution I of reference substances (1 mg/ml): Parathion-methyl, pirimiphos-methyl, quintozene, and triazophos. Each substance (25 mg) is dissolved in some milliliters of dichloromethane and made up to 25 ml with dichloromethane in a volumetric flask (25 ml).

Stock solution II of reference substances (1 mg/ml): Azinphos-methyl, carbofuran, methidathion, and parathion-ethyl. Each substance (25 mg) is dissolved in some milliliters of dichloromethane and made up to 25 ml with dichloromethane in a volumetric flask (25 ml).

Working solution I of reference substances (10 μg/ml): 100 μl of each stock solution I of a reference substance is transferred to a volumetric flask (10 ml) and made up to 10 ml with dichloromethane.

Working solution II of reference substances (10 μg/ml): proceed with stock solution II as described for working solution I.

Negative control sample: drug-free stomach content.

Positive control sample: drug-free stomach content spiked with reference substances. Final concentration: 10 times the detection limit (Table 28.4).

Sample preparation

1. About 0.5 g of the specimen is mixed with anhydrous sodium sulfate.
2. A 30 ml of dichloromethane is added and the mixture is shaken mechanically for 20 min for extraction.
3. The organic phase is separated and evaporated cautiously. Final volume: 0.1 ml.

TLC analysis

1. On each of the four TLC plates, 25 μl of the extract is applied as a spot.
2. Working solution I (10 μl) is applied as a spot on TLC plates 1 and 2.
3. Working solution II (10 μl) is applied as a spot on TLC plates 3 and 4.
4. The TLC plates are developed in a glass tank (saturated):

   Plates 1 and 2: solvent mixture S1
   Plates 3 and 4: solvent mixture S2

   until distance of solvent front from origin is 15 cm.
   The front is marked with a pencil and the plate is air dried.
   Detection:

(a) *UV absorption:* All plates are examined at UV light (254 nm) and absorptions are marked with a pencil.

(b) *Palladium chloride:* Plates 1 and 3 are sprayed with palladium chloride solution and air dried for 3 min. They are stored in a drying oven at 110 °C for 20 min. Some insecticide carbamates become visible as yellow spots.

(c) *Enzymatic reaction:* Plates 2 and 4 are sprayed evenly with enzyme working solution until they are slightly moist. They are incubated for 30–60 min in a water-saturated dry oven at 37 °C. After incubation, the two plates are sprayed with the freshly prepared substrate solution.

Table 28.2 Carbamates and organophosphorus compounds: hR_f^c values [18].

Mobile phase	Reference substance	$hR_f^{c\,a}$
S1	Triazophos	20
	Parathion-methyl	30
	Pirimiphos-methyl	49
	Quintozene	84
S2	Carbofuran	20
	Azinphos-methyl	46
	Methidathion	60
	Parathion-ethyl	85

S1: mobile phase, *n*-hexane/acetone 80 + 20 (v/v); S2: mobile phase, toluene/acetone 95 + 5 (v/v).
[a] hR_f values, corrected (see text).

After some minutes, insecticide carbamates become visible as persistent white spots on a pink-violet background.

The R_f values of the spots pertaining to the chromatograms of the samples are calculated in relation to the reference substances (Table 28.2) and compared with the R_f values of Table 28.3.

By simultaneously analyzing four reference substances on the same plate, the corrected R_f value (R_f^c value) (Section 4.2) can be calculated to improve the reliability of the identification [18].

Negative control sample: No characteristic spots are visible.

Positive control sample: Spots are visible at respective R_f values (Table 28.2).

Table 28.3 Insecticide carbamates: hR_f values [18].

Substance	hR_f		UV	$PdCl_2$
	S1	S2		
Butocarboxim sulfoxide	0	13	–	Yellow
Formetanate	1	1	+	–
Butoxycarboxim	3	2	–	–
Methomyl	6	6	+	–
Phenmedipham	11	17	+	–
Butocarboxim	15	11	+	Yellow
Carbofuran	17	20	+	–
Carbaryl	18	25	+	–
Propoxur	20	21	+	–
Thiofanox	20	17	–	–
Ethiofencarb	21	25	+	Yellow
Pirimicarb	26	17	+	–
Propham	39	57	+	–
Carbosulfan	49	66	+	–

S1: mobile phase, *n*-hexane/acetone 80 + 20 (v/v); S2: mobile phase, toluene/acetone 95 + 5 (v/v); UV, ultraviolet.

Analytical assessment
Sensitivity:

(a) *UV absorption:* UV absorption is only observed, if carbamates with an aryl group are present, for example, carbaryl (Table 28.3).
(b) *Palladium chloride:* Some carbamates become visible as yellow spots (Table 28.3)
(c) *Enzymatic reaction:* The detection limits of some carbamates are provided in Table 28.4.

Specificity: The procedure detects cholinesterase inhibiting compounds such as insecticide carbamates and organophosphorus compounds. The specificity is rooted in separation by two different, poorly related solvent mixture systems and comparison with (at best) four reference substances, which are analyzed simultaneously in the procedure on the same plates and are used for calculation of the corrected R_f value [18].

Practicability
The procedure does not need special, expensive equipment. Technician time is about 35 min if the enzyme stock solution is already prepared. Time of analysis is about 3 h.

Table 28.4 Insecticide carbamates: thin-layer chromatography and enzymatic detection.

Substance	Detection limit (μg/spot)
Carbaryl	3
Dimetan	80
Dimetilan	60
Isolan	10
Methiocarb	400
Minacide	6
Pyramat	200
Zectram	80

28.2.3 Thin-Layer Chromatography (Procedure 2)

Outline
The sample is extracted at low pH and analyzed by TLC with two different mobile phases. The hR_f values and several color reactions applied in sequence are used for a reliable identification of 16 insecticide carbamates.

Specimens

Urine 10 ml
Stomach content, gastric lavage, vomitus
Suspect materials, for example, residues in drinking glass, vials.

Storage: 4 °C protected from light; long long-term storage at −20 °C.
Transport: Separated from blood and urine (to avoid contamination).

Equipment

TOXI-LAB A basic equipment (Ansys Diagnostics, ToxiLab Division, Lake Forrest, CA)
TOXI-GRAMS Blank A (TLC plates with silica gel layer impregnated with ammonium vanadate (NH_4VO_3): Mandelin's reagent)
TOXI-DISKS Blank A
TOXI-TUBES B.

Extraction tubes with screw caps and PTFE-lined seal (e.g., TOXI-LAB)
Glass tank for TLC
Heating block (37 °C) for the evaporation of solvents with nitrogen flow
UV lamp (366 nm).

Chemicals (p.a. grade)

Acetone
Ammonium sulfate
Bismuth(III) nitrate pentahydrate ($Bi(NO_3)_3 \cdot 5H_2O$)
Chloroform
Cyclohexane
Dichloromethane
Diethyl ether
Glacial acetic acid
Formaldehyde
Furfural
n-Hexane
n-Heptane
Hydrochloric acid, fuming (37%)
Iodine
Methanol
Nitrogen, purified
Potassium iodide
Sodium chloride
Sulfuric acid (95–97%)
Toluene.

Reference substances (e.g., Pestanal, Riedel de Haën):

Dimetilan
Furathiocarb
Propoxur.

Reagents
Solvent for sample extracts: toluene/methanol 90 + 10 (v/v).
Mobile phase:

L1: chloroform
L2: *n*-hexane/diethyl ether/acetone 70 + 20 + 10 (v/v/v).

Stock solution of reference substances (250 mg/l): Dimetilan, furathiocarb, and propoxur. 25 mg of each reference substance is dissolved in 100 ml methanol. Stable 12 weeks if stored at 5–8 °C.

Reagents for detection: can be obtained ready-for-use from manufacturer (TOXI-DIP A reagents).

Reagent A1: formaldehyde 37% (20 ml) in a special ToxiLab incubation tank to expose developed chromatograms to formaldehyde vapor. The tank is equipped with a perforated tray at the bottom to prevent the plates from dipping into the formaldehyde solution.

Reagent A2: sulfuric acid concentrated (150 ml).

Dipping solution: aqua bidest (to be replaced daily).

Reagent A3 (modified Dragendorff's reagent): iodine 2 g, potassium iodide 5 g, $Bi(NO_3)_3{\cdot}5H_2O$, glacial acetic acid 10 ml, aqua bidest 150 ml.

Negative control sample: drug-free urine.

Positive control sample (5 µg/ml): 200 µl of each stock solution is transferred to a volumetric flask and made up with drug-free urine to 10 ml. Stable for 1 week if kept at 5–8 °C.

Sample preparation
Samples are extracted as described by the manufacturer using TOXI-TUBES B extraction tubes (see TOXI-LAB Instruction Manual).

In essence, a liquid/liquid extraction at pH 2.2 is performed using extraction tubes that are prepared ready for use with buffers and an organic solvent mixture.

TLC analysis
After application of the extracts, the TL plates are developed with the mobile phases L1 and L2.

Further details of the procedure are described in the TOXI-LAB A manual. Shortly before the expected end of development, the tanks are lighted up indirectly to make localization of the front easier. It is marked with a pencil.

Detection: The TL plates are exposed for 2 min to formaldehyde vapor (reagent A1). They are dipped into Mandelin's reagent (reagent A2, detection state I) and then into aqua bidest (detection state II). They are examined with UV lamp (366 nm) (detection state III) and dipped into modified Dragendorff's reagent (reagent A3, detection state IV). The spots visible in the different detection states are documented with regard to R_f value and color on special data sheets provided by the manufacturer or on self-made forms.

Evaluation of chromatograms: The hR_f values of the spots for both development systems related to the reference substances are calculated and used for their

Table 28.5 Insecticide carbamates: hR_f values and color reactions.

Substance	State I	State II	State III	State IV	hR_f L 1	hR_f L 2
Dimetilan	—	—	—	Brown	42	24
Ethiofencarb	—	—	Pale blue	Brown	68	51
Pirimicarb	—	—	—	Gray	45	48
Carbofuran	Red violet	Gray violet	Absorption	Brown	62	45
Formetanate	Violet	Orange	Green	Brown	7	10
Carbaryl	Gray green	Gray brown	Absorption with orange halo	Brown	63	44
Aminocarb	—	—	—	Brown	52	46
Propoxur	Pink	Gray pink	Absorption	Brown	61	49
Dioxacarb	—	—	Pale green	Pale brown	41	—
Promecarb	Gray violet	Gray	Absorption/pale green	Brown	77	60
Furathiocarb	Red	Red	Absorption	Brown	82	72
Desmedipham	Gray violet	Gray violet	Absorption/pale blue	Brown	64	33
Thiobencarb	Pale blue	—	—	Brown	86	81
Phenmedipham	Violet	Violet	Absorption/pale blue	Brown	61	33
Mercaptodimethur	Pale blue	—	—	Brown	77	60
Propham	Yellow orange	Yellow	Pale yellow	Pale brown	85	83

L1: mobile phase, chloroform; L2: mobile phase, *n*-hexane/diethyl ether/acetone 70 + 20 + 10 (v/v/v). Detection state: I, Mandelin's reagent; II, aqua bidest; III, UV; IV, Dragendorff's reagent, modified.

identification, as well as their color in the four detection states. Relevant information for identification is presented in Table 28.5.

Negative control sample: It should not show any of the absorptions and colors given in Table 28.5.

Positive control sample: It should show the absorptions and colors at the typical hR_f values of the chosen reference substances as documented in Table 28.5.

Analytical assessment

Sensitivity: The detection limit of the procedure is in the range of 0.5–1.0 mg/l according to experiments with spiked urine samples. Acute intoxications can be easily detected. During the early stages of acute poisoning with propoxur, only the parent compound could be detected but not the two main metabolites (2-hydroxypropoxur and 2-isopropoxyphenol). Their detection limit is higher because they are more polar and therefore less extractable and produce less distinctly colored spots than propoxur.

Recovery: Recovery of insecticide carbamates ranged from 60 to 100%.

Specificity: The specificity of this TLC procedure including a dedicated detection sequence is by no doubt superior to procedures using universal means of detection.

In addition, the extraction at acidic pH removes many possibly interfering compounds. When urine samples spiked with frequently used drugs such as analgesics, anticonvulsants, neuroleptics, or illicit drugs were examined by the procedure, no interfering spots were observed. Nevertheless, it is recommended to confirm positive findings with a chromatographic procedure of superior selectivity and specificity, for example, GC–MS. The procedure is suitable to dispel or fortify a suspicion of carbamate poisoning. It is superior to procedure 1 (Section 28.2.1) with regard to specificity for its special detection sequence, whereas in procedure 1, samples are examined only for substances exhibiting cholinesterase-inhibiting properties.

Practicability
Technician time is about 20 min, and time needed for total analysis is about 45 min for one sample, a positive and a negative control sample, and the standard mixture of reference substances.

28.2.4
Medical Assessment and Clinical Interpretation

Medical assessment
Toxic and lethal doses depend on

Individual insecticide carbamate involved
Special industrial product used (e.g., solid, liquid, kind of solvent)
Health status and constitution of the patient
Route of application and so on.

A rough estimate of the doses can be made by using LD_{50} values for rats (Table 28.6, the ADI values [20], and MAC values [21] (Table 28.7). The difference between toxic and lethal doses is bigger compared to organophosphorus compounds [22]. The small number of published single observations does not allow them to be used as reliable, generally applicable data but can be used for orientation.

Toxicokinetics: Insecticide carbamates are absorbed quickly by the lung (inhalation of spray or dust), by the skin, or by the gastrointestinal tract and are usually

Table 28.6 Insecticide carbamates: LD_{50} values [19].

Substance	Classification by hazard[a]	Physical state	LD_{50} (mg/kg)
Aldicarb	Ia	S	0.93
Bendiocarb	II	S	55
Carbaryl	II	S	300
Carbofuran	Ib	S	8
Methomyl	Ib	S	17
Oxamyl	Ib	S	8
Propoxur	II	S	95

S, solid; LD_{50}, mg/kg body mass; rat, oral application.
[a]See Section 28.1.

Table 28.7 Insecticide carbamates: ADI [20] and MAK values [21].

Substance	ADI	MAK
Aldicarb	0–0.003	—
Bendiocarb	0–0.004	—
Benomyl	0–0.1	—
Carbaryl	0–0.008	5
Carbendazim	0–0.03	—
Carbofuran	0–0.002	—
Ethiofencarb	0–0.1	—
Methiocarb	0–0.02	—
Methomyl	0–0.02	—
Oxamyl	0–0.009	—
Pirimicarb	0–0.02	—
Propoxur	0–0.02	2

ADI, acceptable daily intake (mg/kg body mass); MAK, maximum workplace concentration in air (mg/m^3).

metabolized rather rapidly by hydrolysis. They are excreted mainly through the kidney. In propoxur poisoning, 2-isopropoxyphenol and 2-hydroxyphenylmethylcarbamate were identified in the urine.

Half-life was estimated to be 1.3 h for carbaryl during poisoning of a patient with 7.5 g with an elimination following first-order kinetics [22].

Clinical interpretation

Aldicarb [23]

In four patients, after ingestion of aldicarb, serum or blood concentrations of 0.1–3.2 mg/l were measured 3–13 h after admission. Two adults who died after ingesting large amounts of aldicarb had postmortem blood concentrations of 4.8 and 11 mg/l, respectively. The urine aldicarb concentration of the first patient was 9.7 mg/l.

Carbaryl [23]

Volunteers who ingested doses of carbaryl up to 0.13 mg/kg daily for 6 weeks were asymptomatic. Workers exposed to air concentrations of the chemicals up to 31 mg/m^3 were also asymptomatic but did exhibit occasional depression of blood cholinesterase activity. Ingestion of 250 mg of carbaryl has been found to cause severe poisoning in an adult who recovered after administration of atropine. A man who ingested 27 g of carbaryl was comatose for 24 h but survived with treatment. He experienced severe peripheral neuropathy that persisted for 9 months [23].

Three adults who died within hours after intentionally ingesting an unknown amount of carbaryl had an average carbaryl blood concentration of 16 mg/l (range 6–27 mg/l) and an average urine concentration of 31 mg/l [23]. According to Ref. [24], concentrations in blood/plasma/serum exceeding 5 mg/l are considered as toxic and concentrations more than 6 mg/l are life threatening.

Carbofuran

A 20-year-old woman (with 12 weeks of pregnancy) was classified as severely poisoned by carbofuran on admission to the hospital. The highest level of carbofuran in the blood of the expectant mother was 9.71 mg/kg. A spontaneous abortion was reported on the 27th day of poisoning [25].

More cases of intoxications have been reported by Baselt [23]. A 17-year-old pregnant woman who ingested carbofuran had a carbofuran blood concentration of 2.6 mg/l 9 h later. She survived but the 4.5-month-old fetus died. A woman who died after several hours had a postmortal carbofuran blood concentration of 1.9 mg/l. A 26-year-old man died after ingestion of 345 ml of a 45% carbofuran solution. The carbofuran blood concentration was 29 mg/l [23]. In four cases of suicidal ingestion, postmortem blood carbofuran concentrations of 0.3–12 mg/l were observed [23].

Methomyl

A 39-year-old woman patient applied about 4.5 mg/kg body mass. She showed typical clinical symptoms and recovered. Toxic symptoms were observed in 11 patients after intake of 2–16 g (30–200 mg/kg body mass) but all patients survived [22].

According to Baselt [23], an individual who survived the acute ingestion of 2.25 g of methomyl had a methomyl blood concentration of 1.6 mg/l 6 h after intake. Methomyl blood concentrations of 10 adults who died after the accidental or intentional ingestion averaged 26 mg/l, with a range of 8.0–57 mg/l. In one of these cases, a woman patient with a postmortem methomyl blood concentration of 28 mg/l had a urine concentration of 33 mg/l. A man who developed severe poisoning after accidentally ingesting several grams of methomyl was successfully treated with atropine and pralidoxime. Three men died after accidentally ingesting approximately 1 g of methomyl [23]. A lethal dose is about 12–15 mg/kg body mass [22].

Propoxur

A 36-year-old man had ingested 200 ml of Unden (equivalent to 40 g propoxur). He died 4 days later from pneumonia (by aspiration) and renal failure [26].

In cases of propoxur poisoning, concentrations in blood ranged from 0.31 to 41.1 mg/l [27].

Clinical symptoms of carbamate poisoning: Insecticide carbamates inhibit the acetylcholinesterase (EC 3.1.1.7) more quickly than organophosphorus compounds, but for a shorter time period. In contrast to organophosphates, the spontaneous reactivation of the enzyme is rapid and fully effective. Symptoms of an acute intoxication are similar to those of organophosphate poisoning, but develop and disappear more quickly. Atropine is administered as soon as possible to treat poisoning. The application of oximes as an antidote is not advisable [16].

Biochemical findings: The determination of cholinesterase in serum (EC 3.1.1.8) and acetylcholinesterase in erythrocytes (EC 3.1.1.7) is only meaningful in samples taken early after intoxication, because the enzymes are reactivated quickly (Section 33.2).

28.3
Chlorinated Hydrocarbons

M. Geldmacher-von Mallinckrodt

28.3.1
Introduction

Several groups of substances belong to the cyclic chlorinated hydrocarbons that exhibit insecticide effects against, for example, mosquitoes, ants, moths, mites, or lice [28, 29] (Figure 28.2).

Group	Typical examples
Dichlorodiphenylethanes	DT, methoxychlor
Cyclodienes	Aldrin, heptachlor, chlordane, dieldrin
Chlorinated benzenes and cyclohexanes	Hexachlorobenzene, γ-hexachlorocyclohexane (lindane)
Chlorinated norbornenes	Toxaphene

Lindane (γ-isomer of hexachlorocyclohexane) is contained in several pharmaceutical formulations used externally to treat head lice, crab lice and their nits, and itch mites and their eggs [29, 30].

The above-mentioned chlorinated hydrocarbons are well soluble in lipids but not in water. They are sprayed as dry powder or as aqueous solutions with added

DDT

Methoxychlor

Hexachlorcyclohexane

Figure 28.2 Insecticide cyclic chlorinated hydrocarbons.

emulsifier or dissolved in organic solvents. The chlorinated hydrocarbons have in common a high stability against physical or chemical influences. Technical products are contaminated usually with many substances that may be more toxic than the pure hydrocarbon itself. In addition, the vehicle may play an important role in toxicity.

28.3.2
International Agreements on the Use of Chlorinated Hydrocarbons

In the years 1950–1980, cyclic chlorinated hydrocarbons played an important role as pesticides. But their use has been cut down increasingly all over the world for their high stability in the environment and thus danger of global contamination.

According to Ref. [31], especially in developing countries, chlorinated hydrocarbons contaminated the natural environment and destroyed many wild animal species as well as humans. Therefore, governments started to address this problem in the 1980s by establishing a voluntary "prior informed consent" (PIC) procedure. PIC required importers trading in listed hazardous substances to obtain the prior informed consent of importers before proceeding with the trade. In 1998, many governments decided in Rotterdam to strengthen the procedure by adopting the "Rotterdam Convention," which makes PIC legally binding. The convention establishes a first line of defense by giving importing countries the tools and information they need to identify potential hazards and exclude chemicals they cannot manage safely. If a country agrees to import chemicals, the convention promotes their safe use through standards, technical assistance, and other forms of support. The convention also ensures that exporters comply with the requirements. The "Rotterdam Convention" came into force on February 24, 2004 [31]. The PIC procedure of "Rotterdam Convention" applies to (a) banned or severely restricted chemicals and (b) severely hazardous pesticide formulations.

"Banned chemical" means a chemical all uses of which within one or more categories have been prohibited by final regulatory action, to protect human health or environment. It includes a chemical that has been refused approval for first-time use or has been withdrawn by industry either from the domestic market or from further consideration in the domestic approval process, and where there is clear evidence that such action has been taken to protect human health or environment.

"Severely restricted chemical" means a chemical virtually all use of which within one or more categories has been prohibited by final regulatory action to protect human health or environment, but for which certain specific uses remain allowed. This category includes a chemical that has for virtually all use been refused for approval or been withdrawn by industry either from the domestic market or from further consideration in the domestic approval process, and where there is clear evidence that such action has been taken to protect human health or environment.

"Severely hazardous pesticide formulation" means a chemical formulated for pesticidal use that produces severe health or environmental effects observable within a short period of time after single or multiple exposures, under conditions of use.

There are about 40 respective chemical groups currently subject to the interim PIC procedure. Among these chemicals are the following chlorinated hydrocarbons mentioned in this chapter [32]:

Aldrin
Chlordane
DDT(dichlorodiphenyltrichloroethane)
Dieldrin γ-HCH (lindane)
HCH (hexachlorocyclohexane), mixed isomers
Heptachlor
Hexachlorobenzene
Toxaphene (camphechlor).

According to the PIC Convention, export of a chemical can only take place with the prior informed consent of the importing party. The PIC procedure is a means for formally obtaining and disseminating the decisions of importing countries as to whether they wish to receive future shipments of a certain chemical and for ensuring compliance to these decisions by exporting countries. The aim is to promote a shared responsibility between exporting and importing countries in protecting human health and environment from harmful effects of such chemicals.

Some of the above-mentioned pesticides also belong to the "List of Persistent Organic Pollutants (POPs)," banned or severely restricted by the "Stockholm Treaty" in 2001 (see Ref. [32]). Among these are

Aldrin
Chlordane
DDT
Dieldrin
Heptachlor
Hexachlorobenzene.

Finally, there is a third group of active ingredients believed to be *obsolete* or discontinued for use as pesticides mentioned by IPCS (see Ref. [32]). To this group belong

Aldrin
Dieldrin
Endrin.

Although it is difficult in some cases to be sure whether all commercial use of a pesticide of this group has ceased, some of the materials described as "obsolete" may still be in use for nonagricultural purposes.

28.3.3 Screening Procedure (General Unknown)

Most of the chlorinated hydrocarbons discussed in this chapter, and their metabolites, can be detected in case of clinical symptoms of poisoning by the GC–MS screening procedure, which is described in Chapter 12.

28.3.4 Thin-Layer Chromatography

Reagents
Reference substances (e.g., Riedel de Haën).

Spraying reagent: $AgNO_3$ (1 g) is dissolved in 5 ml aqua bidest and 2.5 ml ammonia solution (25%) in a volumetric flask. The solution is made up with acetone ad 100.0 ml.

Samples are extracted by, for example, *n*-hexane and analyzed by TLC. The plates are developed with the mobile phases S1 and S2 as used for insecticide carbamates (see Section 28.2.1). For detecting the chlorinated hydrocarbons, the chromatograms are exposed to UV light to document absorption and sprayed with silver nitrate reagent.

The sprayed plate is dried at 150 °C for 15 min. In the presence of chlorinated hydrocarbons, dark spots on light brown background become visible, which can be identified by their R_f values according to Table 28.8.

Annotation
A positive finding can be confirmed by GC–MS or GC equipped with electron capture detector [34].

Table 28.8 Thin-layer chromatography: hR_f values of chlorinated hydrocarbons.[a]

Substance	UV	hR_f S1	hR_f S2
Aldrin	±	89	98
p,p'-DDT	±	76	98
Dieldrin	–	65	87
Endrin	–	71	90
γ-HCH (lindane)	–	51	92
Methoxychlor	–	43	84

S1: mobile phase, *n*-hexane/acetone 80 + 20 (v/v); S2: mobile phase, toluene/acetone 95 + 5 (v/v); UV, ultraviolet; γ-HCH, γ-hexachlorocyclohexane; DDT, dichlorodiphenyltrichloroethane.
[a]According to Ref. [33].

Table 28.9 Chlorinated hydrocarbons: ADI[a] and MAK[b].

Substance	ADI (mg/kg)	MAK (mg/m^3)
Aldrin	0–0.0001	0.25 I
Chlordane	0.0005	0.5 I
DDT	0.01	1 I
Dieldrin	0.0001	0.25 I
Endrin	0.0002	0.1 I
HCH	0–0.005	0.5 I
γ-HCH (lindane)	0–0.005	0.1 I
Methoxychlor	0–0.1	15 I

I, measured as the inhalable fraction of the aerosol; ADI, acceptable daily intake (mg/kg body mass); MAK, maximum allowable concentration at the workplace in air (mg/m^3); HCH, hexachlorocyclohexane; DDT, dichlorodiphenyltrichloroethane.
[a]According to Ref. [35].
[b]According to Ref. [36].

28.3.5
Medical Assessment and Clinical Interpretation

Medical assessment

Toxic dose: The ADI values of the IPCS as well as the MAK values can be used for estimating toxic doses (Table 28.9). Published toxic and lethal doses are shown in Table 28.10.

Toxicokinetics: Hexachlorocyclohexane, DDT, and other chlorinated diphenylethanes such as methoxychlor are well absorbed from the gastrointestinal tract but poorly from the skin. On the contrary, chlorinated cyclodienes such as aldrin, heptachlor, chlordane, and dieldrin are well absorbed also from the skin. After resorption, chlorinated hydrocarbons are mainly stored in the body fat. DDT and other cyclodienes are eliminated very slowly and the elimination half-lives range from one to several years. However, lindane and toxaphene are metabolized and excreted rather quickly [29]. The chlorinated hydrocarbons, which were mentioned before, pass the placenta and come into mother's milk [30].

Metabolism: Main steps in metabolism are dehydrogenation, cleavage of HCl or Cl, and hydroxylation to produce phenols, which are excreted by the kidneys

Table 28.10 Chlorinated insecticide hydrocarbons: toxic and lethal doses (oral).

Substance	Toxic	Lethal	References
Aldrin	1–3 g	5 g	[37]
Chlordane	>1 g	>2 g	[37]
DDT	n.d.	>30 g	[37]
Dieldrin	>10 mg/kg	>5 g	[30, 37]
Endrin	n.d.	6–12 g	[30, 37]
γ-HCH (lindane)	n.d.	>200 mg/kg	[37]
		10–30 g	[30]

n.d., no data; γ-HCH, γ-hexachlorocyclohexane; DDT, dichlorodiphenyltrichloroethane.

Table 28.11 Chlorinated insecticide hydrocarbons: normal, toxic, and lethal concentrations in plasma.

	Blood/plasma/serum (mg/l)			
Substance	**Normal**	**Toxic from**	**Comatose/ fatal from**	**References**
Aldrin	<0.0015	0.0035	n.d.	[46, 47]
Chlordane	<0.001	0.0025	1–7	[46]
	0.001	0.0025	1.7–4.9	[47]
DDT	<0.005–0.038	0.2–0.5	n.d.	[38]
	0.013	n.d.	n.d.	[47]
	n.d.	0.84	1.3	[29]
Dieldrin	<0.0015	0.15–0.3	n.d.	[46]
	0.0015–0.02	0.15–0.303	0.5–1.16	[47]
Endrin	0.003	0.01–0.03	n.d.	[46]
	0.003	0.007–0.032	0.45	[47]
γ-HCH	0.001–0.031	0.5	1.3	[47]
Methoxychlor	n.d.	0.67	n.d.	[37]

n.d., no data; γ-HCH, γ-hexachlorocyclohexane (lindane); DDT, dichlorodiphenyltrichloroethane.

nonconjugated or conjugated (with glucuronic acid, sulfuric acid, or glutathione) often rather slowly. An alternative route is the elimination by the feces. Normal, toxic and fatal levels in blood, serum and plasma are shown in Table 28.11.

Toxicodynamics: Insecticide organochlorine derivatives are neurotoxins. The susceptibility of neurons to stimuli is increased and even low-level stimuli will cause a depolarization that finally becomes permanent. The effects are caused by a change of permeability of the sodium channel of the nerve cell and an inhibition of enzymes, which support the active transport of sodium, potassium, and calcium ions. Cyclodienes such as aldrin, dieldrin, chlordane, and heptachlor not only inhibit sodium–potassium ATPase but also act as GABA antagonists [30].

Clinical interpretation

Acute poisoning: Insecticide chlorinated hydrocarbons are neurotoxins. Their toxicity focuses on motor and sensory nerves leading to overstimulation and paralysis. At first, about 0.5–1 h after intake of a considerable amount, the tongue is affected. Later on, paresthesia of arms and legs, restlessness, dizziness, and irritability are usually observed, whereas nausea and vomitus seldom occur. Starting with trismus and tremor, tonic–clonic seizures develop, which may lead finally to total paralysis. Motor or sensory paralysis may remain permanent after acute poisoning [30].

Aldrin

Aldrin is rapidly metabolized in the body by epoxidation to dieldrin. Aldrin is slowly eliminated from the body primarily as hydrophilic metabolites in feces and to some extent in urine [37, 38].

Chlordane
Oxychlordane and nonachlor, ingredients of technical chlordane, have been found in the adipose tissue of a general population as well as in the breast milk of women [39, 40].

DDT
DDT is converted to some extent to the much less toxic DDE (dichlorodiphenyl dichloroethylene) by dehydrochlorination. DDE apparently does not undergo further biotransformation. The major detoxification pathway of DDT is dechlorination to DDD (dichlorodiphenyl dichloroethane), an active insecticide, followed by degradation to DDA (dichlorodiphenylacetic acid), which is water soluble and is rapidly excreted with urine [37].

In human adipose tissue of the Japanese, tris-(4-chlorophenyl) methane (TCPMe) and tris-(4-chlorophenyl) methanol (TCPMOH) were identified. Significant correlation between these two metabolites with concentrations of DDT in human adipose tissues suggests that exposure to DDT is the source [41].

Hexachlorobenzene
A set of 53 individuals highly exposed to airborne hexachlorobenzene (HCB) excreted HCB and two main metabolites of HCB, pentachlorophenol and pentachlorobenzenethiole, not only in urine but also in feces [42, 43].

Lindane (γ-hexachlorocyclohexane)
From the three major metabolites 2,3,5-, 2,4,5-, and 2,4,6-trichlorophenol, each accounted for about 18% of the total metabolites in the urine of exposed workers [44]. Only 2,3,4,6-tetrachlorophenol was identified in the urine of a child who died of lindane intoxication [28, 29].

Methoxychlor
Methoxychlor is less persistent than DDT. It is stored in the fat to a limited extent and is eliminated mainly in feces. Small amounts are excreted in urine [37].

Toxaphene
The major metabolic degradation mechanisms for toxaphene in all organisms from bacteria to primates are believed to be reductive dechlorination, reductive dehydrochlorination, and in some cases oxidative dechlorination to produce hydroxyl derivatives, acids, and ketones [45].

28.4 Paraquat

T. Daldrup and C. Köppel

Introduction
Usually, the contact herbicide paraquat (Figure 28.3) is provided commercially as dichloride. The pure substance has a white or light yellow color. It is odorless and

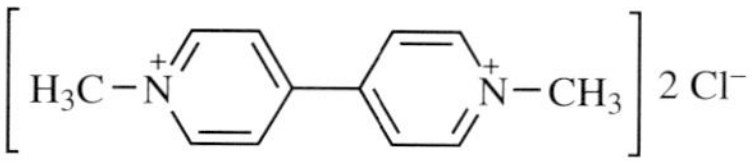

Paraquat

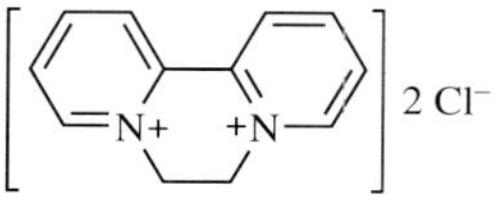

Diquat

Figure 28.3 Paraquat and diquat.

hygroscopic. Industrial products, however, are often colored dark brown, possibly by dyes, and exhibit a distinct smell. Paraquat is not soluble in nonpolar organic solvents, but is well soluble in water shifting pH to strongly alkaline.

Paraquat is one of the most toxic herbicides that is commercially available and causes multiorgan failure in case of poisoning. The prognosis of intoxication, which is not treated appropriately early after intake, is often very serious. A fatal outcome is preceded by the damage to the lungs during 1 or 2 weeks, which do not respond to therapy. Therefore, diagnosis should be available quickly to confirm or to rule out paraquat poisoning.

To a certain extent, there is an inverse correlation between the chance of survival (outcome) and the concentration of paraquat in serum (plasma) during the first 24 h after intake. The herbicide can be determined quantitatively by several methods such as immunochemistry [49–51], chromatography [52, 53], and spectrophotometry [48, 54–57]. More procedures have been published using capillary electrophoresis [58, 59], HPLC/UV spectrophotometry [60–63, 79], or HPLC/tandem mass spectrometry [64]. In the following sections, a simple assay for the detection of paraquat in urine and a photometric procedure for the determination of paraquat in serum are described.

28.4.1
Screening Procedure (General Unknown)

It should be emphasized that paraquat cannot be detected by the HPLC and GC–MS screening procedures described in Chapter 12.

28.4.2
Color Test

Outline
A reducing agent (sodium dithionite) is added to the alkaline sample. Paraquat forms a radical that has a characteristic dark blue color [48].

Specimen
Urine, 10 ml.

Chemicals (p.a. grade)

Paraquat dichloride (e.g., Sigma)
Sodium dithionite, anhydrous. Store in a refrigerator dry and well closed in a container (degradation during storage is possible in the presence of moisture and may cause spontaneous combustion)
Sodium hydroxide 0.1 mol/l.

Reagents
Dithionite solution: Sodium dithionite (0.1 g) is dissolved in 10 ml sodium hydroxide solution. The solution should be prepared fresh and is stable for several hours in ice water. Final concentration: 10 g/l.

Paraquat stock solution: Paraquat dichloride (0.1 g) is dissolved in aqua bidest and made up to 1000 ml with aqua bidest (to be prepared fresh). Final concentration: 100 mg/l.

Negative control sample: Urine of an individual not exposed to paraquat.

Positive control sample: Paraquat stock solution (10 µl) and negative control sample (950 µl) are mixed (to be prepared fresh). Final concentration: 5 mg paraquat dichloride/l equivalent to 3.68 mg paraquat/l.

Sample preparation
Not applicable.

Color test
Sample, negative control sample, and positive control sample (each 1.0 ml) are mixed with 1.5 ml of dithionite solution in a test tube. For comparison, a blank (1 ml sample and 1.5 ml sodium hydroxide solution) is investigated. All the tubes are shaken and the color of the four different solutions is evaluated with white paper as background. In the presence of paraquat, the patient's sample is blue colored.

Evaluation: Negative control sample: Its color should not change to bluish. Positive control sample: A distinct blue color should develop.

Analytical assessment
Sensitivity (Section 4.1.4):

Maximum sensitivity (E_{10}): 1.0 mg/l ($n = 30$)
Practical sensitivity (E_{90}): 2.0 mg/l ($n = 30$).

Specificity: The dark blue color is specific for paraquat. If diquat is involved, urine color changes to faintly yellow up to green.

False positive findings should not be obtained. Negative findings (even though paraquat was absorbed) may occur after intake of a small amount or when the sampling was performed rather a long time after administration of the herbicide. False negative findings can be ruled out by mixing paraquat stock solution (50 µl) and patient's sample (950 µl) and repeating the analytical procedure. The solution shall develop a blue color.

Practicability

Technician time is about 10 min. The total procedure will take 12–15 min.

Annotations

1. The procedure may be applied to all colorless aqueous solutions, such as gastric lavage, in the same way as described for urine. Industrial products should be diluted sufficiently with water to make a change of color visible.
2. A more rapid test can be performed in emergency situations: 5 ml urine is adjusted to pH 9–10 with a spatula point of sodium bicarbonate (or a small volume of diluted sodium hydroxide solution). A spatula point of sodium dithionite is added. A blue color develops in the presence of paraquat. There are no data available with regard to maximum and practical sensitivity.

28.4.3 Spectrophotometry

Outline

The concentration of the blue-colored paraquat radical is determined spectrophotometrically in the sample, which is deproteinized and at the same time extracted by a mixture of chloroform and ethanol containing sodium thiosulfate. The chemicals do interfere neither with the following reaction producing the radical nor with the photometric measurement. The solvent mixture chloroform/ethanol proved to be most effective for extraction, even for postmortem blood [79].

A reducing agent (sodium dithionite) is added to the extract at alkaline pH to produce the blue paraquat radical. The radical is protected against degradation by a covering hexane layer. For calculation of the paraquat concentration, the standard addition method is used [65].

Specimens

Serum, plasma
Urine, gastric lavage, and other aqueous solutions can be investigated alike.

Equipment

Recording spectrophotometer: range 370–430 nm

Centrifuge tubes (10 ml) with standard ground joint and glass stopper or with thread and screw cap (silicone rubber seals, PTFE coated).

Chemicals (p.a. grade)

Chloroform
Ethanol
n-Hexane
Paraquat dichloride (e.g., Sigma)

Sodium dithionite, anhydrous. Store in a refrigerator dry and well closed in a container (degradation during storage is possible in the presence of moisture and may cause spontaneous combustion)
Sodium hydroxide 0.1 mol/l
Sodium thiosulfate pentahydrate.

Reagents
Sodium dithionite solution: Sodium dithionite (0.1 g) is dissolved in 10 ml sodium hydroxide solution. The solution is prepared shortly before use and is stable for several hours if stored in ice water. Final concentration: 10 g/l.

Extraction solution: chloroform/ethanol 80 + 20 (v/v).

Paraquat stock solution: Paraquat dichloride (100 mg) is dissolved in aqua bidest and made up to 100 ml with aqua bidest. Final concentration: 1 g/l. The stock solution is stable for several weeks if stored light protected (e.g., flask wrapped with black paper) at 4 °C.

Paraquat dichloride is dried at 100 °C in a drying oven for 2 h to remove any moisture. The anhydrous paraquat dichloride is stored light protected in a desiccator with drying beads.

Paraquat working solution: Paraquat stock solution (5 ml) is made up with aqua bidest ad 100 ml. Final concentration: 50 mg/l. The solution is prepared fresh shortly before use.

Positive control sample: A serum sample of a nonexposed individual (4.9 ml) is spiked with 100 μl paraquat working solution. The control sample is to be prepared fresh just before running the procedure.

Sample preparation
Not applicable.

Spectrophotometric analysis
See Table 28.12.

Four hundred microliters of the transparent upper layer is transferred to a glass cuvette (semimicro) to which 600 μl sodium dithionite is added. The cuvette is tightly closed with a lid and the solution is mixed cautiously by waving. The content is covered with 50 μl *n*-hexane immediately afterward. Sample and controls (tubes 2–7) are measured by spectrophotometry at 396 nm with tube 1 as reagent blank. In addition, the spectrum between 370 and 430 nm can be recorded for qualitative evaluation.

Calculation of results: The quantitative determination is founded on the standard addition method using the absorption values of the extracts from nonspiked and spiked samples. The *x*-axis shows the mass of paraquat dichloride (0.00, 2.50, and 5.00 μg) used for spiking and the *y*-axis shows the respective absorbance values. A line well fitting to the data points is drawn. Its point of intersection with the *x*-axis gives the mass concentration of paraquat dichloride in the extracted sample (e.g., 1 ml serum). The mass concentration of paraquat is obtained by multiplying the result

Table 28.12 Paraquat in serum: spectrophotometry; pipetting scheme.

	Reagent blank	Sample	Sample spiked 1	Sample spiked 2	Control sample	Control sample spiked 1	Control sample spiked 2
Tube	1	2	3	4	5	6	7
Serum (µl)	—	1000	1000	1000	—	—	—
Control sample (µl)	—	—	—	—	1000	1000	1000
Aqua bidest (µl)	1100	100	50	—	100	50	—
Working standard solution (µl)	—	—	50	100	—	50	100
Extraction solution (µl)	1000	1000	1000	1000	1000	1000	1000
Each tube is shaken vigorously for 1 min							
Sodium thiosulfate (g)	1.0	1.0	1.0	1.0	1.0	1.0	1.0
Shake until salt is dissolved. Centrifuge for 10 min at 1500 × *g*							
Supernatant must be transparent, otherwise filtrate							
Supernatant (µl)	400	400	400	400	400	400	400
Na-dithionite solution (µl)	600	600	600	600	600	600	600
Transfer into cuvette, close, mix, and cover with 50 µl *n*-hexane							
Absorbances are measured at 396 nm in comparison to reagent blank							
Record spectrum in comparison to reagent blank in the range of 370–430 nm							
Graphical evaluation according to method of standard addition (Figure 28.4)							

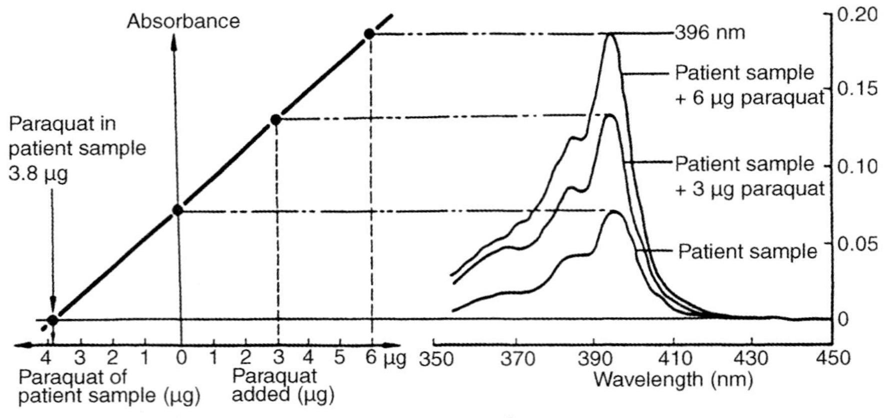

Figure 28.4 Fatal paraquat poisoning. Determination of paraquat in serum: 760 mg paraquat/l serum (sample diluted with aqua bidest before analysis; undiluted: out of range).

with 0.736 (ratio of M_r of paraquat and M_r of paraquat dichloride). A typical example is shown in Figure 28.4.

If the sample has been diluted before extraction, the result has to be multiplied by the respective dilution factor. If a straight line touching the three data points cannot be drawn, the determination failed and has to be repeated.

Analytical assessment

Precision: The spectrophotometer model 24 (Beckman-Coulter) was used for the measurements, which were taken for the estimation of imprecision. The results are presented in Table 28.13.

Serum samples were prepared by spiking (final concentration 0.5 and 2.0 mg/l, respectively) as described (see positive control sample) and proceeded without applying the standard addition method. The extracts were measured by spectrophotometry at 396 nm immediately after addition of sodium dithionite solution.

Table 28.13 does not show the calculated concentrations but the absorption values to give an estimate how sensitive a spectrophotometer should be for this procedure. It may be striking that the imprecision within the series was bigger than between series.

Table 28.13 Paraquat in serum: spectrophotometry; imprecision.

	n	c	$\bar{x}(a)$	$x_{min}(a)$	$x_{max}(a)$	$s(a)$	CV (%)
Imprecision within series	10	0.5	0.0271	0.025	0.030	0.0017	6.3
	10	2.0	0.0953	0.093	0.100	0.0026	2.7
Imprecision between series	10	0.5	0.0274	0.026	0.029	0.0008	2.9
	10	2.0	0.0944	0.090	0.098	0.0020	2.1

(a), absorbance; c, paraquat concentration (mg/l); n, number of determinations; $\bar{x}$, arithmetic mean; x_{min}, minimum value; x_{max}, maximum value; s, standard deviation; CV, coefficient of variation (%).

Table 28.14 Paraquat in serum: spectrophotometry; trueness.

Number of measurements	Target concentration (mg/l)	Concentration, measured (mg/l)	Bias %
4	1.00	1.03	3
4	5.00	5.40	8

That is why reagents deteriorate during the time-consuming within-series measurements, whereas the measurements between series were performed always with freshly prepared reagents.

Trueness: Trueness was evaluated by using control samples and spectrophotometer LS 500 (B. Lange, Berlin) as a measuring device (Table 28.14).

Linearity: The calibration curve is linear at least in the range of 0.1–10 mg/l. The slope is 0.049 absorption values/mg paraquat dichloride in 1000 ml serum.

Specificity: The measurement is considered as specific if a spectrum as shown in Figure 28.5 is obtained. Diquat, also used as a herbicide, forms a greenish radical with a spectrum different from paraquat (Figure 28.5) [66, 67]. At very low paraquat concentrations, characteristic fingerprints of spectrum may be no longer visible and a sure statement on specificity may not be possible. Then, it is recommended to investigate for confirmation qualitatively (see 28.4.2) urine or gastric lavage, which often contain a higher concentration of paraquat than the respective serum.

Bile pigments in the serum can interfere, especially if it is hemolytic. The pigments do not absorb at the wavelengths typical for the paraquat radical but shift the bottom line. However, if the standard addition method is used, paraquat will be determined still sufficiently accurate.

Detection limit: Detection limit was estimated as 0.1 mg/l serum.

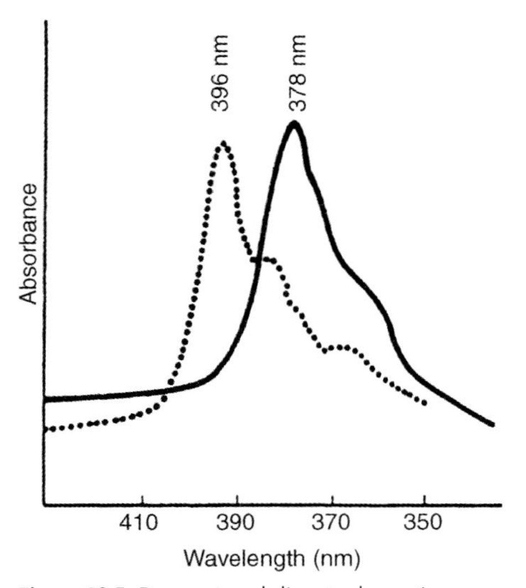

Figure 28.5 Paraquat and diquat: absorption curves.

Practicability
Technician time is about 6 min, and time for total analysis ($n = 1$) is 26 min.

Annotations
Determination of paraquat in other materials than serum: Cloudy liquids such as gastric lavage must be filtrated before use. Materials that may contain a high paraquat concentration should be investigated simultaneously undiluted and diluted with aqua bidest.

28.4.4 Medical Assessment and Clinical Interpretation

Medical assessment

Paraquat is registered in Class II (moderately hazardous) of WHO classification. LD_{50} (rat, oral) is 150 mg/kg body mass [68], the ADI value for the paraquat ion is 0.000–0.004 mg/kg body mass [69], and the maximum concentration at the workplace for paraquat dichloride is 0.1 mg/m^3 air [70].

Proudfoot *et al.* [71] were the first to point to the correlation of the chance of survival and time interval between intake and sampling and the corresponding plasma concentration of paraquat. However, the correlation is valid only if a sample is investigated, which has been obtained, before therapeutic measures to lower the paraquat concentration are taken. A detailed presentation of the correlation was published by Hart [72], who evaluated the medical history of 218 adult patients suffering from paraquat poisoning (Figure 28.6).

The intake of not more than 10 ml of a 20% paraquat solution could be fatal, whereas usually the ingestion of 50 ml was lethal. The chances of survival are said to be rising if the stomach is not empty and there are no lesions in the upper gastrointestinal tract [73]. The knowledge on the toxicokinetics of paraquat in man

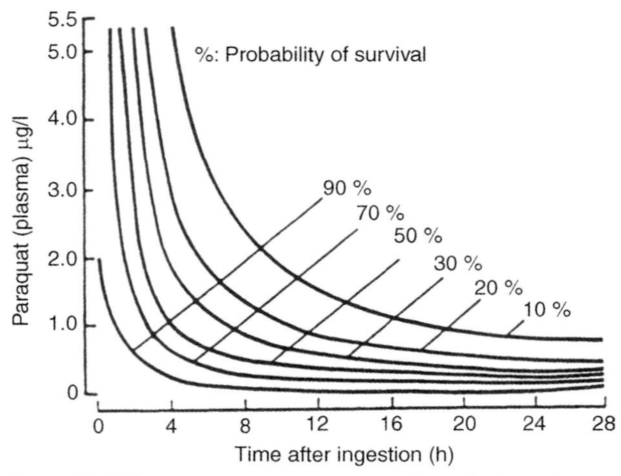

Figure 28.6 Plasma paraquat concentration, time after ingestion, and probability of survival [72].

is poor. The half-life as published ranges from 12 to 120 h, and the absorbed fraction after ingestion is about 5–10% [74]. A considerable biotransformation is not described. After deadly intoxications, paraquat could be detected in all organs of the corpses. The highest concentrations were detected in the kidneys and the liver, followed by the lungs and the brain. In urine, paraquat may be present for several weeks after ingestion [75].

Clinical interpretation

The detection of paraquat in urine is a strong indication that a medically relevant amount of the herbicide has been absorbed. At once, a plasma sample for urgent further investigations should be taken and appropriate measures of detoxification should be started. The mechanism of paraquat toxicity is not known in detail [76].

Paraquat is reduced with consumption of $NADPH_2$ to a metabolite radical, which causes the production of very reactive compounds such as superoxide anions, hydroxy radicals, and singlet oxygen molecules. The available activity of superoxide dismutase, catalase, and peroxidase is not sufficient in severe paraquat poisoning to detoxify the cytotoxic intermediates (see above). They will react with structural proteins, unsaturated fatty acids, and DNA and lead to fatal consequences for cell function. The lungs are the main focus of paraquat toxicity. Their function has to be supported by artificial respiration that unfortunately favors the production of the reactive intermediates.

Even during a rather long time period, there may be scarcely any symptoms or no symptoms at all. Locally, the herbicide may cause severe mucosa irritations, causticizing, and necrosis by colliquation. Hemorrhagic diarrhea may also occur. In the course of time, damage to liver and kidneys becomes obvious: in serum the activity of the transaminases rises, so does the concentrations of bilirubin, creatinine, and urea. Proteinuria or oliguria may develop. After a latent period of some days, a damage to the lung tissue is present, which often cannot be treated successfully, and an ARDS (adult respiratory distress syndrome) develops. The fatal outcome of a severe paraquat intoxication cannot be averted either by harsh measures of detoxification, hyperventilation, and application of antioxidants, cytostatics, immunosuppressants, or high doses of glucocorticoids. Possibly, inhalation of nitrogen monoxide can improve the highly compromised gas exchange and pulmonary hypertension of severe paraquat intoxication [77, 78]. It is doubtful if the administration of Fuller's earth, bentonite [74], or soil (in emergency) will inhibit paraquat absorption and improve outcome [73].

28.5 Organophosphorus Compounds

M. Geldmacher-von Mallinckrodt

Introduction

At the moment, over 100 OPCs with cholinesterase-inhibiting properties are used as pesticides. Most of them are esters, amides, or sulfur-containing derivates of

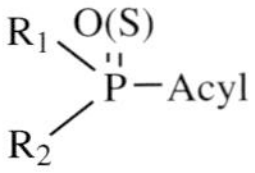

Schrader-Formula

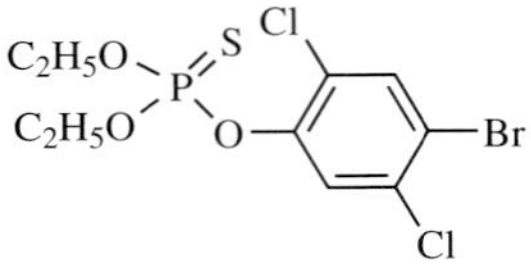

Bromophos-ethyl

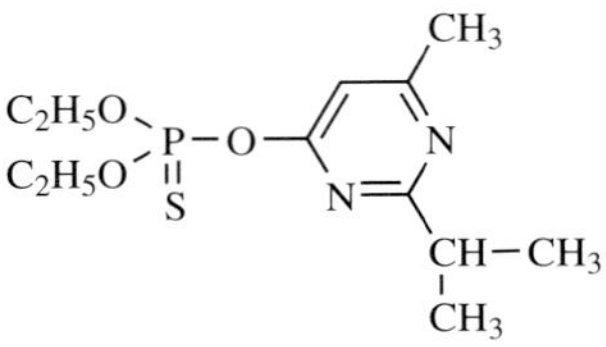

Diazinon

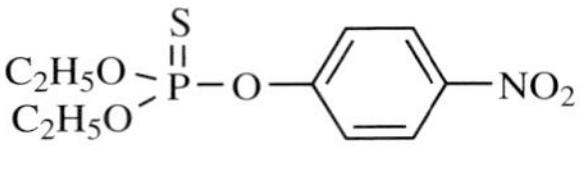

Parathion-ethyl

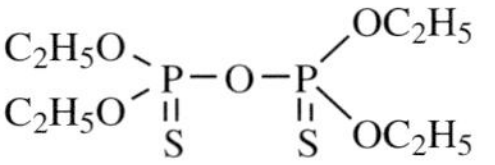

Sulfotepp

Figure 28.7 Insecticide organophosphorus compounds.

phosphoric acid (H_3PO_4), phosphorous acid (H_3PO_3), or their respective "thio" derivatives. Therefore, the term "organophosphates" is not fully correct, if used for all OPCs. Almost all of them can be traced back to a common structure, the so-called "Schrader formula" (Figure 28.7). The OPC pesticides are very toxic for warm-blooded animals including man.

If an intoxication by OPC is suspected, the cholinesterase (EC 3.1.1.8) activity in serum can be determined as a first screening (Section 33.2). More reliable would be the measurement of the acetylcholinesterase activity of erythrocytes [80] (Section 33.2). The assessment of the results is hampered by the fact that the cholinesterase activity of the pertinent individual is usually not known and the results have to be compared with the reference intervals, which are rather broad (Section 33.2).

The parent organophosphorus compounds can be detected qualitatively in acute poisoning after ingestion in vomitus, gastric lavage, and blood by GC or GC–MS. The severity of poisoning, however, could only be estimated by the quantitative determination of the OPCs in blood or serum. Unfortunately, toxicological data are available only for frequently abused OPCs. For others, clinical assessment is restricted to comparisons with single observations and case reports. Usually, only degradation products of the mother compounds are detectable in urine, because OPCs are metabolized rather rapidly.

In one group reaction, the remaining phosphates are detected, such as dimethylphosphate (DMP), diethylphosphate (DEP), dimethylthiophosphate (DMTP), diethylthiophosphate (DETP), dimethyldithiophosphate (DMDTP), and diethyldithiophosphate (DEDTP), as indicators of organophosphate poisoning [81]. If the involved OPC is known, more specific investigations can be undertaken, for example, a search for an acidic degradation product, for example, 4-nitrophenol from parathion.

28.5.1 Screening Procedure (General Unknown)

Many insecticide OPCs can be detected by the general screening procedures (HPLC and GC–MS), which are described in Chapter 12.

28.5.2 Thin-Layer Chromatography

The *procedure* is described in detail in Section 28.2 on carbamates and is suitable for the examination of vomitus, gastric lavage, and exhibits. For the detection of OPCs, however, bromine is needed in addition. The detectability of many OPCs is improved considerably if bromine oxidation precedes the enzyme application. The developed plates are examined especially for OPCs as follows:

(a) *UV absorption:* The plates are exposed to UV light (254 nm) and absorbing bands or spots are marked with a pencil.
(b) *Palladium chloride:* Plates 1 and 3 are sprayed with palladium chloride and air dried for 3 min. They are stored in a dry oven at 110 °C for 20 min.
(c) *Enzymatic reaction:*
  (1) Plates 2 and 4 are exposed to bromine vapor for 30 s in a glass tank. (A glass beaker with 0.1 ml bromine is put 20 min before in the glass tank.)
  (2) The plates are removed from the tank to let bromine evaporate for 25–30 min in a hood.
  (3) The plates are sprayed evenly with enzyme working solution (Section 28.2) until they are moist. They are kept in a dry oven, which is saturated with water, at 37 °C for 30–60 min.
  (4) The plates are sprayed with substrate solution.

After a few minutes, OPC pesticides become visible as white spots on a pink-violet background.

Evaluation of the chromatograms: The R_f values of the samples are calculated in relation to the reference substances (Table 28.3) and compared with the R_f values in Table 28.15. If always four reference substances were examined simultaneously, the so-called corrected R_f value (R_f^c value) (Section 4.2) can be calculated to further improve the reliability of the identification [82].

Table 28.15 Organophosphorus compounds: thin-layer chromatography hR_f values [82].

Substance	S1	S2	UV	$PdCl_2$
Acephate	0	0	–	Yellow
Omethoate	0	0	–	Yellow
Methamidophos	1	0	–	Yellow
Demeton-*S*-methylsulfon	1	3	–	Yellow
Etrimfos	3	0	+	Yellow
Trichlorfon/trichlorphos	4	2	–	–
Dimethoate	4	4	–	Yellow
Phosphamidon	7	2	+	Yellow
Mevinphos/phosdrin	12	10	+	–
Demeton sulfoxide[a]	17	17	–	Yellow
Oxydemeton-methyl	18	0	–	Yellow
Heptenophos	20	18	–	–
Dichlorvos/DDVP	20	20	–	–
Azinphos-methyl	20	42	+	Yellow
Triazophos	21	38	+	Yellow
Temephos	23	75	+	Yellow
Azinphos-ethyl/ethylguthion	24	48	+	Yellow
Tetrachlorvinphos	25	29	+	–
Chlorfenvinphos	26	26	+	–
Coumaphos	27	61	+	Yellow
Methidathion	29	56	+	Yellow
Parathion-methyl	30	73	+	Yellow
Malathion	31	53	–	Yellow
Phosalone	31	67	+	Yellow
Chlorthion	31	71	+	Yellow
Pyrazophos	32	47	+	Yellow
Fenitrothion	32	76	+	Yellow
Ethoprophos/ethoprop	33	28	–	Yellow
Isofenphos	41	67	+	Yellow
Fenthion	41	81	+	Yellow
Parathion-ethyl	41	84	+	Yellow
Phoxim	42	86	+	Yellow
Diazinon/dimpylate	47	50	+	Yellow
Pirimiphos-methyl	50	75	+	Yellow
Sulfotep	50	84	–	Yellow
Etrimfos	52	67	+	Yellow
Tolclofos-methyl	52	89	–	Yellow
Chlorthiophos	55	91	+	Yellow
Chlorpyrifos-methyl	56	89	–	Black
Disulfoton	58	89	–	Yellow
Bromophos	58	92	+	Yellow
Fonofos	59	89	+	Yellow
Terbufos	63	90	–	Red
Chlormephos	64	91	–	Yellow
Chlorpyrifos	64	95	±	Yellow
Bromophos-ethyl	69	93	±	Yellow
Demeton[a]	83	81	–	Yellow

S1: mobile phase, *n*-hexane/acetone 80 + 20 (v/v); S2: mobile phase, toluene/acetone 95 + 5 (v/v); UV, ultraviolet light.

[a]Demeton (D) is oxidized easily to demeton sulfoxide (DS) after application to the plate. D (hR_f 83/81) is separated from DS (hR_f 17/17) in the presented thin-layer chromatography system (Schütz H., personal communication, 2005).

Negative control sample: Extract (see above) from, for example, drug-free stomach content. No spots indicating OPCs shall be visible.

Positive control sample: Reference substances. Respective spots from detection reactions according to Tables 28.2 and 28.15 should become visible.

Analytical assessment

Sensitivity: detection limits

(a) *UV absorption:* 20–200 ng OPC/spot.
(b) *Palladium chloride:* 10–25 ng OPC/spot.
(c) *Enzymatic reaction:*
For most OPCs: 0.1–1 ng/spot.
For some OPCs, for example, guthion, parathion-methyl, and trithion-methyl, the detection limit ranges from 20 to 50 ng/spot.

Specificity

(a) *UV absorption:* UV absorption is rather unspecific. Furthermore, most OPCs that lack an aromatic group are not detectable (Table 28.15).
(b) *Palladium chloride:* In the presence of sulfur-containing compounds, yellow to brownish colors appear. OPCs without sulfur become invisible (Table 28.15).
(c) *Enzymatic reaction:* Not only cholinesterase-inhibiting compounds, for example, OPCs, but also insecticide carbamates are detected.

Practicability

The procedure can be carried easily and no special, sophisticated, or expensive equipment is needed. Technician time is about 35 min (if enzyme stock solution, deep frozen, was prepared before), and time for total analysis is about 3 h.

28.5.3 Spectrophotometry

The determination of the serum cholinesterase activity can be considered as a simple screening for OPCs. CHE activity is determined routinely in many laboratories for clinical chemistry at any time of the day for diagnosis of liver diseases. Ready-for-use reagents for manual or mechanized determination are available from many manufacturers (e.g., Roche) along with a detailed description of the procedure (as package insert) (Section 33.2).

28.5.4 Medical Assessment and Clinical Interpretation

Medical assessment

Toxic and lethal doses: Toxic and lethal doses depend not only on the OPC itself but also on the health and constitution of the patient, on the route of application, as well as on the involved special industrial product (solvents, emulsifiers, and vehicles). LD_{50} values for rats (Table 28.16) and ADI and MAK values (Table 28.17) can be used for

Table 28.16 Organophosphorus compounds: LD_{50} [94].

Substance	Classification by hazard[a]	Physical state	LD_{50} (mg/kg[b])
Azinphos-ethyl	Ib	S	12
Azinphos-methyl	Ib	S	16
Chlorfenvinphos	Ib	L	31
Chlorpyrifos	II	S	135
Demeton-*S*-methyl	Ib	L	40
Diazinon	II	L	1000
Dichlorvos	Ib	L	56
Dimethoate	II	S	150
Disulfoton	Ia	L	2.6
Ethion	II	L	208
Fenitrothion	II	L	503
Malathion	III	L	2100
Mevinphos	Ia	L	4[c]
Omethoate	Ib	L	50
Parathion	Ia	L	13
Parathion-methyl	Ia	L	14
Thiometon	Ib	Oil	120

L: liquid; S: solid.
[a]See Section 28.1.
[b]Rat, oral: mg/kg body mass.
[c]Dermal (more hazardous than oral route).

Table 28.17 Organophosphorus compounds: ADI and MAK values.

Compound	ADI (mg/kg bm) [95]	MAK (mg/m^3) [96]
Acephate	0–0.01	—
Azinphos-methyl	0–0.005	0.2[a]
Bromophos	0–0.04	—
Bromophos-ethyl	0–0.003	—
Carbophenothion	0–0.0005	—
Chlorfenvinphos	0–0.0005	—
Chlorpyrifos-methyl	0–0.01	—
Diazinon	0–0.002	0.1[a]
Dichlorvos	0–0.004	1.0
Dimethoate	0–0.002	—
Disulfoton	0–0.0003	—
Ethion	0–0.002	—
Fenitrothion	0–0.005	—
Fenthion	0–0.007	0.2[a]
Malathion	0–0.3	15[a]
Mevinphos	0–0.0008	0.093
Parathion	0–0.004	0.1[a]
Parathion-methyl	0–0.003	—
Pirimiphos-methyl	0–0.03	—
Thiometon	0–0.003	—

–: No data; bm: body mass; ADI: acceptable daily intake; MAK: maximum workplace concentration in air (mg/m^3).
[a]Measured as the inhalable fraction of the aerosol.

Table 28.18 Organophosphorus compounds: lethal dose for man (estimated) [85].

Substance	Age	Route of application	Lethal dose (g)
Diazinon	Adult	Oral	25
Malathion	n.d.	n.d.	60
Parathion	Adult	Oral or inhaled	0.01–0.3

n.d.: no data.

rough estimates of the severity of poisoning. The values collected in Table 28.18 were published for man. Higher doses can be survived when the patient can be treated appropriately in time.

Toxic and lethal concentrations of some organophosphoros compounds found in serum, plasma and blood of man are shown in Table 28.19.

Toxicokinetics: Organophosphorus compounds are very lipophilic and therefore are absorbed quickly after inhaling, through the skin or after oral intake. Main mechanisms of metabolism are hydrolysis and oxidation, which can produce metabolites more toxic than the mother compound (e.g., parathion paraoxon). The metabolites are eliminated rather quickly in urine or feces [83, 85]. Generally, OPCs are not accumulated in the body. They can be detected

Table 28.19 Organophosphorus compounds: toxic and lethal concentrations.

		Concentration			
Substance	**Material**	**Toxic (mg/l)**	**Lethal (mg/l)**	**Sampling**	**References**
Bromophos	B	1.7	n.d	1 h p.i.	[97]
	P	0.8	n.d.	n.d.	[97]
Chlorfenvinphos	P	0.15–10.6	n.d.	n.d.	[98]
	B	n.d.	0.30–15.0	n.d.	[98]
Chlorpyrifos	n.d.	0.2	27	n.d.	[99]
	S	2.1	n.d.	n.d.	[85]
Diazinon	n.d.	0.05–0.1 (–0.5)	n.d.	n.d.	[99]
	B	n.d.	0.7–277	n.d.	[85]
Dichlorvos	B	n.d.	29	Postmortal[a]	[85]
Fenitrothion	P	0.096–0.35	n.d	n.d.	[98]
	B	n.d.	1.9–14.80	Postmortal[a]	[98]
Isofenphos (i.m.)	P	1.5	n.d.	3 h p.i.	[101]
Malathion	n.d.	0.5	n.d.	n.d.	[99]
	n.d.	n.d.	0.5–3.5	n.d.	[100]
Monocrotophos	S	n.d.	24.0	44 h p.i.	[102]
Parathion	n.d.	0.01–0.05	0.05–0.08	n.d.	[99, 103]
	n.d.	n.d.	0.5–34.0	n.d.	[100]
Phosalone	P	0.005–0.39	n.d.	n.d.	[98]
	B	n.d.	0.024–0.190	Postmortal[a]	[98]
Profenofos	B	n.d.	1.2	Postmortal[a]	[104]

S: serum; P: plasma; B: blood; p.i.: postingestion; i.m.: intramuscular; n.d.: no data.
[a]Postmortal concentration; to be interpreted with caution.

in the body fat, however, several weeks after absorption of very high, acute toxic doses. On release from body fat, clinical symptoms can aggravate after initial recovery.

Toxicodynamics: OPCs are toxic for their inhibition of acetylcholinesterase (ACHE) (EC 3.1.1.7) and serum (pseudo-) cholinesterase (EC 3.1.1.8) by phosphorylation, which becomes irreversible rather rapidly. Severity and duration of the effect depend on the involved individual OPC. Acetylcholine will accumulate at the receptors leading to acetylcholine intoxication with an extreme stimulation of the nervous system.

Clinical interpretation

Typical clinical symptoms

1. *Muscarinic effects on postganglionic parasympathetic nerves:* Stimulation of lacrimal, sweat, and salivary glands, increased bronchial secretion (→ lung edema) and bronchoconstriction (dyspnea), increased secretion of stomach and intestine as well as increased motility and tone (colic, diarrhea, and vomitus), miosis, and inhibition of accommodation (visual disorder), decrease in heart rate and contractility (hypotension).
2. *Nicotinic effects on autonomic ganglia and the neuromuscular function:* Stiffness of muscles, especially face and neck, tremor, fasciculations, tonic–clonic seizures, speech disorder, paresthesia, psychic disorders, confusion, disturbance of consciousness, and respiratory paralysis (central or/and peripheral).

The mnemonic DUMBELS helps to remember some signs of cholinergic excess: diarrhea, urination, mitosis, bronchospasm, emesis, lacrimation, and salivation.

Sometimes, acute pancreatitis was observed [86]. The acute symptoms can be followed by a neuromuscular intermediate syndrome (IMS) [87]. Typically, 10–14 days after acute poisoning, severe chronic neurological symptoms (organophosphorus ester induced delayed neuropathy (OPIDN)) can develop, possibly resulting from a damage of the neurotoxicant sensitive esterase (NTE) [88, 89] or neuropathy target esterase [90].

Typical findings in clinical chemistry: In addition to the enumerated clinical symptoms, a striking low serum cholinesterase activity may be an indicator of OPC intoxication. The interpretation of the result can be difficult; usually, the CHE activity of the individual before poisoning is not known and the value can only be compared to the broad reference interval. Some persons have a congenitally low CHE activity [91], whereas some have an increased activity [92]. The size of serum cholinesterase inhibition and the severity of clinical symptoms are not closely related. A better correlation is given for the acetylcholinesterase activity [80]: a more than 20% decrease below the reference interval is a good indicator of a possibly serious intoxication [93]. It can be helpful for differential diagnosis to follow the duration of serum cholinesterase activity.

In contrast to OPCs, carbamates will inhibit the enzyme only for a short time. Already after some hours, the original activity can be restored. OPCs lead to an irreversible inhibition, and the original serum cholinesterase activity will be regained only weeks later by *de novo* synthesis.

28.6 Pyrethroids

M. Geldmacher-von Mallinckrodt

28.6.1 Introduction

The blossoms of chrysanthemum (*Tanacetum cinerarifolium*) and other flowers contain insecticide pyrethrins. Synthetic analogous compounds, the pyrethroids, are among the most efficient insecticides and are used worldwide for a broad spectrum of applications.

These commercial products consist usually of a mixture of several stereoisomers and optical isomers. Typical compounds of this group of insecticides are shown in Figure 28.8.

Pyrethroids are used on a broad scale not only in agriculture but also for the preservation of woollen carpets and wood, protection from insect (gnats, flies, wasps, etc.) indoors, and as drugs (ectoparasiticides) applied to many animals, because they are by far less toxic for warm-blooded animals than organophosphorus compounds and chlorinated hydrocarbons [105].

Several procedures for the determination of the concentration of respective metabolites in human urine after exposure at the workplace or for studies were published [106–111]. Their detection limits range from 0.5 to 1.0 μg/l, which is, however, not sufficiently sensitive for the determination in environmental medicine.

28.6.2 Gas Chromatography – Mass Spectrometry

Schettgen *et al.* [112] developed a rather sensitive GC–MS procedure for the determination of the most relevant metabolites of insecticide pyrethroids in urine:

cis-Cl_2CA: *cis*-3-(2,2-dichlorovinyl)-2,2-dimethylcyclopropane-1-carboxylic acid
trans-Cl_2CA: *trans*-3-(2,2-dichlorovinyl)-2,2-dimethylcyclopropane-1-carboxylic acid
cis-Br_2CA: *cis*-3-(2,2,-dibromovinyl)-2,2-dimethylcyclopropane-1-carboxylic acid
3-PBA: 3-phenoxybenzoic acid
F-PBA: 4-fluoro-3-phenoxybenzoic acid.

If present, they are indicative of the intake of the pyrethroids listed in Table 28.20.

Cyluthrin

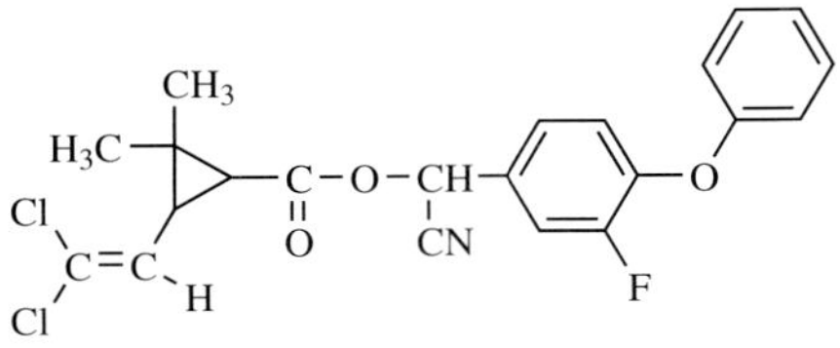

Cypermethrin

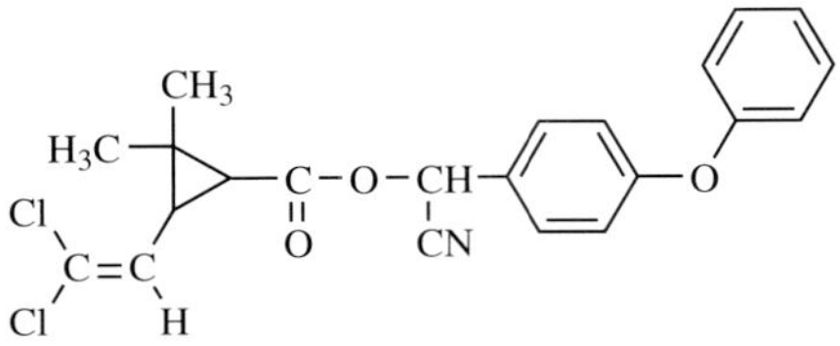

Deltamethrin

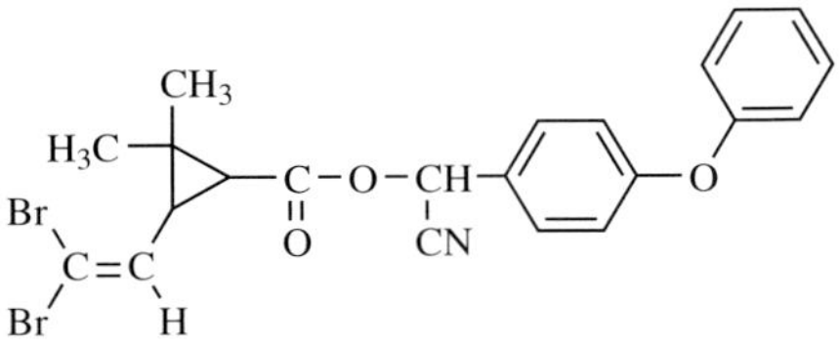

Fenpropathrin

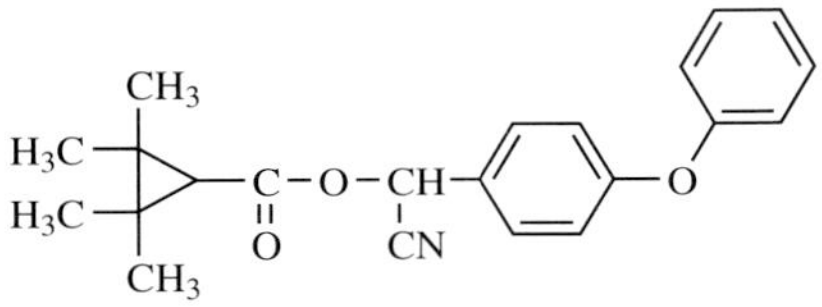

Fenvalerate

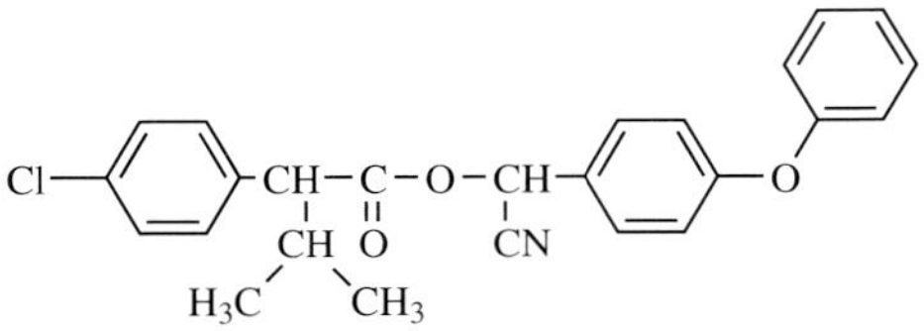

λ-Cyhalothrin

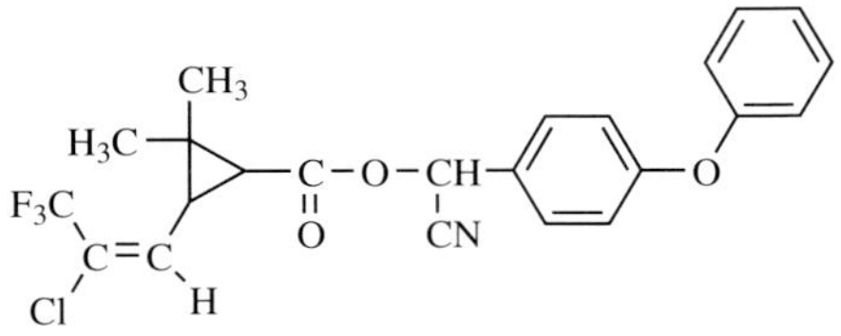

Permethrin

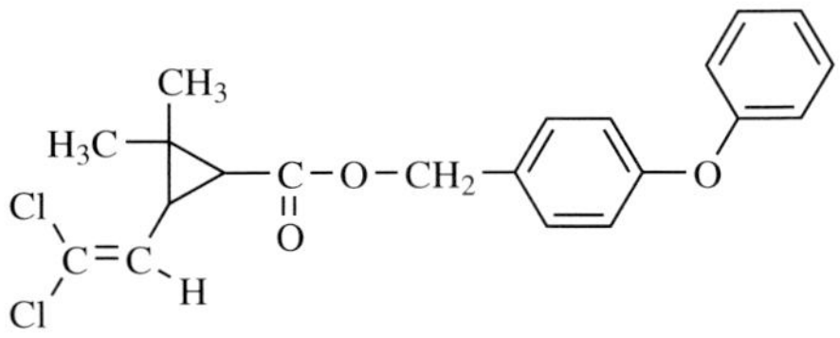

Figure 28.8 Pyrethroids.

Table 28.20 Metabolites of pyrethroids in urine.[a]

Compound (urine)	Respective mother compound(s)
3-PBA	Cypermethrin
	Deltamethrin
	Permethrin
F-PBA	Cyfluthrin
Br_2CA	Deltamethrin
Cl_2CA (*cis* and *trans*)	Cyfluthrin
	Cypermethrin
	Permethrin

[a]Determination according to Ref. [112].

Outline

After acid hydrolysis, a liquid–liquid extraction of the metabolites is performed with *n*-hexane. The extract is derivatized with MTBSTFA (*N-tert*-butyldimethylsilyl-*N*-methyltrifluoroacetamide) and analyzed by GC–MS (Figure 28.9).

The detection limit of 0.05 μg/l for the metabolites is sufficiently sensitive for the investigation of urine from nonexposed persons. Excellent selectivity is achieved by using the mass selective detector. For details see Ref. [112].

28.6.3 Medical Assessment and Clinical Interpretation

Medical assessment

The WHO classifies the pyrethroids of Figure 28.8 as moderately hazardous (class II) [113]. Because of their rather low toxicity for man, no reliable data on toxic or lethal

Conjugated carboxylic acids in urine (10 ml)

| Acidic hydrolysis with hydrochloric acid

Free carboxylic acids

| Extraction with *n*-hexane

Metabolites in *n*-hexane

| Reextraction into aqueous phase

Metabolites in NaOH (0.1 mol/l)

| Acidification, extraction with *n*-hexane

Metabolites in *n*-hexane

| Evaporation and derivatization with MTBSTFA

Derivatives in toluene

|

GC–MS analysis of the derivatives

Figure 28.9 Metabolites of pyrethroids in urine. Determination by gas chromatography–mass spectrometry [112].

Table 28.21 Pyrethroids: LD_{50} values and ADI values.

Pyrethroid	LD_{50}[a] [113]	ADI[b] [114]
Cyfluthrin	C 250	0–0.02
Cypermethrin	C 250	0–0.05
Cyhalothrin	C 144	0–0.002
Deltamethrin	C 135	0–0.01
Fenpropathrin	C 66	0–0.03
Fenvalerate	C 450	0–0.02
Permethrin	C 500	0–0.05

C: very variable. Always the lowest value is given.
[a]Pyrethroids in oil, oral. Data valid for rats (mg/kg body mass).
[b]Allowable daily intake (mg/kg body mass).

doses are available. The ADI values as published by IPCS [114] and the LD_{50} values for rats (Table 28.21) may give a clue.

It should be emphasized that the toxicity data for pyrethroids are highly variable according to isomer ratios, the vehicle used for oral administration, and the husbandry of the test animals. The variability is reflected in the prefix "c" before LD_{50} values. The single LD_{50} value now chosen for classification purposes is based on administration in corn oil and is much lower than that in aqueous solutions. This has resulted in considerable changes in the classification of some products and also underlines the need for classification by formulation if the classification is to reflect true hazard [113].

Toxicokinetics

Pyrethroids are lipophilic and are well absorbed if applied to skin or mucous membrane. Often, they are inhaled and incorporated from suspended particles as vehicles. In the organism, they are detoxified by hydrolysis of the ester bond and hydroxylation [115]. All the pyrethroids, listed in Table 28.20, are metabolized to 3-phenoxybenzoic acid (3-PBA) or the respective fluoric derivative (F-PBA), which can be measured by the procedure.

All but fenvalerate form derivatives of cyclopropanecarboxylic acid. The metabolites are not toxic and are excreted after conjugation as glucuronides, sulfates, or acetates mainly in urine as well as with feces.

Pyrethroids are accumulated less than organophosphorus compounds in the body fat. The half-life time in blood/plasma ranges from 2.5 to 12 h and in fat tissue from 1 to 30 days as estimated from studies with animals [116].

Recent investigations detected interindividual differences in pyrethroid metabolism, which are based on differences in the activity of the carboxyl esterases involved. These variations are especially meaningful in case of a combined exposure, for example, to organophosphorus compounds, which by themselves extend the elimination half-life of pyrethroids [117].

Frequently used pyrethroids and the respective metabolites are listed in Table 28.20. The concentration of the metabolites in urine exceeds by far the concentration of the mother compounds, which are degraded quickly. The metabolite concentration in urine is used as a suitable measure of pyrethroid exposure. As an upper reference limit for a nonexposed population, 0.7 μg/l urine is suggested regarding the metabolites of permethrin, *cis*- and *trans*-Cl_2CA as well as 3-PBA [118]. Schettgen *et al.* [119] investigated urine samples of 1177 inhabitants of a residential area in Frankfurt/Main (Germany) without any known contact with pyrethroids. The 95th percentile for the urinary metabolites of permethrin and cypermethrin (*cis*- and *trans*-CI_2CA) was determined to be 0.5 and 1.4 μg/l, respectively. The 95 percentile of Br_2CA, a specific metabolite of deltamethrin, and F-PBA, a metabolite of cyfluthrin, was 0.3 and 2.7 μg/l, respectively. The metabolic pattern, found for these samples, points out that pyrethroids are probably ingested orally with daily diet. In the United States, Baker *et al.* [120] found 3-PBA in about 12% of 130 urine samples containing concentrations exceeding the limit of detection of 0.5 μg/l with maximal values of 20 μg/l. This shows that the burden of the general population to pyrethroids may be a problem of international concern. The concentrations of metabolites ranged from below 0.5 (detection limit) to 277 μg/l [121] in individuals in charge of pest control working indoors. In the urine of farm and forest laborers, the metabolite concentration could get up to 300 μg/l depending on the quality of the protective equipment [122–124].

Toxicodynamics

The toxicity of the pyrethroids is based on an inhibitory effect on the potential-dependent sodium channels during the excitation phase of the neural cells [125]. There are short-acting type I pyrethroids such as permethrin (TR: tremor syndrome) and long-acting type II pyrethroids such as cypermethrin and fenvalerate (CS: choreoathetosis, salivation) [116].

Clinical interpretation

In case of poisoning, the following symptoms have been observed [125, 126]. After oral intake, at first unpleasant sensations of the skin especially of the head occur, such as burning, pruritus, swelling, feeling of tension, and insensibility, as well as nausea and vomitus. After some minutes or within a few hours, tonicoclonic convulsions may follow. The symptoms disappear completely usually within days or some weeks. A comprehensive report on symptoms and diagnosis of 573 acute intoxications by pyrethroids in China in the years 1983–1989 is given by He [125]. Among the 1580 acute intoxications in China during the years 1983–1997, only very few had a fatal outcome [126].

The diagnosis is based on the anamnesis (Any exposure to pyrethroids during the last 2 days?) and the described symptoms that cannot be attributed to other causes. As hypersalivation and convulsions may be due to poisoning by pyrethroids as well as organophosphorus compounds, serum cholinesterase should always be determined. In case of pure pyrethroid intoxication, serum cholinesterase activity should be within the reference interval if a liver disease is excluded (see Chapter 33.2). If the serum cholinesterase activity is considerably decreased under these circumstances, poisoning by organophosphorus compounds is probable but does not exclude simultaneous intoxication with pyrethroids.

Treatment is guided by the symptoms. In case of intoxication with pyrethroids and organophosphorus compounds, therapy will be focused primarily on the latter [126].

References

References for Section 28.1

1 Mgeni, A.Y. (1991) Toxic effects of pesticides – Africa's health dilemma. Proceedings of the African Workshop on Health Sector Managements in Technological Disasters, November 26–30, 1990, Adis Abeba (eds M. Holopainen, P. Kurttio and P. Tuomiso), National Public Health Institute, Kuopio, Finland, pp. 88–108.

2 IPCS (International Programme on Chemical Safety) (2005) The WHO Recommended Classification of Pesticides by Hazard and Guidelines to Classification 2004. Corrigenda published by 12th April 2005 incorporated. Corrigenda published by 28th June 2006 incorporated. World Health Organization, Geneva, ISBN 92-4-154663-8.

3 WHO (World Health Organization) (2006) Guidelines for Drinking Water Quality, 3rd edn, incorporating 1st addendum. Volume 1: Recommendations. WHO, Geneva, ISBN 92-4-154696-4.

4 WHO (World Health Organisation) Expert Consulation for the 4th Edition of the Guidelines for Drinking-water Quality (GDWQ) Berlin, 7–11 May 2007, WHO/HSE/WSH/07/11, Geneva 2007.

5 IPCS (International Programme on Chemical Safety) (2002) Inventory of IPCS and Other WHO Pesticide Evaluations and Summary of Toxicological Evaluations Performed by the Joint Meeting on Pesticide Residues (JMPR). Evaluations through 2002, 7th edn. WHO/PCS/02.3, Distr.: General.

6 IPCS (International Programme on Chemical Safety) INCHEM Environmental Health Criteria (EHCs) 104 Principles for the Toxicological Assessment of Pesticide Residues in Food World Health Organisation, Geneva 1990, ISBN 92-4-157104-7.

7 ACGIH (American Conference of Governmental Industrial Hygienists) (2001) Threshold Limit Values for Chemical Substances and Physical Agents and Biological Exposure Indices. ACGIH Inc., Cincinnati, OH.

8 ACGIH (American Conference of Governmental Industrial Hygienists) (2001) Documentation of the Threshold Limit Values and Biological Exposure Indices, 7th edn. ACGIH Inc., Cincinnati, OH.

9 IPCS (International Programme on Chemical Safety) (1996) *User's Manual for the IPCS Health and Safety Guides*, WHO, Geneva.

10 DFG (Deutsche Forschungsgemeinschaft) (2007) *List of MAK and BAT Values 2007*. Commission for the Investigation of Health Hazards of Chemical Compounds in the Work Area. Report No. 43. Wiley-VCH Verlag GmbH, Weinheim.

11 DFG (Deutsche Forschungsgemeinschaft) (1991–2002) *Occupational Toxicants. Critical Data Evaluation for MAK Values and Classification of Carcinogens*. Vols 1–17, Wiley-VCH Verlag GmbH, Weinheim.

12 Morgan, M.S. and Schaller, K.H. (1999) An analysis of criteria for biological limit values developed in Germany and in the United States. *International Archives of Occupational and Environmental Health*, **72**, 195–204.

13 DFG (Deutsche Forschungsgemeinschaft) (1994–1999) *Biological Exposure Levels for Occupational Toxicants and Carcinogens. Critical Data Evaluation for BAT and EKA Values*. Vols 1–3, Wiley-VCH Verlag GmbH, Weinheim.

14 Swiss Toxicological Information Centre. Annual Report 2002. CH 8028 Zürich, Freiestrasse 16.

15 Litovitz, T.B., Klein-Schwartz, W., Rodgers, G.C., Cobaugh, D.J., Youniss, J., Omslaer, J.C., May, M.E., Woolf, A.D. and Benson, B.E. (2002) 2001 Annual Report of the American Association of Poison Control Centers – Toxic Exposure Surveillance System. *American Journal of Emergency Medicine*, **20**, 391–452.

References for Section 28.2

16 Bowden, C.A. and Krenzelok, E.P. (1997) Clinical applications of commonly used contemporary antidotes. *Drug Safety*, **16**, 9–47.

17 Erdmann, F., Brose, C. and Schütz, H. (1990) A TLC screening program for 170 commonly used pesticides using the corrected R_f value (R_f^c value). *International Journal of Legal Medicine*, **104**, 25–31.

18 DFG/TIAFT (Deutsche Forschungsgemeinschaft/The International Association of Forensic Toxicologists) (1992) *Thin-Layer Chromatographic R_f-Values of Toxicologically Relevant Substances on Standardized Systems*, 2nd edn, Wiley-VCH Verlag GmbH, Weinheim.

19 IPCS (International Programme on Chemical Safety) (2004) The WHO Recommended Classification of Pesticides by Hazard and Guidelines to Classification. WHO-IPCS, Geneva, ISBN 92-4-154663-8.

20 IPCS (International Programme on Chemical Safety) (2002) Inventory of IPCS and other WHO Pesticide Evaluations and Summary of Toxicological Evaluations Performed by the Joint Meeting on Pesticide Residues (JMPR). Evaluations through 2002. WHO/PCS/02.3, Geneva.

21 Deutsche Forschungsgemeinschaft (2007) *List of MAK and BAT Values 2007.* Commission for the Investigation of Health Hazards of Chemical Compounds. Report No. 43. Wiley-VCH Verlag GmbH, Weinheim.

22 Erdmann, A.R. (2004) Insecticides, in *Medical Toxicology*, 3rd edn (ed. R.C. Dart), Lippincott, Williams and Wilkins, Philadelphia, PA.

23 Baselt, R.C. (2004) *Disposition of Toxic Drugs and Chemicals in Man*, 7th edn, Biomedical Publications, Foster City, CA.

24 Schulz, M. and Schmoldt, A. (2003) Therapeutic and toxic blood concentrations of more than 800 drugs and other xenobiotics. *Die Pharmazie*, **58**, 447–474.

25 Sancewicz-Pach, K., Groszek, B., Pach, D. and Klys, M. (1997) Acute pesticide poisonings in pregnant women. *Przgl-Lek*, **54**, 741–744.

26 Pfordt, J., Magerl, H. and Vock, R. (1987) Tödliche Vergiftungen mit Propoxur. *Zeitschrift fur Rechtsmedizin*, **98**, 43–48.

27 Pragst, F., Correns, A., Sporkert, S., Prügel, M. and Strauch, H. (1998) Morphologische und toxikologisch-chemische Befunde bei einer letalen Intoxikation mit dem Insektenspray Baygon. *Rechtsmed*, **8**, 105–108.

References for Section 28.3

28 Hayes, W.J. Jr (1982) Chlorinated hydrocarbon insecticides, in *Pesticides Studied in Man* (ed. W.J. Hayes Jr), Williams & Wilkins, Baltimore, MD, pp. 172–283.

29 Smith, A.G. (1991) Chlorinated hydrocarbon insecticides, in *Handbook on Pesticide Toxicology* (eds W.J. HayesJr and E.R. Laws), Academic Press, San Diego, CA.

30 Erdmann, A.R. (2004) Insecticides, in *Medical Toxicology*, 3rd edn (ed. R.C. Dart) Lippincott, Williams and Wilkins, Philadelphia, PA.

31 The Rotterdam Convention, Conference of Plenipotentiaries on the Convention on the Prior Informed Consent Procedure for Certain Hazardous Chemicals and Pesticides in International Trade, September 10, 1998, Rotterdam, entered into force on 24 February 2004. Information see www.pic.int.

32 International Programme on Chemical Safety (IPCS) (2005) The WHO Recommended Classification of Pesticides by Hazard and Guidelines to Classification 2004. Corrigenda published by 12th April 2005 incorporated Corrigenda published by 28th June 2006 incorporated. World Health Organization, Geneva, ISBN 92-4-154663-8.

33 DFG/TIAFT (Deutsche Forschungsgemeinschaft/The International Association of Forensic Toxicologists) (1992) *Thin-Layer Chromatographic R_f Values of Toxicologically Relevant Substances on Standardized Systems,* 2nd edn, Wiley-VCH Verlag GmbH, Weinheim.

34 Brock, J.W., Burse, V.W., Ashley, D.L., Najam, A.R., Green, V.E., Korver, M.P. *et al.* (1996) An improved analysis for chlorinated pesticides and polychlorinated biphenyls (PCBs) in human and bovine sera using solid-phase extraction. *Journal of Analytical Toxicology,* **20**, 528–536.

35 International Programme on Chemical Safety (IPCS) (2002) Inventory of IPCS and Other WHO Pesticide Evaluations and Summary of Toxicological Evaluations Performed by the Joint Meeting on Pesticide Residues (JMPR). Evaluations through 2002, 7th edn. WHO/PCS/02.3, Distr.: General.

36 DFG (Deutsche Forschungsgemeinschaft) (2007) *List of MAK and BAT Values 2007.* Commission for the Investigation of Health Hazards of Chemical Compounds. Report No. 43. Wiley-VCH Verlag GmbH, Weinheim.

37 Moffat, A.C., Osselton, M.D. and Widdop, B.(eds) (2004) Clarke's Analysis of Drugs and Poisons in Pharmaceuticals, Body Fluids and Post Mortem Material, 3rd edn, Pharmaceutical Press, London.

38 Baselt, R.C.(ed.) (2004) *Disposition of Toxic Drugs and Chemicals in Man,* 7th edn Biomedical Publications, Foster City, CA.

39 Biros, F.J. and Enos, H.F. (1973) Oxychlordane residues in human adipose tissue. *Bulletin of Environmental Contamination and Toxicology,* **5**, 257–260.

40 Miyazaki, T., Akiyama, K., Kaneko, S., Horii, S. and Yamagishi T. (1980) Chlordane residues in human milk. *Bulletin of Environmental Contamination and Toxicology,* **25** 518–523.

41 Minh, T.B., Watanabe, M., Tanabe, S., Yamada, T., Hata, J. and Watanabe, S. (2000) Occurrence of tris-(4-chlorophenyl) methane, tris-(4-chlorophenyl) methanol and some other persistent organochlorines in Japanese human adipose tissue. *Environ Health Perspectives,* **108**, 599–603.

42 To Figueras, J., Sala, M., Otero, R., Barrot, C., Silva, M., Rodamilans, M., Herrero, C., Grimalt, J. and Sunyer, J. (1997) Metabolism of hexachlorobenzene in humans: association between serum levels and urinary metabolites in a highly exposed population. *Environ Health Perspectives,* **105**, 78–83.

43 To Figueras, J., Barrot, C., Sala, M., Otero, R., Silva, M., Ozalla, M.D., Herrero, C., Corbella, J., Grimalt, J. and Sunyer, J. (2000) Excretion of hexachlorobenzene and metabolites in feces in a highly exposed human population. *Environmental Health Perspectives,* **108**, 595–598.

44 Angerer, J., Maass, R. and Heinrich, R. (1983) Occupational exposure to hexachlorocyclohexane. VI. Metabolism of gamma-hexachlorocyclohexane in man. *International Archives of Occupational and Environmental Health,* **52**, 59–67.

45 Saleh, M.A. (1991) Toxaphene: chemistry, biochemistry, toxicity and environmental fate. *Reviews of Environmental Contamination and Toxicology,* **118**, 1–85.

46 Schulz, M. and Schmoldt, A. (2003) Therapeutic and toxic blood concentrations of more than 800 drugs and other xenobiotics. *Pharmazie,* **58**, 447–474.

47 Winek, C.L., Wahba, W.W., Winek, CL., Jr and Winek-Balzer, T. (2001) Drug and chemical blood level data 2001. *Forensic Science International*, **122**, 107–123.

References for Section 28.4

48 Knepil, J. (1977) A short, simple method for the determination of paraquat in plasma. *Clinica Chimica Acta*, **79**, 387–390.

49 Coxon, R.E., Rae, C., Gallacher, G. and Landon, J. (1988) Development of a simple fluoroimmunoassay for paraquat. *Clinica Chimica Acta*, **175**, 297–306.

50 Tomita, M., Suzuki, K., Shimosato, K., Kohama, A. and Ijiri, I. (1988) An enzyme-linked immunosorbent assay for plasma paraquat levels of poisoned patients. *Forensic Science International*, **37**, 11–18.

51 Levitt, T. (1979) Determination of paraquat in clinical practice using radioimmunoassay. *Proceedings of the Analytical Division of the Chemical Society*, **16**, 72–76.

52 Ikebuchi, J., Yuasa, I. and Kotoku, S. (1988) A rapid and sensitive method for the determination of paraquat in plasma and urine by thin-layer chromatography with flame ionization detection. *Journal of Analytical Toxicology*, **12**, 80–83.

53 Baselt, R.C.(ed.) (1980) *Analytical Procedures for Therapeutic Drug Monitoring and Emergency Toxicology*, Biomedical Publications, Davis, CA.

54 Jarvie, D.R., Fell, A.F. and Stewart, M.J. (1981) A rapid method for the emergency analysis of paraquat in plasma using second derivative spectroscopy. *Clinica Chimica Acta*, **117**, 153–165.

55 Jarvie, D.R. and Stewart, M.J. (1979) The rapid extraction of paraquat from plasma using an ion-pairing technique. *Clinica Chimica Acta*, **94**, 241–251.

56 de Zeeuw, R.A., Wijsbeek, J., Franke, J.P., Bogusz, M. and Klys, M. (1986) Multicentre evaluation of ultrafiltration, dialysis, and thermal coagulation as sample pretreatment methods for the colorimetric determination of paraquat in blood and tissues. *Journal of Forensic Sciences*, **31**, 504–510.

57 Kuo, T.L. (1995) Determination of paraquat in biologic materials by a simplified solid-phase extraction and spectrophotometry. *Journal of the Formosan Medical Association*, **94**, 243–247.

58 Tomita, M., Okuyama, T. and Nigo, Y. (1992) Simultaneous determination of paraquat and diquat in serum using capillary electrophoresis. *Biomedical Chromatography*, **6**, 91–94.

59 Vinner, E., Stievenart, M., Humbert, L., Mathieu, D. and Lhermitte, M. (2001) Separation and quantification of paraquat and diquat in serum and urine by capillary electrophoresis. *Biomedical Chromatography*, **15**, 342–347.

60 Arys, K., van Bocxlaer, J., Clauwaert, K., Lambert, W., Piette, M., van Peteghem, C. and de Leenheer, A. (2000) Quantitative determination of paraquat in a fatal intoxication by HPLC–DAD following chemical reduction with sodium borohydride. *Journal of Analytical Toxicology*, **24**, 116–121.

61 Taylor, P.J., Salm, P. and Pillans, P.I. (2001) A detection scheme for paraquat poisoning: validation of a five-year experience in Australia. *Journal of Analytical Toxicology*, **25**, 456–460.

62 Paixão, P., Costa, P., Bugalho, T., Fidalgo, C. and Pereira, L.M. (2002) Simple method for determination of paraquat in plasma and serum of human patients by high-performance liquid chromatography. *Journal of Chromatography B*, **775**, 109–113.

63 Fuke, C., Arao, T., Morinaga, Y., Takaesu, H., Ameno, K. and Miyazaki, T. (2002) Analysis of paraquat, diquat and two diquat metabolites in biological materials by high-performance liquid chromatography. *Legal Medicine Tokyo*, **4**, 156–163.

64 Lee, X.P., Kumazawa, T., Fujishiro, M., Hasegawa, C., Arinobu, T., Seno, H., Ishii, A., Sato, K. *et al.* (2004) Determination of paraquat and diquat in human body fluids by high-performance liquid chromatography/tandem mass spectrometry. *Journal of Mass Spectrometry*, **39**, 1147–1152.

65 Daldrup, T. and Fowinkel, C. (1991) Determination of substances using the known addition technique in forensic and clinical toxicology. Forensic Toxicology. Proceedings of the 29th International Meeting of TIAFT (The International Association of Forensic Toxicologists), June 24–27, Copenhagen, Denmark (ed. B. Kaempe), Mackenzie, Copenhagen, pp. 379–385.

66 Minakata, K., Suzuki, O. and Asano, M. (1989) A new colorimetric determination of diquat produced with several moderate reductants. *Forensic Science International*, **42**, 231–237.

67 Minakata, K., Suzuki, O., Saito, S. and Harada, N. (2000) A new diquat derivative appropriate for colorimetric measurements of biological materials in the presence of paraquat. *International Journal of Legal Medicine*, **114**, 1–5.

68 IPCS (International Programme on Chemical Safety) (2005) The WHO Recommended Classification of Pesticides by Hazard and Guidelines to Classification 2004. WHO-IPCS, Geneva, ISBN 92-4-154 663-8.

69 IPCS (International Programme on Chemical Safety) (2002) Inventory of IPCS and Other WHO Pesticide Evaluations and Summary of Toxicological Evaluations Performed by the Joint Meeting on Pesticide Residues (JMPR). Evaluations through 2002. WHO/PCS/02.3, Geneva.

70 Deutsche Forschungsgemeinschaft (2007) *List of MAK and BAT Values 2007*. Commission for the Investigation of Health Hazards of Chemical Compounds. Report No. 43. Wiley-VCH Verlag GmbH, Weinheim.

71 Proudfoot, A.T., Stewart, M.J., Levitt, T. and Widdop, B. (1979) Paraquat poisoning: significance of plasma paraquat concentrations. *Lancet*, **II**, 330–332.

72 Hart, T.B., Nevitt, A. and Whitehead, A. (1984) A new statistical approach to the prognostic significance of plasma paraquat concentrations. *Lancet*, **II**, 1222–1223.

73 Bronstein, A.C. (2004) Herbicides, in *Medical Toxicology*, 3rd edn (ed. R.C. Dart), Lippincott, Williams and Wilkins, Philadelphia, PA.

74 van Dijk, A., Maes, R.A.A., Drost, R.H., Douze, J.M.C. and van Heijst, A.N.P. (1975) Paraquat poisoning in man. *Archives of Toxicology*, **34**, 129–136.

75 Baselt, R.C.(ed.) (2004) *Disposition of Toxic Drugs and Chemicals in Man*, 7th edn Biomedical Publications, Foster City, CA.

76 Bismuth, C. and Hall, A.H.(eds) (1995) *Paraquat Poisoning: Mechanism, Prevention, Treatment*, Marcel Dekker, New York, Basel.

77 Köppel, C. and von Wissmann Ch. Barckow, D., Rossaint, D.R., Falke, K., Stoltenburg-Didinger, G. *et al.* (1994) Inhalative nitric oxide in serious paraquat intoxication. *Clinical Toxicology*, **32**, 205–214.

78 Eisenman, A., Armali, Z., Raikhlin-Eisenkraft, B., Bentur, L., Bentur, Y., Guralnik, L. *et al.* (1998) Nitric oxide inhalation for paraquat-induced lung injury. *Clinical Toxicology*, **36**, 575–584.

79 Baeck, S.K., Shin, Y.S., Chung, H.S. and Pyo, M.Y. (2007) Comparison study of the extraction methods of paraquat in post-mortem human blood samples. *Archives of Pharmacal Research*, **30**, 235–239.

References for Section 28.5

80 Lewalter, J., Domik, C. and Schaller, K.H. (1991) Acetylcholinesterase (ACHE; acetylcholine-acetylhydrolase EC 3.1.1.7) and cholinesterase (CHE; acylcholine-acylhydrolase EC 3.1.1.8), in *Analyses of Hazardous Substances in Biological Materials*, Vol. 3 (eds J. Angerer and K.H. Schaller), Wiley-VCH Verlag GmbH, Weinheim.

81 Hardt, J. and Angerer, J. (2000) Determination of dialkylphosphates in human urine using gas chromatography–mass spectrometry. *Journal of Analytical Toxicology*, **24**, 678–684.

82 DFG/TIAFT (Deutsche Forschungsgemeinschaft/The International Association of Forensic Toxicologists) (1992) *Thin-Layer Chromatographic R_f Values of Toxicologically Relevant Substances on Standardized Systems*, 2nd edn, Wiley-VCH Verlag GmbH, Weinheim.

83 Hayes, W.J.(ed.) (1982) *Pesticides Studied in Man*, Williams and Wilkins, Baltimore, MD.

84 Hayes, W.R., Jr and Laws, E.R.(eds) (1991) *Handbook on Pesticide Toxicology*, Academic Press, San Diego, CA.

85 Baselt, R.C. (2004) *Disposition of Toxic Drugs and Chemicals in Man*, 7th edn, Biochemical Publications, Foster City, CA.

86 Lee, W.C., Yang, C.C. and Deng, J.F. (1997) The clinical implications of acute pancreatitis in organophosphate poisoning. *Clinical Toxicology*, **35**, 516.

87 de Bleecker, J.L. (1995) The intermediate syndrome in organophosphate poisoning: an overview of experimental and clinical features. *Clinical Toxicology*, **33**, 683–686.

88 Johnson, M.K. (1975) The delayed neuropathy caused by some organophosphorus esters: mechanism and challenge. *CRC Critical Reviews in Toxicology*, **3**, 289–316.

89 Zech, R. and Chemnitius, J.M. (1987) Neurotoxicant sensitive esterase. Enzymology and pathophysiology of organophosphorus ester-induced delayed neuropathy. *Progress in Neurobiology*, **29**, 193–218.

90 Meredith, T.J. (1999) Mechanisms of toxicity of organophosphorus insecticides abstract. *Clinical Toxicology*, **37**, 366.

91 Ostergaard, D., Samsoe Jensen, F. and Voby-Mogensen, J. (1992) Pseudocholinesterase deficiency and anticholinesterase toxicity, in *Clinical and Experimental Toxicology of Organophosphates and Carbamates* (eds B. Ballantyne and T.C. Marrs), Butterworth-Heinemann, Oxford.

92 Akizuki, S., Ohnishi, A., Kotani, K. and Sudo, K. (2004) Genetic and immunological analyses of patients with increased serum butyrylcholinesterase activity and its C5 variant form. *Clinical Chemistry and Laboratory Medicine*, **42**, 991–996.

93 Jokanovic, M. and Maksimovic, M. (1997) Abnormal cholinesterase activity: understanding and interpretation. *European Journal of Clinical Chemistry and Clinical Biochemistry*, **35**, 11–16.

94 International Programme on Chemical Safety (IPCS) (2005) The WHO Recommended Classification of Pesticides by Hazard and Guidelines to Classification 2004. Corrigenda published by 12th April 2005 incorporated. Corrigenda published by 28th June 2006 incorporated. World Health Organization, Geneva, ISBN 92-4-154663-8.

95 International Programme on Chemical Safety (IPCS) (2002) Inventory of IPCS and Other WHO Pesticide Evaluations and Summary of Toxicological Evaluations Performed by the Joint Meeting on Pesticide Residues (IMPR). Evaluations through 2002, 7th edn. WHO/PCS/02.3, Geneva.

96 DFG (Deutsche Forschungsgemeinschaft) (2007) *List of MAK and BAT Values 2007*. Commission for the Investigation of Health Hazards of Chemical Compounds. Report No. 43. Wiley-VCH Verlag GmbH, Weinheim.

97 Köppel, C., Thomsen, T., Heinemeyer, G. and Roots, I. (1991) Acute poisoning with bromofosmethyl (bromophos). *Clinical Toxicology*, **29**, 203–207.

98 Kala, M. (2004) Pesticides, in *Clarke's Analysis of Drugs and Poisons*, 3rd edn (eds A.C. Moffat, M.D. Osselton and B. Widdop), Pharmaceutical Press, London.

99 Schulz, M. and Schmoldt, A. (2003) Therapeutic and toxic blood concentrations of more than 800 drugs and other xenobiotics. *Pharmazie*, **58**, 447–474.

100 Winek, C.L., Wahba, W.W., Winek, C.L., Jr and Winek-Balzer, T. (2001) Drug and chemical blood-level data 2001. *Forensic Science International*, **122**, 107–123.

101 Zoppelari, R., Borron, S.W., Chieregato, A., Targa, L., Scaroni, I. and Zatelli, R. (1997) Isofenphos poisoning: prolonged intoxication after intramuscular injection. *Clinical Toxicology*, **35**, 401–404.

102 Papa, P., Rocci, L., Riverso, P. and Polettini, A. (1998) A fatal case of monocrotophos poisoning. *TIAFT Bulletin*, **XXVII** (2), 5–6.

103 TIAFT (The International Association of Forensic Toxicologists) (2005) Reference Blood Level List of Therapeutic and Toxic Substances. Update 2005-03-03.

104 Seno, H., Hattori, H., Kumazawa, T., Ishii, A., Watanabe, K. and Suzuki, O. (1998) Quantitation of postmortem profenofos levels. *Clinical Toxicology*, **36**, 63–65.

References for Section 28.6

105 Roberts, T.R., Hutson, D.H., Jewess, P.J., Lee, P.W., Nicholls, P.H. and Plimmer, J.R.(eds) (1999) *Metabolic Pathways of Agrochemicals. Part 2. Insecticides and Fungicides*, Royal Society of Chemistry, Cambridge.

106 Angerer, J. and Schaller, K.H.(eds) (1999) *Analyses of Hazardous Substances in Biological Materials*, Vol. 6. Deutsche Forschungsgemeinschaft, Wiley-VCH Verlag GmbH, Weinheim. pp. 231–254.

107 Angerer, J. and Ritter, A. (1997) Determination of metabolites of pyrethroids in human urine using solid-phase extraction and gas chromatography–mass spectrometry. *Journal of Chromatography B*, **695**, 217–226.

108 Kuhn, K.H., Leng, G., Bucolski, K.A., Dunemann, L. and Idel, H. (1996) Determination of pyrethroid metabolites in human urine by capillary gas chromatography–mass spectrometry. *Chromatographia*, **43**, 285–292.

109 Tuomainen, A., Kangas, J., Liesivuori, J. and Manninen, A. (1996) Biological monitoring of deltamethrin exposure in greenhouses. *Environmental Health*, **69**, 62–64.

110 Woolen, B.H., Marsh, J.R., Laird, W.J.D. and Lesser, J.E. (1992) The metabolism of cypermethrin in man – differences in urinary metabolite profiles following oral and dermal administration. *Xenobiotica*, **22**, 983–991.

111 Eadsforth, C.V., Bragt, P.C. and Vansittert, N.J. (1988) Human dose-excretion studies with pyrethroid insecticides cypermethrin and alpha-cypermethrin – relevance for biological monitoring. *Xenobiotica*, **18**, 603–614.

112 Schettgen, T., Koch, H.M., Drexler, H. and Angerer, J. (2002) New gas chromatographic–mass spectrometric method for the determination of urinary pyrethroid metabolites in environmental medicine. *Journal of Chromatography B*, **778**, 121–130.

113 International Programme on Chemical Safety (IPCS) (2005) The WHO Recommended Classification of Pesticides by Hazard and Guidelines to Classification 2004. Corrigenda published by 12th April 2005 incorporated. Corrigenda published by 28th June 2006 incorporated. World Health Organization, Geneva, ISBN 92-4-154663-8.

114 IPCS (International Programme on Chemical Safety) (2002) Inventory of IPCS and Other WHO Pesticides Evaluations and Summary of Toxicological Evaluations Performed by the Joint Meeting on Pesticide Residues (JMPR). Evaluations through 2002, 7th edn. WHO/PCS/02.3, Geneva.

115 Dorman, D.C. and Beasley, R. (1991) Neurotoxicology of pyrethrin and the pyrethroid insecticides. *Veterinary and Human Toxicology*, **33**, 238–243.

116 Aldridge, W.N. (1990) An assessment of the toxicological properties of pyrethroids and their neurotoxicity. *Critical Reviews in Toxicology*, **21**, 89–104.

117 Leng, G., Lewalter, J., Röhrig, B. and Idel, H. (1999) The influence of individual susceptibility in pyrethroid exposure. *Toxicology Letters*, **107**, 123–130.

118 Butte, W., Walker, G. and Heinzow, B. (1998) Referenzwerte der Konzentration von Permethrin-Metaboliten CI_2CA (3-(2,2-Dichlorvinyl)-2,2-dimethylcyclopropancarbonsäure) und 3-PBA (3-Phenoxybenzoesäure) im Urin. *Umweltmedizin in Forschung Und Praxis*, **3**, 21–26.

119 Schettgen, T., Heudorf, U., Drexler, H. and Angerer, J. (2002) Pyrethroid exposure of the general population – is this due to diet? *Toxicology Letters*, **134**, 141–145.

120 Baker, S.E., Barr, D.B., Driskell, W.J., Beeson, M.D. and Needham, L.L. (2000) Quantification of selected pesticide metabolites in human urine by isotope dilution high-performance liquid chromatography/tandem mass spectrometry. *Journal of Exposure Science and Environmental Epidemiology*, **10**, 789–798.

121 Leng, G., Kühn, K.H. and Idel, H. (1997) Biological monitoring of pyrethroids in blood and pyrethroid metabolites in urine: applications and limitations. *The Science of the Total Environment*, **199**, 173–181.

122 Kolmodin-Hedmann, B., Swensson, A. and Akerblom, M. (1982) Occupational exposure to some synthetic pyrethroids (permethrin and fenvalerate). *Archives of Toxicology*, **50**, 27–33.

123 Kolmodin-Hedmann, B., Akerblom, M., Flato, S. and Alek, G. (1995) Symptoms in forestry workers handling conifer plants treated with permethrin. *Bulletin of Environmental Contamination and Toxicology*, **55**, 487–493.

124 Prinsen, G.H. and van Sittert, N.J. (1980) Exposure and medical monitoring study of a new synthetic pyrethroid after one season of spraying on cotton in Ivory Coast, in *Field Worker Exposure During Pesticide Application* (eds W.F. Tordoir and E.A.H. van Heemstra-Lequin), Elsevier, Amsterdam, pp. 105–120.

125 He, F., Wang, S., Liu, L., Chen, S., Zhang, Z. and Sun, J. (1998) Clinical manifestations and diagnosis of acute pyrethroid poisoning. *Archives of Toxicology*, **63**, 54–58.

126 He, F. (1999) Pyrethroid exposure – clinical features. *Clinical Toxicology*, **37**, 362.

29
Antidiabetics: Proinsulin, Insulin, C-Peptide, and Oral Antidiabetics

Introduction

J. Hallbach

In particular, if hypoglycemia is suspected from surreptitious application of insulin, it is indicated to determine insulin, C-peptide, and possibly also proinsulin in blood plasma. The abuse of antidiabetics such as sulfonylurea derivatives can be detected by LC–MS.

29.1
Insulin, Proinsulin, and C-Peptide

J. Hallbach

Outline
Heterogeneous immunochemical procedures are employed for the determination of insulin, C-peptide, and proinsulin.

Specimens
Serum or heparin plasma, at least 1.0 ml.

The analytes are not stable in blood at room temperature. Insulin, in particular, is stable only for 15 min [1]. Therefore, samples should be centrifuged immediately after sampling. Plasma (serum) should be stored at −20 °C, if not analyzed at once [2].

Equipment

Insulin, C-peptide: devices to carry out heterogeneous immunoassays, at best fully mechanized (e.g., Immulite, Siemens).
Proinsulin: devices to perform RIA.

Chemicals
Not applicable.

Clinical Toxicological Analysis: Procedures, Results, Interpretation. Edited by Wolf-Rüdiger Külpmann

ISBN: 978-3-527-31890-2

Table 29.1 Insulin and C-peptide: immunoassay imprecision.

	Mean (pmol/l)	*s* (pmol/l)	CV (%)	*n*
Insulin	76.8	10.7	13.96	13
C-peptide	794	105	13.2	20
	2460	190	7.7	20

n, number of contributing results; *s*, standard deviation; CV, coefficient of variation.

Reagents

Insulin: luminescence immunoassay (e.g., Siemens)
C-peptide: luminescence immunoassay (e.g., Siemens)
Proinsulin: RIA (e.g., Linco Research).

Sample preparation
Not applicable.

Immunochemical analysis
Immunoassays according to the instructions from the manufacturer.

Analytical assessment
Precision: see Table 29.1.

Trueness: The results depend on the procedure employed.

Specificity: The Siemens insulin assay does not show any cross-reaction with proinsulin, C-peptide, or glucagon. In general, recombinant insulins such as insulin aspart, insulin glargine, or insulin lispro are not detected by the Siemens assay or other insulin immunoassays [3].

The *C-peptide test* from Siemens does not show any cross-reaction with insulin and glucagon. Thirteen percent cross-reaction with proinsulin can be neglected, as its plasma concentration is only a hundredth of insulin.

The *proinsulin assay* from *Linco* does not show relevant cross-reaction with insulin.

Sensitivity: According to the manufacturers, the lower limit of detection is as follows:

Insulin	15 pmol/l
Proinsulin	2 pmol/l
C-peptide	99 pmol/l

Practicability
The Immulite analyzer provides good practicability for the determination of *insulin* and *C-peptide*:

Technician time: 2 min
Time of analysis: 60 min.

The system has access to a stored calibration curve.
The determination of *proinsulin* requires much more time:

Technician time: 2 h
Time of analysis: 3.5 days.

The shelf life of the RIA kit is rather limited.

29.2 Oral Antidiabetics: Sulfonylureas

K. Rentsch

Outline
After solid-phase extraction, the samples are separated by HPLC and identified by mass spectrometry [4].

Specimens
Serum or heparin plasma, at least 3 ml.
The samples should be stored at −20 °C.

Equipment

HPLC equipped for gradient elution
Column: Nucleosil 100-5, C 18 HD; 125 mm × 2 mm i.d., 5 μm film thickness, particle size 10 nm (Macherey-Nagel)
Mass spectrometer (e.g., LCQ, Thermo) with atmospheric pressure chemical ionization.

Chemicals

Acetone
Acetonitrile
Ammonia (25%)
Ammonium carbonate
Hydrochloric acid (1 and 0.1 mol/l)
Methanol
Sulfonylureas: chlorpropamide, glibenclamide, glibornuride, gliclazide, glimepiride, glipizide, tolazamide, and tolbutamide (Figures 29.1–29.7)
C18 SPE columns (e.g., Varian).

Reagents
Ammonium carbonate buffer pH 9.3 (20 mmol/l): 1.93 g ammonium carbonate is dissolved in 1000 ml aqua bidest. pH is adjusted to 9.3 by ammonia (25%). Stable for 6 months if stored in a refrigerator.

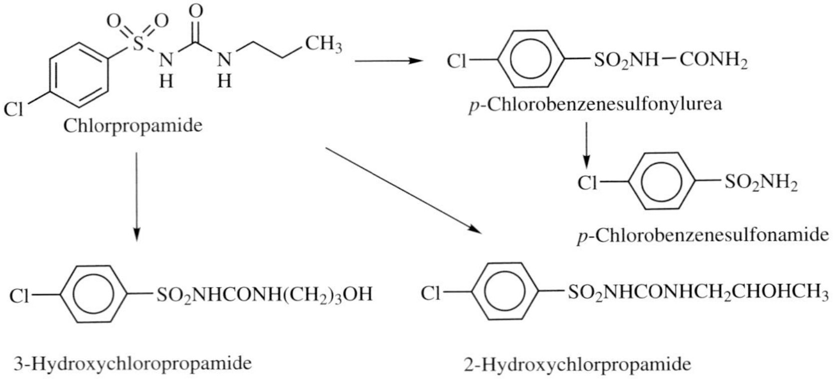

Chlorpropamide

Cl—C6H4—$SO_2NH-CONH_2$

p-Chlorobenzenesulfonylurea

Cl—C6H4—SO_2NH_2

p-Chlorobenzenesulfonamide

Cl—C6H4—$SO_2NHCONH(CH_2)_3OH$

3-Hydroxychloropropamide

Cl—C6H4—$SO_2NHCONHCH_2CHOHCH_3$

2-Hydroxychlorpropamide

Figure 29.1 Chlorpropamide metabolism.

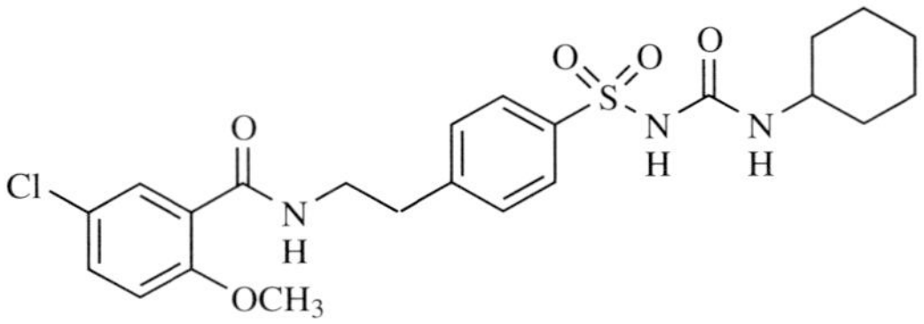

Figure 29.2 Glibenclamide.

Liquid chromatography:

Mobile phase A: ammonium carbonate buffer/acetonitrile 95 + 5 (v/v)
Mobile phase B: acetonitrile.

Stock solutions of the various sulfonylureas (1 mg/ml):

Acetone/acetonitrile 1 + 1 (v/v) as solvent for glibenclamide
Acetone/ethanol 1 + 1 (v/v) as solvent for glipizide and glimepiride
For the other sulfonylureas, acetonitrile is used as solvent.

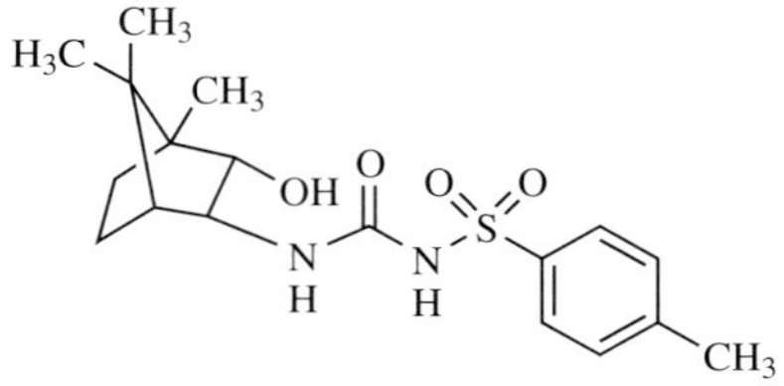

Figure 29.3 Glibornuride.

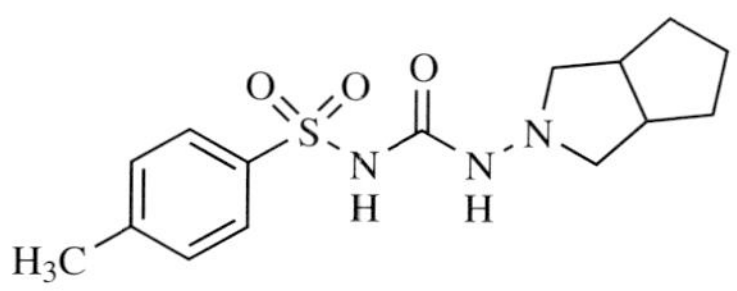

Figure 29.4 Gliclazide.

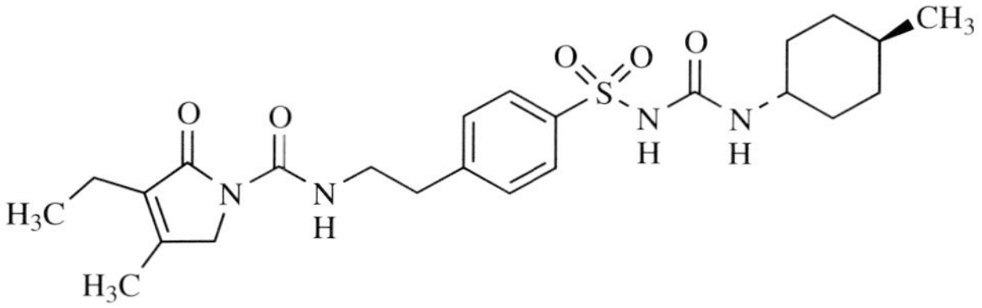

Figure 29.5 Glimepiride.

The stock solutions are stable for 1 year, if stored at −20 °C:

Working solution 1 of sulfonylureas (0.1 mg/ml): 1 ml of each stock solution is transferred to one 10 ml volumetric flask (except for tolazamide). The mixture is made up to 10 ml with acetonitrile.
Working solution 2 of sulfonylureas (0.01 mg/ml): 1 ml of working solution 1 is transferred to a volumetric flask (10 ml) and made up to 10 ml with acetonitrile.
Working solution 3 of sulfonylureas (0.001 mg/ml): 1 ml of working solution 2 transferred to a volumetric flask (10 ml) and made up to 10 ml with acetonitrile.

Glipizide → *m*- and *p*-Hydroxyglipizide

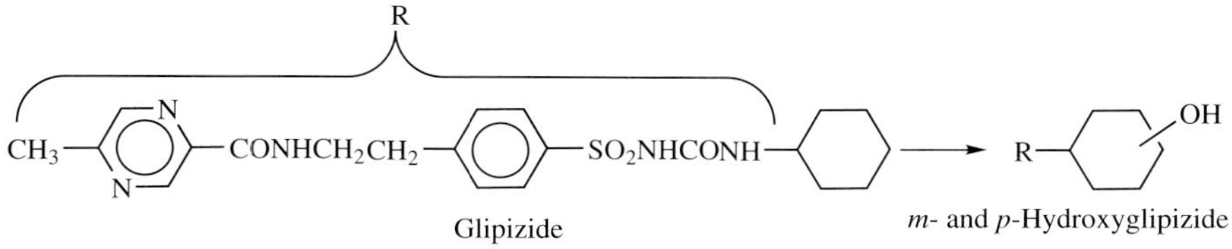

Figure 29.6 Glipizide metabolism.

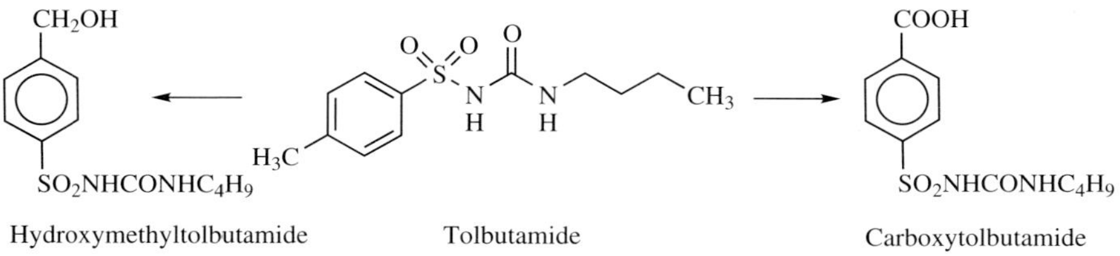

Hydroxymethyltolbutamide ← Tolbutamide → Carboxytolbutamide

Figure 29.7 Tolbutamide metabolism.

The working solutions are stable for 6 months if stored at −20 °C:

Calibrator 1 (20 µg/ml): 20 µl of working solution 3 is added to 1 ml of drug-free serum.
Calibrator 2 (50 µg/l): 50 µl of working solution 3 is added to 1 ml of drug-free serum.
Calibrator 3 (200 µg/l): 20 µl of working solution 2 is added to 1 ml of drug-free serum.
Calibrator 4 (500 µg/l): 50 µl of working solution 2 is added to 1 ml of drug-free serum.
Calibrator 5 (1000 µg/l): 10 µl of working solution 1 are added to 1 ml of drug-free serum.

All calibrators should be prepared just before use.

Tolazamide solution (internal standard, 0.01 mg/ml): 100 µl of the respective stock solution is transferred to a volumetric flask (10 ml) and made up to 10 ml with acetonitrile. The solution is stable for 6 months if stored at −20 °C.

Sample preparation
The plasma samples are extracted by solid-phase extraction on C18 columns. The columns are conditioned with 3 ml methanol, 3 ml aqua bidest, and 1.5 ml hydrochloric acid (0.1 mol/l). Thirty microliters of tolazamide solution (0.01 mg/ml), 0.5 ml hydrochloric acid (1 mol/l), and 0.5 ml aqua bidest are added to 1 ml of the sample. The samples are mixed thoroughly and centrifuged at (2000 × g, 20 °C, 10 min). The supernatant is transferred to the conditioned columns. They are rinsed first with 1.5 ml hydrochloric acid (0.1 mol/l) and then with 6 ml of aqua bidest and later sucked dry (vacuum ~20 mbar, 5 min). The sulfonylureas are eluted with 2.5 ml methanol/acetonitrile 1 + 1 (v/v). The solvents are vaporized with a rotary evaporator and the residues are dissolved in a 200 µl mixture of aqua bidest and acetonitrile 3 + 1 (v/v).

LC–MS analysis
The chromatographic separation is carried out by a C18 column (e.g., Nucleosil) and gradient elution. Starting with 80% of eluent A and 20% of eluent B, the separation is carried out with a linear gradient elution: after 2 min of isocratic operation, the eluent B fraction is increased to 45% within 10 min. The flow rate is 0.25 ml/min.

The mass spectrometer must be equipped with a device for atmospheric pressure chemical ionization, which is used in the negative ion mode. For optimum ionization of the sulfonylureas, the settings of the vaporizer, the corona current, the capillary temperature, and the capillary voltage have to be adjusted accordingly.

Evaluation: The following m/z ratios are used for quantification: chlorpropamide, 275; glibenclamide, 492; glibornuride, 365; gliclazide, 322; glimepiride, 489; glipizide, 444; tolbutamide, 269; tolazamide (IS), 310.

Analytical assessment
Precision: see Table 29.2.

Table 29.2 Oral antidiabetics: LC–MS imprecision.

	Mean(μg/l)	s (μg/l)	CV (%)	n
Chlorpropamide	69.9	3.5	5.1	6
	696.0	6.9	1.0	6
Glibenclamide	70.6	3.8	5.4	6
	689.0	53.5	7.8	6
Glibornuride	72.3	2.1	2.8	6
	708.0	48.5	6.9	6
Gliclazide	71.7	3.2	4.5	6
	684.0	36.0	5.3	6
Glimepiride	70.1	4.6	6.5	6
	691.0	55.3	8.0	6
Glipizide	69.6	2.9	4.2	6
	701.0	22.6	3.2	6
Tolbutamide	69.1	4.1	6.0	6
	703.0	27.6	3.9	6

n, number of contributing results; *s*, standard deviation; CV, coefficient of variation; LC–MS, liquid chromatography–mass spectrometry.

Trueness: Bias: −2 up to 3% in case of an extraction yield >81%.

Specificity: In case of mass spectrometric detection, the measurement of the sulfonylureas is specific.

Sensitivity: Lower limits of detection: glimepiride, 1 μg/l; glipizide and glibenclamide, 2.5 μg/l; chlorpropamide, 5 μg/l; glibornuride, gliclazide, and tolbutamide, 10 μg/l.

Practicability

Technician time is 1.5 h for the analysis of one sample and five calibrators. The total analysis requires approximately 3 h.

Annotations

The calibration curves for the measurement of the sulfonylureas are well below their therapeutic range to achieve good sensitivity. Low detection limits are needed because the compounds have a short half-life and samples are, in general, not obtained soon after administration. In case of a concentration exceeding the upper calibration point, the sample has to be diluted (e.g., 1 + 99 (v/v)).

29.3 Medical Assessment and Clinical Interpretation

J. Hallbach and K. Rentsch

Medical assessment

Insulin is metabolized mainly in the liver and to a minor extent in the skeletal muscles and the kidneys. Its biological half-life is only a few minutes.

Table 29.3 Reference intervals.[a,b]

Insulin	58–172 pmol/l (8–24 mU/l) [5]
C-peptide	up to 1200 pmol/l (up to 3.6 μg/l) [5]
Proinsulin	3.4–12.4 pmol/l

[a] Fasting.
[b] Depending on procedure.

Under physiological conditions, the β-cells secrete mainly insulin and C-peptide in equimolar portion and also secrete proinsulin and its split products but in only small amounts. The reference values of C-peptide are approximately five times higher than those of insulin, a difference that results from the shorter half-life of insulin in plasma compared to that of C-peptide (Table 29.3).

A factitious hypoglycemia will develop if either insulin is injected or sulfonylureas are administered in the absence of hyperglycemia. In diabetics, inadequately high doses of insulin or oral antidiabetics also cause serious hypoglycemia. Hypoglycemia from the injection of insulin is characterized by markedly increased insulin concentrations (>600 pmol/l) on the one hand and distinctly decreased C-peptide concentrations (<200 pmol/l) on the other hand [5]. As the antibodies of most immunoassays do not react with recombinant insulins, these may escape determination, even though they are present at high concentration.

Insulin antibodies may be detectable if insulin has been injected repeatedly. In hypoglycemia resulting from sulfonylureas administration (Figures 29.1–29.7), the concentration of insulin (>600 pmol/l), C-peptide (>200 pmol/l), and proinsulin (>5 pmol/l) is increased.

Table 29.3 shows the reference values for proinsulin, insulin, and C-peptide. The therapeutic ranges of sulfonylureas are listed in Table 29.4 [6].

Clinical interpretation

An increased concentration of insulin and concomitantly a decreased concentration of proinsulin and C-peptide when insulin is applied is a typical finding. However, this constellation can also be found in hypoglycemia by stimulating autoantibodies. Although insulin is never administered, insulin antibodies are present in such patients.

Table 29.4 Sulfonylureas: therapeutic range.[a]

Compound	Therapeutic range (mg/l)
Chlorpropamide	30–250 [6]
Glibenclamide	0.05–0.2 [6]
Glibornuride	1–2
Gliclazide	0.3–8.0
Glimepiride	~0.3
Glipizide	0.1–1.5 [6]
Tolbutamide	50–100 [6]

[a] Steady state.

In principle, it is impossible to differentiate between the analytical constellation characteristics of an intake of sulfonylureas (increased concentration of insulin, C-peptide, and proinsulin) on the one hand and those of insulinoma or multiple endocrine neoplasia with an uncontrolled release of insulin on the other hand. Indeed, insulinoma or multiple endocrine neoplasia increases the concentration of insulin, C-peptide, and proinsulin similarly as sulfonylureas. For differential diagnosis, the concentration of sulfonylureas should be determined. In addition, it should be considered that the concentration of insulin is increasing during supervised fasting only in case of an insulinoma.

References

References for Section 29.1

1 Guder, W.G., Narayanan, S., Wisser, H. and Zawta, B. (1996) List of analytes. Preanalytical variables, in *Samples: From the Patient to the Laboratory*, GIT, Darmstadt.

2 Bolner, A., Lomeo, L. and Lomeo, AM. (2005) Method-specific stability of serum C-peptide in a multicenter clinical study. *Clinical Laboratory*, **51**, 153–155.

3 Owen, W.E. and Roberts, W.L. (2004) Cross-reactivity of three recombinant insulin analogues with five commercial insulin immunoassays. *Clinical Chemistry*, **50**, 257–259.

References for Section 29.2

4 Rentsch, K.M., Gutteck, U. and von Eckardstein, A. (2003) Determination of 7 sulfonylurea drugs with LC–MS. *Therapeutic Drug Monitoring*, **25**, 58.

References for Section 29.3

5 Thomas, L. (1998) Insulin, C-peptide, proinsulin, in *Clinical Laboratory Diagnostics* (ed. L. Thomas) TH-Books Verlagsgesellschaft, Frankfurt/M.

6 Schulz, M. and Schmoldt, A. (2003) Therapeutic and toxic blood concentrations of more than 800 drugs and other xenobiotics. *Pharmazie*, **58**, 447–474.

30
Dyshemoglobins

H.J. Gibitz

Introduction

Carboxyhemoglobin (CO-Hb), methemoglobin (Met-Hb), and sulfhemoglobin are called dyshemoglobins, as they cannot carry oxygen. As the fraction of dyshemoglobins increases, the transport capacity of oxygen decreases and hypoxemia develops in tissues. CO-Hb and, less markedly, Met-Hb shift the oxygen/hemoglobin dissociation curve to the left and impede additionally the release of oxygen in the tissues.

The fractions of CO-hemoglobin and methemoglobin in blood in relation to total hemoglobin are determined by difference spectrophotometry based on the fundamentals from Vierodt and Hüfner (1899): the absorption spectra of different dyes in a solution add up, if the components do not influence each other optically.

30.1
Carboxyhemoglobin

30.1.1
Introduction

Carbon monoxide is produced from combustion of organic materials such as fuels, charcoal, gas, and wool. Oxygen supply to the organism is impaired on inhalation of carbon monoxide, because its affinity to hemoglobin is about 260 times higher than that of oxygen. Poisoning by carbon monoxide is observed rather often and in many countries it is still the most frequent cause of deaths from accidental and nondrug-induced intoxications. Nevertheless, in some countries the frequency decreased considerably by replacing carbon monoxide containing gas by natural gas.

The serious hazards of carbon monoxide are based on the following:

(1) Carbon monoxide is an odorless, tasteless, and colorless gas and therefore not detectable by people. It is rapidly absorbed after inhalation and the person may die within minutes.

Clinical Toxicological Analysis: Procedures, Results, Interpretation. Edited by Wolf-Rüdiger Külpmann

ISBN: 978-3-527-31890-2

(2) Symptoms of carbon monoxide poisoning are not specific. It may be confused with influenza, apoplexy, or myocardial infarction. Often the correct diagnosis is made only after repeated poisoning or after development of chronic intoxication.

The determination of CO-Hb in blood gives the crucial clue for diagnosis. It should be carried out as soon as possible at best by a point-of-care procedure. The fraction of CO-Hb should be known to decide on start or continuation of hyperbaric oxygen therapy (Section 30.1.4).

At present, the fraction of CO-Hb is usually determined by oximeters often attached to blood gas analyzers, whereas manual procedures are seldom applied in hospitals.

30.1.2
Oximetry

Specimens
Preferably arterial blood or capillary blood, anticoagulated by heparin.

Sampling: as soon as possible and before treatment.
Storage: in the dark at 4 °C after addition of sodium fluoride to prevent production of carbon monoxide by bacteria.

Equipment
Oximeter (e.g., Radiometer).

Reagents
Reagents are provided by the manufacturers of the instruments.

Spectrophotometrical analysis
The mechanized procedure includes hemolysis. The blood sample is analyzed spectrophotometrically at different wavelengths. In older oximeters, six wavelengths have been used, whereas in more recent devices 128 wavelengths are involved, which record an almost continuous spectrum ranging from 478 to 672 nm of the hemolyzed blood sample. Compounds such as bilirubin that might interfere are taken into account and do not falsify the results of oxyhemoglobin, carboxyhemoglobin, or methemoglobin measurements.

Practicability
The oximeters measure the carboxyhemoglobin fraction within some minutes. If they are attached to a blood gas analyzer, the other quantities needed for interpretation, such as pH, pCO_2, and pO_2 as well as actual bicarbonate concentration and base excess, are determined simultaneously.

30.1.3
Spectrophotometry

Procedure 1

Heparinized blood is hemolyzed and subdivided into three aliquots; two of them are used for calibration:

Calibrator 1: Hemoglobin of hemolysate is converted completely to CO-hemoglobin by bubbling through carbon monoxide for 5–10 min.
Calibrator 2: Hemoglobin of hemolysate is converted completely to O_2-hemoglobin by bubbling through oxygen for 5–10 min.

After addition of sodium dithionite, the absorbance (A) of the two calibrators and the unchanged aliquot is measured at 540 and 579 nm by a spectrophotometer (Figure 30.1). The ratio A_{540}/A_{579} is linearly related to the fraction of carboxyhemoglobin [1]. The calibration procedure is a rather sophisticated and hazardous procedure and cannot be recommended for routine use [1, 2].

Procedure 2

Heparinized blood is hemolyzed and the absorbance (A) at Hg 546 and Hg 578 nm is measured by a spectrophotometer with a mercury lamp. The fraction of

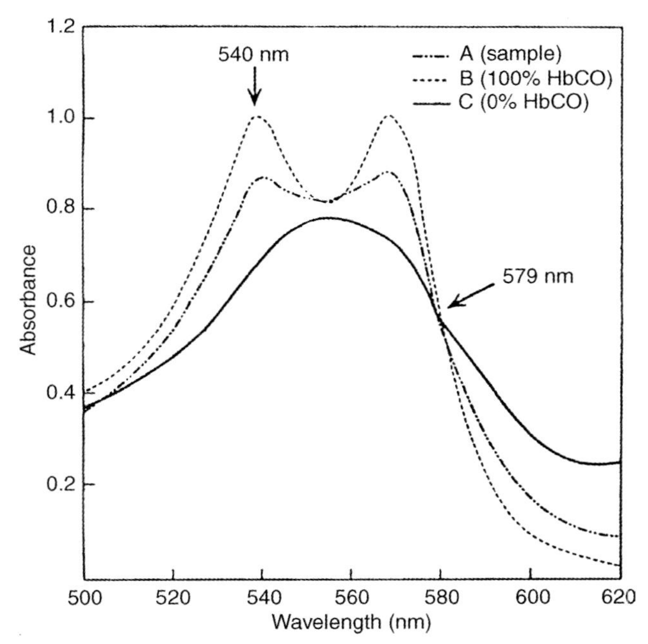

Figure 30.1 Carboxyhemoglobin: spectrophotometry. (A) blood sample of patient; (B) carboxyhemoglobin (100% CO-hemoglobin); (C) oxyhemoglobin (0% CO-hemoglobin). Taken from Ref. [2].

carboxyhemoglobin corresponding to the ratio A_{546}/A_{578} is tabulated. The procedure has been developed by Hartmann and modified by Gibitz [3]. This procedure is less sensitive than procedure 1 because the measurements are not carried out at maximum absorbance. Fractions of carboxyhemoglobin <10% are not determined accurately. However, in the range 10–90%, the results (x) are correlated well to results from an oximeter (y) (model 282 radiometer): $y = 1.031 + 0.994x$, $r = 0.968$ ($n = 76$). The procedure is appropriate to meet the requirements in case of acute carbon monoxide poisoning.

30.1.4 Medical Assessment and Clinical Interpretation

Medical assessment

In healthy people, the fraction of carboxyhemoglobin is approximately 0.4–0.8% from physiological carbon monoxide production [4]. It is increased to 4–6% in case of enhanced heme metabolism, such as hemolytic anemia [5]. In neonates, fractions up to 12% have been reported [6].

The fraction of carboxyhemoglobin from smoking varies considerably and may attain 10–12%, for heavy smokers up to a maximum of 21%. The biological tolerance value (BAT) for carbon monoxide exposure is a fraction of 5% of total hemoglobin [7]. It is not known if this value is exceeded by passive smoking.

For equilibrium, the following equation applies:

$$\frac{\text{CO-Hb}}{\text{O}_2\text{Hb}} = \frac{p\text{CO}}{p\text{O}_2} \times M,$$

where p is the partial pressure and M is not a constant but depends on pH.

The maximal value of 300 is achieved at pH 7.35. According to Roughton [8], as an average, a value of 260 may be used, that is, usually the affinity of carbon monoxide to hemoglobin exceeds the affinity of oxygen 260 times. A carbon monoxide number concentration of $50/10^6$ (50 ppm) of the inhaled air corresponds to 8% carboxyhemoglobin and a number concentration of $500/10^6$ (500 ppm) to almost 50% at equilibrium.

By replacing oxyhemoglobin by carboxyhemoglobin, the oxygen dissociation curve is shifted to the "left," that is, the release of oxygen to the tissues is impeded. An equivalent reduction of erythrocytes as in anemia is less dangerous: the dissociation curve is not shifted to the left, but possibly to the right from increased glycerol-2,3-diphosphate and the release of oxygen is even facilitated.

Clinical interpretation

Clinical symptoms and fraction of carboxyhemoglobin are associated as follows:

0–10%	No substantial symptoms
10–15%	Possibly dyspnea during hard manual work
15–25%	Dyspnea during manual work, not at rest; possibly dizziness and headache; dilation of skin capillaries

(Continued)

25–35%	Dizziness, vomitus, headache, fatigue, visual disorders, irritability, reduced competence to judge, and tachycardia
35–45%	As for 25–35%, but more marked. In addition, confusion, signs of paralysis, and transient loss of consciousness already from little strain
45–55%	Reduced vigilance up to prolonged loss of consciousness, tachypnea, tachycardia, collapse, life threatening if effective during extended time period
55–65%	In addition to 45–55%: convulsion, respiratory depression
>65%	Imminent death

However, the relationship between symptoms and fraction of carboxyhemoglobin is rather variable. For example, symptoms are aggravated in anemic patients, small children, elderly persons, and people suffering from heart failure.

For the treatment of severe carbon monoxide intoxication, hyperbaric oxygen is applied. As a competitive antidote, oxygen replaces carbon monoxide according to its partial pressure. The elimination rate is proportional to the alveolic oxygen pressure. Elimination half-life for carboxyhemoglobin at atmospheric pressure of ambient air comes to 3 h and it is reduced to 20 min by inhalation of oxygen at 250 kPa (2.5 bar). Further therapeutic measures are guided by symptoms, such as osmotic therapy in case of brain or pulmonary edema and hemodialysis in renal failure. Monitoring of circulation and acid–base balance including blood gas analysis is mandatory. It should be considered that other toxic agents, such as cyanide, methemoglobin inducing agents, ethanol, or drugs, may also be involved in addition.

Annotations

(1) Carbon monoxide is formed *in vivo* from dichloromethane [9] as proven by studies in animals [10]. The fraction of carboxyhemoglobin was increased up to 15%.

(2) In victims from fire and suicides from car exhaust, methemoglobinemia may be involved and aggravate the severity of poisoning.

(3) Apart from carbon monoxide, cyanide (Section 31.4) may be formed in addition from combustion of synthetic material, as well as wool, feathers and so on.

(4) If a blood sample is altered from combustion or putrefaction, a spectrophotometric measurement may be impossible. As an alternative, gas chromatography can be applied [11].

(5) Blood gas analyzers (without oximeter) calculate oxygen saturation falsely, if dyshemoglobins are present. Therefore, these results are misleading if carboxyhemoglobin, methemoglobin, or sulfhemoglobin are present at clinically relevant concentrations.

30.2 Methemoglobin

Introduction

Methemoglobin is produced by oxidation of iron(2^+) to iron(3^+) of hemoglobin. It cannot carry oxygen (O_2) to tissues. The fraction of methemoglobin is increased most often by oxidizing agents and rarely by inheritance. It can be determined by oximetry or manual spectrophotometry.

30.2.1 Oximetry

Specimen
See Section 30.1.2.

Analysis
See Section 30.1.2. Oximetry is the method of choice to determine the fraction of methemoglobin in a hospital.

30.2.2 Spectrophotometry

Outline
Methemoglobin exhibits an absorbance maximum at 630 nm in a weakly acidic solution. After addition of cyanide, methemoglobin cyanide is formed. This complex does not exhibit absorbance at 630 nm. Total hemoglobin is determined in this procedure concomitantly: hemoglobin and its derivatives are oxidized completely by potassium hexacyanoferrate(III) ($K_3Fe(CN)_6$) to methemoglobin that is converted to methemoglobin cyanide [12].

Specimen
Blood, anticoagulated (EDTA, potassium oxalate, and lithium heparin) 3.0 ml.

Hemoglobin is considered as stable during 2 h in these specimens. (Later, methemoglobin is reduced by methemoglobin reductase to hemoglobin and results are false low.)

Equipment
Spectrophotometer (630 nm).

Chemicals (p.a. grade)

Phosphate buffer pH 6.8 (66.7 mmol/l)
Potassium cyanide
Potassium hexacyanoferrate(III)
Sodium carbonate.

Table 30.1 Methemoglobin: spectrophotometric determination.

Cuvette no.	A_1	A_2	A_3	A_4
Hemolysate buffer (µl)	1500	1500	1500	1500
Aqua bidest (µl)	100	50	50	0
Potassium cyanide solution (µl)	0	50	0	50
Ferric cyanide solution (µl)	0	0	50	50

The solutions are mixed thoroughly. After 5 min, the absorbance is read against water at 630 nm.

Reagents

Ferric cyanide solution: 0.5 g potassium hexacyanoferrate(III) and 50 mg sodium carbonate are dissolved in 10 ml aqua bidest. The solution is stable several months if stored in the refrigerator.

Potassium cyanide solution: 0.05 g potassium cyanide is dissolved in 10 ml aqua bidest. The solution is very hazardous. It is stable for 1 week if stored in the refrigerator.

Control sample: Lyophilized concentrate of erythrocytes (e.g., Instrumentation Laboratory or Combitrol Plus B, Roche).

Sample preparation

After sampling, 2 ml blood or control sample is hemolyzed by addition of 18 ml aqua bidest.

Spectrophotometrical analysis

To 10 ml hemolysate, 10 ml phosphate buffer is added. The mixture of hemolysate and buffer is centrifuged. For pipetting scheme, see Table 30.1. Absorbances are measured at 630 nm.

The fraction of methemoglobin f_M (%) in relation to hemoglobin is calculated from (see Table 30.2)

$$\frac{A_1 - A_2}{A_3 - A_4} \times 100 = f_M.$$

Analytical assessment

Precision: Coefficient of variation for the imprecision within the series was 0.3%. For the imprecision between series, it ranged from 0.3 to 6.9% (Table 30.2).

Table 30.2 Methemoglobin: imprecision between series.

n	$\bar{x}$	s	CV
20	0.49	0.034	6.9
20	8.07	0.075	0.9
20	71.91	0.232	0.3

n, number of measurements; $\bar{x}$, mean methemoglobin fraction (%) of total hemoglobin; s, standard deviation; CV, coefficient of variation (%).

Trueness: Trueness was investigated by measurement of hemolysates that contained different portions of methemoglobin by mixing. The average bias was 0.5%.

Specificity: The procedure is rather specific and not interfered by sulfhemoglobin (see annotation).

Detection limit: A blood sample of a non-exposed person was analyzed 20 times. The fraction of methemoglobin came up to $\bar{x} = 0.495\%$, and the standard derivation to $s = 0.034\%$. The detection limit was estimated as 0.6% (mean + 3s). It is recommended to report quantitative results only for fractions exceeding 1.0%.

Practicability

Time of analysis is 22.5 min, including a technician time of 12.5 min.

Annotations

(1) Sulfhemoglobin also shows an absorbance maximum at 630 nm, which does not disappear after addition of cyanide. This phenomenon may be considered as an indication of the presence of sulfhemoglobin.

(2) Cyanide poisoning may be treated with 4-(*N*,*N*-dimethylamino)phenol (4-DAMP). The antidote produces methemoglobin that binds cyanides rather quickly. The methemoglobin cyanide complex escapes detection as it does not show absorbance at 630 nm and the portion of methemoglobin is underestimated.

(3) At a portion of methemoglobin exceeding 15%, the blood shows a brownish color, which does not change, if exposed to air.

(4) The described manual procedure should only be used, if oximetry is not available.

30.2.3 Medical Assessment and Clinical Interpretation

In methemoglobinemia, usually a partial oxidation of iron in the four subunits predominates. Oxygen carried by a remaining nonoxidized subunit is more tightly bound as compared to oxyhemoglobin and poorly released to tissues. Therefore, the oxygen dissociation curve is shifted to the left, even though the shift is only half in relation to the corresponding portion of carboxyhemoglobin. Methemoglobinemia should be considered, if cyanosis is associated with normal partial pressure of oxygen in blood.

Methemoglobin is also a natural compound of blood and its portion may come up to 1–2%. It is produced in metabolism, but is reduced steadily by methemoglobin reductase in a first-order reaction proceeding at approximately 15%/h. Cytochrome b_5 reductase plays a major role in this process by transferring electrons from NADH to methemoglobin [13].

An increased portion of methemoglobin is most often caused by oxidizing agents (Table 30.3). Direct oxidizers react with hemoglobin to form methemoglobin. Indirect oxidizers reduce oxygen to the free radical O_2^- or water to hydrogen peroxide, which in turn oxidize hemoglobin to methemoglobin.

Table 30.3 Methemoglobinemia inducing agents.

Aniline derivatives, for example, acetanilide, acetaminophen, indometacin, and dipyrone
Aniline dyes
Chlorate
Clofazimine
Local anesthetics, for example, benzocaine, lidocaine, and prilocaine
Nitrate and nitrite, for example, amyl nitrite, nitroglycerin, sodium nitroprusside, and sodium nitrite
Nitroalkanes (e.g., nail polish remover)
Phenacetin
Phenazopyridine
Sulfonamides, for example, dapsone, sulfamethoxazole
Sulfones
Industrial nitrogen containing compounds, for example, dyes, fertilizers
Food: beets, cabbage, nitrite/nitrate preservatives, nitrogen-rich foods, preserved meats, spinach, and well water

Selected from Ref. [13].

In victims of combustion or suicides from car exhaust, methemoglobin may be produced by the influence of nitrogen oxides or denaturing heat. The methemoglobin fraction may attain 30%. Life-threatening concentrations have been reported for nitrogen oxides released from combustion of fertilizers.

Administration of drugs, nitrate containing drinking water, or vegetables may cause methemoglobinemia (Table 30.4). Nitrate may be reduced to nitrite inducing methemoglobin. The activity of the methemoglobin reductase is low in neonates that are very susceptible to methemoglobinemia [14].

Congenital methemoglobinemia

The disease is founded on cytochrome b_5 reductase deficiency and causes cyanosis and neurological disorders. Methemoglobinemia develops also in carriers of hemoglobin M, which is detectable by hemoglobin electrophoresis [15].

Clinical interpretation

The clinical symptoms of methemoglobinemia depend on the state of health of the individual and on the size of fraction (Table 30.4).

Table 30.4 Symptoms associated with methemoglobinemia [13].

Fraction of methemoglobin (%)[a]	Symptoms
10–20	Cyanosis, skin discoloration
20–30	Anxiety, light headedness, headache, and tachycardia
30–50	Fatigue, confusion, dizziness, tachypnea, and tachycardia
50–70	Coma, seizures, dysrhythmia, and acidosis
>70	Imminent death

[a] In relation to total hemoglobin.

For treatment, oxygen is insufflated and as antidote methylene blue or toluidine blue is administered to reduce methemoglobin to hemoglobin.

Note: Blood gas analyzers (without oximeter) calculate oxygen saturation falsely, if dyshemoglobins are present. Therefore, the results are misleading, if methemoglobin, carboxyhemoglobin, or sulfhemoglobin occur at clinically relevant concentrations.

30.3 Sulfhemoglobin

30.3.1 Introduction

Sulfhemoglobin is a greenish hemoglobin derivative with a sulfur atom irreversibly bound between two carbon atoms in the periphery of the porphyrin ring. Sulfhemoglobinemia is observed during chronic application of phenacetin and sulfonamides. It is controversial, if it is also produced in poisoning from hydrogen sulfide [16–19].

30.3.2 Spectrophotometry

A special manual procedure is applied for the determination of the fraction of sulfhemoglobin as compared to total hemoglobin [20]:

Hemoglobin of a blood sample of a nonexposed individual is oxidized to methemoglobin and saturated with sodium disulfide to form sulfhemoglobin. Aliquots are used for spiking a "normal" blood sample to obtain calibrators in the range from 1 to 25% sulfhemoglobin. The portion of sulfhemoglobin in the patient's sample is calculated from the calibrating function.

More accurate results are obtained by isoelectric focusing [21].

Annotation

(1) The portion of sulfhemoglobin cannot be determined by conventional oximetry. However, it interferes with the measurement of other hemoglobin derivatives such as carboxyhemoglobin and methemoglobin. CO-hemoglobin is increased and methemoglobin is decreased. OSM3 hemoximeter (radiometer) detects the presence of sulfhemoglobin and corrects the results for carboxyhemoglobin and methemoglobin accordingly as long as the fraction of sulfhemoglobin does not exceed 10%. In this case, sulfhemoglobin is reported ">0.5%." Otherwise, it is stated that oximetry cannot be carried out.

(2) The presence of sulfhemoglobin is suspected, when in the spectrophotometric determination of methemoglobin, a marked absorbance persists after addition of cyanide (Section 30.2.2).

(3) Sulfhemoglobin is suspected, if treatment of assumed "methemoglobinemia" with methylene blue is not effective. Sulfhemoglobin, indeed, persists for the lifetime of the erythrocyte.

(4) Sulfhemoglobin is less dangerous than methemoglobin with regard to the oxygen supply to tissues, because oxyhemoglobin dissociation curve is shifted to the right facilitating oxygen release to the tissues.

References

References for Section 30.1

1 Moffat, A.C., Osselton, M.D. and Widdop, B.(eds) (2004) *Clarke's Analysis of Drugs and Poisons*, 3rd edn, Pharmaceutical Press, London, p. 19.

2 Flanagan, R.J., Braithwaite, R.A., Brown, S.S., Widdop, B. and de Wolff, F.A. (eds) (1995) *Basic Analytical Toxicology*, World Health Organization, Geneva, pp. 96–100.

3 Thomas, L. (1998) Hemoglobins, in *Clinical Laboratory Diagnostics* (ed. L. Thomas), TH-Books, Frankfurt/M, pp. 477–479.

4 Werner, B. (1978) Inter- and intra-individual variation of carbon monoxide production in young healthy males. *Scandinavian Journal of Clinical and Laboratory Investigation*, **38**, 199–202.

5 Coburn, R.F., Williams, W.J. and Kahn, S.B. (1966) Endogenous carbon monoxide production in patients with hemolytic anemia. *The Journal of Clinical Investigation*, **45**, 460–468.

6 Goldsmith, J.R. and Landaw, S.A. (1968) Carbon monoxide and human health. *Science*, **162**, 1352–1359.

7 Deutsche Forschungsgemeinschaft (DFG) (2007) *List of MAK and BAT Values 2007*. Commission for the Investigation of Health Hazards of Chemical Compounds. Report 43. Wiley-VCH Verlag GmbH, Weinheim.

8 Roughton, F.J.W. (1972) Comparison of the O_2Hb and COHb dissociation curves of human blood, in *Oxygen Affinity of Hemoglobin and Red Cell Acid–Base Status* (eds M. Rørth and P. Astrup), Munksgaard, Copenhagen.

9 Astrand, I. Ovrum, P. and Carlsson, A. (1975) Exposure to methylene chloride. I. Its concentration in alveolar air and blood during rest, and exercise and its metabolism. *Scandinavian Journal of Work Environment & Health*, **1**, 78–94.

10 Kubic, V.L. Anders, M.W. Engel, R.R. Barlow, C.H. and Caughey, W.S. (1974) Metabolism of dihalomethanes to carbon monoxide. I. *In vivo* studies. *Drug Metabolism and Disposition*, **2**, 53–57.

11 Angerer, J. and Zorn, H. (1985) Carboxyhemoglobin, in *Analyses of Hazardous Substances in biological Materials*, Vol. 1 (eds J. Angerer and K.H. Schaller), Wiley-VCH Verlag GmbH, Weinheim.

References for Section 30.2

12 Evelyn, K.A. and Malloy, H.T. (1938) Microdetermination of oxyhemoglobin, methemoglobin and sulfhemoglobin in a single sample of blood. *The Journal of Biological Chemistry*, **126**, 655–662.

13 Woolf, A.D. and Wright, R.O. (2004) Methemoglobinemia, in *Medical Toxicology*, 3rd edn (ed. R.C. Dart), Lippincott, Williams and Wilkins, Philadelphia, PA.

14 Shearer, L.A., Goldsmith, J.R., Young, C., Kearns, O.A. and Tamplin, B.J. (1972) Methemoglobin levels in infants in an

area with high nitrate water supply. *American Journal of Public Health*, **62**, 1173–1180.

15 Keitt, A.S. (1972) Hereditary methemoglobinemia with deficiency of NADH-methemoglobin reductase, in *The Metabolic Basis of Inherited Disease* (eds J.B. Stanbury, J.B. Wyngaarden and D.S. Fredrickson), McGraw-Hill, New York, pp. 1389–1397.

References for Section 30.3

16 Smith, R.P. and Gosselin, R.E. (1979) Hydrogen sulphide poisoning. *Journal of Occupational Medicine*, **21**, 93–97.

17 Caravati, E.M. (2004) Hydrogen sulfide, in *Medical Toxicology*, 3rd edn (ed. R.C. Dart), Lippincott, Williams and Wilkins, Philadelphia, PA.

18 Adelson, L. and Sunshine, I. (1966) Fatal hydrogen sulfide intoxication. *Archives of Pathology*, **81**, 375–380.

19 Baselt, R.C. (2004) *Disposition of Toxic Drugs and Chemicals in Man*, 7th edn, Biomedical Publications, Foster City, CA.

20 Nørgaard-Pedersen, B., Siggaard-Andersen, O. and Rem, J. (1972) Hemoglobin pigments. Spectroscopic determination of oxy-, carboxy-, met- and sulfhemoglobin in capillary blood. *Clinica Chimica Acta*, **42**, 85–100.

21 Park, C.M. and Nagel, RL. (1984) Sulfhemoglobinemia: clinical and molecular aspects. *The New England Journal of Medicine*, **310**, 1579–1584.

31
Various Drugs and Toxic Agents

31.1
Chloroquine

M. Geldmacher-von Mallinckrodt and H. Käferstein

Introduction

Chloroquine is available in formulations such as tablets, syrup, and solutions for injection. It is used as an antimalarial agent, for the basic treatment of chronic and juvenile polyarthritis, systemic lupus erythematodes, porphyria cutanea tarda, polymorphous light eruption, solar urticaria, and chronic cutaneous vasculitis unresponsive to other therapy [1].

Chloroquine poisoning is rather rare in Europe, but nevertheless, reports on intoxications even with fatal outcome have been published repeatedly in the past. Thirteen deadly intoxications from suicides have been documented by the poison information center in Marseille for the southeast of France during January 1993 and December 2002 [2]. In Africa and the Pacific region, however, chloroquine is the second most frequent taken drug for suicide following pesticides [3]. If chloroquine poisoning is suspected from symptoms or setting, it should be proven by toxicological analysis as quickly as possible. A screening procedure for general unknown should include chloroquine [4].

31.1.1
Chromatographic Procedures: HPLC and GC

The HPLC and GC procedures as described in Chapter 12 are appropriate for the detection of chloroquine.

31.1.2
Thin-Layer Chromatography

Outline

Chloroquine is extracted from urine or stomach contents at alkaline pH by chloroform. After thin-layer chromatography, the plates are exposed to UV light (366 nm)

Clinical Toxicological Analysis: Procedures, Results, Interpretation. Edited by Wolf-Rüdiger Külpmann

ISBN: 978-3-527-31890-2

and evaluated for the presence of fluorescent bands and sprayed with iodoplatinate and evaluated for colored bands.

Specimens

Urine 10 ml, stomach contents 5 ml.
Storage: approximately 5 °C (refrigerator), −20 °C for prolonged period.

Equipment

Centrifuge tubes with standard ground joint and conical bottom, 10 and 30 ml
UV lamp (366 nm)
Heating block
Devices for thin-layer chromatography.

Chemicals

Ammonia solution (25%)
Chloroform
Chloroquine diphosphate (e.g., Sigma-Aldrich)
Ethyl acetate
Hexachloroplatinic (IV) acid solution (10%)
Hydrochloric acid (1 mol/l)
Methanol
Nitrogen
Potassium iodide
Sodium hydroxide (10 mol/l)
Sodium sulfate, anhydrous.

Reagents

Silica gel thin-layer plates, layer thickness 0.25 mm. Not to be activated before use.
Mobile phase: ethyl acetate/methanol/ammonia solution 85 + 10 + 5 (v/v/v). To be prepared just before use.

Iodoplatinate spray solution: Hexachloroplatinic acid (1 ml) and potassium iodide (2 g) are dissolved in aqua bidest (60 ml) by shaking the mixture vigorously in an Erlenmeyer flask. If stored in the dark at 4 °C, the solution is stable for at least 2 weeks.

Chloroquine solution 1 g/l: 16 mg chloroquine diphosphate is dissolved in 10 ml methanol.

Negative control sample: Drug-free pooled urine.

Positive control sample (chloroquine 10 mg/l): Drug-free pooled urine/chloroquine solution 99 + 1 (v/v).

Sample preparation

The sample (e.g., 5 ml urine) – and the control samples alike – is adjusted to pH 11–12 (pH strip test) by adding a few drops of sodium hydroxide solution. It is extracted with

25 ml chloroform. After centrifugation, the organic phase is transferred to a centrifuge tube, dried with sodium sulfate and evaporated with a stream of nitrogen or by rotary evaporator.

TLC analysis

Thin-layer chromatography: The dry residues are dissolved with 100 µl methanol. Different volumes are applied separately to a plate (5, 10, and 50 µl) and additionally 10 and 20 µl of the chloroquine solution. The plates are developed with the mobile phase in a saturated chamber. The distance of solvent front from origin should be approximately 10 cm at the end of development.

Detection: The still humid plate is immediately exposed to UV light (366 nm), and fluorescent zones are marked with a pencil. The procedure is repeated after air-drying of the plate. Then, it is sprayed with iodoplatinate solution (use hood). Blue/gray/brown colored zones are marked with a pencil [5].

Evaluation: A violet fluorescent zone at hR_f 46, identical for sample and standard lane present on the humid plate, is an indication of the presence of chloroquine. Typically, the zone changes to a blue/gray/brown color after spraying [5].

Negative control sample: At hR_f 46, a fluorescent and colored zone should be absent.

Positive control sample: At hR_f 46, a fluorescent (humid plate) and a colored zone should be present markedly, similar to the lane of the standard.

Analytical assessment

Sensitivity: The detection limit is 0.1 µg chloroquine/spot for investigation by fluorescence and 0.2 µg chloroquine/spot for iodinate spray solution and practical sensitivity (Section 6.2) is 1 and 2 µg chloroquine/spot, respectively.

Specificity: If a fluorescent zone at hR_f 46, which is typically colored after spraying, is observed, the presence of chloroquine is probable. The hR_f value of quinine is 45 [5] and exhibits light blue fluorescent only on the dry plate. Several other basic compounds show a color reaction with iodoplatinate. A comprehensive collection of hR_f values is available including other mobile phases [5]. Positive findings should be confirmed by a different method (see Section 31.1.1).

Practicability

The TLC procedure is an easy approach for the detection of chloroquine. Analysis time is approximately 2 h, and technician time is 35 min.

Annotations

(1) Chloroquine is a white or slightly yellow powder that is scarcely soluble in water, but well soluble in chloroform or diethyl ether. Chloroquine diphosphate is well soluble in water and poorly soluble in chloroform. Chlorochin, resochin, and weimerchin contain chloroquine diphosphate, and also for other preparations, water-soluble salts, for example, sulfate or dihydrochloride are used. Scene residues, such as tablets can be dissolved in water and examined as described

for urine and stomach contents. Other suspicious material should be extracted by chloroform in addition.

(2) The given hR_f values refer to 20 °C. For markedly different temperatures, the hR_f value of the chloroquine standard solution should be used for comparison. Alternatively, the corrected hR_f value hR_f^c (Section 4.2) that is related to four reference substances may be applied for identification.

(3) Chloroquine concentration in different parts of hairs reflects period of application and respective doses [6].

31.1.3 Medical Assessment and Clinical Interpretation

Twenty milligrams of chloroquine/kg body mass is considered a toxic dose for adults, 30 mg/kg may be lethal, and 40 mg/kg is usually fatal, if not treated very soon [7].

A 25-year-old man ingested 10 g but survived by early and effective treatment [8]. The IV injection of 250 mg in a 42-year-old man had a fatal outcome [9].

For children, the toxic dose is 20 mg/kg body mass, too. Several children below 3 years died from doses between 300 and 2000 mg [10–12].

Serum concentrations [13, 14]	
Therapeutic range	0.02–0.2 mg/l
Toxicity	from 0.5–1 mg/l
Comatose/lethal	from 3 mg/l
Plasma elimination half-life (dose-dependent)	72–576 h [15–18]
Volume of distribution	132–261 l/kg [15]

Urine: In acute chloroquine intoxication with fatal outcome, a chloroquine concentration of 462 mg/l and a monodesethylchloroquine concentration of 140 mg/l were measured (by NMR spectroscopy) [19].

The bioavailability of chloroquine diphosphate is approximately 100% for oral ingestion as well as for subcutaneous and intramuscular application. Also, 55–60% of plasma chloroquine is bound to proteins.

Concentrations in erythrocytes are in the range of 1–23 times plasma concentration. The concentration in several tissues exceeds plasma concentration 200–700 times: eyes, liver, kidneys, spleen, lungs, and leukocytes. Chloroquine is metabolized mainly to mono-desethylchloroquine, di-desethylchloroquine, and 4-amino-7-chloroquinoline. Fifty percent of a dose is excreted in the urine, partly unchanged (70%) or metabolized (e.g., desethylchloroquine 23%) [1, 15–19]. Chloroquine crosses the placenta and is present in human breast milk. When chloroquine administration is discontinued after application of therapeutic doses during 14 days, plasma concentrations decreases gradually at an elimination half-life of approximately 9

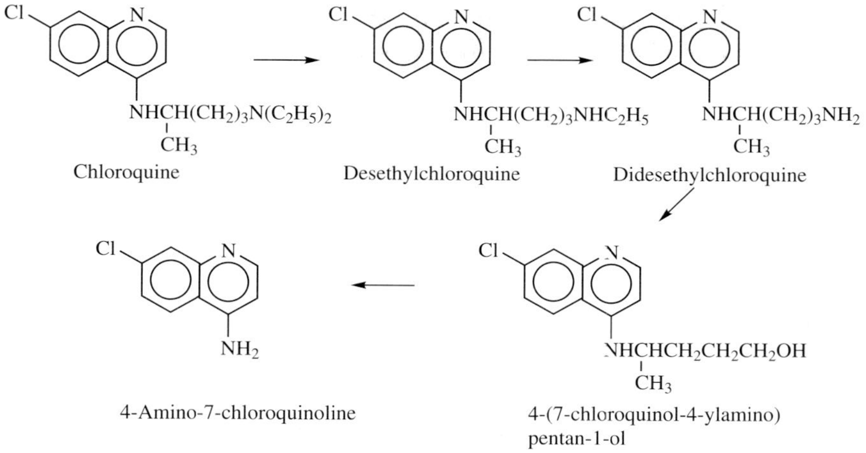

Cl N NHCH(CH2)3N(C2H5)2 CH3 Chloroquine → Cl N NHCH(CH2)3NHC2H5 CH3 Desethylchloroquine → Cl N NHCH(CH2)3NH2 CH3 Didesethylchloroquine

Cl N NH2 4-Amino-7-chloroquinoline ← Cl N NHCHCH2CH2CH2OH CH3 4-(7-chloroquinol-4-ylamino) pentan-1-ol

Figure 31.1 Chloroquine metabolism.

days during 4 weeks. Afterward, half-life is considerably increased and small amounts of chloroquine may be observed in the urine even after years [15] (Figure 31.1).

Chloroquine is a very frequently used effective drug for the prevention and therapy of malaria. It is a blood schizonticide and a gametocide possibly by inhibition of nucleic acid synthesis. Unfortunately, plasmodium falciparum seems to become increasingly resistant to chloroquine [15].

For intoxications, the cardiotoxic effects of chloroquine are most important. A dose of 2–5 g chloroquine can be lethal within 2–3 h after application by developing ventricular fibrillation, cardiac arrest, and apnea. Poisoning may be associated with hypokalemia [20]. Methemoglobinemia is observed especially in the presence of glucose-6-phosphate deficiency [7].

Already 30 min after oral ingestion of a toxic dose, the first symptoms of acute poisoning may be present: nausea, vomitus, diarrhea, and headache. Later on, the cardiotoxic effects are observed as well as respiratory depression or arrest. Visual disorders are observed frequently [7]. Intravenous injection of chloroquine may cause a possibly lethal drop in blood pressure [18]. In hypokalemia, the severity of poisoning and potassium concentration are related inversely.

Therapy is guided by the clinical symptoms and possibly application of diazepam [21].

31.2 Nicotine

H. König

Introduction

Nicotine is the major toxic alkaloid of the tobacco plant (*Nicotiana tabacum* L.) that has been indigenous to North America, Central America, and South America, but is now cultivated worldwide. In the beginning, tobacco has been confined mainly to ritual purposes. Meanwhile, huge amounts are being smoked daily probably supported by the addictive properties of tobacco and the use of tobacco has become a serious problem in health care. About one third of the adult population of Germany is smoking (Chapter 1). Apart from nicotine, several other toxic compounds are released and inhaled during smoking, for example, carbon monoxide, ammonia, cyanide, heavy metals, and polycyclic aromatic hydrocarbons with cancerigenic properties. It should be realized that nonsmokers may be exposed to these hazards by passive smoking.

31.2.1 Immunoassay

An immunoassay for the determination of cotinine in urine is commercially available (e.g., Abbott). The procedure is described in detail in the package inserted. Interferences from endogenous compounds or other toxicologically relevant drugs and their metabolites have not been described.

31.2.2 High-Performance Liquid Chromatography

Nicotine and its metabolite cotinine can be detected and determined by the HPLC procedure of Section 13.1.1. For identification, the retention times related to 5-*p*-methylphenyl-5-phenylhydantoin can be used: nicotine: 0.022 and cotinine: 0.036 [23].

31.2.3 Gas Chromatography – Mass Spectrometry

Nicotine and cotinine can be detected by the procedure of Section 13.1.3. The compounds are identified unambiguously by the attached mass spectrometer (Table 31.1).

Table 31.1 Nicotine: gas chromatography – mass spectrometry.

Substance	RRT_{OV-1} [a]	RRT_{OV-17} [b]	Retention index [26]	Principal ions *m/z* [27]
Cotinine	0.645	0.655	1715	98, 176
Nicotine	0.301	0.301	1380	84, 133

[a] Retention time related to methaqualone on column OV-1. Retention time of methaqualone: 7.6 min.
[b] Retention time related to methaqualone on column OV-17. Retention time of methaqualone: 9.3 min.

31.2.4
Medical Assessment and Clinical Interpretation

Serum concentrations [24]	
Nonexposed persons	
Nicotine	0.005–0.02 (–0.03) mg/l
Nicotine + cotinine	0.025–0.35 mg/l
Toxicity	
Nicotine	from 0.4 (–1.0) mg/l
Nicotine + cotinine	from 0.3–1.0 mg/l
Comatose/lethal	from 5 mg/l
Plasma elimination half-life	
Nicotine	1–4 h
Cotinine	16–20 h

Nicotine is easily accessible from tobacco, for example, cigarettes, cigars, and snuff. It may be used in agriculture and gardening as a pesticide and in medicine as transdermal patches. Nicotine is absorbed quickly from the gastrointestinal tract after oral ingestion. More than 90% is absorbed in the respiratory tract during smoking. Nicotine is oxidized to cotinine (Figure 31.2) that may be further degraded. Only 4–10% of a dose is excreted unchanged in urine (the lesser the percentage at alkaline pH, the higher the percentage at acidic pH), the major part as cotinine and the rest as metabolites. Forty milligrams of nicotine (equivalent to four cigarettes) peroral is considered a lethal dose for adults, 10 mg nicotine peroral for children.

The symptoms of nicotine poisoning are markedly dose dependent. Mild intoxications are associated with nausea, agitation, headache, confusion, salivation, and

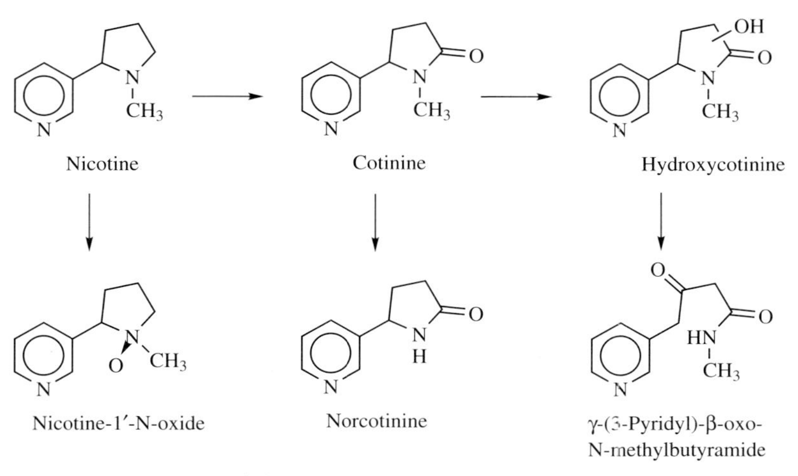

Figure 31.2 Nicotine metabolism.

Table 31.2 Cotinine: concentration in urine.

	Cotinine (μg/l)
Smoker	1000–2500
Nonsmoker	
Nonexposed	1.7
Exposed	2.6
Restaurant stuff	≤5.6
Discotheque stuff	≤24
Nightclub stuff	≤45

possibly tremor. A further sign may be miosis that later changes to mydriasis. In serious poisoning, severe abdominal pain, diarrhea, and diaphoresis are observed. Tachycardia and hypertension are followed by bradycardia and hypotension and possibly cardiac collapse. Fasciculations and convulsions occur as well as coma and possibly cardiac or respiratory arrest.

Convulsions should be prevented or treated by, for example, i.v. diazepam. Intensive care measures including artificial respiration and possibly resuscitation should be applied [25].

The chronic exposure to smoke from tobaccos is considered hazardous for smokers and nonsmokers. Some authors claim that passive smoking may be even more problematic than active smoking, because nonsmokers inhale cold smoke (500 °C) that contains more toxic compounds and more particles as cold smoke does not pass through the tipped cigarette.

In Germany, 110 000 deaths/year are attributed to active smoking (total population 82 million including children) and 3000 deaths/year to passive smoking (roughly twice the deaths from abuse of illicit drugs). In addition, the teratogenic properties of smoking during pregnancy have to be considered.

Smokers can be distinguished from nonsmokers by the amount of cotinine excreted in urine [22] (Table 31.2).

31.3 Strychnine

M. Geldmacher-von Mallinckrodt and H. Käferstein

Introduction

Strychnine (Figure 31.3) is a potent central nervous stimulant, and strychnine preparations have been used as a tonic to improve circulation and muscle tone. It has been applied for the treatment of the rare nonketotic hyperhyperglycinemia [28]. Currently, strychnine is considered as obsolete in many countries except for homeopathic preparations. In some countries, it is, however, still in use. Some preparations that are applied in veterinary medicine contain this alkaloid and furthermore, it is present in baits for the extermination of rodents. In the past, cocaine, heroin, LSD,

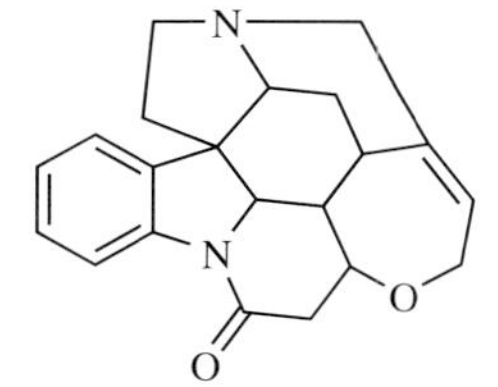

Figure 31.3 Strychnine.

and other drugs have been adulterated occasionally with strychnine [29]. Reports on poisoning have been published in recent years, too [30–32].

31.3.1 Chromatographic Procedures: HPLC and GC

Strychnine can be detected in urine by the GC–MS procedure, which is described in detail in Chapter 12.

31.3.2 Thin-Layer Chromatography

Outline

Urine or stomach contents are extracted with chloroform at alkaline pH. The extract is applied to the plate, and after development, strychnine is identified by its color after spraying Mandelin's reagent and by the respective R_f value.

Specimens

Urine 10 ml; stomach contents 5 ml.
Storage: approximately 5 °C (refrigerator), long-term −20 °C.

Equipment

Centrifuge tubes with standard ground joint and conical bottom, 10 and 30 ml
Heating block
Devices for thin-layer chromatography
Silica gel plates (layer thickness 0.25 mm), not to be activated before use
UV lamp (254 nm).

Chemicals

Ammonia solution (25%)
Ammonium vanadate
Chloroform
Ethyl acetate
Methanol
Nitrogen
Sodium hydroxide (10 mol/l)
Sodium sulfate, anhydrous

Strychnine (e.g., Fluka, Sigma)
Sulfuric acid (95–97%).

Reagents
Mobile phase: Ethyl acetate/methanol/ammonia solution 85 + 10 + 5 (v/v/v). To be prepared just before use.

Mandelin's reagent: 50 mg ammonium vanadate is dissolved with 50 ml sulfuric acid.

Strychnine standard solution 2 g/l: 20 mg strychnine is dissolved with 10 ml chloroform.

Negative control sample: Drug-free pooled urine.

Positive control sample: 4 mg strychnine/l.

10 ml drug-free pooled urine is spiked with 20 µl strychnine standard solution and mixed thoroughly.

Sample preparation
The sample, for example, 5 ml urine, and the control samples alike are adjusted to pH 11–12 by adding a few drops of sodium hydroxide solution (pH strip test). It is extracted with 20 ml chloroform. After centrifugation, the chloroform layer is transferred to a centrifuge tube, dried with sodium sulfate and evaporated with a stream of nitrogen.

TLC analysis
Thin-layer chromatography: The dry residues are dissolved with 250 µl chloroform. Different volumes are applied separately to a plate: 20 and 50 µl of the dissolved residues and 2 µl of the strychnine standard solution. The plate is developed with the mobile phase in a saturated chamber. The distance of solvent front from origin should be approximately 10 cm at the end of development. The plate is removed from the tank and air-dried.

Detection: The plate is exposed to UV light (254 nm) and those zones are marked with a pencil, which show marked fluorescence quenching. Then, it is sprayed with Mandelin's reagent (use hood).

Evaluation: If strychnine is present, a zone with fluorescence quenching at hR_f 32 is observed [33], which shows at first violet color with Mandelin's reagent and later changes to orange/crimson [34]. Metabolites of strychnine do not move as far as strychnine.

Negative control sample: A colored zone at hR_f 32 should not be visible.

Positive control sample: A colored zone at hR_f 32 should be visible changing from violet to orange/crimson.

Analytical assessment
Sensitivity: According to the practical sensitivity of the procedure, 0.05 µg strychnine/spot is detected in 90% of the investigated samples.

Specificity: A violet colored zone at hR_f 32 after spraying with Mandelin's reagent is a strong indication for the presence of strychnine [34], which, however, should be confirmed by another method (e.g., HPLC or GC–MS). A comprehensive collection of hR_f values including other mobile phases is available [33].

Practicability

The TLC procedure is an easy approach for the detection of strychnine. Analysis time is approximately 2 h, and technician time 35 min.

Annotations

(1) Strychnine is the major alkaloid in the ripe seeds of *Strychnos nux-vomica*, a tree indigenous to India, but it is also present in some other plants. It is a white, crystalline, and odorless powder with a rather bitter taste still at a dilution of 1 : 130 000.
(2) Scene residues, such as tablets or suspicious materials can be investigated as described after extraction with chloroform at alkaline pH.
(3) Chloroform is most appropriate for extraction, because the solubility for strychnine is 1 g/5 ml, as compared to methanol 1 g/260 ml and water 1 g/6400 ml. Diethyl ether or petrol ether are not at all suitable for dissolution [34].
(4) A positive finding in urine can be expected only 1 or 2 h after oral ingestion. After intake of 4.8 g and fatal outcome, the drug could not be detected in urine sample obtained 1 h after intake [35].

31.3.3
Medical Assessment and Clinical Interpretation

Strychnine is very toxic and an oral dose of 10–15 mg may be lethal for children and 30–120 mg for adults. However, poisoning with higher doses (3.5–4.8 g) has been survived, if treated appropriately and timely [36].

Serum concentrations [37]	
Therapeutic range	no data
Toxicity	from 0.075–0.100 mg/l
Comatose/lethal	from 0.2–2.0 mg/l
Plasma elimination half-life	10–16 h [31, 35, 38, 39]

In deadly poisoning, 30 min after ingestion serum concentration came up to 3.8 mg/l [35]. In a nonfatal intoxication, 4.73 mg/l was measured at 1.5 h after application, which had decreased to 0.38 mg/l 72.5 h later [31]. In another report [38], a peak serum concentration of 2.1 mg/l was attained 3 h after ingestion of 2.25 g strychnine. A serum concentration of 11 mg/l was measured after parenteral application [40].

Strychnine is absorbed rapidly from gastrointestinal tract and nasal mucosa. Only a small portion is bound to plasma proteins in blood. Strychnine is metabolized in the liver [41, 42] following probably first-order kinetics [38]. Indeed, little is known about degradation in man, but according to studies in rats, major metabolites are strychnine-21,22-epoxide and 21,22-dihydrostrychnine [43]. A small part of a dose is excreted unchanged by the kidneys; the major part is eliminated with the urine after metabolic degradation.

Clinical interpretation: Strychnine is a potent central nervous stimulant and convulsant agent. The first symptoms of intoxication occur most often 15–30 min after oral ingestion. Initially, ache, stiffness, and fasciculations of the muscles for chewing and in the neck are reported, which is quickly followed by painful systemic tonic convulsions including opisthotonus and risus sardonicus at full consciousness. The convulsions are subsiding after some minutes, but are revoked already by minor stimuli. The symptoms may be associated with hyperthermia and acidosis. The patient may die from suffocation during a fit, from exhaustion after several hits, or from cardiac failure [44]. A dedicated antidote is not known and treatment is guided by symptoms. Gastric lavage is not recommended anymore and may only be carried out in the absence of convulsions.

For differential diagnosis, tetanus, rabies, meningitis, hysteria, and intoxications from phenothiazines, cocaine, phencyclidine, chlorinated hydrocarbons, and isoniazid and other compounds, which may induce myoclonus and convulsions, have to be considered.

31.4 Cyanide

31.4.1 Introduction

M. Geldmacher-von Mallinckrodt

Cyanide is an extremely toxic and rapidly acting agent. The primary sources of medical significance comprise hydrogen cyanide, inorganic cyanide salts, metallic cyanides, and organic cyanogens.

Hydrogen cyanide is used in industry for synthesis of resins (e.g., acrylate) or is produced as a by-product. It is applied as a fumigant in ships and storerooms for the extermination of rodents. It is liberated with the combustion of organic materials, for example, polyurethanes, polyacrylnitriles, wool, silk, horsehair, and tobacco. In case of fire in closed rooms, cyanide will be often present concomitantly with carbon monoxide. Hydrogen cyanide has been used as a warfare agent in the two World Wars and as Zyklon B in the concentration camps during the Holocaust.

Inorganic cyanide salts are used for the extraction of precious metals from ore or wastes (e.g., circuit boards) and for electroplating as well as metal cleaning. They liberate HCN after addition of acid.

Organic cyanogens comprise synthetic nitriles, for example, acetonitrile [45] that are metabolized and release cyanide as well as naturally occurring cyanogens. Among these are amygdalin from, for example, apricot pits, cherry pits, and bitter almonds. Cassava (see Chapter 35) contains linamarin, which may cause poisoning if not prepared appropriately. Laetrile is a semisynthetic derivative of amygdalin that may release cyanide during metabolism. Sodium nitroprusside, which is applied as an antihypertensive agent, liberates cyanide *in vivo*, too.

31.4.2
Indicator Tube Method

M. von Clarmann and M. Geldmacher-von Mallinckrodt

Outline
Sulfuric acid is added to a blood sample. Hydrogen cyanide is liberated and sucked by a pump to an indicator tube. The presence of HCN is indicated by a color change of the pertinent reaction zone.

Specimens
Venous blood, 5–10 ml anticoagulated by EDTA or heparin. Samples should be investigated as quickly as possible. During storage, cyanide concentration may increase or decrease unpredictably already within few hours [46] (Figure 31.4).

Equipment
Indicator tube for the detection of cyanide (e.g., Dräger) (Figure 31.5), to be stored in a refrigerator, maximum 2 years (see expiry date). Dedicated pump for indicator tube (e.g., Dräger) (Figure 31.6).

Glass reaction tube, height 6 cm, diameter 2 cm; fitting rubber stopper with two holes; fitting glass tube, length 2–3 cm (Figure 31.6).

Chemicals

Potassium cyanide (be aware of decomposition)
Sulfuric acid 100 g/l
Sodium hydroxide 1 mol/l.

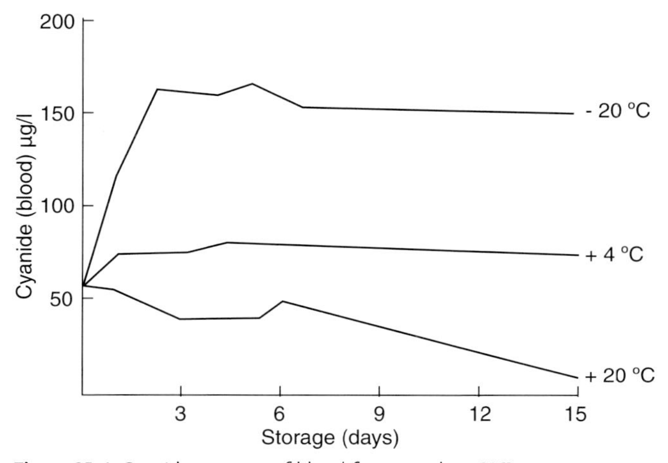

Figure 31.4 Cyanide: storage of blood from smokers [46].

Figure 31.5 Cyanide: indicator tube. 1: Melted end; 2: melted end; 3: documentation zone; 4: precedent zone (white); 5: indicator zone (yellow) and scale (number concentration 10^{-6}, equivalent to ppm); 6: cover (red); 7: indicating arrow.

Reagents

Cyanide solution 100 mg/l: 25 mg potassium cyanide (equivalent to 10 mg cyanide) is dissolved with sodium hydroxide solution. The solution is stable for 3 months if stored in a refrigerator.

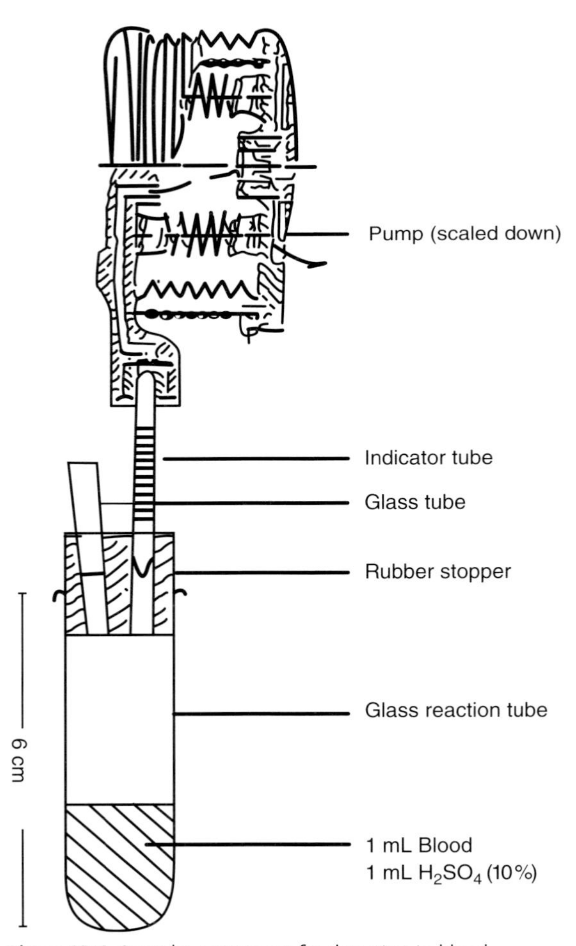

Figure 31.6 Cyanide: equipment for detection in blood.

Negative control sample: Blood of a person not exposed to cyanide.

Positive control sample 5 mg/l: 1 ml blood of a nonexposed person in a glass reaction tube is spiked with 50 μl cyanide solution.

Sample preparation
Not applicable.

Indicator tube analysis
The procedure is described in detail in the package insert of the manufacturer. The pump is checked for tightness by attaching an original indicator tube.

Afterward, the two ends of the original indicator tube are broken off. One end is put tightly in the free hole of the rubber stopper, whereas the other end is attached to the pump with the indicating arrow pointing to the pump (Figure 31.6).

One milliliter sample and 1 ml sulfuric acid are pipetted into the glass reaction tube that is immediately closed with a finger, followed by pumping once. The finger is removed and the pump moves back to its starting position. The whole procedure is repeated 14 times. The mixture in the glass reaction tube may not be sucked. Otherwise, false-positive results are obtained.

Immediately after the last pumping, the indicator zone of the indicator tube is evaluated. In the presence of cyanide, the following reaction takes place:

$$2HCN + HgCl_2 \rightarrow 2HCl + Hg(CN)_2.$$

HCl induces a change of color of the indicator methyl red from yellow to red.

Evaluation *Negative control sample:* Color of the indicator zone should not change.

Positive control sample: Color of the indicator zone should change markedly to red.

Analytical assessment
Sensitivity: The sensitivity of the procedure was estimated for EDTA blood:

Maximal sensitivity: 1.0 mg cyanide/l blood.
Practical sensitivity: 2.3 mg cyanide/l blood.

Specificity: The zone preceding the indicator zone brings about that hydrochloric acid, SO_2, and dihydrogen sulfide, even if present at a considerable concentration, do not cause a change of the indicator color. Sodium azide or the mixture in the reaction tube, if sucked in, will produce false-positive results.

Practicability
The total procedure takes approximately 10 min.

Annotations

(1) The procedure is appropriate for the detection of cyanide in blood. It can be used in the field by the fire brigades of a community or in hazardous environments for example, industry, mining, and shipping.

(2) A rough estimation of the cyanide concentration can be carried out after calibration with spiked blood samples and use of the tube scale.

(3) The procedure should be applied to samples that have been obtained before treatment of cyanide poisoning with methemoglobin-inducing agents. Sulfuric acid that is used in the procedure liberates cyanide from methemoglobin. Although bound cyanide does not exhibit toxic effects, it will contribute to the reaction. Dosage of methemoglobin-inducing agents should be guided only by the concentration of unbound cyanide.

31.4.3 Paper Strip

A. Scholer

Outline

A dedicated test strip is used for the detection of hydrogen cyanide and cyanide in aqueous solutions. Hydrogen cyanide is liberated from the sample by addition of sulfuric acid.

$$2NaCN + H_2SO_4 \rightarrow 2HCN + Na_2SO_4.$$

HCN reacts with *N,N,N′,N′*-tetramethyl-4,4′-diaminodiphenylmethane of the test strip. The strip changes its color from slightly green to slightly blue or even markedly blue depending on the hydrogen cyanide concentration. The strip shows highest sensitivity at the border of a weak sulfuric acid solution and the gas phase. It does not detect cyanide in alkaline solutions.

Specimen

EDTA blood, 2 ml.
Sampling tube should be filled rather completely and closed gastight. The examination should be carried out immediately; otherwise, the sample should be stored at 4 °C.

Equipment

Polyethylene tubes (3.5 ml) and appropriate stoppers (e.g., Sarstedt, part no. 55 484) alternatively polypropylene tubes (Figure 31.7).
Hood.

Chemicals

Sodium cyanide (e.g., Merck)
Sulfuric acid (3.6 mol/l), at least 6 months stable at room temperature.

Reagents

Test strip CYANTESMO (Macherey and Nagel), to be stored gastight.

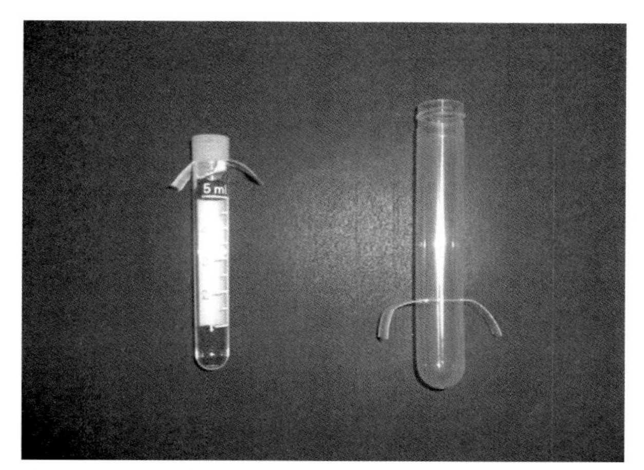

Figure 31.7 Cyanide: paper strip procedure. Left: Polyethylene tube with stopper and test strip in place, right: polypropylene tube with cut and test stripe in place.

Cyanide solution 50 mg/l: 9.42 mg sodium cyanide (equivalent to 5.00 mg cyanide) is dissolved with 100 ml aqua bidest. The solution is stable at room temperature for 6 months. During the preparation of the solution under the hood, the technician should wear protective gloves and mask.

Negative control sample: EDTA blood from nonexposed person or quality control material.

Test strip analysis

The procedure is carried out under the hood according to Table 31.3. The reaction mixtures are transferred to the reaction tubes. The following reaction tubes can be used alternatively: A) Polyethylene tubes (3.5 ml). The bottoms of the stoppers are cut and discarded. B) Polypropylene tubes (Figure 31.7). They are partly cut at 0.5 cm above the level of the reaction mixture.

After addition of sulfuric acid, the solutions are mixed vigorously (Vortex). After transfer of the reaction mixture 3–4 cm of the CYANTESMO test strip is put to every reaction tube, either under the stopper (A) or in the cut (B) (Figure 31.7). The tubes

Table 31.3 Test strip procedure: pipetting scheme.

	Negative control 0 mg/l	**Positive control 1 0.5 mg/l**	**Positive control 2 1.0 mg/l**	**Positive control 3 3.3 mg/l**	**Sample**
Negative control sample (μl)	1500	1500	1500	1500	0
Cyanide solution (μl)	0	15	30	100	0
Mix thoroughly					
Sample (μl)	0	0	0	0	1500
Sulfuric acid (μl)	1500	1500	1500	1500	1500

Mix vigorously (Vortex). For further procedure, see text.

are kept in a water bath at 37 °C. The color of the strips is evaluated after 15, 30, 60, and possibly 120 min (if the results are still ambiguous after 60 min).

Samples producing a strip color as control 1 or more intensive are considered positive. The concentration of a sample may be estimated by comparison to the three controls. Positive findings should be confirmed by another method.

Evaluation: No change of color should be observed for the negative control sample. Positive control samples should exhibit a marked color change increasing from control 1 to control 3.

Analytical assessment

Sensitivity: A positive finding is obtained for cyanide concentrations $\geq$1.5 mg/l within 30 min. For concentrations in the range of 1.0–1.5 mg/l, a positive result can be expected for more than 90% of the samples.

In a comparative study with a spectrophotometric procedure ($n = 22$), usually a positive finding was obtained with the test strip for cyanide concentrations exceeding 1 mg/l [47].

Practicability

The procedure can be carried out within 75 min.

Annotations

(1) The major part of cyanide is in the erythrocytes. Therefore, blood is preferred to plasma, which should be used only exceptionally.

(2) According to our experiments, the cyanide portion liberated from urine is higher than the portion from blood.

(3) The given volumes for pipetting (Table 31.3) fit the size of the polyethylene tubes. For reduced volumes, the size of the tubes has to be scaled down accordingly. The volume of the gas phase above the blood–sulfuric acid mixture shall be sufficiently small to allow an intensive reaction with the test strip. However, if it is too small, the test strip may come into contact with the mixture and invalidate the result (see next item).

(4) Sulfuric acid solutions in touch with the test strip may destroy the reagent. Hexacyanoferrate(II), thiocyanate, thiosulfate, and chlorine interfere with the reaction.

31.4.4
Absorption Spectrophotometry

T. Daldrup

Introduction

Several methods are available for the determination of cyanide. Among these are fluorescence spectrometry [48], gas chromatography with NP sensitive detector [49], gas chromatography–mass spectrometry following formation of chlorocyane [50], and potentiometry by cyanide selective electrode [51–54] (see Section 31.4.5). In the following, absorption spectrometry following color reaction is used for cyanide determination.

Outline

Hydrogen cyanide is released from the sample by sulfuric acid and absorbed by sodium hydroxide solution after microdiffusion. Diffusion is enhanced by addition of zinc granules to generate hydrogen. The cyanide concentration of the sodium hydroxide solution is determined by absorption spectrophotometry after reaction with pyridine/barbituric acid [55] using standard addition.

Specimen

Blood, 5 ml, anticoagulated by heparin or EDTA.

Minimum blood volume required for a determination including two standard additions is 1.5 ml.

Sampling should precede antidote therapy (Section 31.4.2: annotation 3).

Storage temperature: 4 °C (Figure 31.4) [56], but preferably determination should be carried out immediately.

Equipment

Spectrophotometer or spectral line photometer (Hg 578 nm)
Cuvettes: special optical glass or polystyrene, 3 ml, light path 1 cm
Widmark flasks (Figure 31.8).

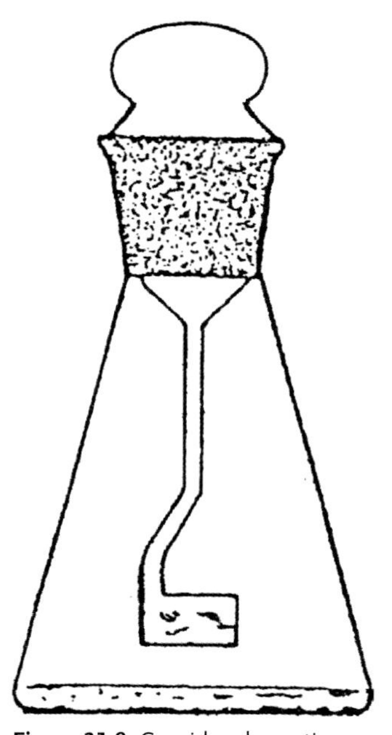

Figure 31.8 Cyanide: absorption spectrophotometry. Widmark flask for HCN absorption.

Chemicals

Barbituric acid
Chloramine-T trihydrate
Hydrochloric acid, concentrated
Potassium cyanide
Pyridine: If not colorless, to be distilled. Storage in the dark.
Sodium dihydrogen phosphate dihydrate
Sodium hydroxide solution 0.1 mol/l
Sulfuric acid 100 g/l
Zinc granules

Reagents

Phosphate solution 1 mol/l: 156 g sodium dihydrogen phosphate are dissolved with 1000 ml aqua bidest. Stable at room temperature for at least 4 weeks.

Chloramine-T solution: 250 mg chloramine-T is dissolved with 100 ml aqua bidest. To be prepared just before use.

Pyridine–barbituric acid reagent: In a 50 ml volumetric flask, 3 g barbituric acid is dissolved in 15 ml pyridine and 3 ml hydrochloric acid. The mixture is made up to 50 ml by aqua bidest, mixed well and filtrated. To be prepared just before use.

Cyanide solution 50 mg/l: 12.5 mg potassium cyanide (equivalent to 5 mg cyanide) is dissolved with 100 ml sodium hydroxide solution. Stable for several days at 4 °C.

Negative control sample: Blood anticoagulated by heparin or EDTA of a nonsmoker not exposed to cyanide or cyanogenic compounds.

Positive control sample 1 mg cyanide/l: Negative control sample (4.9 ml) spiked with 100 µl cyanide solution. To be prepared just before use.

Spectrophotometrical analysis

See pipetting scheme in Table 31.4.

The standard ground joint of the Widmark flask shall be greased with silicone grease before starting.

After closing, the stopper is fixed with two springs. During all handling, the sodium hydroxide solution in the small bowl of the Widmark flask may not be sloped.

In the presence of cyanide, a red-violet color develops after addition of the pyridine–barbituric acid reagent.

Calculation of results: A graph is constructed:

x-axis: Cyanide added to blood (0.5 and 1.0 µg).
y-axis: Absorbance.

The data points are connected by a line that crosses the *x*-axis at the point reflecting the negative value for cyanide mass in 0.5 ml blood.

The value is multiplied with −2 to give the cyanide concentration in the blood sample (mg/l).

If a straight line connecting the data points cannot be drawn, the assay failed and has to be repeated.

Table 31.4 Cyanide: absorption spectrophotometry; pipetting scheme.

	Reagent blank (µl)	Sample (µl)	Sample + addition 1 (µl)	Sample + addition 2 (µl)	Control sample (µl)	Control sample + addition 1 (µl)	Control sample + addition 2 (µl)
Flask number	1	2	3	4	5	6	7
Into flasks:							
Blood sample	—	500	500	500	—	—	—
Positive control sample	—	—	—	—	500	500	500
Aqua bidest	3500	3000	3000	3000	3000	3000	3000
Zinc granules, number	3	3	3	3	3	3	3
Sodium hydroxide solution	20	20	10	—	20	10	—
Cyanide solution	—	—	10	20	—	10	20
Into bowl of stoppers:							
Sodium hydroxide solution	500	500	500	500	500	500	500
Into flasks:							
Sulfuric acid	500	500	500	500	500	500	500
Immediately after the addition of sulfuric acid, the flask is closed tightly. The solutions are mixed so thoroughly that the sodium hydroxide solution in the bowls does not slop over. The flasks are kept for 3 h at room temperature.							
Into centrifuge tubes:							
Bowl solution	400	400	400	400	400	400	400
Phosphate solution	800	800	800	800	800	800	800
Chloramine-T solution	400	400	400	400	400	400	400
After 2–3 min							
Pyridine–barbituric acid reagent	1200	1200	1200	1200	1200	1200	1200
After 10 min absorbance is measured against reagent blank at 585 nm or Hg 578 nm.							

Table 31.5 Determination of cyanide by absorption spectrophotometry: imprecision in the series.

	n	$\bar{x}$ (mg/l)	s (mg/l)	CV (%)
Aqueous solution	5	1.80	0.04	2.2
Blood	5	1.58	0.10	6.3

n, number of measurements; $\bar{x}$, mean; s, standard deviation; CV, coefficient of variation.

Analytical assessment

Precision: Coefficient of variation for imprecision within the series was 6.3% (Table 31.5).

Trueness: The linearity ranges from 0 to 4 mg/l for blood. Cyanide added to blood was recovered by approximately 100% considering the sample blank.

Specificity: Specificity of the procedure is increased as compared to the procedure given in Section 31.4.3 by separation of hydrogen cyanide by microdiffusion from the sample before color reaction. However, hydrogen sulfide and SO_2 may interfere if present in markedly increased concentration. They may be absorbed by the sodium hydroxide solution and reduce chloramine-T in the same way as hydrogen cyanide.

Sensitivity: The detection limit is below 0.1 mg cyanide/l.

Practicability

Time of analysis is about 4 h, and technician time is approximately 1 h.

Annotations

(1) If the absorption exceeds 2 due to the presence of high cyanide concentrations, the solution ready for measurement should not be diluted. However, 50 µl of the remaining sodium hydroxide solution in the little bowl is used for repetition of the color reaction.

(2) As the procedure is rather time consuming, it is recommended to carry out the screening procedures as described in Sections 31.4.2 and 31.4.3, if a serious acute intoxication is suspected. Any delay in the treatment of acute poisoning should be avoided.

(3) Similar procedures use Conway dishes instead of Widmark flasks. Conway dishes are difficult to obtain commercially nowadays.

31.4.5
Potentiometry

F. Pragst and M. Geldmacher-von Mallinckrodt

Outline

Cyanide is released from the sample by sulfuric acid and absorbed by sodium hydroxide solution after microdiffusion. Cyanide concentration is determined by potentiometry by the use of cyanide-selective electrodes.

Specimen
Blood, 5 ml, anticoagulated by heparin or EDTA. Sampling should precede antidote therapy (Section 31.4.2, annotation 3). Storage temperature 4 °C (Figure 31.4) [56], but preferably determination should be carried out at once.

Equipment

Widmark flask (Figure 31.8)
Magnetic stirrer
Magnetic stirrer bars, PTFE coated, length 6 mm
Cyanide-selective electrode (Ag_2S/AgI) with integrated reference electrode (e.g., Model 9606 Sure-Flow Combination Cyanide Electrode, Orion)
pH meter or (preferable) ion meter (e.g., PerpHecT Meter 370, Orion)
Sample containers, polypropylene, diameter 15 mm, capacity 13 ml, round bottom (e.g., Sarstedt no. 60.541.004PP).

Chemicals

Sodium hydroxide solution 0.1 mol/l
Sulfuric acid 1.0 mol/l
Potassium cyanide.

May decompose during storage. In case of doubt, to be analyzed by argentometric titration or to be replaced.

Reagents
Cyanide solution 100 mg/l: 25 mg potassium cyanide (equivalent to 10 mg cyanide) is dissolved with 100 ml sodium hydroxide solution. Stable for at least 6 months if kept in the refrigerator at 4 °C (Figure 31.9).

Sample preparation and potentiometric analysis
One milliliter sodium hydroxide solution is pipetted into the bowl, 2 ml blood is transferred to the bottom of the flask. Immediately after the addition of 2 ml of sulfuric acid, the flask is closed tight with the silicon-greased stopper that is fixed by springs. The flask stands 3 h at room temperature or in urgency 20 min at 40 °C (water bath). Every sample is analyzed twice. For further details, see Section 31.4.4.

The liquid in the bowl is transferred completely to a polypropylene sample container and adjusted to 2 ml by sodium hydroxide solution. A magnetic stirrer bar is added to the container that is put on to the magnetic stirrer. The cyanide-selective electrode is dipped into the mixture to a constant depth identical for all the measurements. The potentiometer is attached and is at always the same constant stirrer velocity and constant temperature, and the potential difference is recorded when a stable signal is achieved.

Linearity check A linearity check is carried out with every new electrode to define the linear range of the semi-logarithmic potential activity curve. For this purpose, the

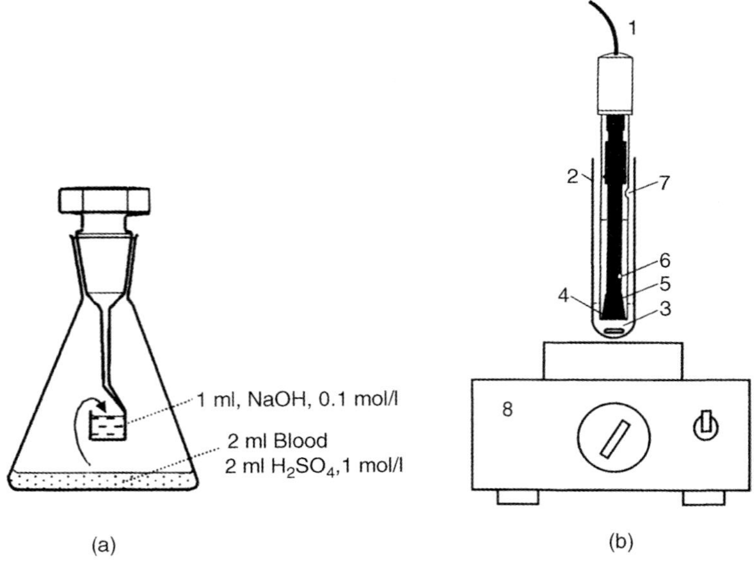

Figure 31.9 Cyanide: potentiometric determination by cyanide selective electrode (a) Sample preparation, (b) potentiometric analysis. 1: Cyanide-selective electrode; 2: polypropylene sample container; 3: magnetic stirrer bar; 4: liquid junction; 5: silver sulfide membrane; 6: reference electrode; 7: opening for filling solution; 8: magnetic stirrer.

following cyanide concentrations (μg/ml NaOH (0.1 mol/l)) are investigated: 0.01, 0.03, 0.1, 0.3, 1, 3, 10, and 30 μg/ml.

In general, there is a linear relationship at concentrations more than 0.1 μg/ml, that is, in the toxicologically most relevant range.

Two-point calibration

Measurement: 2 ml sodium hydroxide solution is spiked with 20 μl cyanide solution (concentration 0.99 μg/ml) and measured by potentiometry (CN-selective electrode).
Measurement: 2 ml sodium hydroxide solution is spiked with 200 μl cyanide solution (concentration 9.9 μg/ml) and measured by potentiometry (CN-selective electrode).

A graph is constructed on a semi-logarithmic paper:

x-axis (linear): Concentration
y-axis (logarithmic): Millivolt.

The data points should be connected and show a slope of 56–59 mV/decade, depending on the temperature.

The calibration is carried out before measurement to eliminate electrode drift. A constant signal may be attained rather slowly at low concentrations. Readings may be taken at fixed time, for example, after 60 s without awaiting constancy to save time.

Analytical assessment

Specificity: The measurement is interfered by sulfides present in decomposed blood samples.

Practicability

The procedure is as difficult as pH measurement. The technician should be knowledgeable about possible interferences in potentiometry. Time of analysis is 3.5 h and the technician time approximately 30 min. In urgency, time can be reduced considerably by incubation at 40 °C (see text) and concomitant calibration.

Annotations

(1) Sulfuric acid may be replaced by lead acetate (0.1 mol/l in 4% perchloric acid, 40 ml concentrated $HClO_4$ + 163 g lead acetate adjusted to 1000 ml with aqua bidest) to eliminate hydrogen sulfide. As an alternative, sulfide may be precipitated by a drop of saturated lead acetate after microdiffusion.

(2) Cyanide concentrations exceeding 30 μg/ml considerably deteriorate the AgI electrode and should be diluted beforehand. In general, the contact of the electrode with cyanide containing solutions should be as short as possible.

(3) The measuring area should be protected against mechanical damage. It is rinsed with aqua bidest for cleaning and dabbed cautiously with tissue wiper. It may be necessary to polish the electrode thoroughly before use.

(4) For measurement of concentrations less than 0.1 μg/ml, the dedicated commercial Ag_2S/AgI electrode is replaced by an Ag_2S electrode (e.g., Model 9616 Sure-Flow Combination Silver/Silver sulfide, Orion). The Ag_2S electrode is sensitized for cyanide by adding potassium dicyanoargentate as silver ion buffer to all the solutions before potentiometry (final concentration 10 mmol/l).

31.4.6 Medical Assessment and Clinical Interpretation

M. Geldmacher-von Mallinckrodt

Medical assessment

The gaseous hydrogen cyanide is an extremely toxic and very rapidly acting poison. Symptoms may occur within seconds and death within minutes after inhalation of sufficiently high contaminated air. The lethal number concentration is approximately 200–300/1000 000 (200–300 ppm) [57]. The typical odor of bitter almonds cannot be smelled by half of the people. Inorganic cyanide salts also exhibit extreme toxicity and are very fast acting poisons after ingestion. Already several minutes after administration, symptoms of intoxication are appearing. The lethal oral dose of potassium cyanide is 200–300 mg for adults and 1.2–5 mg/kg body mass for children [57]. After ingestion of naturally occurring cyanogens, symptoms develop less rapidly and are obvious only after hours [57].

Cyanide concentrations in blood [58]	
Nonsmoker	0.005–0.04 mg/l
Smoker	0.04–0.07 mg/l
Toxic symptoms	from 0.1–1.0 mg/l
Mild cyanide intoxication	from < 2.0 mg/l
Moderate cyanide intoxication	from 2–3 mg/l
Severe cyanide intoxication	from > 3–4 mg/l

0.95–7.68 mg/l cyanide in blood was measured in samples from nine patients suffering from acute cyanide poisoning, comatose for a short while, and recovering quickly in fresh air without application of antidotes [59].

Cyanide in the blood of not exposed nonsmokers stems from vitamin B_{12} metabolism, ingestion of cyanogenic food (Chapter 35), and air pollution. It is accumulated in the erythrocytes with only 1% of total in plasma.

Cyanide concentration in blood during treatment with sodium nitroprusside did not exceed 1.8 mg/l and was not associated with any symptoms of poisoning [60]. A good 80% of cyanide is detoxified to thiocyanate in the liver by rhodanese. The remaining cyanide is exhaled unchanged or bound to vitamin B_{12}. A small portion is metabolized to formic acid and CO_2.

Clinical interpretation

Cyanide binds avidly to the ferric ion of cytochrome oxidase. The enzyme becomes inactivated, oxidative phosphorylation is blocked and production of energy is inhibited and may lead to cell death. Acute cyanide poisoning is associated with anxiety, headache, nausea, vomitus, dizziness, hyperventilation, dyspnea, cyanosis, increase of blood pressure, bradycardia, and cardiac dysrhythmia. Seizures, hypotension, and coma may follow.

The procedures described in Sections 31.4.2 to 31.4.5 are sufficiently sensitive to detect moderate or severe cyanide poisoning.

For therapy of acute cyanide intoxication, symptomatic measures and the administration of antidotes are recommended. IPCS (International Programme on Chemical Safety of the WHO) and the CEC (Commission of the European Communities) listed the following antidotes in volume 2 “Antidotes for poisoning by cyanide” of the “IPCS/CEC Evaluation of antidotes series” [61]:

sodium thiosulfate, to provide sulfur for detoxification by rhodanese;
hydroxocobalamin, converted to vitamin B_{12} by binding cyanide;
dicobalt EDTA (Kelocyanor) to form a stable complex with cyanide;

methemoglobinemia-inducing agents (e.g., sodium nitrite, amyl nitrite, and 4-dimethylaminophenol): Fe^{3+} of methemoglobin avidly binds cyanide and inhibits competitive binding to cytochrome oxidase.

Inducing methemoglobinemia is risky (particularly if carbon monoxide is involved in addition), because it reduces the number of oxygen transporting molecules. Therefore, this treatment should be only applied if cyanide intoxication is rather probable and a blood sample for the quantitative measurement of cyanide concentration should be taken before methemoglobinemia is induced.

It should be appreciated that the procedures given in Sections 31.4.2 and 31.4.3 easily and rapidly detect clinically relevant cyanide poisoning and prevent the unnecessary induction of risky therapy. In case of a positive finding, the cyanide concentration is determined, for example, by absorption spectrometry (Section 31.4.4) or potentiometry (Section 31.4.5) to estimate the severity of poisoning and prognosis of sequelae.

In case of cyanide intoxication from combustion, the additional determination of the carbon monoxide hemoglobin fraction is mandatory.

In general, metabolic acidosis is present in clinically significant acute cyanide poisoning associated with an increased lactate concentration in plasma and increased anion gap, reflecting the degree of cellular hypoxia. The extraction of oxygen by the tissues is markedly decreased and the arterial-venous difference of blood oxygen pressure is reduced. In case of artificial respiration, the respiratory ratio as a measure of oxygen uptake should be recorded during respiration monitoring.

31.5 Fluoride

F. Pragst

31.5.1 Outline

Potentiometry, by use of a fluoride-selective electrode, is the only technique to measure easily the fluoride concentration [62–71].

The basics of potentiometric measurements are described in Section 4.8.

For fluoride determination, a LaF_3 electrode is used. It exhibits a slope according to theory of 57 mV/decade for fluoride concentrations exceeding 50 µg/l. At lower concentrations, the slope decreases steadily, and the time required to attain a constant potential increases concomitantly.

31.5.2 Potentiometry

Sample
2 ml of blood, urine, or stomach contents.

Equipment

Fluoride-selective electrode with integrated reference electrode including filling solution (e.g., Model 9609 Sure-Flow Combination Fluoride Electrode, Orion).

The fluoride electrode is stored dipped into sodium fluoride solution or conditioned by dipping into the solution some hours before measurement.

For further details on equipment, see Section 31.4.5 (cyanide).

Chemicals

EDTA disodium concentrate (Komplexon III) (Fluka, Taufkirchen, Germany)
Sodium acetate
Sodium fluoride.

Reagents

Buffer solution: Komplexon III is dissolved in 900 ml aqua bidest at room temperature until saturated. 136.1 g sodium acetate is added and the solution is made up to 1000 ml by aqua bidest.

Sodium fluoride solution I 100 mg/l: 221.1 mg sodium fluoride are dissolved in 1 l aqua bidest.

Sodium fluoride solution II 10 mg/l: Aqua bidest/sodium fluoride solution I 900 + 100 (v/v).

Fluoride calibration solutions: See Table 31.6.

Potentiometric analysis [70]

One milliliter sample is transferred to a polypropylene tube and 2 ml buffer solution is added. Fluoride electrode and reference electrode are dipped shortly below the surface of the respective solutions for measurement. An appropriate volume for the proposed tubes is 1.5 ml. The electrodes are rinsed for cleaning after every measurement and dabbed cautiously with tissue wiper. Depending on fluoride concentration, a constant potential is attained after 5–30 min. Time of measurement can be reduced if all readings are recorded, for example, after 60 s (stopwatch) without stirring.

Practicability

The procedure is as difficult as pH measurement. The technician should be knowledgeable about possible interferences in potentiometry.

One measurement takes 5–10 min depending on fluoride concentration, total analysis (sample + two-point calibration) approximately 30 min.

Table 31.6 Fluoride: calibration solutions.

	Fluoride concentration (mg/l)		
	0.1	**0.5**	**1.0**
Sodium fluoride solution II (ml)	0.5	0.0	0.0
Sodium fluoride solution I (ml)	0.0	0.25	0.5
Aqua bidest (ml)	16.0	16.0	16.0
Buffer solution, till up to (ml)	50.0	50.0	50.0

Annotations

(1) The buffer solution adjusts pH to 5–6. At alkaline pH, OH^- would compete with F^- and at lower pH, HF_2^- and HF are present and escape measurement. The buffer solution provides a constant ionic strength and binds cations that might otherwise bind to fluoride.

(2) For the measurement of the fluoride concentration in urine or of high concentrations in serum in case of poisoning, two-point calibration is sufficiently accurate. At high concentrations, a linear relationship exists between activity (concentration) and logarithm of the respective potential. The pertinent slope is constant.

(3) At low concentrations, for example, fluoride concentration in serum of nonexposed people, deviation from the linear relationship (see annotation 2) occurs. Nevertheless, accurate measurements may be achieved by two alternative procedures:

(a) Standard addition repeated twice to establish the slope,
(b) Slope by dilution method: The potential of the sample is measured as described, after one standard addition and after dilution with buffer solution (1 + 1; v/v), to establish the slope for the present sample concentration [63].

31.5.3 Medical Assessment and Clinical Interpretation

Medical assessment

Plasma/serum concentrations [73]	
Nonexposed/therapeutic below	0.5 mg/l
Toxicity from	2 mg/l
Comatose/lethal from	3 mg/l
Plasma elimination half-life	2–9 h
Urine concentration [64, 65]	
Nonexposed/therapeutic	0.156–1.99 mg/24 h

The plasma concentration of nonexposed people ranges from 0.001 to 0.047 mg/l and partly depends on the fluoride concentration in drinking water [62–65]. It is increased in hemodialyzed patients to 0.08 mg/l. After application of fluorine containing anesthetics (e.g., Sevofluran), plasma concentrations between 0.37 and 0.65 mg/l were measured [66].

In spot urine of nonexposed people, concentrations ranged from 0.2 to 2.0 mg/l. In the urine of workers in aluminum production, 1.7–17.7 mg fluoride/l urine was found [67]. Concentrations in urine between 5.3 and 6.1 mg/l were observed after an inadvertent addition of fluoride to drinking water. In lethal poisoning from ingestion of $ZnSiF_6$ (Fluat-Vogel), hydrogen fluoride 30%, sodium fluoride or SO_2F_2, and plasma concentrations up to 60 mg/l were measured [72].

Clinical interpretation
Toxic effects of ingested fluoride are based on

(a) the formation of hydrogen fluoride in the stomach;
(b) chelation of calcium and magnesium associated with inhibition of the activity of several enzymes.

Most of the symptoms of poisoning may be attributed easily to (a) or (b). One to five hours after ingestion, abdominal pain, salivation, sweating, nausea, bloody diarrhea, and thirst occur and so also tremor, tetanic contractions, elevated temperature, decreased blood pressure, tachycardia, ventricular dysrhythmia, and dyspnea. The symptoms subside within 24 h. In severe poisoning after ingestion of 5–10 g sodium fluoride (sometimes even from lower amounts), fatal outcome is observed already after some minutes or within 3 h from cardiac failure or respiratory arrest. Survivors may suffer from liver or kidney disorders.

Chronic ingestion of 10–80 mg sodium fluoride daily may cause fluorosis, which is characterized by weakness, marked increase in bone density, and calcification of soft tissues.

31.6 Thallium

Introduction

F. Degel and H. J. Gibitz

Severe poisoning by thallium is still life threatening, but has become rare. Most often thallium sulfate is ingested, a white crystalline, odorless and tasteless powder that still exists as a rodenticide in many households. Usually, thallium concentration in blood or urine is determined by atomic absorption spectrometry (AAS). Alternatively, inverse voltammetry and spectrophotometry following color reaction may be used for determinations in urine. These methods can be carried out easily and may be less expensive than AAS. In general, the determination of the urinary thallium concentration is sufficient to detect and monitor poisoning from thallium.

31.6.1 Atomic Absorption Spectrometry

F. Degel

Basic information on atomic absorption spectrometry and its applications are given in Ref. [74].

Thallium can be determined in urine by flame AAS after wet digestion or extraction. Lower limit of quantification comes to approximately 0.1 mg/l for these procedures.

For the determination of thallium in blood, a more sensitive method is needed. For this purpose, flameless AAS with graphite furnace after wet digestion of the sample is most appropriate. It is a demanding method that is rather accurate, specific, and sensitive (lower limit of measurement 0.002 mg/l) [75, 76].

31.6.2
Inverse Voltammetry

F. Degel

Outline
Anodic stripping voltammetry with hanging mercury drop electrode is used for the determination of thallium in urine. Basic information on the method is given in Section 4.8.

Specimens

Spot urine 50 ml.
24 h urine, aliquot of 50 ml.

Note: In case of forced diuresis, 24 h urine may come up to 10–20 l.
Samples are stable for 3 days, if stored at 4–8 °C. For prolonged periods, samples should be kept at −18 °C.

Equipment

Voltammeter (e.g., 757 VA Computrace with multimode electrode, Deutsche Metrohm)
Digestion device (e.g., 705 UV Digester, Deutsche Metrohm).

Chemicals

Acetic acid 100% (suprapur)
EDTA disodium salt dihydrate (Titriplex III) (p.a.)
Hydrogen peroxide 30% (p.a.)
Sodium acetate (suprapur)
Sulfuric acid 95–97% (p.a.)
Thallium standard solution 1.000 g/l (p.a.) (e.g., Sigma-Aldrich).

Reagents
Thallium solution for standard addition procedure 1 mg/l: Thallium standard solution/aqua bidest 1 + 999 (v/v). It is stored in a glass container that has been cleaned with nitric acid (suprapur). It can be used for 1 year.

EDTA solution 25 mmol/l: 9.3 g EDTA disodium salt is dissolved in aqua bidest ad 1000 ml.

Sodium acetate solution 200 mmol/l: 16.4 g is dissolved with aqua bidest, adjusted to pH 4.5 with acetic acid and made up to 1000 ml with aqua bidest. The solution is stable for 6 months, if kept at 4–8 °C.

Control samples: Urine Metals Control levels I and II (BioRad).

Sample preparation
One milliliter urine (or control samples), 9 ml aqua bidest, 100 μl hydrogen peroxide, and 100 μl sulfuric acid are transferred to a quartz tube of the digestion device and mixed. All remaining quartz tubes are filled with 10 ml aqua bidest.

Cooling water flow rate is adjusted to 400–500 ml/min to adjust the sample temperature to 80–90 °C. Then, the digestion device is started. Sample temperature is monitored and adjusted to 80–90 °C if necessary. After 1 h, the device is switched off.

Analysis by voltammetry

(1) *Step:* Thallium is deposited with formation of amalgam at the cathodic mercury drop and simultaneously accumulated.
(2) *Step:* Thallium is solubilized again by oxidation, passing a voltage ramp of −0.7 to + 0.2 V.

The current required for oxidation of thallium to Tl^+ is recorded and the time of analysis/voltage ramp is documented. The peak height is proportional to the Tl^+ concentration of the prepared sample and the position of the oxidation potential (−0.5 V) is specific for the transition of Tl to Tl^+.

Before initiating measurement, nitrogen is supplied (1.7–1.9 bar), the reference electrode is checked, for example, for filling level (filling solution KCl 4 mol/l), and the formation of the mercury drop is monitored.

For measurement, 13 ml sodium acetate solution and 200 μl EDTA solution are added to 3 ml of the digested solution in the measuring cell that is bubbled with nitrogen. Accumulation is effected at −0.7 V. For the measurement, a voltage ramp from −0.7 to + 0.2 V is passed. For further details on settings, see Table 31.7.

The method of standard addition is applied for determination. All steps including two standard additions (in duplicates) are controlled by the computer and carried out automatically according to the stored dedicated program file. A linear relationship between peak height and thallium concentration exists in the range of 10–50 000 μg/l. The chosen volumes of standard addition solution should always approximately double the peak height of the previous measurement. The three data

Table 31.7 Thallium: inverse voltammetry; settings.[a]

Electrode: HMDE; mode: DP; standard addition: manual; drop size: 4; stirrer speed (rpm): 2000			
Initial purge time (s)	600	Deposition potential (V)	−0.7
Deposition time (s)	300	Initial potential (V)	−0.7
Equilibration time (s)	5	End potential (V)	+0.2
Addition purge time (s)	10	Voltage step (V)	0.0060
Number of additions	2	Pulse amplitude (V)	0.050
Number of replications	2	Pulse time (s)	0.040
		Voltage step time (s)	0.40
		Sweep rate (V/s)	0.0149

DP, differential pulse; HMDE, hanging mercury drop electrode; rpm, revolutions per minute.
[a] 757 VA Computrace.

points (x-axis: concentration; y-axis: peak height) should be in a straight line. Otherwise the measurement is not valid and should be repeated.

After measurements, the measuring cell is emptied into a waste container for thallium/mercury and cleaned with nitric acid. The aqueous layer of the waste container is collected separately and treated as hazardous waste. The contaminated mercury should be recycled. The measuring cell is filled with aqua bidest to prevent drying of the electrodes. The reference electrode is stored separately dipped into filling solution (KCl 4 mol/l).

Analytical assessment

Imprecision: Coefficients of variation ranged from 7% for high concentrations to 20% for low concentration (Table 31.8).

Linearity: Urine was spiked with thallium to evaluate linearity. A linear relationship for peak height versus concentration was obtained in the range of 10–50 000 μg/l, which exceeds considerably the concentrations observed in acute poisoning. Nevertheless, higher concentrations can be measured by reduction of the volume of the digested sample taken for voltammetry.

Trueness: Two control materials with different thallium concentrations were investigated repeatedly. Bias below 10% for both control samples was obtained.

Detection limit: Detection limit is about 10 μg/l, but can be reduced by increasing the volume of the digested sample taken for voltammetry up to 6 ml (detection limit 5 μg/l) and by prolongation of accumulation time.

Specificity: Interference from other heavy metals with similar precipitation potential, such as lead, is suppressed by EDTA complexation of the compound. Supplied EDTA is not sufficient for very high lead concentrations. In these cases or in case of doubt, the procedure should be repeated by addition of the double volume of EDTA and the results should be compared.

Practicability

Five samples are prepared by UV digestion within 65 min. Voltammetry of one sample including standard additions (2) in duplicates takes 45 min. In acute poisoning, one may carry out single determinations to save about 20 min. However, data points should be in straight line. To rule out intoxication, at first the urine sample of the patient is analyzed. In case of a negative finding, a control sample is

Table 31.8 Thallium (urine): inverse voltammetry; imprecision.

	n	Coefficient of variation %	
		In the series	**Between series**
Thallium 8.58 μg/l[a]	6	11.8	20.1
Thallium 193 μg/l[b]	6	9.6	7.1

n: number of measurements.
[a]Lyphochek Urine Metals Control, level 1 (BioRad).
[b]Lyphochek Urine Metals Control, level 2 (BioRad).

investigated. It should contain a thallium concentration of 100–250 μg/l, which is slightly below the toxic range, to demonstrate sufficient sensitivity. In case of a positive result, two control samples of different concentrations are analyzed to demonstrate adequate accuracy.

Annotations

(1) In the past, a cation exchanger has been used successfully for time-saving sample preparation. Unfortunately, the respective reagent (Chelite N Serva, Heidelberg, Germany) is not available at present.
(2) Occasionally, interference occurs when urine is analyzed without preceding sample preparation and suppressed or/and peaks out of shape are observed. Therefore, voltammetry of native urine is not recommended.
(3) As usual, in the analysis of rare elements, all tubes, pipettes, and so on shall be cleaned thoroughly, for example, with nitric acid (suprapur) to avoid erroneous results from contamination with thallium.

31.6.3 Absorption Spectrophotometry

R. Aderjan, T. Daldrup and H. J. Gibitz

Outline
Complexes of thallium halogenides are formed. After extraction with diisopropyl ether (DIPE), they react with rhodamine B. Maximum absorbance of the thallium halogenide–rhodamine B complex is at 550 nm.

Specimens

Spot urine 50 ml
24 h urine, aliquot of 50 ml (Section 31.6.2).

Equipment

Spectrophotometer or spectral line photometer (Hg 546 nm)
Cuvettes 1 cm light path, special optical glass (do not use plastic cuvettes)
Pipette 5 ml with long tips
Centrifuge tubes, 12 ml (polypropylene or glass) with polypropylene stopper.

Chemicals (p.a. grade)

Diisopropyl ether
Hydrochloric acid 25%
Hydroxylammonium chloride (hydroxylamine hydrochloride)
Nitric acid 1 mol/l
Rhodamine B
Sulfuric acid 1 mol/l
Thallium solution 1.000 g/l (e.g., Sigma-Aldrich).

Reagents
Hydroxylammonium chloride solution: 0.2 g hydroxylammonium chloride is dissolved in 100 ml aqua bidest. To be prepared shortly before use.

Rhodamine B solution: Ten milligrams of rhodamine B is dissolved with 100 ml sulfuric acid by warming. To be prepared shortly before use.

Calibrator solution 3 mg/l: Thallium solution/aqua bidest 3 + 997 (v/v).

Control sample: For example, Lyphochek Urine Metals Control, level I and II (BioRad).

Spectrophotometrical analysis
See pipetting scheme in Table 31.9.

It is recommended to analyze simultaneously the native urine and diluted urine (urine/aqua bidest 1 + 9 (v/v)) if a thallium concentration >3 μg/l is suspected. After addition of rhodamine B solution, the color of the upper organic layer changes to pink in the presence of a considerably increased thallium concentration, otherwise it is colorless or slightly yellow. The absolutely transparent organic layer of a tube is transferred completely to the measuring cuvette, but without any water (not even drops). The absorbance is measured against reagent blank at 550 nm or Hg 546 nm.

Calculation of results
The calibration line *x*-axis: Concentration *y*-axis: Absorbance is constructed from the absorbance values of the four calibrators. It should be straight and go through the origin. A thallium concentration of 1 mg/l should correspond to an absorbance of 0.4, and the absorbance of the reagent blank should be below 0.015 when measured against diisopropyl ether. The concentration of the samples is calculated from the respective linear equation of the calibration line.

Analytical assessment
Imprecision: Coefficient of variations ranged from 3.1 to 10.1% (Table 31.10).

Linearity: A linear relationship for absorbance and thallium concentration is established in the range of 0.1–3.0 mg/l.

Recovery: 1000 ml urine of nonexposed people was spiked with 1 mg thallium. Recovery ranged from 92 to 107% (mean: 0.98 mg/l, $n = 11$).

Detection limit: The detection limit was estimated from the analyses of 10 urine samples of nonexposed people. It comes to 0.1 mg/l (mean + 3 times standard deviation).

Specificity: For possible interference from other metals, see Table 31.11. The only metal of this presentation that may lead to a false elevated result at a clinically relevant concentration is gold. The interference from gallium, indium, platinum, and zirconium, which in principle can also form colored complexes, has not been investigated.

Practicability
Technician time for one sample is approximately 40 min and for five samples within a series of measurements is approximately 85 min. Total time of analysis comes to 105 min for one sample and 150 min for five samples.

Table 31.9 Thallium (urine): absorption spectrophotometry; pipetting scheme.

	Reagent blank	Calibrator 1 (0.75 mg/l)	Calibrator 2 (1.50 mg/l)	Calibrator 3 (2.25 mg/l)	Calibrator 4 (3.00 mg/l)	Sample[a]	Sample[a]	Control sample	Control sample
Centrifuge tube number	1	2	3	4	5	6	7	8	9
Urine sample (μl)	—	—	—	—	—	4000	4000	—	—
Control sample (μl)	—	—	—	—	—	—	—	4000	4000
Calibrator solution (μl)	—	1000	2000	3000	4000	—	—	—	—
Aqua bidest (μl)	4000	3000	2000	1000	—	—	—	—	—
Nitric acid (μl)	500	500	500	500	500	500	500	500	500
Hydrochloric acid (μl)	2000	2000	2000	2000	2000	2000	2000	2000	2000
The solutions are mixed and kept (tubes not closed) 40 min at 95–100 °C (water bath or heating block). Cooling to room temperature.									
DIPE (μl)	3000	3000	3000	3000	3000	3000	3000	3000	3000
The tubes are closed and the mixture is shaken vigorously (2 min). After centrifugation (2000 × g, 5 min), the aqueous lower layer is removed (pipette 5 ml with long tip) almost completely and discarded, avoiding loss of organic layer.									
Hydroxylammonium chloride solution (μl)	2000	2000	2000	2000	2000	2000	2000	2000	2000
Hydrochloric acid (μl)	1000	1000	1000	1000	1000	1000	1000	1000	1000
DIPE (μl)	1000	1000	1000	1000	1000	1000	1000	1000	1000
The tubes are closed and processed as before.									
Rhodamine B solution (μl)	2000	2000	2000	2000	2000	2000	2000	2000	2000
The tubes are closed and the mixture is shaken (2 min). After centrifugation (2000 × g, 5 min), the organic upper layer is transferred to a measuring cuvette. The absorbance is measured against reagent blank at 550 nm (or Hg 546 nm).									

DIPE: diisopropyl ether.

[a] If a thallium concentration >3 mg/l is suspected, diluted urine (1 + 9) is used instead of native urine.

Table 31.10 Thallium: absorption spectrophotometry; imprecision.

	n	$\bar{x}$ (mg/l)	*s* (mg/l)	CV (%)
Within series	11	0.98	0.03	3.06
Between series	26	1.09	0.11	10.09

n, number of measurements; $\bar{x}$, mean; *s*, standard deviation; CV, coefficient of variation.

Annotations

(1) The procedure can be performed with an internal standard instead of external calibration to minimize interference attributable to the matrix (Aderjan, personal communication).
(2) Another spectrophotometric procedure based on the absorbance of a red thallium–dithizone complex [90] is not sufficiently specific. The results must always be confirmed by atomic absorption spectrometry [91].

31.6.4
Medical Assessment and Clinical Interpretation

F. Degel and H.J. Gibitz

Thallium is a rather toxic metal. It is present in semiconductor industry, optical systems, photoelectric cells, alloys, and smelters, and may cause chronic poisoning to exposed workers. Although it is banned in many countries for this purpose, most often thallium sulfate used as rodenticide is still involved in acute poisoning

Table 31.11 Thallium: absorption spectrophotometry; specificity.

	Concentration (mg/l urine)			
Metal	**50**	**5**	**0.5**	**0.05**
Gold (Au)	+	+	+	+
Iron (Fe)	–	–	–	–
Mercury (Hg)	+	–	–	–
Lead (Pb)	–	–	–	–
Antimony (Sb)	+	+	–	–
Tin (Sn)	+	–	–	–
Chromium (Cr III)	–	–	–	–
Chromium (Cr VI)	+	+	–	–
Vanadium (V)	+	+	–	–

Measurement of absorbance of a urine sample: thallium concentration 1 mg/l. Spiked with metal (concentration after adding: 0.05 to 50 mg/l. + denotes increase of absorbance >0.02, and – denotes increase of absorbance <0.02.

(including homicides and suicides). But also all other thallium (I) and (III) salts, which are soluble at physiological pH, are toxic, including the less soluble thallium halogenides. One gram or 13–15 mg/kg body mass is considered as lethal doses for adults. For children, already a dose exceeding 1 mg/kg body mass is hazardous.

Concentration of thallium in urine of nonexposed people comes to 0.002 mg/l [76]; toxic symptoms are associated with concentrations exceeding 0.25 mg/l. In the urine of workers suffering from chronic poisoning, concentrations ranged from 0.2 to 1.0 mg/l [77]. Details on the distribution of thallium in several organ systems are given in Ref. [84].

In severe acute poisoning and lethal intoxications, urine concentrations from 1.8 to 20 mg/l were measured [78–83]. Urine is the appropriate material for the diagnosis of thallium poisoning after absorption and distribution especially since the first unspecific toxic effects occur only after a latent period of hours or days. In case of a negative finding early after ingestion, a second sample voided 6 h later should be analyzed. Treatment should be monitored by regular determination of urinary thallium concentration during 2–3 months because of the long elimination half-life (up to 30 days), which depends on the time elapsed since ingestion and the period of intake [92]. For forensic purposes, the following materials should be stored if available: blood 5–10 ml (anticoagulated with heparin or EDTA), stomach contents (10–50 ml), gastric lavage (100–200 ml) hair, and scene residues (e.g., drinking vessel, packaging, and left over).

Several organ systems are affected in thallium poisoning. Gastrointestinal disturbances such as gastroenteritis, diarrhea, and constipation are observed associated with abdominal pain. Peripheral and central nervous system are involved resulting not only in sensory neuropathy but also in ataxia, tremor, and convulsions, as may be observed in lead or arsenic poisoning.

Hair loss is a rather specific symptom and may occur already 8 days after exposure. Damage of pulmonary alveoli and myocardium has been observed. Impaired renal function will slow down urinary excretion. Further details and clinical symptoms of thallium poisoning are described in Refs [85, 86]. After 24–48 h of ingestion before the end of distribution, hemodialysis or hemoperfusion may be applied to enhance elimination if the thallium concentration in blood exceeds 1 mg/l [87] (measurement by AAS, see Section 31.4.3). Later on, these measures are used only in case of impaired renal function.

In Europe, Prussian blue (ferric III hexacyanoferrate II) (K Fe (III) [Fe $(CN)_6$]) is used as an antidote that binds thallium. Gastrointestinal absorption is inhibited and fecal elimination is increased. The antidote is administered until urinary excretion of thallium is less than 0.5 mg/24 h. One milligram of thallium/24 h is eliminated by forced diuresis. Elimination half-life can be reduced to 2–3 days by administration of antidote and concomitant forced diuresis. Further details on the treatment of thallium intoxication are described in Refs [88, 89].

References

References for Section 31.1

1 USPDI (2004) *Drug Information for the Health Care Professional,* Vol. 1, 24th edn, Micromedex, Englewood, Colorado. Content reviewed by the United States Pharmacopoeia Convention.

2 De Haro, L., Menut, A., Tichadou, L., Pommier, P., Arditti, J. and Hayek-Lanthois, M. (2004) Lethal chloroquine self poisonings: 13 cases of the Marseilles Poison Centre. *Clinical Toxicology,* **42**, 503.

3 Eddleston, M. (2000) Patterns and problems of deliberate self-poisoning in the developing world. *Quarterly Journal of Medicine (QJM)*, **93**, 715–731.

4 Lambert, W., Meyer, E., Van Bocxlaer, J. and De Leenheeer, A. (1997) Relevance of toxicological screening for chloroquine in non-malarious areas. *Journal of Analytical Toxicology,* **21**, 321–322.

5 DFG/TIAFT (Deutsche Forschungsgemeinschaft/The International Association of Forensic Toxicologists) (1992) *Senatskommission für klinisch-toxikologische Analytik. Report XVII/Special Issue of the TIAFT Bulletin: thin-layer chromatographic* R_f values of toxicologically relevant substances on standardized systems, 2nd edn, VCH Verlagsgesellschaft, Weinheim.

6 Runne, U., Ochsendorf, F.R., Schmidt, K. and Raudonat, H.W. (1992) Sequential concentration of chloroquine in human hair correlates with ingested dose and duration of therapy. *Acta Dermato-Venereologica*, **72**, 355–357.

7 Jaeger, A., Sauder, P.H., Kopferschmitt, J. and Nothdurft, H. (1987) Clinical features and management of poisoning due to antimalarial drugs. *Medical Toxicology,* **2**, 242–273.

8 Messant, I., Jeremie, N., Lenfant, F. and Freysz, M. (2004) Massive chloroquine intoxication: importance of early treatment and prehospital treatment. *Resuscitation,* **60**, 343–346.

9 Abu-Aisha, H., Abu-Sabaa, H.M.A. and Nur, T. (1979) Cardiac arrest after intravenous chloroquine injection. *Journal of Tropical Medicine and Hygiene,* **82**, 36–37.

10 Cann, H.M. and Verhulst, H.M. (1961) Fatal acute chloroquine poisoning in children. *Pediatrics,* **27**, 95–102.

11 Markowitz, H.A. and McGinley, J.M. (1964) Chloroquine poisoning in a child. *The Journal of the American Medical Association,* **1**, 22–25.

12 Whyte, I.M. (2004) Antimalarial agents, in *Medical Toxicology,* 3rd edn (ed. R.C. Dart), Lippincott Williams and Wilkins, Philadelphia.

13 The International Association of Forensic Toxicologists (TIAFT). TIAFT reference blood level list of therapeutic and toxic substances. Update 2007 (www.tiaft.org/).

14 Schulz, M. and Schmoldt, A. (2003) Therapeutic and toxic blood concentrations of more than 800 drugs and xenobiotics. *Pharmazie,* **58**, 447–474.

15 Hardman, J.G. and Lombard, L.E. (2001) *Goodman & Gilman's The Pharmacological Basis of Therapeutics,* 10th edn (eds L.S. Goodman and A. Gilman), MacGraw-Hill, New York.

16 Gustaffson, L.I., Walker, O., Alvan, G., Beerman, B., Estevez, F., Gleisner, L. *et al.* (1983) Disposition of chloroquine in man after single intravenous and oral doses. *British Journal of Clinical Pharmacology,* **15**, 471–479.

17 Houze, P., de Reynies, A., Baud, F.J., Benatar, M.F. and Pays, M. (1992) Simultaneous determination of chloroquine and its three metabolites in human plasma, whole blood and urine by ion pair high-performance liquid chromatography. *Journal of Chromatography,* **574**, 305–312.

18 Krishna, S. and White, N.J. (1996) Pharmacokinetics of quinine, chloroquine

and amodiaquine. *Clinical Pharmacokinetics*, **30**, 263–299.
19 Maschke, S., Azaroual, N., Wieruszeski, J.M., Lippens, G., Imbenotte, M., Mathieu, D. *et al.* (1997) Diagnosis of a case of acute chloroquine poisoning using H-1 NMR spectroscopy: characterisation of drug metabolites in urine. *NMR in Biomedicine*, **10**, 277–284.
20 Clemessy, J.L., Favier, C., Borron, S.W., Hantson, P.E., Vicaut, E. and Baud, F.J. (1995) Hypokalaemia related to acute chloroquine ingestion. *Lancet*, **346**, 877–880.
21 Yanturali, S. (2004) Diazepam for treatment of massive chloroquine intoxication. *Resuscitation*, **63**, 347–348.

References for Section 31.2

22 http://de.Wikipedia.org/wiki/Passivrauchen.
23 Pragst, F., Herzler, M., Herre, S., Erxleben, B.-T. and Rothe, M. (2001) *UV Spectra of Toxic Compounds. Database of Photodiode Array Spectra of Illegal and Therapeutic Drugs, Pesticides, Ecotoxic Substances and Other Poisons. Manual and Database on CD-ROM*, Dieter Helm, Heppenheim.
24 Schulz, M. and Schmoldt, A. (2003) Therapeutic and toxic blood concentrations of more than 800 drugs and other xenobiotics. *Pharmazie*, **58**, 447–474.
25 McGuigan, M.A. (2004) Nicotine, in *Medical Toxicology*, 3rd edn (ed. R.C. Dart), Lippincott Williams and Wilkins, Philadelphia, pp. 601–604.
26 De Zeeuw, R.A., Franke, J.P., Maurer, H.H. and Pfleger, K. (1992) *Gas Chromatographic Retention Indices of Toxicologically Relevant Substances on Packed and Capillary Columns with Dimethylsilicone Stationary Phases* (Report XVIII of the DFG Commission for Clinical–Toxicological Analysis, Special Issue of the TIAFT Bulletin), 3rd revised and enlarged edn, Wiley-VCH Verlag GmbH, Weinheim, p. 407.
27 Maurer, H.H., Pfleger, K. and Weber, A.A. (2007) *Mass Spectral Library of Drugs, Poisons, Pesticides, Pollutants and their Metabolites*, 4th revised edn, Wiley-VCH Verlag GmbH, Weinheim.

References for Section 31.3

28 Gitzelmann, R., Steinmann, B., Otten, A., Dumermuth, G., Herban, M., Reubi, J.C. and Cuenod, M. (1977) Non-ketotic hyperhyperglycinemia treated with strychnine, a glycine receptor antagonist. *Helvetica Paediatrica Acta*, **32**, 517–525.
29 O'Callaghan, W.G., Joice, N., Counihan, H.E., Ward, M., Lavelle, P. and O'Brien, E. (1982) Unusual strychnine poisoning and its treatment: report of eight cases. *British Medical Journal (Clinical Research Ed.)*, **285**, 478.
30 Scheffold, N., Heinz, B., Albrecht, H., Pickert, A. and Cyran, J. (2004) Strychnine poisoning. *Deutsche Medizinische Wochenschrift*, **129**, 2236–2238.
31 Shadnia, S., Moiensadat, M. and Abdollahi, M. (2004) A case of acute strychnine poisoning. *Veterinary and Human Toxicology*, **46**, 76–79.
32 Wood, D.M., Webster, D., Martinez, D., Dargan, P.I. and Jones, A.L. (2002) Case report: Survival after deliberate strychnine self-poisoning with toxicokinetic data. *Critical Care*, **6**, 456–459.
33 DFG/TIAFT (Deutsche Forschungs gemeinschaft/The International Association of Forensic Toxicologists) (1992) Senatskommission für klinisch-toxikologische Analytik. Report XVII/ Special Issue of the TIAFT Bulletin: thin-layer chromatographic R_f values of toxicologically relevant substances on standardized systems, 2nd edn, VCH Verlagsgesellschaft, Weinheim.
34 Dijkhuis, I.C., Uges, D.R.A., Eerland, J.J. and Harteveld, A. (1990) *Eurotox – 1990, KKGT Proficiency Study in Clinical Toxicology January–May 1990*, The Hague

Hospital Pharmacy, The Hague (Netherlands).
35 Heiser, J.M., Daya, M.R., Magnussen, A.R., Norton, R.L., Spyker, D.A. and Allen, D. (1992) Massive strychnine intoxication: serial blood levels in a fatal case. *Clinical Toxicology*, **30**, 269–283.
36 Dart, R.C.(ed.) (2004) *Medical Toxicology*, 3rd edn, Lippincott Williams and Wilkins, Philadelphia.
37 TIAFT (The International Association of Forensic Toxicologists). TIAFT reference blood level list of therapeutic and toxic substances. Update 2007 (www.tiaft.org/).
38 Palatnick, W., Meatherall, R., Sitar, D. and Tenenbein, M. (1997) Toxicokinetics of acute strychnine poisoning. *Clinical Toxicology*, **35**, 617–620.
39 Edmunds, M., Sheehan, T.M. and Hoff, W. (1986) Strychnine poisoning: clinical and toxicological observations on a non-fatal case. *Clinical Toxicology*, **24**, 245–255.
40 Decker, W.J., Baker, H.E. and Tamulinas, S.E. (1982) Two deaths resulting from apparent parenteral injection of strychnine. *Veterinary and Human Toxicology*, **24**, 161–162.
41 Adamson, R.H. and Fouts, J.R. (1959) Enzymatic metabolism of strychnine. *The Journal of Pharmacology and Experimental Therapeutics*, **127**, 87–91.
42 Tanimoto, Y., Kaneko, H., Ohkuma, T., Oguri, K. and Yoshimura, H. (1991) Site-selective oxidation of strychnine by phenobarbital inducible cytochrome P-450. *Journal of Pharmacobio-Dynamics*, **14**, 161–169.
43 Oguri, K., Tanimoto, Y. and Yoshimura, H. (1988) *Identification of Urinary Metabolites of Strychnine in Rats by Gas Chromatography–Mass Spectrometry*. Proceedings of the 24th International Meeting of TIAFT, July 28–31, 1987 (eds G.R. Jones and P.P. Singer), University of Alberta Printing Service, Banff (Alberta), Canada.
44 Yamarick, W., Walson, P. and DiTraglia, J. (1992) Strychnine poisoning in an adolescent. *Clinical Toxicology*, **30**, 141–148.

References for Section 31.4.1

45 Michaelis, H.C., Clemens, C.H., Kijewski, H., Neurath, H. and Eggert, A. (1991) Acetonitrile serum concentrations and cyanide blood levels in a case of suicidal oral acetonitrile ingestion. *Clinical Toxicology*, **29**, 447–458.

References for Section 31.4.2

46 Ballantyne, B. (1983) Artifacts in the definition of toxicity by cyanides and cyanogens. *Fundamental and Applied Toxicology*, **3**, 400–408.

References for Section 31.4.3

47 Fligner, C.L., Lüthi, R., Linkaityle, E. and Raisys, V. (1992) Paper strip screening method for detection of cyanide in blood using CYANTESMO test paper. *The American Journal of Forensic Medicine and Pathology*, **13**, 81–84.

References for Section 31.4.4

48 Felscher, D. and Wulfmeyer, M. (1998) A new specific method to detect cyanide in body fluids, especially whole blood, by fluorometry. *Journal of Analytical Toxicology*, **22**, 363–366.
49 Zamecnik, J. and Tam, J. (1987) Cyanide in blood by gas chromatography with NP detector and acetonitrile as internal standards. Application on air accident fire victims. *Journal of Analytical Toxicology*, **11**, 47–48.
50 Asselborn, G. and Wennig, R. (2000) Program and abstracts of the 38th International Meeting TIAFT 2000 (ed. I. Rasanen), Helsinki, p. 40.

51 Egekeze, J.O. and Oehme, F.W. (1979) Direct potentiometric method for the determination of cyanide in biological materials. *Journal of Analytical Toxicology*, **3**, 119–124.
52 Ikeda, S., Schweiss, J.F., Frank, P.A. and Homan, S.M. (1987) *In vitro* cyanide release from sodium nitroprusside. *Anesthesiology*, **66**, 381–385.
53 Steinecke, H. and Heesen, U. (1989) Direct potentiometric determination of the level of cyanide in plasma following an infusion of sodium nitroprusside. *Anesthesiology and Reanimation*, **14**, 333–337.
54 Yagi, K., Ikeda, S., Schweiss, J.F. and Homan, S.M. (1990) Measurement of blood cyanide with a microdiffusion method and an ion-specific electrode. *Anesthesiology*, **73**, 1028–1031.
55 Feldstein, L. and Klendshoj, N.C. (1954) The determination of cyanide in biological fluids by microdiffusion analysis. *The Journal of Laboratory and Clinical Medicine*, **44**, 166–170.
56 Ballantyne, B. (1983) Artifacts in the definition of toxicity by cyanides and cyanogens. *Fundamental and Applied Toxicology*, **3**, 400–408.

References for Section 31.4.6

57 Erdman, A.R. (2004) Cyanide, in *Medical Toxicology*, 3rd edn (ed. R.C. Dart), Lippincott Williams and Wilkins, Philadelphia.
58 Meredith, T.J., Jacobsen, D., Haines, J.A., Berger, J.C. and Van Heijst, A.N.P. (eds) (1993) *Antidotes for Poisoning by Cyanide*, Cambridge University Press, Cambridge, NY.
59 Peden, N.R., Taha, A., McSorley, R.D., Bryden, G.T., Murdoch, I.B. and Anderson, J.M. (1986) Industrial exposure to hydrogen cyanide: implications for treatment. *British Medical Journal*, **239**, 538–541.
60 Bogusz, M., Moroz, J., Karski, J., Gierz, J., Regieli, A., Witkowska, R. and Golabek, A. (1979) Blood cyanide and thiocyanate concentrations after administration of sodium nitroprusside as hypotensive agent in neurosurgery. *Clinical Chemistry*, **25**, 60–63.
61 IPCS/CEC (International Programme on Chemical Safety/Commission of the European Communities) (1993) *Evaluation of Antidotes Series, Volume 2: Antidotes for Poisoning by Cyanide* (eds T.J. Meredith, D. Jacobsen, J.A. Haines and A.N.P. van Heijst), World Health Organization, Geneva and ECSC-EEC-EAEC, Brussels-Luxembourg, Cambridge University Press, Cambridge, NY.

References for Section 31.5

62 Czarnowski, W. and Krechniak, J. (1990) Fluoride in the urine, hair and nails of phosphate fertiliser workers. *British Journal of Industrial Medicine*, **4**, 349–351.
63 Fuchs, C., Dorn, D., Fuchs, C.A., Henning, H.B., McIntosh, C. and Scheler, F. (1975) Fluoride determination in plasma by ion selective electrodes: a simplified method for the clinical laboratory. *Clinica Chimica Acta*, **60**, 157–167.
64 Torra, M., Rodamilans, M. and Corbella, J. (1998) Serum and urine ionic fluoride: normal range in a non-exposed population. *Biological Trace Element Research*, **60**, 67–71.
65 Torra, M., Rodamilans, M. and Corbella, J. (1998) Serum and urine fluoride concentration: relationships to age, sex and renal function in a non-fluoridated population. *The Science of the Total Environment*, **220**, 81–85.
66 Wiesner, G., Wild, K., Schwürzer, S., Merz, M. and Hobbhahn, J. (1996) Serum fluoride concentrations and exocrine kidney function with sevoflurane and enflurane. An open, randomized, comparative phase III study of patients

with healthy kidneys. *Anaesthesist*, **45**, 31–36.

67 Neefus, J.D., Cholak, J. and Saltzmann, B.E. (1970) The determination of fluoride in urine using a fluoride-specific ion electrode. *American Industrial Hygiene Association Journal*, **3**, 96–99.

68 Penman, A.D., Brackin, B.T. and Embrey, R. (1997) Outbreak of acute fluoride poisoning caused by a fluoride overfeed. *Public Health Reports*, **112**, 403–409.

69 Scheuerman, E.H. (1986) Suicide by exposure to sulfuryl fluoride. *Journal of Forensic Sciences*, **31**, 1154–1158.

70 Steinecke, H. and Fischer, W. (1983) Fluoride determination in human organs and body fluids using an ion-sensitive electrode. *Zeitschrift für Medizinische Laboratoriumsdiagnostik*, **24**, 304–307.

71 Duly, E.B., Luney, S.R., Trinick, T.R., Murray, J.M. and Comer, J.E.A. (1995) Validation of an ion selective electrode system for the analysis of serum fluoride ion. *Journal of Automatic Chemistry*, **17**, 219–223.

72 Poklis, A. and Mackell, M. (1989) Disposition of fluoride in a fatal case of unsuspected sodium fluoride poisoning. *Forensic Science International*, **41**, 55–59.

73 Stead, A.H. and Moffat, A.C. (1983) A collection of therapeutic, toxic and fatal blood drug concentrations in man. *Human Toxicology*, **2**, 437–464.

References for Section 31.6

74 Skoog, D.A. and Leary, J.J. (1997) *Principles of Instrumental Analysis*, 5th edn, Saunders College Publishing.

75 Paschal, D.C. and Bailey, G.G. (1986) Determination of thallium in urine with Zeeman effect graphite furnace atomic absorption. *Journal of Analytical Toxicology*, **10**, 252–254.

76 Kubasik, N.P. and Volosin, T. (1973) A simplified determination of urinary cadmium, lead and thallium with the use of carbon rod atomization and atomic absorption spectrometry. *Clinical Chemistry*, **19**, 954–958.

77 Richeson, E.M. (1958) Industrial thallium intoxication. *Industrial Medicine and Surgery*, **27**, 607–619.

78 Stein, D.M. and Perlstein, M.A. (1959) Thallium poisoning. *American Journal of Diseases of Children*, **98**, 80–85.

79 Taber, P. (1964) Chronic thallium poisoning: rapid diagnosis and treatment. *The Journal of Pediatrics*, **65**, 461–463.

80 Grunfeld, O. and Hinostroza, G. (1964) Thallium poisoning. *Archives of Internal Medicine*, **114**, 132–138.

81 Koshy, K.M. and Lovejoy, F.H. (1981) Thallium ingestion with survival: ineffectiveness of peritoneal dialysis and potassium chloride diuresis. *Clinical Toxicology*, **18**, 521–525.

82 Hologgitas, J., Ullucci, P. and Driscoll, J. (1980) Thallium elimination kinetics in acute thalliotoxicosis. *Journal of Analytical Toxicology*, **4**, 68–73.

83 Chandler, H.A., Archbold, G.P.R. and Gibson, J.M. (1990) Excretion of a toxic dose of thallium. *Clinical Chemistry*, **36**, 1506–1509.

84 Baselt, R.C. (2004) *Disposition of Toxic Drugs and Chemicals in Man*, 7th edn, Biomedical Publications, Foster City, CA.

85 Mulkey, J.P. and Oehme, F.W. (1993) A review of thallium toxicity. *Veterinary and Human Toxicology*, **35**, 445–453.

86 Herrero, F., Fernandez, E., Gomez, J. *et al.* (1995) Thallium poisoning presenting with abdominal colic paresthesia and irritability. *Clinical Toxicology*, **33**, 261–264.

87 de Groot, G. (1982) Haemoperfusion in clinical toxicology. Thesis. Utrecht.

88 Malbrain, J.M.L.N.G., Lambrecht, G.L.Y., Zandijk, E. *et al.* (1997) Treatment of severe thallium intoxication. *Clinical Toxicology*, **35**, 97–100.

89 Nogue, S., Mas, A., Pares, A. *et al.* (1982) Acute thallium poisoning: an evaluation of different forms of treatment. *Clinical Toxicology*, **19**, 1015–1021.

90 Flanagan, R.J., Braithwaite, R.A., Brown, S.S., Widdop, B. and Wolff, F.A. (1995) *Basic Analytical Toxicology*, World Health Organization, Geneva, pp. 222–223.

91 Moffat, A.C., Osselton, M.D. and Widdop, B.(eds) (2004) *Clarke's Analysis of Drugs and Poisons*, 3rd edn, Pharmaceutical Press, London.

92 Moore, D., House, I. and Dixon, A. (1993) Thallium poisoning. *British Medical Journal*, **306**, 1527–1529.

32
Chemical Warfare Agents

M. Koller and L. Szinicz

32.1
Overview

Definition
Chemical warfare agents (CWA) are highly toxic compounds synthesized for warfare purposes. They are used to cause a temporary incapacitation, a lasting damage, or even the death of enemy soldiers. If these chemical warfare agents are stockpiled in form of grenades or missile warheads, they are classified as chemical weapons.

Classification
Chemical warfare agents are mainly classified according to their target organ. There are nerve, vesicant, blood, lung-damaging, and psychotomimetic agents as well as eye, nose, and throat irritants. Table 32.1 provides a detailed list of the relevant classes of agents.

Toxicity
Table 32.2 provides the toxicity of selected chemical warfare agents. A compilation of the AEGL (Acute Exposure Guideline Levels) in tabular form is given by Kluge and Szinicz [1].

Deployment of chemical warfare agents
Chemical warfare agents can be released in the form of liquids, aerosols, or vapors and gases [3]. In case of the so-called binary chemical warfare agents, two components of minor toxicity are reacted at the moment of their deployment to release the highly toxic compound. Suitable weapons for chemical warfare agents are missiles, bombs, and grenades that cause, in addition, mechanical and/or thermal injuries. Various types of aircraft including drones and agricultural aircraft with spraying devices can be also used. To impede diagnosis and therapy, the various chemical agents are mixed or chemical agents are used in combination with nuclear or biological warfare agents.

Clinical Toxicological Analysis: Procedures, Results, Interpretation. Edited by Wolf-Rüdiger Külpmann

ISBN: 978-3-527-31890-2

Table 32.1 Chemical warfare agents (Szinicz, 2004).

Classification	Substance	Common name	Abbreviation (official)
Blood agent	Hydrogen arsenide	Arsine	SA
Blood agent (cyanogen)	Cyanogen chloride	—	CK
Blood agent (cyanogen)	Hydrogen cyanide		AC
Eye irritant (lacrimator)	Bromoacetone	—	BA
Eye irritant (lacrimator)	Chloroacetophenone	—	CN
Eye irritant (lacrimator)	2-Chlorobenzylidenemalonodinitrile	—	CS
Lung-damaging agent (choking agent)	Carbonyl dichloride	Phosgene	CG
Lung-damaging agent (choking agent)	Chloroformic acid trichloromethyl ester	Diphosgene	DP
Lung-damaging agent (choking agent)	Trichloronitromethane	Chloropicrin	PS
Nerve agent	Ethyl-*N,N*-dimethylphosphor-amido cyanidate	Tabun	GA
Nerve agent	Isopropylmethylphosphonofluoridate	Sarin	GB
Nerve agent	Pinacolylmethylphosphonofluoridate	Soman	GD
Nerve agent	Cyclohexylmethylphosphonofluoridate	Cyclosarin	GF
Nerve agent	*O*-Ethyl-*S*-[2-diisopropylamino] ethylmethylphosphonothioate	VX	VX
Nose–throat irritant	Diphenylarsine chloride	Clark I	DA
Nose–throat irritant	Diphenylarsine cyanide	Clark II	DC
Nose–throat irritant	Phenarsazine chloride	Adamsite	DM
Psychotomimetic agent	3-Quinuclidinyl benzilate	BZ	BZ
Vesicant (blistering agent)	Bis-(2-chloroethyl)sulfide	Sulfur mustard (yperite)	HD
Vesicant (blistering agent)	Tris-(2-chloroethyl)amine	Nitrogen mustard	HN-3
Vesicant (blistering agent)	Dichloro(2-chlorovinyl)arsine	Lewisite	L

Exposure to CWA may occur in case of military conflicts, terrorist or criminal activities, checks in the framework of the Chemical Weapons Convention (CWC), destruction of stockpiles, and release from CWA waste.

The most important routes of poisoning are inhalation and skin absorption, followed by the ingestion of contaminated food or water ("sabotage poisons"). Liquids with different evaporation rates are the most frequently used CWA.

Although substances of low volatility show a persistence of days or weeks, the persistence of volatile substances is limited to minutes or hours. CWA in ammunition or in containers can retain their toxicity for decades, even if they have been buried or sunk.

Analysis of chemical warfare agents in biomedical samples

Overview

For various reasons, the determination of CWA in biomedical samples is a great analytical challenge and will therefore be restricted to a few highly specialized

Table 32.2 Chemical warfare agents: toxicity [2] (Klimmek, 1983).

Dose (man)	Tabun	Sarin	Soman	VX	Sulfur mustard	Nitrogen mustard	Lewisite
LD_{50} p.o. (mg/kg)[a] estimated	5	0.14	0.14	0.07	—	2	—
LD_{50} percutaneous (mg/kg)[a] estimated	14–21	25	5	0.04	40–60	15	38–54
LCt_{50} inhaled at rest (mg min/m^3) estimated	200–400	100	70	36	1500	400	1200–1500
Minimal dose required for							
Miosis (mg min/m^3)	0.9	2–4	0.5	0.5	—	—	—
Tremor (mg min/m^3)	—	4	—	1.6	—	—	—
Erythema (mg/cm^2)	—	—	—	—	0.01	0.001	0.05–0.1
Skin vesicles (mg/cm^2)	—	—	—	—	0.1–0.15	0.1	0.2
Conjunctival irritation (mg/l)	—	—	—	—	0.0012 (30 min)	0.0007 (15 min)	0.01 (15 min)
No effect dose (mg min/m^3)	—	0.5	0.3	0.02	2	—	—
Minimal lethal dose (inhaled, 30 min) (mg/m^3)	—	—	—	—	—	—	48

LCt_{50}, mean lethal concentration (inhalation). LD_{50}, mean lethal dose.
[a]mg/kg body mass.

laboratories. At the moment, there is no laboratory worldwide that is accredited for that kind of analysis at the Organization for the Prohibition of Chemical Weapons (OPCW) in The Hague. It is the objective of the OPCW to recruit laboratories for the analysis of biomedical samples in the near future, besides the ones that have already been accredited for the analysis/verification of CWA in environmental samples. The extreme toxicity of the compounds constitutes the main problem in the analysis of these agents in biomatrix. The analytical procedures must be very sensitive to detect even minute amounts in the body. That is why the analysis is mostly carried out at the limit of detection of gas chromatographic or liquid chromatographic procedures. In case of biomedical samples, the application of simple rapid tests is rarely possible. Although test tubes for the detection of CWA are available from Dräger, they are only applicable to environmental samples to have a first overview of the contamination grade of an area. In a similar way, detection devices such as the Raid-M (Bruker-Daltonik), which are also used in the German "Fuchs" tank, are used to reconnoiter an area and to raise an alarm if necessary. Moreover, the measurement of the parent agents in biomatrix is usually of minor importance as compared to the products from reaction and degradation, for which no simple tests are available. An exception to that is the determination of the cholinesterase activity, which enables a fast detection of the presence of cholinesterase inhibitors (see Section 33.2). The activity allows an estimate of the severity of poisoning with a nerve agent and prognosis but is not able to identify the poison, because an organophosphorus or a carbamate pesticide (see Chapter 28) may be responsible for the inhibition of the cholinesterase activity as well. Moreover, vesicant or other CWA are not detected. For more detailed information about the cholinesterase status, the reader is referred to Section 33.2.

Although a reliable identification of a nerve agent would be desirable with regard to therapy with oximes, it is rather unlikely that this highly sophisticated analysis can be carried out fast enough to give guidance for treatment. In case of exposure, measures have to be taken immediately. Therefore, soldiers are equipped with two autoinjectors, of which one contains atropine only and the other atropine and obidoxime. These injectors have been developed especially for self and buddy aid. A later identification of the CWA can possibly explain the ineffectiveness of oxime as an antidote. Soman poisoning can serve as an example: due to the rapid aging of the acetylcholinesterase–soman complex, the oxime treatment is not effective.

The unambiguous identification of a CWA is of utmost importance for the verification of the deployment in a bilateral conflict. Such an accusation has to be very well supported by analyses, and therefore, more than one test shall be carried out to verify the presence of a particular CWA. For a valid verification, the analytical spectrum should be employed in full, including free CWA, the respective metabolites, and protein and possibly DNA adducts. The selectivity of the methods must be sufficiently high to rule out, for example, organophosphorus pesticides.

The clarification of entitlements for state benefit of soldiers claiming that they have been exposed to at least small doses of CWA constitutes another reason for the development of such sophisticated analytical methods.

The poor availability of reference substances constitutes a further obstacle to the analysis of CWA, because only a few metabolites can be obtained commercially. All other compounds have to be specifically synthesized. Sophisticated security measures and a staff that is absolutely trustworthy are of paramount importance when producing the active CWA themselves. In addition, the synthesis of all CWA of schedule 1 of the Chemical Weapons Convention has to be registered at OPCW. As a prerequisite, the national authorities of the home country have to give the official permission for synthesis, which is only granted to very few dedicated laboratories.

A further item concerns the development of appropriate procedures. Apart from the terrorist attack on the Tokyo subway in 1995 [4] and the Iran–Iraq war of 1980 (sulfur mustard), there are (fortunately) hardly any samples available from human individuals. Therefore, the development of new procedures is solely based on spiked samples. Practically, an evaluation of the procedures would only be possible by samples from poisoned animals. However, due to ethical reasons, respective experiments can hardly be performed in Germany. As a consequence, a clinical toxicological assessment of the measurements is almost impossible, because there is a lack of experience concerning the correlation between the severity of the poisoning and the analytical results.

32.2 Nerve Agents

Introduction

Nerve agents such as the derivatives of methylphosphonic acid (sarin, soman, cyclosarin, VX, VR, and CVX) and phosphoric acid (tabun) (see Figure 32.1 for

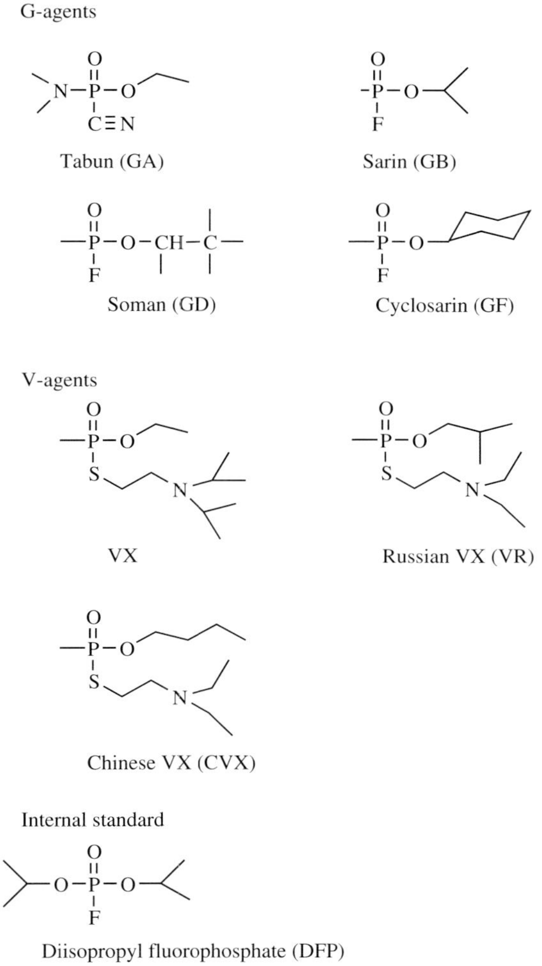

G-agents

Tabun (GA)

Sarin (GB)

Soman (GD)

Cyclosarin (GF)

V-agents

VX

Russian VX (VR)

Chinese VX (CVX)

Internal standard

Diisopropyl fluorophosphate (DFP)

Figure 32.1 Nerve agents.

structural formulas) belong to the group of organophosphorus compounds and are frequently simply termed as organophosphates (OP) (see also Section 28.5). There are two groups: G- and V-agents. "G" stands for Germany, because the first nerve agents such as sarin (GB), tabun (GA), and soman (GD) have been developed in Germany, whereas the so-called V-substances are Swedish advancements. These are more similar to acetylcholine, which is the natural substrate of the acetylcholinesterase (AChE) [130].

After intoxication, the major part of the nerve agents is quickly bound to enzymes (acetylcholinesterase, butyrylcholinesterase, and serinesterase) [5] and to proteins such as (serum) albumin [6, 7]. The remaining minor part (according to Polak and

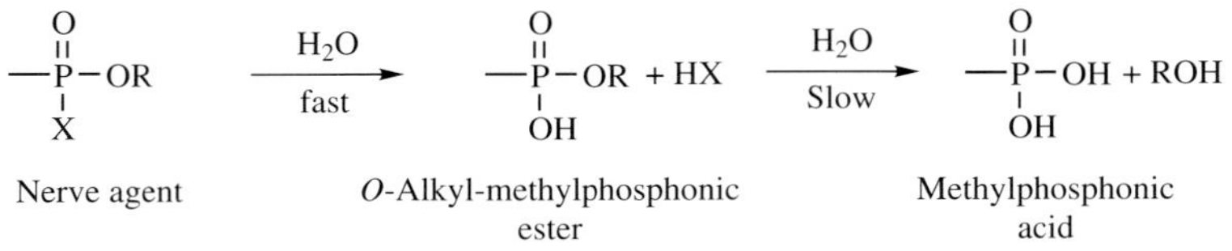

$$-\underset{X}{\overset{O}{\overset{\|}{P}}}-OR \xrightarrow[\text{fast}]{H_2O} -\underset{OH}{\overset{O}{\overset{\|}{P}}}-OR + HX \xrightarrow[\text{Slow}]{H_2O} -\underset{OH}{\overset{O}{\overset{\|}{P}}}-OH + ROH$$

Nerve agent — *O*-Alkyl-methylphosphonic ester — Methylphosphonic acid

Figure 32.2 Nerve agents: metabolic reaction scheme [26].

Cohen [8], 18% for sarin in rat plasma) is subject to spontaneous as well as enzymatic hydrolysis from phosphorylphosphatases [9–12] (see Figure 32.2).

The analytical principles regarding nerve agents have to consider these fast reactions: freely circulating nerve agents are only detectable immediately after exposure during acute phase. VX only remains unchanged for up to 40 h [13]. Products from hydrolysis are eliminated within a few days and can be detected in urine for up to a week. If, however, the exposure has been more than a week ago, one can only prove exposure to nerve agents from a laborious determination of that portion that was bound to proteins.

32.2.1
Free Nerve Agents

Specimen
EDTA plasma (deproteinized), 1 ml.

Equipment

GC–MS unit with cold injection system (e.g., Agilent, Gerstel)
Capillary column (5 MS, length 30 m, ID 0.25 mm, film thickness 0.25 μm; e.g., Factor 4, Varian)
LC–MS unit (e.g., Applied Biosystems, Perkin-Elmer) without separation column
Solid-phase extraction unit
Vacuum system.

Chemicals

Acetonitrile
Ammonia solution (0.5%)
Aqua bidest (chromatographic grade)
n-Hexane
Isopropanol (GC–MS grade)
Methanol
Perchloric acid (1 mol/l)
Potassium acetate solution (1 mol/l)
Solid-phase extraction cartridges (C8, Separtis).

Reagents
Reference stock solutions: 0.01% (v/v) of GA, GB, GD, GF, VX, VR, CVX, and DFP, each in hexane. The calculation of the concentration is based on the specific density (e.g., VX 1.01 g/ml, GF 1.13 g/ml) [130]. There are no data available for VR and CVX [131].

Calibration standard mixture: 100 pg/μl of GA, GB, GD, GF, VX, VR, VCX, and DFP, each in isopropanol. These mixtures are freshly prepared daily from the reference stock solutions.

Internal standard(IS): 535 pg/μl of diisopropyl fluorophosphate (DFP) in isopropanol (at room temperature stable for 6 months).

Sample preparation
For the determination of the free nerve agents, it is necessary to take EDTA blood within the first few hours after exposure and to centrifuge it immediately (EDTA inhibits phosphorylphosphatases by binding their cofactor Ca^{2+}). Seventy-five microliters of perchloric acid are pipetted to 1 ml separated plasma (to precipitate the proteins and impede enzymatic hydrolysis [5]). After mixing, 15 μl potassium acetate solution are added. The supernatant can be either mixed immediately with 20 μl internal standard solution and is then ready for further sample preparation, or frozen until analysis. Deproteinized samples are stable for 1 year at −60 °C.

A C8 cartridge is put into the vacuum system for each sample. The cartridge is preconditioned by successive rinsing with 1 ml methanol and 1 ml aqua bidest. The system may not run dry.

The sample is mixed with 20 μl of the internal standard solution and applied completely onto the respective cartridge. It is rinsed successively with 1 ml aqua bidest and 1 ml hexane and sucked to dryness by maximal vacuum for approximately 3 min. For elution, the solid phase is overlaid by 1 ml isopropanol, which is sucked into the sample vials by moderate vacuum. Five microliters of the eluate (which contains DFP and possibly GA, GB, GD, GF) is injected into the GC–MS system. For eluting VX, the cartridge is reconditioned by applying 1 ml of ammonia solution and afterward 1 ml of hexane. A layer of isopropanol is added, which is sucked through at a moderate rate. Ten microliters of this second eluate are injected into the LC–MS system.

GC–MS analysis
The cold on-column injection system of the gas chromatograph is run in the solvent venting mode and is programmed in such a way that at the initial temperature (e.g., 45 °C, maintained for 1.05 min) and with an open split, only the solvent is vented, whereas the analytes remain in the liner. The mode enables to inject 5 μl without overloading the system. Afterward, the split is closed and the liner is heated at a maximum rate (12 °C/s) to an analyte-dependent final temperature (e.g., 195 °C, maintained for 1 min). All compounds, which are now vaporized in the liner, reach directly the column. The temperature of the column is low (e.g., 50 °C, 1.8 min) to condense all substances vaporized in the liner at the beginning of the column and avoid broadening of peaks.

Column oven temperature program	
Initial temperature	50 °C (1.8 min)
Heating rate	10 °C/min (up to 80 °C)
Heating rate	20 °C/min (up to 180 °C)
Heating rate	30 °C/min (up to 270 °C)
Final temperature	270 °C (0.5 min)

This leads to the following retention times: 5.56 min (GB), 6.65 min (DFP; IS), 7.36/7.41 min (GD as a double peak), 8.08 min (GA), 8.80 min (GF), 12.19 min (VX), 12.34 min (VR), and 12.61 min (CVX).

The mass spectrometer is run in the selected ion monitoring mode (SIM). This mode increases the selectivity by using only a few typical mass ions per substance for detection.

Before determination of the samples, seven standard injections are carried out. The result of the first is always discarded. The mean of the remaining six area values is calculated and used for calibration. In case of long analytical series, three standard injections each are carried out regularly after some samples and at the end of the series. The determination is based on the mean value of two injections per sample.

For the analysis of VX, the LC–MS device is run in the flow injection analysis mode (FIA) (i.e., without a separating column) with acetonitrile as mobile phase. The ionization is effected in the ESI source at +4500 V. The LC–MS detection of VX (mass 267) by means of the molecule ion ($[M\text{-}H]^+$: m/z 268) is approximately 500 times more sensitive than the detection by the GC–MS system. The device itself optimizes the other parameters necessary for the detection. VR and CVX are isomers of VX and not distinguishable under these conditions. If necessary, the second eluate should also be investigated by GC–MS and metabolites or fluoridated derivatives have to be used.

The most frequently occurring nerve agents are detectable by the described procedure for screening. For an accurate determination of a particular nerve agent, the conditions of extraction and the parameters of the devices have to be adjusted accordingly (Table 32.3).

Details about other extraction procedures and methods of determination for free nerve agents are listed in Table 32.4.

An overview of the methods used in the trace analysis of free nerve agents is given by Benschop and de Jong [24].

32.2.2 Nerve Agent Metabolites [25]

Introduction

As mentioned in Section 32.2.1, nerve agents are hydrolyzed *in vivo* nonenzymatically, but predominantly by enzymes such as phosphorylphosphatases. Both processes lead to the same products as can be seen from the reaction pattern (Figure 32.2) [26]. Methylphosphonic acid esters are specific metabolites that give a clear indication to the involved nerve agent. In contrast to that, methylphosphonic

Table 32.3 Nerve agents: measurement by gas chromatography–mass spectrometry.

Substance	Extraction	Temperature program: cold on-column injection	Temperature program: oven	Selected ions (*m/z*)	Retention time (min)
GA	1 ml plasma[a], SPE: IST-C8 (100 mg), 1 ml *i*-propanol, vacuum	60 °C (0.05 min) → 195 °C (5 min); rate: 12 °C/s	50 °C (1 min) → 150 °C (0 min) → 200 °C (1.75 min); rate 1: 18 °C/min; rate 2: 40 °C/min	70, 106, 117, 133, 162	6.14
GB	5 ml plasma, SPE: IST-cyclohexyl (100 mg, 10 ml reservoir), 1 ml aqua bidest, 1 ml hexane, 500 µl *i*-propanol, vacuum	45 °C (1.05 min) → 150 °C (1 min); rate: 12 °C/s	60 °C (1.8 min) → 80 °C (0 min) → 100 °C (0 min) → 140 °C (0 min); rate 1: 10 °C/min; rate 2: 20 °C/min; rate 3: 5 °C/min	79, 81, 99, 125	6.4
GD	1 ml plasma, SPE: IST-C8 (25 mg), 100 µl *i*-propanol, centrifuge	40 °C (0.25 min) → 140 °C (1 min); rate: 12 °C/s	60 °C (1.8 min) → 200 °C (1.2 min); rate: 20 °C/min	126	5.79/5.82 (2 peaks)
GF	1 ml plasma, SPE: Baker C8 (100 mg), 1 ml *i*-butanol, vacuum	50 °C (1.05 min) → 300 °C (1 min); rate: 12 °C/s	60 °C (1.4 min) → 150 °C (3 min) → 200 °C (2 min); rate 1: 20 °C/min; rate 2: 20 °C/min	67, 99, 125, 137	6.2
VX	1 ml plasma, SPE: IST-C8 (100 mg), 1 ml aqua bidest, 1 ml 0.5% ammonia, 1 ml hexane, 1 ml *i*-propanol, vacuum	45 °C (1.05 min) → 195 °C; rate: 12 °C/s	50 °C (1.8 min) → 80 °C (0 min) → 180 °C (0 min) → 270 °C (0.5 min); rate 1: 10 °C/min; rate 2: 20 °C/min; rate 3: 30 °C/min	72, 85, 114, 127, 162	12.78
VR	See VX	45 °C (1.05 min) → 195 °C; rate: 12 °C/s	See VX	71, 86, 99, 139	12.86
DFP (IS)	See respective analyte	See respective analyte	See respective analyte	69, 101, 127, 169	Depending on oven temperature program

IS, internal standard; IST, name of manufacturer; MS, mass spectrometry; SPE, solid-phase extraction.
[a]No appropriate internal standard available.

Table 32.4 Unbound nerve agents: survey of measurement procedures.

Analyte	Method	System	Sample preparation	Detection limit	Reference
GB, GD	GC–MS	Blood (rabbit)	Liquid/liquid extraction: chloroform	n.d.	[5]
GB	Radiometric detection	Serum (rat)	Sephadex filtration	n.d.	[8]
GD	TLC	Tissue (several)	Liquid/liquid extraction: toluene	0.01 ng/g	[10]
GD	GC–MS	Liver: homogenate, perfusate (rat)	Liquid/liquid extraction: dichloromethane	n.d.	[11]
VX	HPLC–ECD	Blood (guinea pig, marmoset)	Liquid/liquid extraction: methanol/hexane 5 + 95 (v/v)	25 pg absolute	[13]
GD	GC–MS and AChE activity	Liver perfusate (rat, guinea pig, mouse)	Not applied	n.d.	[14]
GD	GC–MS	Blood, nerve tissue (mouse)	Liquid/liquid extraction: dichloromethane	<0.5 ng/ml and <0.5 ng/g	[15]
GB, GD	GC–NPD	Blood, stabilized (guinea pig)	SPE: Sep-Pak C18	1–5 pg/ml	[16]
GD	Enzymatic assay	Blood (rat)	Protein precipitation: perchloric acid	18 pg/ml	[17]
GB	HPLC with UV detection	Plasma, urine (rat)	SPE: Sep-Pak C18	10 μg/ml	[18]
GD	Immunoassay	Antibody dilution	Not applied	45.5 ng/ml	[19]
GB	Two-dimensional GC	Blood (guinea pig)	SPE: Sep-Pak C18	8.3 pg absolute	[20]
VX	CIEIA	Serum (rabbit)	Not applied	3.7 nmol/ml	[21]
GD	GC–MS/MS	CSF (pig)	SPE: C18 column	n.d.	[22]
VX	LC–MS (APCI)	Plasma, spiked (human)	Liquid/liquid extraction: hexane	n.d.	[23]

AChE, acetylcholinesterase; APCI, atmospheric pressure chemical ionization; CIEIA, competitive inhibition enzyme immunoassay; CSF, cerebrospinal fluid; ECD, electrochemical detector; GC, gas chromatography; HPLC, high-performance liquid chromatography; MS, mass spectrometry; n.d., no data; NPD, N–P detector; SPE, solid-phase extraction; TLC, thin-layer chromatography.

acid does not allow such an assignment, but is only indicative of the presence of some nerve agent. (Moreover, according to our own experience, extracts from plasma and urine contain a large (endogenous) peak that interferes with methylphosphonic acid.)

If additional information is required, one may search for the corresponding leaving group: fluoride in case of most G-agents [27], aminosulfide in case of V-agents [26], and the groups from ester cleavage (e.g., isopropanol [28]).

Figure 32.3 shows the structural formulas of the metabolites [29] from the first hydrolytic stage (see Figure 32.2) including the internal standards.

In addition to the main metabolite ethylmethylphosphonic acid (EMPA), further degradation products of VX are shown in Figure 32.4 including an extended hydrolysis pattern [31].

This hydrolysis pattern does not apply to tabun. As a phosphoric acid amide, it offers a third step of hydrolytic degradation and, in addition, the sequence, in which the metabolites are formed, is pH dependent [31, 32]. The respective reaction pattern is shown in Figure 32.5.

Two of the metabolites shown in Figure 32.5 still have the characteristic N–P bonding as a clear indicator of tabun. In contrast to that, the ethyl ester could also result from hydrolysis of organophosphate pesticides [32]. Phosphoric acid from the

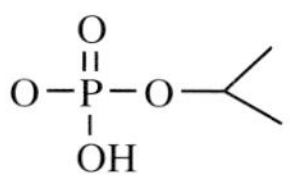

i-Propyl methylphosphonic acid (IMPA)

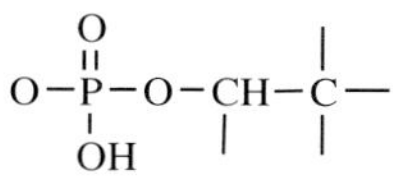

Pinacolyl methylphosphonic acid (PMPA)

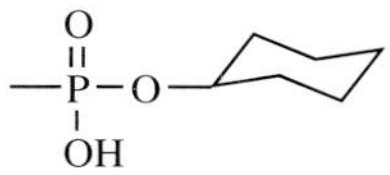

Cyclohexyl methylphosphonic acid (CHMPA)

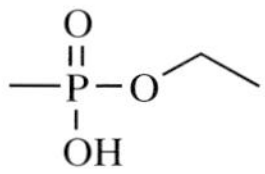

Ethyl methylphosphonic acid (EMPA)

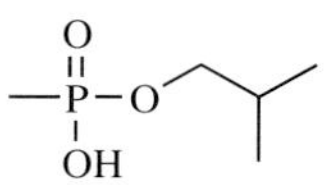

i-Butyl methylphosphonic acid (iBMPA)

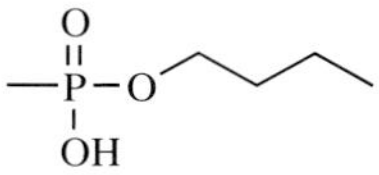

n-Butyl methylphosphonic acid (*n*BMPA)

Internal standard for plasma analysis

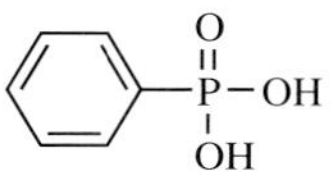

Internal standard for urine [30]

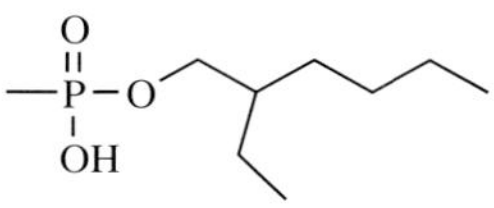

Phenylphosphonic acid (PhPA)

(2-Ethylhexyl) methylphosphonic acid (EHMPA)

Figure 32.3 Methylphosphonic acid derivatives [30].

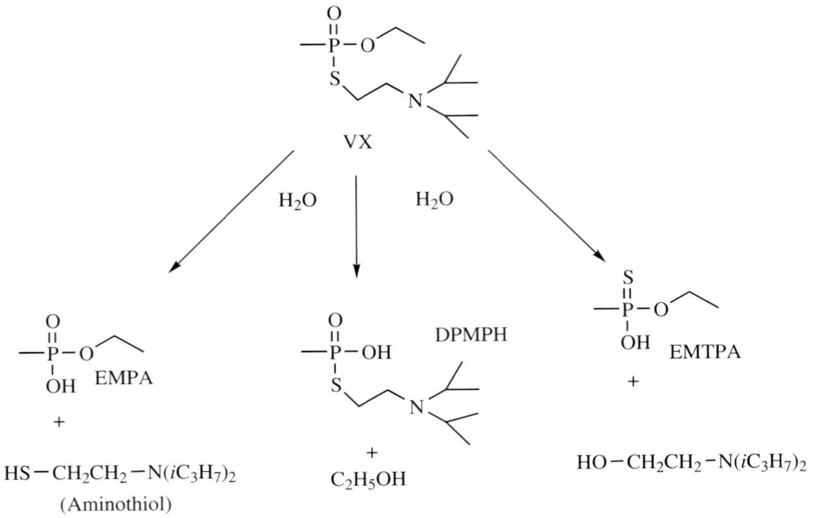

Figure 32.4 VX: hydrolysis [31].

final stage of the hydrolytic process is, of course, as useless for identification as methylphosphonic acid mentioned above.

In principle, there are two strategies for the detection of the metabolites:

(a) Direct determination via liquid chromatography–mass spectrometry (LC–MS). (Other HPLC methods can be neglected, because the molecules carry neither chromophores nor any other groups necessary for a sensitive and selective detection.)

(b) Determination by gas chromatography–mass spectrometry (GC–MS). As the metabolites are very polar and poorly volatile, they have to be converted prior to GC–MS analysis into nonpolar compounds that can be vaporized easily [33].

The following matrices are suitable for extraction: whole blood (hemolyzed) [34], serum or plasma [27, 29, 33–40], urine [35–37, 41–46], and saliva [37, 47].

32.2.2.1 **Phosphonic Acid Esters**

Liquid chromatography–mass spectrometry (plasma)

Specimen
EDTA plasma, 1 ml.

Equipment

LC–MS unit (e.g., Applied Biosystems, Perkin-Elmer)
Column: Hypercarb 100 mm length, 2.1 mm ID (e.g., ThermoHypersil)
Centrifugal evaporator (e.g., RVC, Christ)
Vacuum unit
Solid-phase extraction unit.

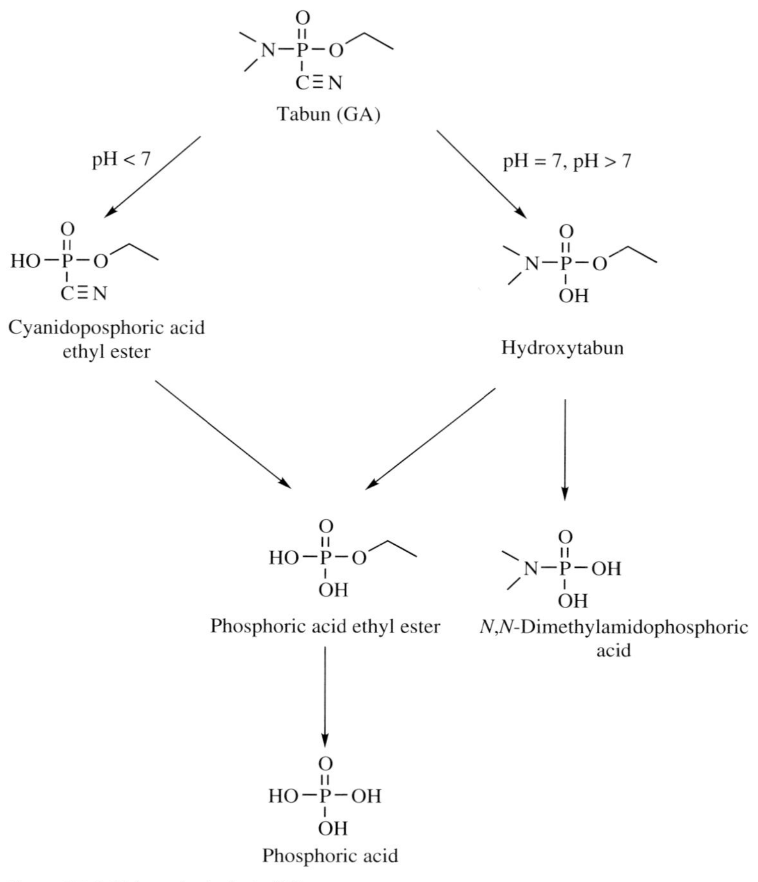

Figure 32.5 Tabun: hydrolysis [32].

Chemicals

Acetonitrile
Ammonia solution (1%)
Aqua bidest (HPLC grade)
Formic acid (2, 5, and 10%)
Methanol
Perchloric acid (1 mol/l)
Solid-phase cartridges ENV+ (25 mg, 1 ml; Separtis).

Reagents
Reference substances.

Stock solutions: EMPA, IMPA, *n*BMPA, *i*BMPA, CHMPA, and PMPA (Figure 32.3): each 10 ng/μl of methanol.

Working solution: Stock solutions of the reference substance and internal standard solution diluted to obtain each 100 pg/μl of formic acid (5%). To be prepared daily for analysis.

Internal standard solution:

Phenylphosphonic acid(PhPA): 10 ng/μl methanol.
Eluent (solid-phase extraction): ammonia solution (1%)/methanol 1 + 1 (v/v).

Gradient elution:

Mobile phase A: formic acid (2%).
Mobile phase B: acetonitrile.

Sample preparation
Ten microliters of the internal standard (phenylphosphonic acid) and 300 ml of perchloric acid are added to 1 ml EDTA plasma. After mixing, the mixture is centrifuged at 9500 × g for 10 min. (The acid is added to precipitate the plasma proteins and to suppress the dissociation of the analytes by strong acidification). In the meantime, a solid-phase cartridge for each sample is fixed to the unit and preconditioned with 1 ml methanol and 1 ml aqua bidest. (The solid phase shall not run dry.) After centrifugation, the supernatant is transferred onto the solid phase of the cartridge and is slowly sucked through by a moderate vacuum. (Should never run dry.) The cartridge is rinsed with 1 ml aqua bidest and sucked dry under maximum vacuum for 3 min. Two hundred fifty microliters of the eluent is applied and slowly sucked into an Eppendorf test tube. In the centrifugal evaporator, ammonia and methanol are removed from the eluate. The remaining 125 μl aqueous solution are thoroughly mixed with 125 μl formic acid (10%). After transferring the mixture into a sample vial, 20 μl are injected into the LC–MS system.

LC–MS analysis
A Hypercarb column is used to carry out the separation of the phosphonic acid esters by gradient elution:

Equilibration time 2 min: 100% A, flow rate 150 μl/min
0–2 min: 100% A at a flow rate of 150 μl/min
2–9 min: 80% A at a flow rate of 175 μl/min
9–16 min: 20% A at a flow rate of 175 μl/min
16–20 min: 100% A at a flow rate of 150 μl/min.

The following retention times are obtained: 3.8 min (EMPA), 4.7 min (IMPA), 9.6 min (*i*BMPA), 10.4 min (PhPA), 10.8 min (*n*BMPA), 12.1 min (CHMPA), and 12.3 min (PMPA).

The ionization is carried out in the cold electrospray source at −4500 V. The parameters for the detection of the following mass ions m/z (corresponding to the molecular mass minus 1) are optimized by the system: 123 (EMPA), 137 (IMPA), 151 (*i*BMPA and *n*BMPA), 157 (PhPA, IS), 177 (CHMPA), and 179 (PMPA).

At the beginning, as mentioned in Section 32.2.1, seven standard injections are carried out and each sample is injected twice.

Liquid chromatography–mass spectrometry (urine) [46]

Specimen
Urine, 4 ml.

Equipment

LC–MS unit (e.g., Applied Biosystems, Perkin-Elmer)
Column: Hypercarb 50 mm length, 2.1 mm ID (e.g., ThermoHypersil)
Centrifugal evaporator (e.g., RVC, Christ)
Solid-phase extraction unit.

Chemicals

Acetonitrile
Aqua bidest (chromatography grade)
Formic acid (2, 5, and 10%)
Hydrochloric acid (concentrated)
Methanol
Solid-phase cartridges ENV+ (25 mg, 10 ml; Separtis).

Reagents
Calibration standard solution and eluent, see the aforementioned procedure for plasma (Section 32.2.2).

Internal standard solution:(2-ethylhexyl)methylphosphonic acid (EHMPA) 10 ng/μl methanol.

Sample preparation
Four milliliters of urine are centrifuged at 2500 × g for 20 min (to remove the urine sediment). In the meantime, the solid-phase cartridges are preconditioned with 1 ml methanol and 1 ml aqua bidest. The separated supernatant is acidified with 300 μl concentrated hydrochloric acid before adding 40 μl of the internal standard solution. For dilution, 12 ml aqua bidest are added, and the mixture is applied stepwise to the solid-phase cartridge and sucked through at moderate vacuum and afterward at maximum vacuum for 3 min to dryness. 1 ml of the eluent is applied and slowly sucked into Eppendorf test tubes. In the centrifugal evaporator, methanol and ammonia are removed by vaporization (1 h). The remaining 500 μl aqueous solution are mixed with 500 μl of formic acid (10%) and transferred into a sample vial. Twenty microliters of the mixture are injected into the LC–MS unit.

LC–MS analysis
The separation of the phosphonic acid esters is carried out by a Hypercarb column and gradient elution as for plasma (see Section 32.2). The deviating column length results in different retention times: 2.4 min (EMPA), 2.9 min (IMPA), 5.8 min (*i*BMPA), 7.7 min (*n*BMPA), 10.2 min (CHMPA), 11.0 min (PMPA), and 13.8 min (EHMPA, IS). For ionization, detection, and calibration, see the procedure for plasma in this section.

Table 32.5 shows further procedures for the determination of phosphonic acid esters.

Table 32.5 Phosphonic acids: survey of LC–MS and CE procedures for measurement.

Analyte	System	Sample preparation	Method	Internal standard	Detection limit	Reference
EMPA, IMPA, BMPA, CMPA, PMPA	Serum	SPE: "molecular imprinted polymer" fibers	CE-UV ($\lambda = 210$ nm)	KH_2PO_4	EMPA: 500 ng/ml; IMPA: 300 ng/ml; BMPA: 100 ng/ml; PMPA: 100 ng/ml; CMPA: 200 ng/ml	[29]
MPA, EMPA, IMPA, PMPA	Serum	Ultrafiltration; dichloromethane: extraction; Ag^+ cartridge	Ion chromatography, indirect photometric detection	—	MPA: 40 ng/ml; EMPA: 80 ng/ml; IMPA: 80 ng/ml; PMPA: 80 ng/ml	[38]
IMPA	Serum	Liquid/liquid extraction	LC–MS	Deuterated IMPA	4 ng/ml	[39]
MPA, EMPA, IMPA, PMPA	Serum	SPE; derivatization: 4-bromophenacyl bromide	LC–MS	—	0.5–3 ng/ml	[40]
EMPA, IMPA, *i*BMPA, *n*BMPA, CMPA, PMPA	Urine	SPE	LC–MS	EHMPA	1–10 ng/ml	[46]
EMPA, IMPA, CMPA, PMPA	Urine (1); saliva (2)	(1) Protein precipitation; (2) protein precipitation and PTFE filter (0.45 μm)	LC–MS/MS	Deuterated IMPA	10–50 ng/ml	[47]

SPE, solid-phase extraction; CE, capillary electrophoresis; LC, liquid chromatography; MS, mass spectrometry.

High-performance liquid chromatography (other specimens)
As described before, it is also possible to extract phosphonic acid esters from other matrices such as whole blood, serum, or saliva (Table 32.5).

Gas chromatography–mass spectrometry

Introduction
If a LC–MS system is not available for the direct determination of phosphonic acid esters, they can also be detected by GC–MS. Prior to the analysis, however, a derivatization of the polar and low volatile compounds is necessary. Several derivatization reagents are available to esterify the hydroxy function of phosphonic acids (Figure 32.6). An overview of the methods for derivatization is given by Black and Muir [31].

In principle, the analytes are extracted first before they are converted with the aid of an appropriate reagent. As each of the derivatization reagents shown in Figure 32.6 is sensitive to hydrolysis, the samples have to be really dry to avoid a premature degradation of the reagent. Therefore, it is necessary to introduce the removal of water from the eluate as an additional step into routine sample preparation. The

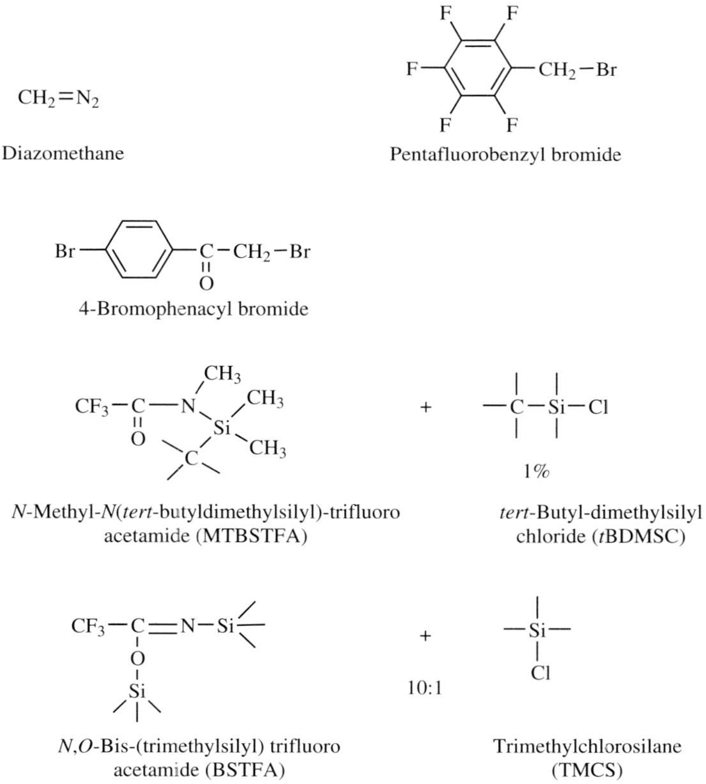

$CH_2{=}N_2$

Diazomethane

Pentafluorobenzyl bromide

4-Bromophenacyl bromide

N-Methyl-*N*(*tert*-butyldimethylsilyl)-trifluoro acetamide (MTBSTFA) + 1% *tert*-Butyl-dimethylsilyl chloride (*t*BDMSC)

N,O-Bis-(trimethylsilyl) trifluoro acetamide (BSTFA) + Trimethylchlorosilane (TMCS) 10:1

Figure 32.6 Phosphonic acids: reagents for derivatization.

conversion with pentafluorobenzyl bromide is described as an example of such a derivatization reaction [34–37, 45]. Regarding the use of the other reagents, the reader is referred to the overview from Ref. [31] and to Table 32.6, which lists further procedures.

Specimens

EDTA plasma, 1 ml
Urine, 4 ml.

Equipment

Oil bath
Heating block.

Chemicals (p.a. grade)

Aqua bidest (chromatography grade)
Carbon tetrachloride
Crown ether: 18-crown-6
Dichloromethane
Nitrogen (gaseous)
Pentafluorobenzyl bromide.

Reagents
Derivatization reagent: 10 μl pentafluorobenzyl bromide and 3 mg 18-crown-6 in 1 ml dichloromethane.

Sample preparation

Plasma: see LC–MS for plasma
Urine: see LC–MS for urine.

After solid-phase extraction, the eluate is evaporated to dryness with nitrogen in an oil bath at 85 °C. The derivatization reagent is added and the closed Eppendorf test tube is kept at 50 °C (1 h) in a heating block and shaken time and again. Then, the mixture is centrifuged and the organic phase is transferred into an Eppendorf test tube and evaporated to dryness. One hundred microliters of aqua bidest and 100 μl carbon tetrachloride are added to the residue and 1 μl of the organic phase is injected into the GC for analysis.

GC–MS analysis
In the splitless mode, 1 μl of the sample is injected at an injection liner temperature of 180 °C.

Column oven temperature program	
Initial temperature	60 °C (1 min)
First heating rate	20 °C/min up to 200 °C (4 min)
Second heating rate	30 °C/min up to 260 °C (not maintained)

Table 32.6 Phosphonic acids: survey of gas chromatographic procedures.

Analyte	System	Sample preparation	Derivatization reagent	Method	Internal standard	Detection limit	References
EMPA	Serum	Liquid/liquid extraction	MTBSTFA/1% BDMSC	GC–MS, GC–MS/MS	Diphenylmethane	<100 ng/ml	[26]
EMPA, IMPA, PMPA	Serum, urine	Anion exchanger (fluoride); after derivatization: Florisil	Pentafluorobenzyl bromide	GC–NICI-MS or GC–NICI-MS/MS	D_3-phosphonic acids	fmol	[31]
EMPA	Serum	Microultrafiltration, liquid/liquid extraction	BSTFA or MTBSTFA/1% BDMSC	GC–MS	Diphenylmethane	3 ng/ml	[33]
IMPA, PMPA, CMPA	Hemolysate, plasma, tissue (lung), urine (rat)	IMPA: SPE C18; PMPA: SPE C2; CMPA: SPE C2	Pentafluorobenzyl bromide	GC–MS	—	PMPA: 1 ng/ml; CMPA: 5 ng/ml; IMPA: 10 ng/ml	[34]
MPA, EMPA, IMPA, PMPA	Serum, urine, saliva	Serum: protein precipitation (acetone), cation exchanger; urine: cation exchanger; saliva: dilution, adjustment to pH 4.5	Pentafluorobenzyl bromide	GC–MS	Dipropylphosphate (before derivatization)	50 ng/ml	[37]
EMPA, IMPA, PMPA	Urine	SPE: three-layer cartridge: Ag^+, Ba^{2+}, H^+; derivatization of the eluate	BSTFA/TMCS 10 : 1	GC–FPD	Ethylphosphonic acid (EPA)	MPA: 60 ng/ml; IMPA: 25 ng/ml; EMPA: 25 ng/ml	[41]

(Continued)

Table 32.6 (*Continued*)

Analyte	System	Sample preparation	Derivatization reagent	Method	Internal standard	Detection limit	References
MPA, IMPA	Urine	SPE: C18 column and cation exchanger; derivatization of the eluate	MTBSTFA	GC–FPD	*p*-Cyanophenyldimethyl-phosphorothioate	10 ng/ml	[42]
EMPA, IMPA, PMPA, CMPA, EDAPA	Urine	Azeotropic distillation with acetonitrile	Diazomethane	GC–MS/MS	Deuterated phosphonic acids	EDAPA: 15–20 ng/ml; others: 2–4 ng/ml	[43]
EMPA, IMPA, *i*BMPA, CMPA, PMPA	Urine	Liquid/liquid extraction	Diazomethane	GC–MS/MS	d_5-EMPA, ^{13}C-derivatives	<1 ng/ml	[44]
EMPA, IMPA, *i*BMPA, CMPA, PMPA	Urine	SPE	Pentafluorobenzyl bromide	GC–MS/MS (ion trap)	Deuterated phosphonic acids	100 pg/ml; EMPA: 500 pg/ml	[45]
IMPA	Erythrocytes; tissue: cerebellum	Isolation of AChE; digestion with trypsin; alkaline phosphatase to split IMPA	1% Trimethylchlorosilane in (bis-trimethylsilyl) trifluoroacetamide	GC–MS (EI and CI)	—	No data	[48–50]

AChE, acetylcholinesterase; CI, chemical ionization; EI, electron impact ionization; FPD, flame photometric detection; GC, gas chromatography; MS, mass spectrometry; NICI, negative ion chemical ionization.

The MS detection is carried out in the SIM mode with the mass ions m/z 80, 181, and 256. The respective derivatized internal standard is taken into account in the evaluation of the results.

32.2.2.2 Other Nerve Agent Metabolites

Introduction
Indications of nerve agent poisoning can also be obtained by the detection of those products that, besides phosphonic acid esters, are formed from hydrolysis: fluoride [27] and cyanide, respectively [51], are produced at the first hydrolytic step of the G-agents and aminothiols (molecular structure: see hydrolytic pattern for VX, Figure 32.4) at the first hydrolytic step of the V-agents [26, 52]. (*In vivo* the aminothiols are subjected to methylation of the mercapto function, which is not shown in Figure 32.4.) The detection of the alcohols from ester cleavage at the second hydrolytic step is a further possibility [28].

Fluoride
Only a fluoride concentration exceeding 70 µmol/g of creatinine in urine is really typical for nerve agent poisoning (G-agent). Smaller amounts of fluoride may also be found in people not exposed to nerve agents [27]. Therefore, the diagnostic sensitivity of the fluoride concentration is insufficient and a respective procedure is not given here (Section 31.5).

Aminothiol [26]

Outline
Aminothiol and ethylmethylphosphonic acid are extracted together but determined separately. Aminothiol is a degradation product of the V-agents.

Specimen
Serum, 1 ml.

Equipment
GC–MS system with chemical ionization (CI).

Chemicals

Acetonitrile
Dichloromethane
Nitrogen, gaseous
Sodium sulfate, anhydrous.

Reagents
Calibrator stock solution: 2-(diisopropylaminoethyl)methylsulfide 100 µg/ml aqua bidest.

Calibrator working solution: to be prepared by spiking drug-free serum with calibrator stock solution.

Internal standard solution: 2-(diisopropylaminoethyl)methoxide 30 μg/ml dichloromethane.

Sample preparation

One milliliter of serum is extracted twice with 1 ml of dichloromethane. The aqueous phase can be further processed separately for the analysis of EMPA (see the aforementioned procedure). The organic phase, which can be easily separated from the aqueous phase after centrifugation, is dried over anhydrous sodium sulfate and evaporated to dryness in a mild nitrogen flow. The residue is dissolved in 100 μl of dichloromethane, and 1 μl is injected into the GC–MS system.

GC–MS analysis

The injection is carried out in the splitless mode at 270 °C.

Column oven temperature program	
Initial temperature	70 °C (2 min)
Heating rate	10 °C/min
Final temperature	300 °C (0 min)

The detection is carried out in the scan mode from *m*/*z* 50 to 800 at a scan rate of 0.5 s/scan. In case of chemical ionization with isobutane as reactant gas, *m*/*z* 176 is used for detection.

Isopropanol [28]

Isopropanol, which is produced from sarin by hydrolytic degradation of the ester, can be detected in urine. A respective procedure has been described by Minami (in little detail, however). The sample is injected into the GC–FID system without any sample preparation via a headspace unit.

Fluoroorganophosphates (from fluoride-induced reactivation of OP-inhibited BuChE)

A meaningful application of the methods described for the analysis of nerve agents is restricted to hours after exposure in case of the free nerve agents and to days in case of the phosphonic acid metabolites. For the detection of nerve agent poisoning after more than 1 week after exposure, the butyrylcholinesterase (BuChE) can be reactivated by fluoride to release intact fluoridated nerve agents [53–59]. The reaction is shown in Figure 32.7.

$$\text{Enzyme}-\text{O}-\overset{\text{O}}{\overset{\|}{\underset{\underset{\text{CH}_3}{|}}{\text{P}}}}-\text{OR} + \text{F}^-_{\text{excess}} \longrightarrow \text{Enzyme-OH} + \text{F}-\overset{\text{O}}{\overset{\|}{\underset{\underset{\text{CH}_3}{|}}{\text{P}}}}-\text{OR}$$

Figure 32.7 Organophosphorus compound–enzyme complex: cleavage by fluoride.

Annotations

(1) The original nerve agent is obtained in case of sarin and cyclosarin.
(2) Fluorotabun is produced in case of tabun-inhibited BuChE.
(3) V-agents: ethylsarin (VX), *iso*-butylsarin (VR), and *n*-butylsarin (CVX) are obtained from the procedure.

The procedure is not applicable to soman. Human enzymes split the pinacolyl group from the agent (so-called "aging") [60].

The corresponding reference compounds must be available for the identification of the reaction products.

Specimen
EDTA plasma, 500 μl.

Equipment

GC–MS system with cold injection system (CIS)
Capillary column: 5 MS 30 m, 0.25 mm ID, 0.25 μm film (e.g., Factor 4, Varian)
Solid-phase extraction unit.

Chemicals (p.a. grade)

Aqua bidest (chromatographic grade)
n-Hexane
Hydrochloric acid (1 mol/l)
Isopropanol
Methanol
Potassium fluoride solution (10 mol/l)
Solid-phase cartridges: Amino (e.g., Separtis)
Solid-phase cartridges: ENV+ (25 mg, 1 ml; e.g., Separtis).

Reagents
Standard stock solutions: 0.01% (v/v) of ethylsarin, GB, isobutylsarin, fluorotabun, *n*-butylsarin, GF, and DFP, each in hexane (see Section 32.2.1).

Calibration standard solution: GB, isobutylsarin, fluorotabun, *n*-butylsarin, GF, and DFP 100 pg/μl each in isopropanol (to be prepared daily from the stock solutions).

Internal standard solution (DFP): 535 ng/ml isopropanol.

Sample preparation
Five hundred microliters of EDTA plasma are mixed with 100 μl potassium fluoride solution and incubated at room temperature for 30 min [54]. The mixture is applied to an "Amino" solid-phase cartridge (preconditioned with 1 ml methanol and 1 ml hydrochloric acid), sucked through, and is directly collected. The cartridge is rinsed with 100 μl aqua bidest, which is added to the first eluate. If the mixture is turbid, the combined eluates are centrifuged. The supernatant is mixed with 20 μl of the internal standard solution and completely applied to the ENV+ cartridges (which have been

preconditioned with 1 ml of methanol and aqua bidest). Prior to the elution with 200 μl isopropanol, the cartridges are sucked dry for 3 min. 5 μl of the eluate are injected into the GC–MS system.

GC–MS analysis

The cold injection system is run in the solvent venting mode and programmed as follows (see also Section 32.2.1)

Initial temperature	40 °C (maintained for 0.25 min)
Heating rate	12 °C/s
Final temperature	190 °C (1 min, 0.8 min of it splitless)

Column oven temperature program	
Initial temperature	60 °C (1.8 min)
Heating rate	20 °C/min up to 120 °C
Heating rate	3 °C/min up to 125 °C
Heating rate	30 °C/min
Final temperature	200 °C (1 min)

The related retention times are: ethylsarin 4.33 min, sarin 4.49 min, *iso*-butylsarin 5.09 min, fluorotabun 5.18 min, DFP (IS) 5.28 min, *n*-butylsarin 5.37 min, and cyclosarin 7.69 min.

The detection is carried out in the SIM mode using the following mass ions: m/z 99, 101, and 126.

Protein–organophosphate complexes

Another possibility for a retrospective diagnosis of an exposure to nerve agents is the direct determination of the adducts from proteins and organophosphates. The technical progress in mass spectrometry has facilitated the determination of such adducts [61–64]. The protein is split into peptide fragments by a proteinase. The fragment carrying the organophosphate is picked out by means of an LC–MS/MS system. Up to now, methods have been described for phosphorylated acetylcholinesterase [61, 64], butyrylcholinesterase [65], as well as phosphorylated α-chymotrypsin [63].

Metabolites bound to acetylcholinesterase

A special extraction procedure has to be applied, if metabolites of nerve agents shall be investigated, which are bound to acetylcholinesterase. This approach was successfully applied several times using samples taken from victims of the terrorist attack in Tokyo in the year 1995. AChE was obtained from erythrocytes [48, 49, 66] and from cerebral tissue [49, 50]. As sample preparation is rather time consuming, the procedure should only be applied for samples that have been taken long time after exposure or if the aforementioned methods have not been able to detect the nerve agents or their metabolites in plasma or urine.

32.3
Vesicants/Blister Agents

Introduction

The group of vesicants comprises two classes of compounds: arsenic-containing vesicants (arsines) and mustard agents. The most important member of the arsines is 2-chlorovinylarsine dichloride or lewisite, named after W.L. Lewis, the person who synthesized it first. Other compounds of this class, such as methyldichloroarsine, ethyldichloroarsine, and phenyldichloroarsine (Klimmek, 1983), are in general not used as warfare agents and no analytical methods have been published. The mustards have in common chlorinated alkyl substituents bound to sulfur (sulfur mustard) or nitrogen (nitrogen mustard). Sulfur mustard (HD) is 2,2′-bis-dichloroethyl sulfide and in case of nitrogen mustard there are three different tertiary amines: HN-1, HN-2, and HN-3. The corresponding structural formulas are shown in Figure 32.8.

Annotation: The following methods are taken from the literature.

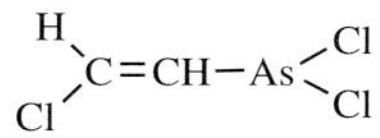

$$ClHC{=}CH{-}AsCl_2$$

2-Chlorovinylarsine dichloride
(Lewisite, L)

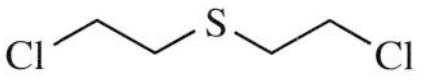

$$Cl{-}CH_2{-}CH_2{-}S{-}CH_2{-}CH_2{-}Cl$$

2,2′-Bis-chloroethyl sulfide
(sulfur mustard, HD)

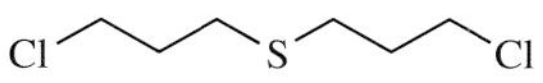

$$Cl{-}CH_2{-}CH_2{-}CH_2{-}S{-}CH_2{-}CH_2{-}CH_2{-}Cl$$

3,3′-Bis-dichloropropyl sulfide
(IS)
[67]

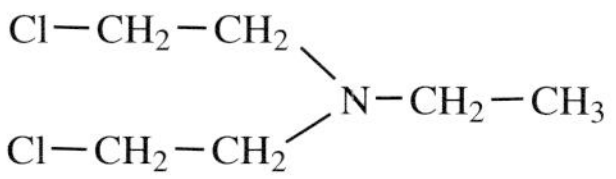

$$(Cl{-}CH_2{-}CH_2)_2N{-}CH_2{-}CH_3$$

N-Ethyl-2,2′-dichloroethyl amine
(nitrogen mustard, HN-1)

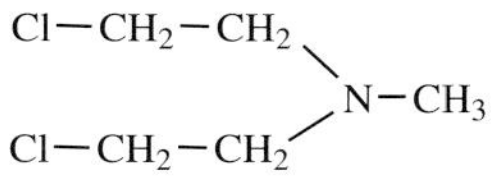

$$(Cl{-}CH_2{-}CH_2)_2N{-}CH_3$$

N-Methyl-2,2′-dichloroethyl amine
(nitrogen mustard, HN-2)

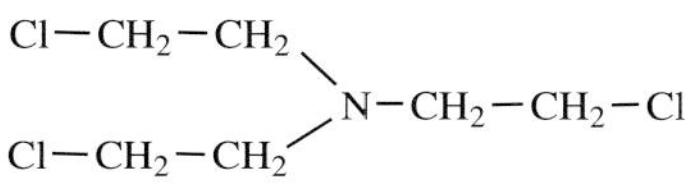

$$(Cl{-}CH_2{-}CH_2)_2N{-}CH_2{-}CH_2{-}Cl$$

Tris-(2-chloroethyl) amine
(nitrogen mustard, HN-3)

Figure 32.8 Vesicants or blistering agents [67].

32.3.1
Sulfur Mustard

Introduction

Sulfur mustard (HD) is regarded as a strong alkylating agent that forms adducts with endogenous substances such as proteins or DNA bases in a characteristic way. Besides this reaction, hydrolysis and the ensuing oxidation of the hydrolytic products [68] lead to a fast progressing metabolic degradation of HD. Nevertheless, there are reports that intact HD was detected in the blood of rats up to 8 h after exposure [67, 69]. In the urine [70] and various tissues of the victims of the Iran–Iraq war in the 1980s, unchanged HD was found even 7 days after exposure [71]. In a study using a ^{35}S-labeled HD, one-fifth of the original amount was found in the blood of rats 6 weeks after exposure [72].

The detection of sulfur mustard can be based on

(1) the unchanged molecule,
(2) the (numerous) metabolites in urine, and
(3) the protein and DNA adducts.

32.3.1.1 **Sulfur Mustard in Blood** [69]

Specimens
Whole blood, 6 ml, or plasma, 1 ml.

Equipment

GC–FID system
Cooling centrifuge
Vacuum concentrator with cooling trap
Solid-phase extraction unit.

Chemicals

Ethyl acetate
Methanol
Nitrogen (gaseous)
Sodium chloride (solid)
Solid-phase cartridge: C18 (100 mg, 3 ml; Baker).

Reagents
Stock solution: Sulfur mustard (HD) and 3,3′-dichloropropyl sulfide (IS), each 20 mg/ml isopropanol.

Calibration standard: HD 1 mg/ml isopropanol (at +4 °C stable for 3 weeks).

Internal standard solution: 2,2′-dichloropropyl sulfide 100 μg/ml isopropanol.

Sample preparation

(a) A 1.08 g sodium chloride and 50 μl internal standard solution are added to 6 ml whole blood. 3 ml of ethyl acetate are added to the mixture twice. After each time, the mixture is shaken for 30 s and then centrifuged at +4 °C and 3000 × g for 3 min. The separated organic phases are combined and evaporated to a volume of 100 µl, first in the vacuum concentrator, followed by a mild nitrogen flow. A 1 µl aliquot of this extract (which may be stored at +4 °C until analysis) is injected into the GC–FID system. The quantitative determination is related to the internal standard.

(b) 180 mg sodium chloride are added to 1 ml *plasma*. The solution is applied to a solid-phase cartridge (which has been equilibrated with 3 ml methanol). The cartridge is sucked dry for 10 min and the analytes are eluted with 2 × 0.5 ml of ethyl acetate. 1 µl of the combined eluates is injected into the GC–FID system.

GC analysis
The injection is carried out at 200 °C in the splitless mode.

Column oven temperature program	
Blood extract	
Initial temperature	70 °C (0.5 min)
Heating rate	30 °C/min up to 140 °C (4 min)
Heating rate	30 °C/min up to 230 °C
Plasma extract	
Initial temperature	70 °C (0.5 min)
Heating rate	30 °C/min up to 140 °C (2 min)
Heating rate	40 °C/min up to 200 °C

The detector temperature is 250 °C.

Annotation
Also, an HPLC method for HD extracted from whole blood has been described [73]. As HD shall be detected via a UV detector at the rather unspecific wavelength of 200 nm, the method may be not appropriate for practical use.

32.3.1.2 **Sulfur Mustard in Urine** [70]

Specimen
Urine, 20 ml.

Equipment
GC–MS system.

Chemicals

Dichloromethane
Diethyl ether
n-Hexane

Nitrogen (gaseous)
Silica gel (0.2–0.5 mm)
Sodium chloride (solid).

Reagents
Stock solution: sulfur mustard 0.01% (v/v) in hexane (density 1.268 g/ml).
Calibration standard solution: sulfur mustard 100 ng/ml dichloromethane (to be prepared daily).

Sample preparation
20 ml of urine are saturated with solid sodium chloride and shaken for 1 h with 5 ml diethyl ether in a vortex mixer. After centrifugation, the organic phase is separated and evaporated with a mild nitrogen flow at room temperature. The residue is dissolved in 1 ml dichloromethane. 100 mg silica gel are added and the mixture is shaken for 1 h. After centrifugation to separate the silica gel, the solvent is evaporated with nitrogen and the residue is redissolved in 10 μl diethyl ether. 1 μl aliquot are injected in the splitless mode into the GC–MS system.

GC–MS analysis
The injection is carried out manually (as the volume is too small for the autosampler) in the splitless mode at an injector temperature of 200 °C.

Column oven temperature program	
Initial temperature	50 °C (2 min)
Heating rate	20 °C/min
Final temperature	230 °C

Corresponding retention time for HD is 7.39 min.

The detection is carried out in the SIM mode using the following mass ions: m/z 63, 95, 109, and 123.

32.3.2 Sulfur Mustard Metabolites

Introduction

Sulfur mustard is mainly converted to water-soluble metabolites, which are eliminated via the kidneys [25, 74–90]. In experiments with rats, 60% of the dose was recovered in urine after 24 h of intraperitoneal administration of HD and several metabolites were detected [76]. Therefore, metabolites are determined preferably in urine.

As shown in Figure 32.9, the first step of HD metabolism in aqueous solution is the formation of the reactive episulfonium ion [77]. This intermediate stage is the starting point of three different metabolic reactions: (a) hydrolysis, (b) conjugation with glutathione, and (c) formation of protein and DNA adducts. Small amounts of thiodiglycol (TDG) and its oxidation product thiodiglycolsulfoxide may be detectable

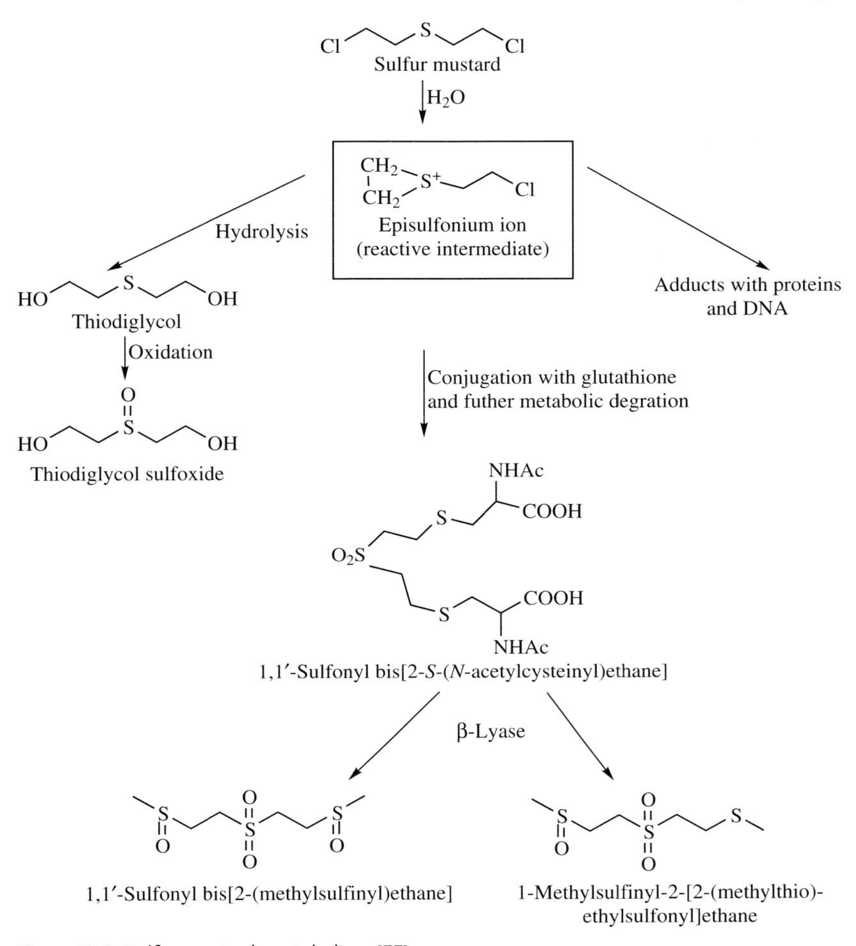

Figure 32.9 Sulfur mustard: metabolism [77].

in healthy people [77–79]. The methods for the detection of HD metabolites are summarized in Table 32.7.

32.3.2.1 **Thiodiglycol (TDG)**

Introduction

Thiodiglycol is a rather polar compound, which is not well amenable to gas chromatography. Therefore, it is converted into a nonpolar and more volatile compound by derivatization [31]. Several reagents appropriate for derivatization are listed in Table 32.7. The minor part of TDG in urine is unbound, whereas the major part is conjugated (glucuronidated). Conjugated TDG may be released by acidic or enzymatic hydrolysis to enable the determination of total TDG.

Table 32.7 Metabolites of sulfur mustard: survey of measurement procedures.

Analyte	System	Sample preparation	Derivatization reagent	Method	Internal standard	Detection limit	References
TDG	Urine[a]	SPE β-glucuronidase (to split glucuronidated TDG)	HFBA	Isotope dilution GC–MS/MS	$^{13}C_4$-TDG	<1 ng/ml	[78]
TDG	Blood, plasma, urine[a]	SPE C18 β-glucuronidase (to split glucuronidated TDG)	Pentafluorobenzoyl chloride	GC–MS (SIM)	d_4-TDG	<1 ng/ml	[80]
TDG	Urine[a]	SPE HCl (to split conjugated TDG)	Pentafluorobenzoyl chloride	GC–MS	d_4-TDG	1 ng/ml	[85]
TDG	Urine[b]	SPE HCl (to split conjugated TDG)	NaCl	GC–MS	Deuterated TDG	1 ng/ml	[87, 92]
TDG	Urine[b]	Liquid/liquid extraction	HFBA	GC–MS	d_8-TDG and thiodipropanol	<1 ng/ml	[88, 89]
TDG-sulfoxide	Urine	SPE (only) or SPE after reduction with $TiCl_3$ (to form TDG)	Pentafluorobenzoyl chloride	GC–MS	d_4-TDG-sulfoxide and d_4-TDG	2 ng/ml	[81, 82, 85]
1,1′-Sulfonyl-bis[2-*S*-(*N*-acetylcysteinyl) ethane]	Urine	SPE	—	ESI-LC–MS (negative mode)	—	0.5–1 ng/ml	[83]
Metabolites from β-lyase catalyzed reaction as SBMTE	Urine	SPE reduction with $TiCl_3$	HFBA	Isotope dilution GC–MS/MS	^{13}C-SBMTE	<1 ng/ml	[78]
Metabolites from β-lyase catalyzed reaction as SBMTE	Urine	SPE reduction with $TiCl_3$	—	GC–MS	—	2 ng/ml	[82, 85, 90]
Metabolites from β-lyase catalyzed reaction as SBMTE	Urine	SPE reduction with $TiCl_3$	—	GC–MS/MS	^{13}C-SBMTE	0.1 ng/ml	[86]
Metabolites from β-lyase catalyzed reaction (without reduction)	Urine	SPE	—	ESI-LC–MS/MS	—	0.1–0.5 ng/ml	[84]

ESI, electrospray ionization; LC, liquid chromatography; SIM, selected ion monitoring; SPE, solid-phase extraction; HFBA, heptafluorobutyric anhydride; SBMTE, 1,1′-sulfonyl-bis[2-methylthio]ethane.

[a]Nonconjugated and conjugated TDG.

[b]Conjugated TDG.

Thiodiglycol (blood plasma) [80]

Specimens
Blood or plasma, 1 ml.

Equipment

GC–MS system with chemical ionization (e.g., Finnigan 4600)
Round bottom flask, 50 ml
Rotary evaporator
Heating block.

Chemicals

Acetone
ClinElut (Analytichem International) and C18 Sep-Pak (Waters) cartridges, directly connected
Ethyl acetate
Methanol
Nitrogen (gaseous)
Pentafluorobenzoyl chloride (derivatization reagent)
Pyridine
Toluene.

Reagents
Stock standard solution: thiodiglycol 0.1, 1, 5, 10, 25, 50, 75, and 100 µg/ml acetone.
Internal standard solution: 1,1,1′,1′-tetradeuterothiodiglycol 1 µg/ml acetone.
Aquasil siliconization solution (for deactivation of glassware) (Pierce and Warriner).

Sample preparation
Annotation: before use, all glassware is deactivated by rinsing with the siliconization reagent.

1 ml blood or plasma is mixed with 10 µl internal standard solution and applied to a 3 ml ClinElut cartridge, which is directly connected to a C18 Sep-Pak cartridge. (Before use, the ClinElut cartridge has been rinsed with 3 × 5 ml of methanol and dried in a vacuum at 60 °C. The Sep-Pak cartridge has been preconditioned with 5 ml methanol and 5 ml ethyl acetate.) The cartridges are eluted five times with 5 ml ethyl acetate. The eluates are collected in a 50 ml round bottom flask and evaporated to 1 ml in a rotary evaporator at 30 °C. The concentrated eluates are transferred completely to an Eppendorf test tube by rinsing the flask with 0.5 ml methanol. The extract is dried at 40 °C with a flow of nitrogen and can be stored at −20 °C.

For derivatization, 50 µl pyridine and 10 µl pentafluorobenzoyl chloride are added to the dry residue. The mixture is shaken and then kept for 5 min at room temperature. It is made up to 500 µl with 440 µl toluene, mixed, and centrifuged. 2 µl of the supernatant are injected into the GC–MS system. At −20 °C, the derivatized sample is stable for 4 weeks.

GC–MS analysis
The injection of 2 µl is carried out in the splitless mode at 265 °C.

Column oven temperature program	
Initial temperature	90 °C (0.5 min)
Heating rate	25 °C/min up to 230 °C
Heating rate	4 °C/min
Final temperature	260 °C (2 min)

Thiodiglycol-bis-(pentafluorobenzoate) has a retention time of 10 min, whereas the (derivatized) deuterated internal standard is eluted 1.5 s earlier. Applying negative chemical ionization and methane as reactant gas, the mass spectrometer is programmed in the SIM mode for the detection of the following mass ions: m/z 510 (thiodiglycol derivate) and 514 (derivative of the internal standard).

The calibration of the system is based on a standard curve obtained from extracting blood or plasma spiked with TDG (concentration 1, 5, 10, 25, 50, 75, and 100 pg/µl).

Thiodiglycol (urine) [85]

Specimen
Urine, 2 ml.

Equipment

GC–MS system with chemical ionization (e.g., Finnigan 4600)
Round bottom flask (50 ml)
Rotary evaporator
Heating block.

Chemicals

Acetone
ClinElut tubes (3 ml)
Ethyl acetate
Florisil Sep-Pak cartridges (normal phase) (Waters)
Hydrochloric acid (concentrated)
Methanol
Nitrogen (gaseous)
Pentafluorobenzoyl chloride (derivatization reagent)
Sodium hydroxide solution (5 mol/l)
Toluene.

Reagents
Thiodiglycol standard: different concentrations in urine.
Internal standard solution: 1,1,1′,1′-tetradeuterothiodiglycol 1 µg/ml acetone.
Aquasil siliconization solution (for deactivation of glassware).

Sample preparation
1 ml urine is pipetted into an Eppendorf test tube and mixed with 50 μl IS solution. The sample is mixed vigorously and applied to a ClinElut tube, which is directly connected to a Florisil Sep-Pak cartridge.

(See the aforementioned blood/plasma procedure for deactivation of glassware, preconditioning of the cartridges, elution, and derivatization.)

For the analysis of total thiodiglycol, 100 μl concentrated hydrochloric acid are added cautiously to 1 ml urine (sample may effervesce) and incubated at 37 °C for 24 h. After incubation, the mixture is neutralized with sodium hydroxide solution. Further steps: see free TDG.

GC–MS analysis
See the blood/plasma procedure.

Annotation
Conjugated fraction can be calculated from the difference of total TDG and its free fraction.

Thiodiglycol (saliva)
A method for the extraction of TDG from saliva is proposed by Hayes *et al.* [47]. This method is probably of little practical relevance.

Thiodiglycol after cleavage from blood proteins

Outline
The measurement of thiodiglycol after release from blood proteins through alkaline hydrolysis is a special approach in the analysis of metabolites [91]. After liberation, TDG is derivatized with pentafluorobenzoyl chloride and analyzed by GC–MS as in the aforementioned procedures.

Specimens
EDTA whole blood (250 μl), erythrocyte concentrate (RBC) (500 μl), or EDTA plasma (1 ml).

Equipment

GC–MS system with chemical ionization (e.g., Agilent 6890 (GC) and Agilent 5973 (MS))
Capillary column (e.g., DB-5MS 30 m × 0.25 mm ID, 0.25 μm film; J&W Scientific)
Heating block
Shaking water bath.

Chemicals

Acetone
BondElut silica cartridges (100 mg; Varian)

Diethyl ether
Ethyl acetate
Hydrochloric acid (3 mol/l)
Pentafluorobenzoyl chloride (derivatization reagent)
Pyridine
Saline
Sodium hydrogen carbonate
Sodium hydroxide solution (1 mol/l)
Sodium sulfate (anhydrous).

Reagents
Hydrochloric acid (1%) (as precipitant for whole blood or RBC samples).

Internal standard solution: D_8-thiodiglycol 1 ng/μl acetone.

Standard solution: sulfur mustard (HD) 40 μmol/l saline.

Calibration solution: 9.9 ml pooled human plasma (anticoagulant: sodium citrate) is mixed with 100 μl HD solution. Aliquots of 1 ml are further diluted with pooled plasma to obtain the following concentrations: 25, 50, 100, 200, and 400 nmol/l. The spiked plasma samples are incubated in a shaking water bath at 37 °C for 18 h. The resulting thiodiglycol is extracted for establishing the standard curves.

Sample preparation [91]
The plasma proteins are precipitated with acetone and washed with diethyl ether. For whole blood or RBC, hydrochloric acid (1%) in acetone is used as precipitant (to remove the heme group from the hemoglobin). After drying the proteins at room temperature, they are weighed (to relate mass of thiodiglycol to protein mass).

A 200 μl sodium hydroxide solution is added to the proteins and the mixture is kept at 70 °C for 90 min. After neutralization with hydrochloric acid (60–70 μl), the sample is dried with 200 mg of anhydrous sodium sulfate and extracted with 1.1 ml ethyl acetate. A 500 μl aliquot of the organic phase is transferred and mixed with 10 μl IS solution and another 200 mg of sodium sulfate. 15 μl of pyridine and 20 μl pentafluorobenzoyl chloride are added and shaken (10 min). Twenty microliters of water and 20 mg sodium hydrogen carbonate are added (to destroy excess of derivatization reagent). The organic phase is separated and an aliquot (200 μl) is applied to the silica solid-phase cartridge. The cartridge is rinsed with 400 μl ethyl acetate. A 1 μl aliquot of the combined eluates is injected into the GC–MS system.

GC–MS analysis
The injection is carried out in the splitless mode at 250 °C using an autosampler.

Column oven temperature program	
Initial temperature	80 °C (1 min)
Heating rate	30 °C/min
Final temperature	225 °C (11 min)

The retention times of the derivatized internal standard and the derivatized thiodiglycol are 13.3 and 13.5 min, respectively. As methane serves as reactant gas for the detection, the following molecule ions $[M\text{-}H]^+$ are used: m/z 510 and 511 for the thiodiglycol derivative and 518 and 519 for the derivative of the internal standard.

Thiodiglycol sulfoxide in urine [81]

Specimen
Urine, 0.5 ml.

Equipment

GC–MS system with chemical ionization (e.g., Finnigan 4600)
Round bottom flask (50 ml)
Round bottom flask (25 ml)
Rotary evaporator
Heating block.

Chemicals

Acetone
ChemElut tubes (3 ml) (Analytichem International)
Chloroform
Ethyl acetate
Florisil Sep-Pak cartridges (Fisons)
Methanol
Nitrogen (gaseous)
Pentafluorobenzoyl chloride (derivatization reagent)
Pyridine
Toluene.

Reagents
Internal standard solution: 1,1,1′,1′-tetradeuterothiodiglycol sulfoxide 1 μg/ml methanol.

Thiodiglycol sulfoxide standard solution: 0.1–100 μg/ml urine.

Aquasil siliconization solution (for deactivation of glassware) (Pierce and Warriner).

Sample preparation
Annotation: prior to use, all glassware is deactivated with siliconization solution. 0.5 ml of urine are mixed with 50 μl of the internal standard solution and applied onto a 3 ml ChemElut tube. Low polar substances are removed by rinsing the ChemElut tube with a mixture of ethyl acetate/methanol 100 + 2 (v/v) 3 × 5 ml. TDG sulfoxide is eluted with 5 × 5 ml ethyl acetate/methanol 100 + 7 (v/v). The eluates are collected in a 50 ml flask and evaporated to dryness at 40 °C with a rotary evaporator. The residue is dissolved in 2 × 2 ml acetone and applied to a Florisil Sep-Pak cartridge. It is rinsed first with 2 × 5 ml chloroform/methanol 100 + 20 (v/v). The analytes are

eluted with 2×4 ml chloroform/methanol 50 + 50 (v/v). The eluates are collected in a 25 ml flask and evaporated to dryness at 40 °C with a rotary evaporator. The residue is redissolved in 2×0.5 ml methanol, transferred to an Eppendorf test tube, and evaporated to dryness at 60 °C with a stream of nitrogen.

For derivatization, 80 µl pyridine and 20 µl pentafluorobenzoyl chloride are added to the dry residue. The mixture is shaken vigorously and kept for 2 min at room temperature. The solution is made up to 500 µl with toluene, mixed again, and centrifuged. A 0.5 µl aliquot of the solution is injected in the splitless mode into the GC–MS system.

GC–MS analysis
The GC–MS analysis can be carried out as described for thiodiglycol, because the sulfoxide is reduced to thiodiglycol by the derivatization reaction [81].

Thiodiglycol sulfoxide in urine after reduction to TDG [81]

Specimen
Urine, 0.5 ml.

Equipment
See thiodiglycol in urine.

Chemicals
See thiodiglycol in urine.

Reagents
Titan(III) chloride: 15% solution in hydrochloric acid (20–30%).
Internal standard solution: 1,1,1′,1′-tetradeuterothiodiglycol 1 µg/ml methanol.

Sample preparation
0.5 ml titan chloride solution are added to 0.5 ml urine and incubated at 75 °C for 1 h. After reduction of the sulfoxide, extraction, derivatization, and measuring of the thiodiglycol thus produced are carried out as described before with regard to thiodiglycol.

1,1′-Sulfonyl-bis[2-S-(N-acetylcysteinyl)ethane [83]

Specimen
Urine, 2 ml.

Equipment

LC–MS/MS system (e.g., Finnigan TSQ 700)
Column: RPR-1 (150 mm × 2 mm ID; e.g., ThermoHypersil)
Solid-phase extraction unit
Evaporating centrifuge (e.g., SpeedVac, ThermoSavant).

Chemicals

Acetonitrile
Aqua bidest (chromatographic grade)
Formic acid (0.05%)
Methanol
Oasis HLB cartridge (3 ml, 60 mg; Waters)
Trifluoroacetic acid (0.1 and 10%).

Reagents

Trifluoroacetic acid 0.1% in methanol/aqua bidest 5 + 95 (v/v)
Trifluoroacetic acid 0.1% in methanol/aqua bidest 30 + 70 (v/v)
Trifluoroacetic acid 0.1% in methanol/aqua bidest 60 + 40 (v/v).

Mobile phase LC–MS:

A: Formic acid 0.05% in aqua bidest
B: Formic acid 0.05% in acetonitrile.

Sample preparation
Oasis HLB cartridges are preconditioned with methanol and 0.1% trifluoroacetic acid (TFA). 2 ml of urine are acidified with 100 μl trifluoroacetic acid (10%) and transferred to a cartridge (no sucking vacuum). The cartridges are successively rinsed with (1 ml each) 0.1% TFA, 0.1% TFA in methanol/water 5 + 95 (v/v), and 0.1% TFA in methanol/water 30 + 70 (v/v). Then they are sucked dry by a mild vacuum. For elution of the analytes, 2 × 0.5 ml 0.1% TFA in methanol/water (60 + 40) are sucked through under a mild vacuum. The collected eluates are evaporated to dryness at 50 °C in an evaporating centrifuge. The dry residue is redissolved in 0.2 ml aqua bidest. A 10 μl aliquot is injected into the LC–MS or LC–MS/MS system.

LC–MS analysis
The injection is performed manually via a Rheodyne valve. The chromatographic analysis is performed with a RPR-1 column with gradient elution: 5% B (0–2 min) to 90% B (15 to 20 min) at a flow rate of 0.2 ml/min. For ionization, an electrospray source (ESI) in the negative mode at – 5000 V is used. It enables to measure m/z 433.2 $[M\text{-}H]^-$ with transition to the daughter ion m/z 162.1.

β-Lyase products

Outline
It has been proven difficult to determine the two β-lyase products directly by GC–MS. Therefore, both are converted by reduction to 1,1′-sulfonyl-bis[2-(methylthio)ethane] (SBMTE) (Figure 32.10), which is easy to extract and amenable to GC–MS analysis [78, 86].

β-Lyase products (GC–MS)

Specimen
Urine, 0.5 ml.

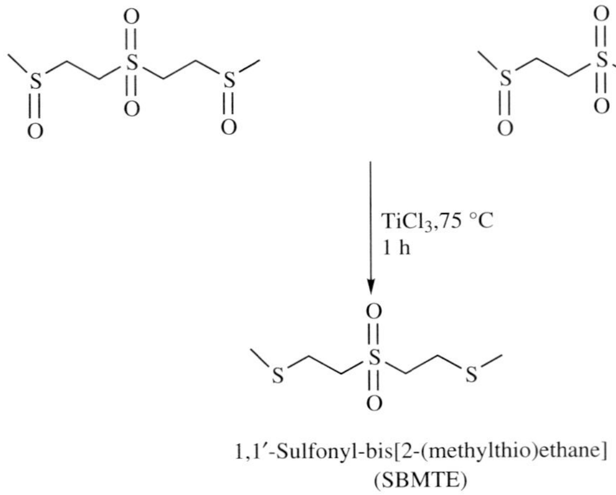

$TiCl_3$,75 °C
1 h

1,1′-Sulfonyl-bis[2-(methylthio)ethane]
(SBMTE)

Figure 32.10 Sulfur mustard: reduction of metabolites (from lyase catalyzed reaction) by titan(III) chloride [78, 86].

Equipment

GC–MS/MS system (e.g., Trace GC, Thermoquest, connected to Finnigan TSQ Triple-Quadrupol)
Capillary column: DB-5MS (30 m, 0.25 mm ID, 0.25 μm film)
Evaporating unit (e.g., TurboVap, Zymark).

Chemicals

Acetonitrile
ChemElut cartridges (3 ml; Varian)
Dichloromethane
Nitrogen (gaseous)
Sodium hydroxide solution (6 mol/l)
Toluene.

Reagents
Internal standard solution: ^{13}C-labeled 1,1′-sulfonyl-bis[2-(methylthio)ethane] 1 μg/ml acetonitrile (National Laboratory, Los Alamos, NM).

Standard solutions: 1,1′-sulfonyl-bis[2-(methylthio)ethane] 50 μg/ml acetonitrile, 1 μg/ml acetonitrile, and 10 ng/ml acetonitrile.

The solutions are used to spike urine: SBMTE 0.1, 0.25, 0.5, 1.5, 20, 50, and 100 μg/ml urine. After mixing thoroughly, the urine standards are frozen at −70 °C and are used for establishing calibration curves.

Reduction agent: titan(III) chloride (10%) in hydrochloric acid.

Sample preparation [86]
0.5 ml urine are mixed with 20 μl of the internal standard solution and 1 ml of the $TiCl_3$ solution in a 15 ml glass tube with screw cap and incubated at 75 °C for 1 h for

reduction of the lyase products. 2 ml of NaOH are added, and the mixture is shaken vigorously (vortex). The major part of $TiCl_3$ is precipitated at alkaline pH. The mixture is centrifuged at 4000 × g for 5 min and the supernatant is transferred to a 3 ml ChemElut cartridge (which has not been preconditioned). The analytes are eluted with 16 ml dichloromethane/acetonitrile 3 + 2 (v/v). The eluate is collected in conical 15 ml tubes and evaporated to dryness in a nitrogen stream at 40 °C (evaporating unit). The residue is redissolved in 20 µl toluene, mixed (vortex), and injected into the GC–MS/MS for analysis.

GC–MS analysis
1 µl of the toluene solution is injected in the splitless mode at 320 °C via autosampler.

Column oven temperature program	
Initial temperature	90 °C (2 min)
Heating rate	70 °C/min
Final temperature	320 °C (3 min)

Analyte and internal standard are eluted at the same time. Isobutane gas at a source pressure of 1.5 Torr is used for chemical ionization. SBMTE is distinguished from the internal standard:

SBMTE: m/z 215 with transition to m/z 75
IS: m/z 219 with transition to m/z 77.

β-Lyase products (LC–MS–MS)

Outline
In a recent approach, the analysis of the lyase products is directly performed via LC–MS/MS [84], obviating reduction.

Specimen
Urine, 2 ml.

Equipment

LC–MS/MS system
Column: Hypercarb (150 × 2.1 mm ID; 5 µm film thickness + Uniguard precolumn (ThermoHypersil)
Solid-phase extraction unit
Evaporating centrifuge (e.g., SpeedVac, ThermoSavant).

Chemicals

Acetonitrile
Aqua bidest (chromatographic grade)
Dithiothreitol (100 µg/ml in aqua bidest)

ENV+ cartridges (3 ml, 200 mg; Separtis)
Methanol.

Reagents
Internal standard solution (IS): mixture of the D_6-labeled derivatives of the β-lyase products (each 100 μg/ml aqua bidest).
Standard solutions of the β-lyase products: each 10 μg/ml urine.

Mobile phases for LC:
A: Ammonium formate (0.02 mol/l aqua bidest)
B: Ammonium formate (0.02 mol/l methanol).

Sample preparation [84]
2 ml of urine are mixed with 100 μl IS solution and applied without vacuum to the ENV+ cartridges (which have been preconditioned with 2 × 1 ml acetonitrile and 2 × 1 ml dithiothreitol). The cartridges are rinsed with 1 ml degassed aqua bidest and 0.5 ml acetonitrile/aqua bidest 20 + 80 (v/v) applying a mild excess pressure. The analytes are eluted with 2 × 0.5 ml acetonitrile/aqua bidest 60 + 40 (v/v) with excess pressure. The combined eluates are concentrated to approximately 250 μl in an evaporating centrifuge at 40 °C. The volume is made up to accurately 250 μl and 10 μl are injected into the LC–MS/MS system.

LC–MS analysis
The following gradient is programmed for separation: 5% B (0.5 min) to 80% B (15–20 min) at a flow rate of 0.2 ml/min.

The bis-sulfoxide is eluted at 14.3 min and the mono-sulfoxide at 18 min. The corresponding internal standards are eluted a few seconds earlier. The ionization is performed in the hot electrospray source (300 °C) at a voltage of +3000 V. Argon at a pressure of 0.8 Torr is used as collision gas. The following transitions are selected for the detection in the Triple-Quadrupol:

Bis-sulfoxide: m/z 247 to 183
Corresponding internal standard: m/z 253 to 186
Mono-sulfoxide: m/z 231 to 75.1
Corresponding internal standard: m/z 237 to 78.1.

Table 32.7 lists further procedures for the detection of the various metabolites of sulfur mustard.

32.3.2.2 **Protein–Sulfur Mustard Adducts**

Introduction
As known for nerve agents, also adducts of sulfur mustard (HD) with macromolecules are detectable for a longer period of time than the free parent compounds and their respective metabolites [93] (see also Section 32.3.1). The adducts of HD with hemoglobin [94–97], with serum albumin [98, 99], as well as with DNA [100–106] are of particular analytical importance. The analysis of globin adducts is more meaningful than the measurement of DNA adducts.

Hemoglobin is present in the body for 120 days [107], whereas alkylated DNA is subject to repair mechanisms of the body and less concentrated than hemoglobin or serum albumin [94, 95, 97].

N-terminal sulfur mustard–valine adduct of hemoglobin [96, 97]

Outline
The complex reaction basic for the detection is shown in Figure 32.11.

Specimen
Whole blood, 5 ml.

Equipment

GC–MS system with chemical ionization
Capillary column: CP-SIL 5 CB 50 m × 0.32 ID, 0.25 μm film (Chrompack) or HP 5 MS 30 m × 0.25 mm ID, 0.25 μm film (J&W Scientific)
Evaporating centrifuge (e.g., Jouan RC 10.10).

Cl–S–Cl
2,2′-Bis-dichlorethyl sulfide
(sulfur mustard)
+
$H_2N-CH(CH(CH_3)_2)-C{=}O$–Globin

↓

HO–S–NH–$CH(CH(CH_3)_2)$–CH=O–Globin

↓ PFPITC

S=C–N–R′; F_5C_6–N–C(=O); CH–$CH(CH_3)_2$

R′: $CH_2CH_2-S-CH_2CH_2OH$ → $CH_2CH_2-S-CH_2CH_2-OC(=O)-C_3F_7$
HFBA [97]
HFBI [96]

Figure 32.11 Isolation of alkylated terminal valine by modified Edman degradation [96, 97].

Chemicals

Acetonitrile
Aqua bidest (chromatographic grade)
Dichloromethane
Florisil cartridges (Waters)
Formamide
Magnesium sulfate
Methanol
Nitrogen, liquid
Pentafluorophenyl isothiocyanate (PFPITC)
Pyridine
Sodium carbonate (0.1 mol/l)
Toluene.

Reagents

Standard stock solution: Sulfur mustard (HD) 1 mol/l acetonitrile.

Internal standard stock solution: D_8-sulfur mustard 1 mol/l acetonitrile.

Calibration standards: 5 ml whole blood are incubated with 50 µl of the HD standard stock solution for 2 h at 37 °C, and 5 ml whole blood is incubated with 50 µl of the IS solution for 2 h at 37 °C.

Sample preparation [96]

Twenty milligrams of globin isolated according to Ref. [108] and alkylated with HD and 20 mg isolated globin alkylated with D_8-HD (see "Reagents") are dissolved with 2 ml formamide. The solution is mixed with 8 µl pyridine and 8 µl PFPITC and incubated in the heating block at 60 °C for 2 h. After cooling, the mixture is extracted three times with 1 ml toluene: each time, it is vigorously mixed for 30 s, centrifuged in the evaporating centrifuge at 15 × g, and frozen in liquid nitrogen (to enhance the separation of the phases). The combined toluene phases are washed with 2 × 0.5 ml aqua bidest, 0.5 ml sodium carbonate solution, and 0.5 ml aqua bidest. The organic phase is dried over magnesium sulfate and evaporated to dryness in the evaporating centrifuge. The residue is redissolved in 100 µl toluene. The Florisil cartridge is preconditioned with 2 ml methanol/dichloromethane 1 + 9 (v/v) and finally with 2 ml dichloromethane. The toluene solution is applied to the cartridge, which is rinsed with 2 ml dichloromethane. Thiohydantoin is eluted with 1.5 ml methanol/dichloromethane 1 + 9 (v/v). The eluate is dried and the residue is dissolved with 100 µl toluene. A 10 µl aliquot of 1-(heptafluorobutyryl) imidazol (HFBI) is added and the mixture is kept at 60 °C for 30 min. After cooling, the reaction mixture is washed twice with 100 µl aqua bidest, then with 100 µl sodium carbonate solution and 100 µl aqua bidest. The toluene phase is dried over magnesium sulfate and concentrated to 30 µl. 1 µl is injected into the GC–MS system.

GC–MS analysis

The injection is performed in the splitless mode.

Column oven temperature program	
Initial temperature	120 °C (5 min)
Heating rate	15 °C/min
Final temperature	275 °C (10 min)

The detection is based on chemical ionization using the following mass ions: m/z 564 (M-3HF, analyte) and 572 (M-3HF, IS).

Sulfur mustard–albumin adduct [99]

Outline

For the detection of sulfur mustard (HD) as an adduct to serum albumin, this protein is first separated by affinity chromatography or extracted from plasma according to the procedure of Bechtold [109]. It is then cleaved into peptide fragments by adding pronase as a digestive enzyme, and the alkylated tripeptide shown in Figure 32.12 is released.

Specimen

Plasma, 1 ml.

Equipment

LC–MS/MS system (e.g., API 4000 Triple-Quadrupol)
Column: Luna C-18 (150 mm × 1 mm ID, Phenomenex)
Affinity chromatography system with UV detector (FPLC)
Column: HiTrap Blue HP column and a PD 10 column (filled with 10 ml Sephadex-G-25) (Amersham, Sweden).

Chemicals

Acetonitrile
Ammonium hydrogen carbonate (50 mmol/l)
Formic acid 1% in aqua bidest
Potassium dihydrogen phosphate (50 mmol/l, pH 7)
Potassium chloride (1.5 mol/l)
Ultrafilter (10 kDa) used in the centrifuge (e.g., Centrex UF-2, Schleicher and Schüll).

O
||
H_2N-CH-C—Pro—Phe
|
CH_2
|
S ~ S ~ OH

Tripeptide adduct of sulfur mustard

Figure 32.12 Sulfur mustard–tripeptide adduct (obtained from alkylated albumin by pronase digestion) [99].

Reagents
Pronase (type XIV from *Streptomyces griseus*; EC 3.4.24.31): 10 mg/ml in ammonium hydrogen carbonate (50 mmol/l).

Affinity chromatography:
Buffer A: potassium dihydrogen phosphate pH 7 (50 mmol/l)
Buffer B: potassium dihydrogen phosphate 50 mmol/l and potassium chloride 1.5 mol/l pH 7.

Calibration standards: plasma incubated with different concentrations of HD.
Internal standard solution: Plasma incubated with D_8–HD 100 μmol/l.

LC–MS/MS:
Mobile phase A: formic acid 1% in aqua bidest
Mobile phase B: formic acid 1% in acetonitrile/aqua bidest 80 + 20 (v/v).

Sample preparation [99]
1 ml plasma is mixed with 50 μl plasma that has previously been incubated with D_8-HD. Then the serum albumin is separated from the plasma via affinity chromatography: the plasma is applied onto a HiTrap Blue HP column (which has been preconditioned with 10 ml of buffer A). The column is first rinsed with 7 ml of buffer A at a flow rate of 1 ml/min (to remove any impurities). Elution of the analytes is carried out with 7 ml of buffer B at a flow rate of 1 ml/min. A 2.5 ml aliquot of the fraction with UV absorption 280 nm is collected and applied to the PD 10 column (which has been previously equilibrated with 50 mmol/l ammonium hydrogen carbonate). The elution is first performed in two steps with 0.5 ml and 2.5 ml ammonium hydrogen carbonate solution. The first eluate is discarded and only the second eluate (2.5 ml) is collected.

(The affinity chromatography can be replaced by extracting the serum albumin according to the procedure of Bechthold [109], which involves a time-consuming multistep precipitation procedure.)

For enzymatic digestion, 0.25 ml of the serum albumin fraction are diluted with 0.5 ml ammonium hydrogen carbonate solution before adding 100 μl of a freshly prepared pronase solution. The mixture is incubated at 37 °C for 2 h and then filtered over a 10 kDa cutoff filter by centrifugation at 2772 × g. A 50 μl aliquot of the filtrate is injected into the LC–MS/MS system.

LC–MS analysis
The following gradient is used for the separation of the analytes: 100% A to 100% B within 25 min at a flow rate of 50 μl/min.

The detection of the tripeptide is based on the mass ion m/z 470 with transition to m/z 105 and on the mass ion m/z 478 of the deuterated tripeptide (IS) and its transition to m/z 113.

32.3.2.3 **Sulfur Mustard–DNA Adducts**

Introduction
Already several years ago, it was suspected that the strong alkylating effect of HD would also attack the DNA. As early as 1960, fragments of DNA were reacted with HD and isolated and characterized [100]. It could be shown that the nitrogen atom in 7-

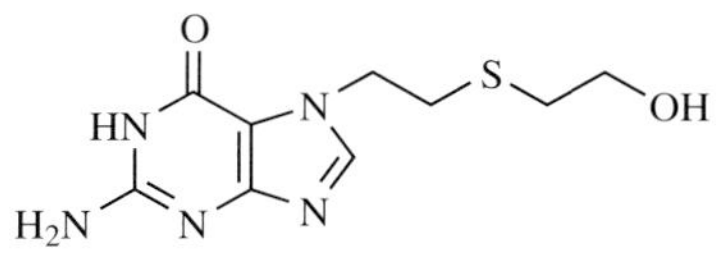

N7-(2-Hydroxyethyl thioethyl) guanine (HETEG)

Figure 32.13 N_7-Guanine–mustard adduct [25].

position of the guanine is most appropriate for the reaction. Later, Ludlum *et al.* also found the resultant 7-(2-hydroxyethylthioethyl)guanine (HETEG; for structural formula, see Figure 32.12) as the main product [105], although a number of other DNA adducts are formed with HD as well [101]. It was deduced from these results that HETEG could be used as a biomarker for the verification of an exposure to HD [102, 104, 106] (Figure 32.13).

7-(2-Hydroxyethylthioethyl)guanine

Outline

Specimen: EDTA blood, 300 μl.

The monoclonal antibodies against HETEG, which are required for this method, seem to be available only from the authors. Moreover, this method is extremely time and labor consuming, so for more details, the reader is referred to the literature [104].

32.3.3 Nitrogen Mustard

Introduction

From their molecular structure, the nitrogen mustards comprise HN-1, HN-2, and HN-3 (Figure 32.8). They are still classified as warfare agents [110], but they are by far less important as vesicants as compared to sulfur mustard [25]. For this reason, there is only one more recent GC procedure for measuring HN-1 and HN-3 in air [111] besides two older colorimetric methods [112, 113] for the determination of nitrogen mustard in aqueous solutions. The insight into the metabolism has been gained from the use of HN-2 [114] and from several metabolites of nitrogen mustard used as cytostatics in the treatment of cancer [25, 114–116]. The knowledge on metabolism provides valuable information for analysis [114, 117, 118].

32.3.4 Nitrogen Mustard Metabolites

32.3.4.1 Nitrogen Mustard Hydrolysis Products

Introduction

Nitrogen mustards are hydrolyzed *in vivo*:

HN1 → ethyldiethanolamine (EDEA)
HN2 → methyldiethanolamine (MDEAN)
HN3 → triethanolamine (TEA) as shown in Figure 32.14.

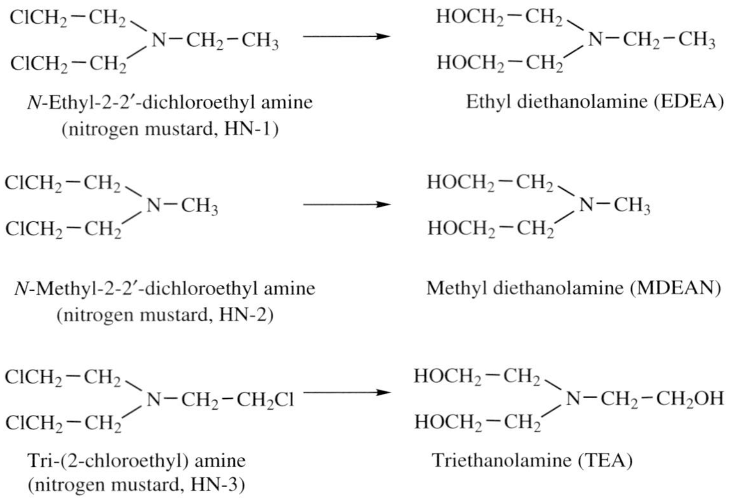

$$\begin{array}{l} ClCH_2-CH_2 \\ ClCH_2-CH_2 \end{array}\!\!>\!N-CH_2-CH_3 \longrightarrow \begin{array}{l} HOCH_2-CH_2 \\ HOCH_2-CH_2 \end{array}\!\!>\!N-CH_2-CH_3$$

N-Ethyl-2-2′-dichloroethyl amine (nitrogen mustard, HN-1) → Ethyl diethanolamine (EDEA)

$$\begin{array}{l} ClCH_2-CH_2 \\ ClCH_2-CH_2 \end{array}\!\!>\!N-CH_3 \longrightarrow \begin{array}{l} HOCH_2-CH_2 \\ HOCH_2-CH_2 \end{array}\!\!>\!N-CH_3$$

N-Methyl-2-2′-dichloroethyl amine (nitrogen mustard, HN-2) → Methyl diethanolamine (MDEAN)

$$\begin{array}{l} ClCH_2-CH_2 \\ ClCH_2-CH_2 \end{array}\!\!>\!N-CH_2-CH_2Cl \longrightarrow \begin{array}{l} HOCH_2-CH_2 \\ HOCH_2-CH_2 \end{array}\!\!>\!N-CH_2-CH_2OH$$

Tri-(2-chloroethyl) amine (nitrogen mustard, HN-3) → Triethanolamine (TEA)

Figure 32.14 Nitrogen mustards: hydrolysis.

The ethanolamines can be analyzed as trimethylsilyl (TMS) or *tert*-butyldimethylsilyl esters by gas chromatography [31]. However, an easier procedure is the determination in urine without derivatization via LC–MS [117, 118].

Specimen

Urine, 1 ml.

Equipment

LC–MS/MS system (Agilent 1100 LC system and API 3000 Triple-Quadrupol, Applied Biosystems)
Column: Xterra RP_{18}, 150 mm × 2.1 mm ID (Waters)
Solid-phase extraction unit
Evaporating unit (e.g., TurboVap LV, Zymark).

Chemicals

Acetic acid (1 mol/l)
Acetonitrile
Ammonia (3 mmol/l and 15 mmol/l)
Ammonia (2% in acetonitrile)
Aqua bidest (chromatographic grade)
Cation-exchange cartridges (e.g., LiChrolut 500 mg, EM Separations)
β-Glucuronidase 2 mg/ml aqua bidest
Methanol
Nitrogen, gaseous.

Reagents
Standard stock solution: A volumetric flask (25 ml) that contains 500 mg EDEA and 500 mg MDEAN is made up to 25 ml with an ammonia solution (15 mmol/l).

Based on this solution, eight different standard solutions in 15 mmol/l ammonia are prepared in such a way that aliquots of 50 µl are added to 1 ml urine to achieve concentrations between 1.6 and 270 ng/ml.

Internal standard solution: The stock solution ($^{13}C_4$-EDEA and $^{13}C_4$-MDEAN 1 µg/ml each in ammonia (15 mmol/l)) is diluted with ammonia (15 mmol/l) in such a way that an aliquot of 20 µl added to urine will give a concentration of 25 ng/ml.

Sample preparation [117]
In a centrifuge tube, 1 ml urine is spiked with 20 µl IS solution (final concentration 25 ng/ml). The mixture is diluted with 1 ml aqua bidest, applied to a preconditioned cation-exchange cartridge, and the liquid is then allowed to pass through the cartridge under gravity. The cartridge is rinsed with 3 ml acetonitrile and the analytes are eluted with 2% ammonia in acetonitrile. The cartridge is sucked dry by application of vacuum. The eluate is evaporated to dryness with nitrogen in an evaporating unit at 70 °C. The dry residue is redissolved in 0.5 ml ammonia (3 mmol/l) and 10 µl are injected into the LC–MS/MS system. The prepared samples are stable for 4 weeks at +4 °C.

To include the conjugated (glucuronidated) metabolites in urine, 1 ml urine is adjusted to pH 5.0 by adding 1 ml acetic acid. 0.5 ml β-glucuronidase solution are added and the mixture is incubated at 37 °C for 4 h. After hydrolysis, the samples are processed as described above [117].

LC–MS analysis
The separation of the analytes is performed by isocratic chromatography at 27 °C with ammonia pH 10.5 (3 mmol/l)/methanol 73 + 27 (v/v) at a flow rate of 200 µl/min. The corresponding retention times are 2.53 min for MDEAN and 2.99 min for EDEA. The respective ^{13}C-labeled internal standards are eluted at the same time. They are distinguished by the mass ions. The molecules are ionized in the electrospray source at 400 °C and +5000 V. The following molecule ions and transitions are obtained: MDEAN, m/z 120 to 102; ^{13}C-MDEAN, m/z 124 to 106; EDEA, m/z 134 to 116; ^{13}C-EDEA, m/z 138 to 120.

32.3.4.2 **Nitrogen Mustard–DNA Adducts**
Similar to sulfur mustard, nitrogen mustard also forms an alkylated guanine (see Figure 32.15), *N*-[2-(hydroxyethyl)-*N*-(2-(7-guaninyl)ethyl]methylamine (*N*-7-G), which can be used as a biomarker for HN-2 [114]:

The extraction of samples artificially produced from DNA taken from the calf thymus gland is described in detail. Unfortunately, an extraction method for blood, which might give more realistic picture for the handling of common samples, is not available. Moreover, the equipment used may be regarded as obsolete: (1) UV absorption (wavelengths 284 and 285 nm) is applied for the

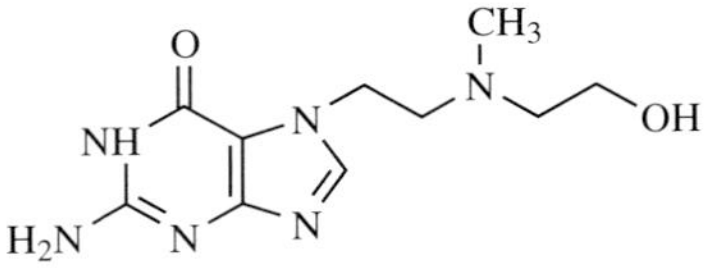

N-[2-(Hydroxyethyl)-*N*-(2-(7-guaninyl)ethyl)] methylamine (N-7-G)

Figure 32.15 Guanine alkylated with HN-2 [114].

detection of the guanine–nitrogen mustard adducts, which would not be sensitive enough for the verification of a nitrogen mustard warfare agent exposure. (2) The detection by a thermospray LC–MS system is not feasible, because these instruments are no more available since several years and have been replaced by the more sensitive electrospray devices. Therefore, the procedure is not described in more detail.

32.3.5
Lewisite (2-Chlorovinylarsine Dichloride)

Overview

Lewisite (L) (Figure 32.8) is a blister agent (vesicant). It is effective immediately after exposure, whereas HD exhibits a latency of several hours (Klimmek, 1983) [25]. It is indeed rather reactive in aqueous solutions:

(1) Lewisite is hydrolyzed to 2-chlorovinylarsonous acid (CVAA).
(2) Lewisite reacts with cysteine-containing proteins such as hemoglobin due to its great affinity to vicinal dithiol compounds [62, 77].

Due to its high reactivity, there is no procedure for the determination of intact lewisite in body fluids, although a procedure has been developed for the quality control of lewisite standards [119]. A hint on the presence of L could be obtained by AAS determination of arsenic, but for this purpose, the compound has first to be converted to arsine (AsH_3) [120]. A more specific proof for lewisite intoxication is based on the measurement of its characteristic hydrolysis product CVAA, which can be extracted from whole blood [62] and urine [121] (Figure 32.16).

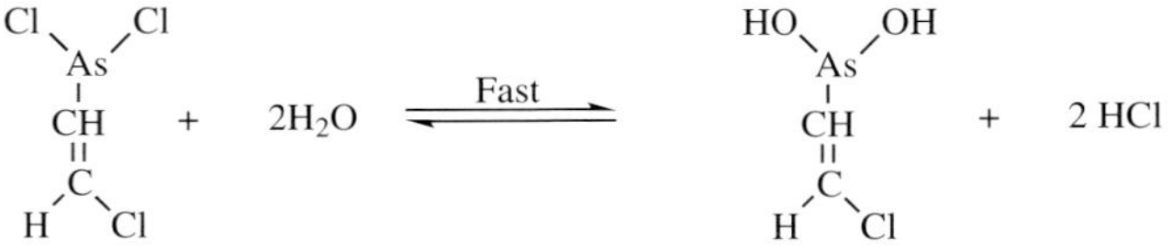

2-Chlorovinylarsine dichloride (Lewisite) + $2H_2O$ ⇌ (Fast) 2-Chlorovinylarsonous acid (CVAA) + 2 HCl

Figure 32.16 Lewisite: hydrolysis [25].

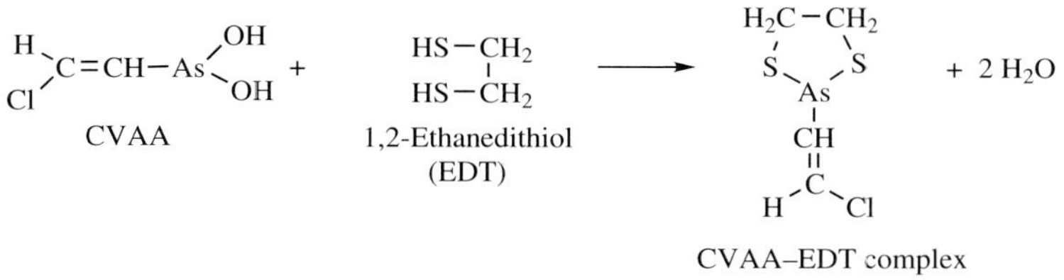

H(Cl)C=CH–As(OH)$_2$ + HS–CH_2–CH_2–SH ⟶ CVAA–EDT complex + 2 H_2O

CVAA | 1,2-Ethanedithiol (EDT) | CVAA–EDT complex

Figure 32.17 CVAA (from lewisite hydrolysis): derivatization with vicinal dithiol [122].

32.3.6 Lewisite Metabolites

32.3.6.1 **2-Chlorovinylarsonous Acid** [62]

Outline

Considering the fast hydrolysis of lewisite, it is assumed that the metabolite CVAA mainly contributes to the actual toxicity [122]. As CVAA has not been found from any other exposure, its detection is considered as a specific indication of the presence of lewisite. The polarity and low volatility of CVAA require the derivatization of the compound prior to the GC–MS analysis [31]. Due to its great affinity to dithiols, 2,3-dimercapto-1-propanol (British antilewisite, BAL) [62] or 1,2-ethanedithiol (EDT) [122] are particularly appropriate. The reaction is shown in Figure 32.17.

Specimen

Whole blood, 2 ml.

Equipment

GC–MS system (e.g., Agilent Technologies)
Column: CP-SIL 5 CB (50 m × 0.32 mm ID, 0.25 μm film thickness; Chrompack)
Solid-phase extraction unit
Heating block
Round bottom flasks (250 ml).

Chemicals

Acetonitrile
Aqua bidest (chromatographic grade)
Dichloromethane
1-(Heptafluorobutyryl)imidazole (HFBI; derivatization reagent)
Magnesium sulfate
Methanol
Solid-phase cartridges: C18 (Sep-Pak, Waters)
Toluene.

Reagents

Lewisite: 0.2–200 µmol/l acetonitrile
Dimercaprol (BAL): 50 mmol/l acetonitrile
Internal standard solution: phenylarsine–BAL complex (PAB) 0.18 mmol/l acetonitrile.

Sample preparation
2 ml whole blood are mixed with 10 µl BAL solution and 10 µl PAB solution (IS) and shaken overnight at room temperature. The solution is diluted with 18 ml water and loaded onto a C18 cartridge (which has been preconditioned with 10 ml methanol and 10 ml aqua bidest). The cartridge is rinsed with 2 × 10 ml aqua bidest. The analytes are eluted with 3 ml dichloromethane/acetonitrile 4 + 1 (v/v). The eluate is evaporated and the residue is suspended in 1 ml toluene. The solution is evaporated again (to eliminate any traces of water). The dried sample is dissolved in 100 µl toluene, mixed with 10 µl 1-(heptafluorobutyryl)imidazole, and incubated at 50 °C for 1 h. Finally, the sample is washed with 4 × 100 µl aqua bidest, dried over magnesium sulfate, and analyzed by GC–MS.

GC–MS analysis
1 µl toluene solution is injected on-column.

Column oven temperature program	
Initial temperature	120 °C (5 min)
Heating rate	8 °C/min
Final temperature	280 °C (5 min)

Retention time of CVAA-BAL-heptafluorobutyryl ester: 15.3 min; retention time of the IS (PAB-heptafluorobutyryl ester): 18.3 min. Each of the substances is present as two isomers at a ratio of 7 : 3. The detection is based on the mass ions m/z 454 (analyte) and m/z 470 (IS).

32.3.6.2 **2-Chlorovinylarsonous Acid** [121]

Specimen
Urine, 1 ml.

Equipment

GC–MS system (e.g., Agilent Technologies)
Column: DB-1, 30 m × 0.25 mm ID, 0.25 µm film thickness (J&W Scientific)
Solid-phase extraction unit.

Chemicals

Ethanol
Hydrochloric acid (1 mol/l)

Methanol
Nitrogen, gaseous
Solid-phase cartridges: C18 BondElut (Varian).

Reagents
Internal standard solution: phenylarsine oxide 3.6 μg/ml ethanol.
1,2-Ethanedithiol (derivatization reagent): 0.027% in ethanol.

Sample Preparation
1 ml of urine is adjusted to pH 6 by adding 1 mol/l hydrochloric acid, mixed with 100 μl internal standard solution, and then applied to a C18 cartridge (which has been preconditioned with methanol and water). The elution is carried out with 0.5 ml methanol, and the collected eluate is dried in a nitrogen stream. The residue is dissolved in the 1,2-ethanedithiol solution and mixed thoroughly. 1 μl is injected into the GC–MS system.

GC–MS Analysis
The injection is carried out in the splitless mode at 280 °C.

Column oven temperature program	
Initial temperature	50 °C (1 min)
Heating rate	50 °C/min
Final temperature	250 °C (6 min)

For the selected ion monitoring mode, the following mass ions are used: m/z 165, 167, and 228 for the derivatized analyte and m/z 215 and 244 for the derivatized internal standard.

32.3.6.3 Lewisite–Hemoglobin Adduct
The presence of the adduct has only been proven by radiometric detection after incubation of whole blood with ^{14}C-labeled lewisite [62]. Unfortunately, a detailed analytical procedure to enable a verification of lewisite intoxication in practice is not yet available.

32.4 Lung-Damaging Agents (Choking Agents)

Introduction
The term lung-damaging agent mainly comprises chloropicrin and phosgene (Figure 32.18), which are both capable of inducing toxic pulmonary edema (Klimmek, 1983). As a highly reactive acid chloride, phosgene reacts preferably with hydroxy, amino, and carboxylic acid groups. As mostly proteins bear numerous groups of this kind, tissues are damaged the most by phosgene. In contrast to that, chloropicrin belongs to the group of substances inducing methemoglobinemia (Klimmek, 1983).

$$\mathrm{Cl{-}C(=O){-}Cl}$$

Phosgene

$$\mathrm{Cl{-}CCl_2{-}NO_2}$$

Chloropicrin

Figure 32.18 Lung-damaging agents.

32.4.1 Phosgene

Overview

Due to its high reactivity, it is impossible to detect phosgene as an intact molecule. The only solution to the problem is to find a biomarker that can emerge exclusively from a reaction with phosgene:

(1) A specific phospholipid–phosgene adduct in mitochondria [123].
(2) An adduct of phosgene with hemoglobin and serum albumin (*in vitro*) [124].

In the following, only the globin and serum albumin adducts are discussed, because the sample preparation for the mitochondrial biomarker is very time consuming and suitable tissue that could serve as a specimen is not always available.

32.4.2 Phosgene Metabolites

32.4.2.1 **Phosgene–Globin Adducts** [124]

Specimen

Whole blood, 2 ml.

Equipment

LC–MS/MS system (e.g., Q-TOF, Micromass)
Column: PepMap C18, 15 cm × 0.3 mm ID (LC Packings).

Chemicals

Acetonitrile
Ammonium hydrogen carbonate (50 mmol/l)
Aqua bidest (chromatographic grade)
Disodium hydrogen phosphate
Formic acid
Methanol
Potassium chloride
Potassium dihydrogen phosphate
Sep-Pak C18 cartridges (Waters)
Sodium chloride
Sodium hydroxide solution (5 mol/l)

Trifluoroacetic acid (0.1% in aqua bidest)
Ultrafilters (10 kDa).

Reagents

Pronase (type XIV from *S. griseus*; EC 3.4.24.31): 10 mg/ml ammonium hydrogen carbonate (50 mmol/l)
Phosphate buffer system (PBS) pH 8.0 (consisting of NaCl 0.14 mol/l, KCl 2.6 mmol/l, Na_2HPO_4 8.1 mmol/l, and KH_2PO_4 15 mmol/l)
^{14}C-labeled phosgene.

LC–MS/MS:

Mobile phase A: 0.2% formic acid in aqua bidest
Mobile phase B: 0.2% formic acid in acetonitrile.

Sample preparation
2 ml whole blood are centrifuged and the erythrocytes are separated. The globin is isolated from the erythrocytes [108] and dissolved in 400 µl H_2O/PBS 1 + 1 (v/v). It is cleaved to peptides by pronase as described above (see Section Sulfur mustard–albumin adduct in 32.3.2.2). The digestion is stopped by filtration over a 10 kDa ultrafilter. The filtrate is applied to a Sep-Pak C18 cartridge (which has been conditioned previously with 5 ml methanol and 5 ml aqueous trifluoroacetic acid solution). The cartridge is rinsed successively with TFA solution, 2 ml of a mixture of TFA solution/acetonitrile 9 + 1 (v/v), and 4 ml of TFA solution/acetonitrile 8 + 2 (v/v). The analytes are eluted with 2 ml of TFA solution/acetonitrile 6 + 4 (v/v). A 10 µl aliquot of the eluate is injected into the LC–MS/MS system.

LC–MS analysis
The peptide fragments are separated by using the following gradient:

95% A (1–5 min; flow rate: 0.1 ml/min)
95% A to 80% B (5–90 min; flow rate: 0.5 ml/min).

The flow rates are reduced to 5 µl/min by a splitter before reaching the mass spectrometer. The particular peptide fragment (five amino acids) is identified by the mass ion m/z 512.3. If ^{14}C-labeled phosgene is involved, the resultant mass ion is m/z 514.2.

32.4.2.2 **Phosgene–Albumin Adduct** [124]

Specimen
Plasma, 1 ml.

Equipment

LC–MS/MS system (e.g., Q-TOF)
Column: PepMap C 18 (15 cm × 0.3 mm ID) (LC Packings).

Chemicals
Aqueous ammonium hydrogen carbonate solution (50 mmol/l).

Reagents
V8 protease solution (0.2 mg/ml ammonium hydrogen carbonate solution).

Sample preparation
Albumin is isolated as proposed by Bechtold *et al.* [109] or according to Section 32.3.1 by affinity chromatography.

It is dissolved in 100 μl ammonium hydrogen carbonate solution to obtain a final concentration of 6 mg albumin/ml. 100 μl of a freshly prepared V8 protease solution are added and the mixture is incubated at 37 °C for 2.5 h. The mixture is now ready for analysis by LC–MS/MS.

LC–MS analysis
The enzymatic hydrolysis releases a peptide with 19 amino acids from albumin, which has been carboxymethylated by reaction with phosgene. (Two lysine molecules of the peptide are bridged by carboxymethylation.) It is not possible to separate the particular peptide distinctly by chromatography and henceforth it can only be identified by the mass ion m/z 741.7 corresponding to $(M\text{-}3H)^{3+}$.

The sophisticated approach may not be feasible in practice and not appropriate for the verification of phosgene exposure in routine use.

Peptides modified from reaction with phosgene and released by other proteinases are even more difficult to identify [124].

32.4.3 Chloropicrin

Only a method for the determination of chloropicrin in air samples by thermodesorption using tenax has been published [125]. No method is available for its determination in body fluids.

32.5 Blood Agents

Hydrogen cyanide and its salts, the so-called cyanides, are termed blood agents. Procedures regarding the analysis of cyanides are described in detail in Section 31.4.

32.6 Psychotomimetic Agents

Psychotomimetic agents, mainly represented by the alkaloid quinuclidinyl benzilate (BZ) (Figure 32.19), do not kill soldiers but incapacitate them temporarily.

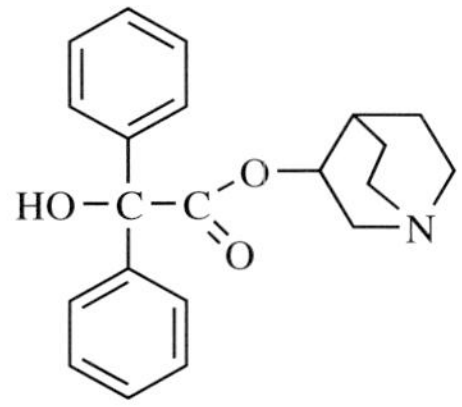

3-Quinuclidinyl benzilate (BZ)

Figure 32.19 Psychotomimetic agent [126].

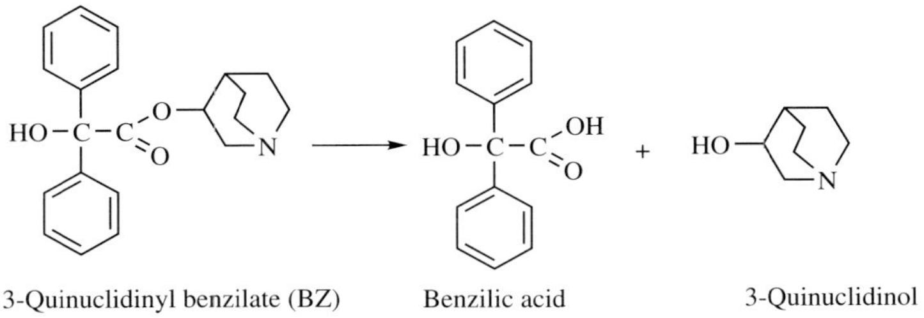

3-Quinuclidinyl benzilate (BZ) Benzilic acid 3-Quinuclidinol

Figure 32.20 3-Quinuclidinyl benzilate (BZ): hydrolysis [31].

In case of BZ, for example, the effects decline after approximately 24 h (Klimmek, 1983).

Literature regarding the analysis of BZ is scarce, because the verification of a BZ exposure is not considered as very important (as compared to nerve or blister agents). Animal studies have shown that BZ is not bound to macromolecules and is exclusively degraded by hydrolysis (Figure 32.20).

The detection of the products from hydrolysis, quinuclidinol and benzilic acid, may be performed according to the comprehensive screening procedure that has been developed by Black and Read for products from hydrolyzed warfare agents. Until now, this method has been applied only to the extract of hydrolysis products from soil samples obtained from bomb craters [127].

32.7 Eye Irritants

Eye irritants (lacrimators) (Figure 32.21) belonging to the "riot control agents" (RCA) have already been known during World War I and were sprayed by the US army in Vietnam. Nowadays, they are mainly used by the police to control public riots.

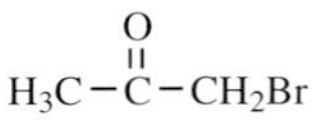

$H_3C-C(=O)-CH_2Br$

Bromoacetone (BA)

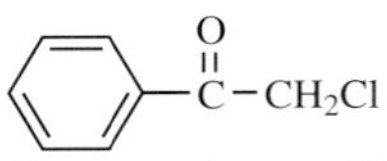

$C_6H_5-C(=O)-CH_2Cl$

Chloroacetophenone (CN)

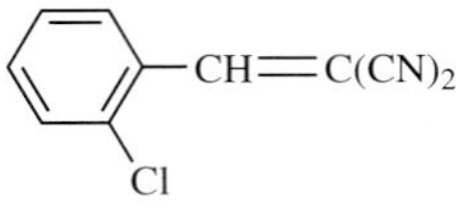

$CH{=}C(CN)_2$ (with Cl on the ring)

o-Chlorobenzylidene malonodinitrile

Figure 32.21 Eye irritants (lacrimators) (Klimmek, 1983).

Only methods are available for the identification of the pure substances, for example, in extracts from environmental samples [128].

32.8 Nose and Throat Irritants

The compounds Clark I and Clark II (Figure 32.22) were used by the German army in World War I as "mask breakers." Adamsite was synthesized in 1918 and is no longer applied today. The symptoms of poisoning with nose and throat irritants are occurring within a few minutes after exposure and decline within 1–3 h (Klimmek, 1983). The need for the verification of an exposure is not considered as very important. Therefore, methods for their extraction from biomatrix can hardly be found and appropriate biomarkers are not known.

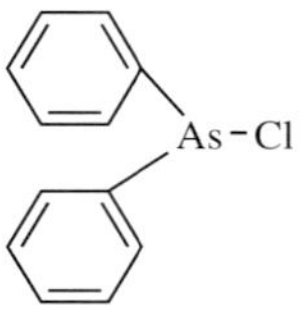

As−Cl

Diphenylarsine chloride (Clark I)

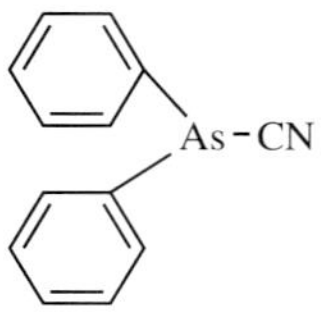

As−CN

Diphenylarsine cyanide (Clark II)

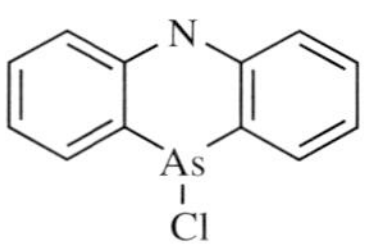

N

As

Cl

Phenarsazine chloride (Adamsite)

Figure 32.22 Nose-throat irritants (Klimmek, 1983).

References

References for Section 32.1

1 Kluge, S. and Szinicz, L. (2005) Acute Exposure Guideline Levels (AEGLs), toxicity and properties of selected chemical and biological agents. *Toxicology*, **214**, 268–270.

2 Robinson, J.P. and Leitenberg, M. (1971) The rise of CB weapons, in *The Problem of Chemical and Biological Warfare*, Vol. I, Stockholm International Peace Research Institute (SIPRI), Almquist and Wiksell, Stockholm.

3 Bismuth, C., Borron, S.W., Baud, F.J. and Barriot, P. (2004) Chemical weapons: documented use and compounds on the horizon. *Toxicology Letters*, **149**, 11–18.

4 Tu, A.T. (1996) Basic information on nerve gas and the use of sarin by Aum Shinrikyo. *Journal of the Mass Spectrometry Society of Japan*, **44**, 293–320.

References for Section 32.2

5 Singh, A.K., Zeleznikar, R.J., Jr and Drewes, L.R. (1985) Analysis of soman and sarin in blood utilizing a sensitive gas chromatography–mass spectrometry method. *Journal of Chromatography*, **324**, 163–172.

6 Black, R.M., Harrison, J.M. and Read, R.W. (1999) The interaction of sarin and soman with plasma proteins: the identification of a novel phosphonylation site. *Archives of Toxicology*, **73**, 123–126.

7 Li, B., Sedlacek, M., Manoharan, I., Boopathy, R., Duysen, E.G., Masson, P. *et al.* (2005) Butyrylcholinesterase, paraoxonase, and albumin esterase, but not carboxyl esterase, are present in human plasma. *Biochemical Pharmacology*, **70**, 1673–1684.

8 Polak, R.L. and Cohen, E.M. (1970) The binding of sarin in the blood plasma of the rat. *Biochemical Pharmacology*, **19**, 877–881.

9 Benschop, H.P. and de Jong, L.P.A. (1991) Toxicokinetics of soman: species variation and stereospecificity in elimination pathways. *Neuroscience & Biobehavioral Reviews*, **15**, 73–77.

10 Reynolds, M.L., Little, P.L., Thomas, B.F., Bagley, R.B. and Martin, B.R. (1985) Relationship between the biodisposition of [^{3}H]soman and its pharmacological effects in mice. *Toxicology and Applied Pharmacology*, **80**, 409–420.

11 Sterri, S.H., Valdal, G., Lyngaas, S., Odden, E., Malthe-Sørenssen, D. and Fonnum, F. (1983) The mechanism of soman detoxification in perfused rat liver. *Biochemical Pharmacology*, **32**, 1941–1943.

12 Katsemi, V., Lücke, C., Koepke, J., Löhr, F., Maurer, S., Fritzsch, G. *et al.* (2005) Mutational and structural studies of the diisopropylfluorophosphatase from *Loligo vulgaris* shed new light on the catalytic mechanism of the enzyme. *Biochemistry*, **44**, 9022–9033.

13 Van der Schans, M.J., Lander, B.J., van der Wiel, H., Langenberg, J.P. and Benschop, H.P. (2003) Toxicokinetics of the nerve agent ($\pm$)-VX in anesthetized and atropinized hairless guinea pigs and marmosets after intravenous and percutaneous administration. *Toxicology and Applied Pharmacology*, **191**, 48–62.

14 Fonnum, F. and Sterri, S.H. (1981) Factors modifying the toxicity of organophosphorus compounds including soman and sarin. *Fundamental and Applied Toxicology*, **1**, 143–147.

15 Beck, O., Holmstedt, B., Lundin, J., Lundgren, G. and Santesson, J. (1981) Quantitation of free soman in nervous tissue and blood. A preliminary communication. *Fundamental and Applied Toxicology*, **1**, 148–153.

16 Benschop, H.P., de Jong, L.P.A., and Langenberg, J.P. (1995) Inhalation toxicokinetics of C($\pm$)P($\pm$)-soman and ($\pm$)-sarin in the guinea pig, in *Enzymes of the Cholinesterase Family* (eds D.M. Quinn,

A.S. Balasubramanian, B.P. Doctor and P. Taylor), Plenum Press, New York, pp. 361–368.

17 Loke, W.K., Karlsson, B., Waara, L., Göransson-Nyberg, A. and Cassel, G.E. (1998) Enzyme-based microassay for accurate determination of soman in blood samples. *Analytical Biochemistry*, **257**, 12–19.

18 Abu-Qare, W. and Abou-Donia, B. (2001) Simultaneous analysis of sarin, pyridostigmine, bromide and their metabolites in rat plasma and urine using HPLC. *Chromatographia*, **53**, 251–255.

19 Miller, J.K. and Lenz, D.E. (2001) Development of an immunoassay for diagnosis of exposure to toxic organophosphorus compounds. *Journal of Applied Toxicology*, **21**, S23–S26

20 Spruit, H.E.T., Trap, H.C., Langenberg, J.P. and Benschop, H.P. (2001) Bioanalysis of the enantiomers of (±)-sarin using automated thermal cold-trap injection combined with two-dimensional gas chromatography. *Journal of Analytical Toxicology*, **25**, 57–61.

21 Ci, Y., Zhou, Y., Guo, Z., Rong, K. and Chang, W. (1995) Production, characterization and application of monoclonal antibodies against the organophosphorus nerve agent VX. *Archives of Toxicology*, **69**, 565–567.

22 Göransson-Nyberg, A., Fredriksson, S.-A., Karlsson, B., Lundström, M. and Cassel, G. (1998) Toxicokinetics of soman in cerebrospinal fluid and blood of anaesthetized pigs. *Archives of Toxicology*, **72**, 459–467.

23 Smith, J.R. (2004) Analysis of the enantiomers of VX using normal-phase chiral liquid chromatography with atmospheric pressure chemical ionization-mass spectrometry. *Journal of Analytical Toxicology*, **28**, 390–392.

24 Benschop, H.P. and de Jong, L.P.A. (2001) Toxicokinetics of nerve agents, in *Chemical Warfare Agents: Toxicity at Low Levels* (eds S.M. Somani and J.A. RomanoJr), CRC Press, Boca Raton, FL, pp. 25–81.

25 Black, R.M. and Noort, D. (2005) Methods for retrospective detection of exposure to toxic scheduled chemicals. Part A. Analysis of free metabolites, in *Chemical Weapons Convention Chemicals Analysis: Sample Collection, Preparation and Analytical Methods* (ed. M. Mesilaakso) John Wiley & Sons, Inc., New York, pp. 403–431.

26 Tsuchihashi, H., Katagi, M., Nishikawa, M., and Tatsuno, M. (1998) Identification of metabolites of nerve agent VX in serum collected from a victim. *Journal of Analytical Toxicology*, **22**, 383–388.

27 Hui, D. and Minami, M. (2000) Monitoring of fluorine in urine samples of patients involved in the Tokyo sarin disaster, in connection with the detection of other decomposition products of sarin and the by-products generated during sarin synthesis. *Clinica Chimica Acta*, **302**, 171–188.

28 Minami, M., Hui, D., Wang, Z., Katsumada, M., Inagaki, H., Li, Q. *et al.* (1998) Biological monitoring of metabolites of sarin and its by-products in human urine samples. *The Journal of Toxicological Sciences*, **23** (Suppl. II), 250–254.

29 Zi-Hui, M. and Qin, L. (2001) Determination of degradation products of nerve agents in human serum by solid phase extraction using molecularly imprinted polymer. *Analytica Chimica Acta*, **435**, 121–127.

30 Chaudot, X., Tambuté, A. and Caude, M. (2000) Selective extraction of hydrocarbons, phosphonates, and phosphonic acids from soils by successive supercritical fluid and pressurized liquid extractions. *Journal of Chromatography A*, **866**, 231–240.

31 Black, R.M. and Muir, B. (2003) Derivatization reactions in the chromatographic analysis of chemical warfare agents and their degradation products (review). *Journal of Chromatography A*, **1000**, 253–281.

32 Tan, S.L., Yong, Y.L., Chua, E., Leong, C.G. and Loke, W.K. (2005) Application of 1D

^{1}H–^{31}P inverse NMR spectroscopy and GC–MS for monitoring tabun urinary metabolites. Poster 9, Medizinische C-Schutz-Tagung, Munich.

33 Katagi, M., Nishikawa, M., Tatsuno, M. and Tsuchihashi, H. (1997) Determination of the main hydrolysis product of *O*-ethyl *S*-2-diisopropylaminoethyl methylphosphonothiolate, ethyl methylphosphonic acid, in human serum. *Journal of Chromatography B*, **689**, 327–333.

34 Shih, M.L., Smith, J.R., McMonagle, J.D., Dolzine, T.W. and Gresham, V.C. (1991) Detection of metabolites of toxic alkylmethylphosphonates in biological samples. *Biological Mass Spectrometry*, **20**, 717–723.

35 Fredriksson, S.-A., Hammarström, L.-G., Henriksson, L. and Lakso, H.-Å. (1995) Trace determination of alkyl methylphosphonic acids in environmental and biological samples using gas chromatography/negative-ion chemical ionization mass spectrometry and tandem mass spectrometry. *Journal of Mass Spectrometry*, **30**, 1133–1143.

36 Shih, M.L., McMonagle, J.D., Dolzine, T.W. and Gresham, V.C. (1994) Metabolite pharmacokinetics of soman, sarin and GF in rats and biological monitoring of exposure to toxic organophosphorus agents. *Journal of Applied Toxicology*, **14**, 195–199.

37 Miki, A., Katagi, M., Tsuchihashi, H. and Yamashita, M. (1999) Determination of alkylmethylphosphonic acids, the main metabolites of organophosphorus nerve agents, in biofluids by gas chromatography–mass spectrometry and liquid–liquid–solid-phase-transfer-catalyzed pentafluorobenzylation. *Journal of Analytical Toxicology*, **23**, 86–93.

38 Katagi, M., Nishikawa, M., Tatsuno, M. and Tsuchihashi, H. (1997) Determination of the main hydrolysis products of organophosphorus nerve agents, methylphosphonic acids, in human serum by indirect photometric detection ion chromatography. *Journal of Chromatography B*, **698**, 81–88.

39 Noort, D., Hulst, A.G., Platenburg, D.H.J.M., Polhuijs, M. and Benschop, H.P. (1998) Quantitative analysis of *O*-isopropyl methylphosphonic acid in serum samples of Japanese citizens allegedly exposed to sarin: estimation of internal dosage. *Archives of Toxicology*, **72**, 671–675.

40 Katagi, M., Tatsuno, M., Nishikawa, M. and Tsuchihashi, H. (1999) On-line solid-phase extraction liquid chromatography–continuous flow frit fast atom bombardment mass spectrometric and tandem mass spectrometric determination on hydrolysis products of nerve agents alkyl methylphosphonic acids by *p*-bromophenacyl derivatization. *Journal of Chromatography A*, **833**, 169–179.

41 Minami, M., Hui, D., Katsumata, M., Inagaki, H. and Boulet, C.A. (1997) Method for the analysis of the methylphosphonic acid metabolites of sarin and its ethanol-substituted analogue in urine as applied to the victims of the Tokyo sarin disaster. *Journal of Chromatography B*, **695**, 237–244.

42 Nakajima, T., Sasaki, K., Ozawa, H., Sekijima, Y., Morita, H. Fukushima, Y. *et al.* (1998) Urinary metabolites of sarin in a patient of the Matsumoto sarin incident. *Archives of Toxicology*, **72**, 601–603.

43 Driskell, W.J., Shih, M.L., Needham, L.L. and Barr, D.B. (2002) Quantitation of organophosphorus nerve agent metabolites in human urine using isotope dilution gas chromatography–tandem mass spectrometry. *Journal of Analytical Toxicology*, **26**, 6–10.

44 Barr, J.R., Driskell, W.J., Aston, L.S. and Martinez, R.A. (2004) Quantitation of metabolites of the nerve agents sarin, soman, cyclohexylsarin, VX, and Russian VX in human urine using isotope-dilution gas chromatography–tandem mass spectrometry. *Journal of Analytical Toxicology*, **28**, 372–378.

45 Riches, J., Morton, I., Read, R.W. and Black, R.M. (2005) The trace analysis of alkyl alkylphosphonic acids in urine using gas chromatography–ion trap negative ion tandem mass spectrometry. *Journal of Chromatography B*, **816**, 251–258.

46 Koller, M., Rehm, J. and Szinicz, L. (2005) Determination of phosphonic acids by LC–MS after their extraction from urine. Poster 9, Medizinische C-Schutz-Tagung, Munich.

47 Hayes, T.L., Kenny, D.V. and Hernon-Kenny, L. (2004) Feasibility of direct analysis of saliva and urine for phosphonic acids and thiodiglycol-related species associated with exposure to chemical warfare agents using LC–MS/MS. *Journal of Medical Chemical Defense*, **2**, 1–23.

48 Nagao, M., Takatori, T., Matsuda, Y., Nakajima, M., Iwase, H. and Iwadate, K. (1997) Definitive evidence for the acute sarin poisoning diagnosis in the Tokyo subway. *Toxicology and Applied Pharmacology*, **144**, 198–203.

49 Nagao, M., Takatori, T., Maeno, Y., Isobe, I., Koyama, H. and Tsuchimochi, T. (2003) Development of forensic diagnosis of acute sarin poisoning (review). *Legal Medicine*, **5**, S34–S40

50 Matsuda, Y., Nagao, M., Takatori, T., Niijima, H., Nakajima, M., Iwase, H. *et al.* (1998) Detection of the sarin hydrolysis product in formalin-fixed brain tissues of victims of the Tokyo subway terrorist attack. *Toxicology and Applied Pharmacology*, **150**, 310–320.

51 Copper, C.L. and Collins, G.E. (2004) Separation of thiol and cyanide hydrolysis products of chemical warfare agents by capillary electrophoresis. *Electrophoresis*, **25**, 897–902.

52 Bonierbale, E., Debordes, L. and Coppet, L. (1997) Application of capillary gas chromatography to the study of hydrolysis of the nerve agent VX in rat plasma. *Journal of Chromatography B*, **688**, 255–264.

53 Van der Schans, M.J., Noort, D., van der Schans, G.P., Langenberg, J.P. and Benschop, H.P. (2000) Verification and diagnosis of exposure to chemical warfare agents. The Second Singapore International Symposium on Protection Against Toxic Substances, Singapore.

54 Van der Schans, M.J., Polhuijs, M., Langenberg, J.P. and Benschop, H.P. (2002) Diagnosis, dosimetry and/or verification of exposure to organophosphorus nerve agents. Conversion of protein-bound phosphyl moieties into phosphofluoridates and methylphosphonic acid. Final Report, TNO.

55 Van der Schans, M.J., Polhuijs, M., van Dijk, C., Degenhardt, C.E.A.M., Pleijsier, K., Langenberg, J.P. *et al.* (2004) Retrospective detection of exposure to nerve agents: analysis of phosphofluoridates originating from fluoride-induced reactivation of phosphorylated BuChE. *Archives of Toxicology*, **78**, 508–524.

56 Polhuijs, M., Langenberg, J.P. and Benschop, H.P. (1997) New method for retrospective detection of exposure to organophosphorus anticholinesterases: application to alleged sarin victims of Japanese terrorists. *Toxicology and Applied Pharmacology*, **146**, 156–161.

57 Degenhardt, C.E.A.M., Pleijsier, K., van der Schans, M.J., Langenberg, J.P., Preston, K.E., Solano, M.I. *et al.* (2004) Improvements of the fluoride reactivation method for the verification of nerve agent exposure. *Journal of Analytical Toxicology*, **28**, 364–371.

58 Jakubowski, E.M., McGuire, J.M., Evans, R.A., Edwards, J.L., Hulet, S.W., Benton, B.J. *et al.* (2004) Quantitation of fluoride ion released sarin in red blood cell samples by gas chromatography–chemical ionization mass spectrometry using isotope dilution and large-volume injection. *Journal of Analytical Toxicology*, **28**, 357–363.

59 Noort, D. and Black, R.M. (2005) Methods for retrospective detection of exposure to toxic scheduled chemicals. Part B. Mass

spectrometric and immunochemical analysis of covalent adducts to proteins and DNA, in *Chemical Weapons Convention Chemicals Analysis: Sample Collection, Preparation and Analytical Methods* (ed. M. Mesilaakso), John Wiley & Sons, Inc., New York, pp. 433–451.

60 Millard, C.B., Kryger, G., Ordentlich, A., Greenblatt, H.M., Harel, M., Raves, M.L. *et al.* (1999) Crystal structures of aged phosphonylated acetylcholinesterase: nerve agents reaction products at the atomic level. *Biochemistry*, **38**, 7032–7039.

61 Elhanany, E., Ordentlich, A., Dgany, O., Kaplan, D., Segall, Y., Barak, R. *et al.* (2001) Resolving pathways of interaction of covalent inhibitors with the active site of acetylcholinesterases: MALDI-TOF/MS analysis of various nerve agent phosphyl adducts. *Chemical Research in Toxicology*, **14**, 912–918.

62 Fidder, A., Noort, D., Hulst, A.G., de Jong, L.P.A. and Benschop, H.P. (2000) Biomonitoring of exposure to lewisite based on adducts of haemoglobin. *Archives of Toxicology*, **74**, 207–214.

63 Tsuge, K. and Seto, Y. (2002) Analysis of organophosphorus compound adducts of serine proteases by liquid chromatography–tandem mass spectrometry. *Journal of Chromatography B*, **776**, 79–88.

64 Jennings, L.L., Malecki, M., Komives, E.A. and Taylor, P. (2003) Direct analysis of the kinetic profiles of organophosphate–acetylcholinesterase adducts by MALDI-TOF mass spectrometry. *Biochemistry*, **42**, 11083–11091.

65 Fidder, A., Hulst, A.G., Noort, D., de Ruiter, R., van der Schans, M.J., Benschop, H.P. *et al.* (2002) Retrospective detection of exposure to organophosphorus anti-cholinesterases: mass spectrometric analysis of phosphorylated human butyrylcholinesterase. *Chemical Research in Toxicology*, **15**, 582–590.

66 Nagao, M., Takatori, T., Matsuda, Y., Nakajima, M., Niijima, H., Iwase, H. *et al.* (1997) Detection of sarin hydrolysis products from sarin-like organophosphorus agent-exposed human erythrocytes. *Journal of Chromatography B*, **701**, 9–17.

References for Section 32.3

67 Maisonneuve, A., Callebat, I., Debordes, L. and Coppet, L. (1993) Biological fate of sulphur mustard in rat: toxicokinetics and disposition. *Xenobiotica*, **23**, 771–780.

68 Brimfield, A.A., Zweig, L.M., Novak, M.J. and Maxwell, D.M. (1998) *In vitro* oxidation of the hydrolysis product of sulfur mustard, 2,2′-thiobis-ethanol, by mammalian alcohol dehydrogenase. *Journal of Biochemical and Molecular Toxicology*, **12**, 361–369.

69 Maisonneuve, A., Callebat, I., Debordes, L. and Coppet, L. (1992) Specific and sensitive quantitation of 2,2′-dichlorodiethyl sulphide (sulphur mustard) in water, plasma and blood: application to toxicokinetic study in the rat after intravenous intoxication. *Journal of Chromatography*, **583**, 155–165.

70 Vycudilik, W. (1985) Detection of mustard gas bis(2-chloroethyl)sulfide in urine. *Forensic Science International*, **28**, 131–136.

71 Drasch, G., Kretschmer, E., Kauert, G. and von Meyer, L. (1987) Concentrations of mustard gas [bis(2-chloroethyl)sulfide] in the tissues of a victim of a vesicant exposure. *Journal of Forensic Sciences*, **32**, 1788–1793.

72 Hambrook, J.L., Howells, D.J. and Schock, C. (1993) Biological fate of sulphur mustard (1,1′-thiobis(2-chloroethane)): uptake, distribution and retention of ^{35}S in skin and in blood after cutaneous application of ^{35}S-sulphur mustard in rat and comparison with human blood *in vitro*. *Xenobiotica*, **23**, 537–561.

73 Dangi, R.S., Jeevaratnam, K., Sugendram, K., Malhotra, R.C. and Raghuveeran, C.D. (1994) Solid-phase extraction and reversed-phase high-performance liquid chromatographic determination of sulphur mustard in blood. *Journal of Chromatography B*, **661**, 341–345.

74 Davison, C., Rozman, R.S. and Smith, P.K. (1961) Metabolism of bis-β-chloroethyl sulfide (sulfur mustard gas). *Biochemical Pharmacology*, **7**, 65–74.

75 Roberts, J.J. and Warwick, G.P. (1963) Studies of the mode of action of alkylating agents. VI. The metabolism of bis-2-chloroethylsulphide (mustard gas) and related compounds. *Biochemical Pharmacology*, **12**, 1329–1334.

76 Black, R.M., Brewster, K., Clarke, J., Hambrook, J.L., Harrison, J.M. and Howells, D.J. (1992) Biological fate of sulphur mustard, 1,1′-thiobis (2-chloroethane): isolation and identification of urinary metabolites following intraperitoneal administration to rat. *Xenobiotica*, **1** (22), 405–418.

77 Noort, D., Benschop, H.P. and Black, R.M. (2002) Biomonitoring of exposure to chemical warfare agents: a review. *Toxicology and Applied Pharmacology*, **184**, 116–126.

78 Boyer, A.E., Ash, D., Barr, D.B., Young, C.L., Driskell, W.J., Whitehead, R.D., Jr *et al.* (2004) Quantitation of the sulfur mustard metabolites 1,1′-sulfonylbis[2-(methylthio)ethane] and thiodiglycol in urine using isotope-dilution gas chromatography–tandem mass spectrometry. *Journal of Analytical Toxicology*, **28**, 327–332.

79 Black, R.M. and Read, R.W. (1995) Biological fate of sulphur mustard, 1,1′-thiobis(2-chloroethane): identification of β-lyase metabolites and hydrolysis products in human urine. *Xenobiotica*, **25**, 167–173.

80 Black, R.M. and Read, R.W. (1988) Detection of trace levels of thiodiglycol in blood, plasma and urine using gas chromatography–electron-capture negative-ion chemical ionization mass spectrometry. *Journal of Chromatography*, **449**, 261–270.

81 Black, R.M. and Read, R.W. (1991) Methods for the analysis of thiodiglycol sulphoxide, a metabolite of sulphur mustard, in urine using gas chromatography–mass spectrometry. *Journal of Chromatography*, **558**, 393–404.

82 Black, R.M. and Read, R.W. (1995) Improved methodology for the detection and quantitation of urinary metabolites of sulphur mustard using gas chromatography–tandem mass spectrometry. *Journal of Chromatography*, **665**, 97–105.

83 Read, R.W. and Black, R.M. (2004) Analysis of the sulfur mustard metabolite 1,1′-sulfonylbis[2-*S*-(*N*-acetylcysteinyl) ethane] in urine by negative ion electrospray liquid chromatography–tandem mass spectrometry. *Journal of Analytical Toxicology*, **28**, 352–356.

84 Read, R.W. and Black, R.M. (2004) Analysis of β-lyase metabolites of sulfur mustard in urine by electrospray liquid chromatography–tandem mass spectrometry. *Journal of Analytical Toxicology*, **28**, 346–351.

85 Black, R.M., Hambrook, J.L., Howells, D.J. and Read, R.W. (1992) Biological fate of sulphur mustard, 1,1′-thiobis(2-chloroethane). Urinary excretion profiles of hydrolysis products and β-lyase metabolites of sulphur mustard after cutaneous application in rats. *Journal of Analytical Toxicology*, **1**, **16**, 79–84.

86 Young, C.L., Ash, D., Driskell, W.J., Boyer, A.E., Martinez, R.A., Silks, L.A. *et al.* (2004) A rapid, sensitive method for the quantitation of specific metabolites of sulfur mustard in human urine using isotope-dilution gas chromatography–tandem mass spectrometry. *Journal of Analytical Toxicology*, **28**, 339–345.

87 Wils, E.R.J., Hulst, A.G. and van Laar, J. (1988) Analysis of thiodiglycol in urine of victims of an alleged attack with mustard gas. Part II. *Journal of Analytical Toxicology*, **12**, 15–19.

88 Jakubowski, E.M., Woodard, C.L., Mershon, M.M. and Dolzine, T.W. (1990) Quantification of thiodigycol in urine by electron ionization gas chromatography–mass spectrometry. *Journal of Chromatography*, **528**, 184–190.

89 Jakubowski, E.M., Sidell, F.R., Evans, R.A., Carter, M.A., Keeler, J.R., McMonagle, J.D. *et al.* (2000) Quantification of thiodiglycol in human urine after an accidental sulfur mustard exposure. *Toxicology Methods*, **10**, 143–150.

90 Black, R.M., Clarke, R.J. and Read, R. (1991) Analysis of 1,1′-sulphonylbis[2-(methyl-sulphinyl)ethane] and 1-methylsulphinyl-2-[2-(methylthio) ethylsulphonyl] ethane, metabolites of sulphur mustard, in urine using gas chromatography–mass spectrometry. *Journal of Chromatography*, **558**, 405–414.

91 Capacio, B.R., Smith, J.R., DeLion, M.T., Anderson, D.R., Graham, J.S., Platoff, G.E. *et al.* (2004) Monitoring sulfur mustard exposure by gas chromatography–mass spectrometry analysis of thiodiglycol cleaved from blood proteins. *Journal of Analytical Toxicology*, **28**, 306–310.

92 Wils, E.R.J., Hulst, A.G., de Jong, A.L., Verweij, A. and Boter, H.L. (1985) Analysis of thiodiglycol in urine of victims of an alleged attack with mustard gas. *Journal of Analytical Toxicology*, **9**, 254–257.

93 Black, R.M., Clarke, R.J., Harrison, J.M. and Read, R.W. (1997) Biological fate of sulphur mustard: identification of valine and histidine adducts in haemoglobin from casualties of sulphur mustard poisoning. *Xenobiotica*, **27**, 499–512.

94 Noort, D., Verheij, E.R., Hulst, A.G., de Jong, L.P.A. and Benschop, H.P. (1996) Characterization of sulfur mustard induced structural modifications in human hemoglobin by liquid chromatography–tandem mass spectrometry. *Chemical Research in Toxicology*, **9**, 781–787.

95 Noort, D., Hulst, A.G., Trap, H.D., de Jong, L.P.A. and Benschop, H.P. (1997) Synthesis and mass spectrometric identification of the major amino acid adducts formed between sulphur mustard and haemoglobin in human blood. *Archives of Toxicology*, **71**, 171–178.

96 Noort, D., Fidder, A., Benschop, H.P., de Jong, L.P.A. and Smith, J.R. (2004) Procedure for monitoring exposure to sulfur mustard based on modified Edman degradation of globin. *Journal of Analytical Toxicology*, **28**, 311–315.

97 Fidder, A., Noort, D., de Jong, A.L., Trap, H.C., de Jong, L.P.A. and Benschop, H.P. (1996) Monitoring of *in vitro* and *in vivo* exposure to sulfur mustard by GC/MS determination of the N-terminal valine adduct in hemoglobin after a modified Edman degradation. *Chemical Research in Toxicology*, **9**, 788–792.

98 Noort, D., Hulst, A.G., de Jong, L.P.A. and Benschop, H.P. (1999) Alkylation of human serum albumin by sulfur mustard *in vitro* and *in vivo*: mass spectrometric analysis of a cysteine adduct as a sensitive biomarker of exposure. *Chemical Research in Toxicology*, **12**, 715–721.

99 Noort, D., Fidder, A., Hulst, A.G., Woolfitt, A.R., Ash, D. and Barr, J.R. (2004) Retrospective detection of exposure to sulfur mustard: improvements on an assay for liquid chromatography–tandem mass spectrometry analysis of albumin/sulfur mustard adducts. *Journal of Analytical Toxicology*, **28**, 333–338.

100 Brookes, P. and Lawley, P.D. (1960) The reaction of mustard gas with nucleic acids *in vitro* and *in vivo*. *The Biochemical Journal*, **77**, 478–484.

101 Fidder, A., Moes, G.W.H., Scheffer, A.G., van der Schans, G.P., Baan, R.A., de Jong, L.P.A. *et al.* (1994) Synthesis, characterization, and quantitation of the major adducts formed between sulfur mustard and DNA of calf thymus and human blood. *Chemical Research in Toxicology*, **7**, 199–204.

102 Fidder, A., Noort, D., de Jong, L.P.A., Benschop, H.P. and Hulst, A.G. (1996) N7-(2-Hydroxyethylthioethyl)-guanine: a novel urinary metabolite following exposure to sulfur mustard. *Archives of Toxicology*, **70**, 854–855.

103 Van der Schans, G.P., Scheffer, A.G., Mars-Groenendijk, R.H., Fidder, A., Benschop, H.P. and Baan, R.A. (1994) Immunochemical detection of adducts of sulfur mustard to DNA of calf thymus and human white blood cells. *Chemical Research in Toxicology*, **7**, 408–413.

104 Van der Schans, G.P., Mars-Groenendijk, R.H., de Jong, L.P.A., Benschop, H.P. and Noort, D. (2004) Standard operating procedure for immunuslotblot assay for analysis of DNA/sulfur mustard adducts in human blood and skin. *Journal of Analytical Toxicology*, **28**, 316–319.

105 Ludlum, D.B., Austin-Ritchie, P., Hagopian, M., Niu, T.-Q. and Yu, D. (1994) Detection of sulfur mustard-induced DNA modifications. *Chemico-Biological Interactions*, **91**, 39–49.

106 Benschop, H.P., van der Schans, G.P., Noort, D., Fidder, A., Mars-Groenendijk, R.H. and de Jong, L.P. (1997) Verification of exposure to sulfur mustard in two casualties of the Iran–Iraq conflict. *Journal of Analytical Toxicology*, **21**, 249–251.

107 Black, R.M., Harrison, J.M. and Read, R.W. (1997) Biological fate of sulphur mustard: *in vitro* alkylation of human haemoglobin by sulphur mustard. *Xenobiotica*, **27**, 11–32.

108 Bailey, E., Brooks, A.G.F., Dollery, C.T., Farmer, P.B., Passingham, B.J., Sleightholm, M.A. and Yates, D.W. (1988) Hydroxyethylvaline adduct formation in haemoglobin as a biological monitor for cigarette smoke intake. *Archives of Toxicology*, **62**, 247–253.

109 Bechtold, W.E., Willis, J.K., Sun, J.D., Griffith, W.C. and Reddy, T.V. (1992) Biological markers of exposure to benzene: *S*-phenylcysteine in albumin. *Carcinogenesis*, **13**, 1217–1220.

110 Technical Secretariat of the Organization for Prohibition of Chemical Weapons (1997) Chemical Convention: Convention on the prohibition of the development, production, stockpiling and use of chemical weapons and on their destruction. Technical Secretariat of the Organization for Prohibition of Chemical Weapons, The Hague.

111 Stuff, J.R., Cheicante, R.L., Durst, H.D. and Ruth, J.L. (1999) Detection of chemical warfare agents bis-(2-chloroethyl) ethylamine (HN-1) and tris-(2-chloroethyl) amine (HN-3) in air. *Journal of Chromatography A*, **849**, 529–540.

112 Trams, E.G. (1958) Determination of bis-(beta-chloroethyl) amines and related compounds with 8-quinolinol. *Analytical Chemistry*, **30**, 256–259.

113 Friedman, O.M. and Boger, E. (1961) Colorimetric estimation of nitrogen mustards in aqueous media. Hydrolytic behaviour of bis(beta-chloroethyl)amine, *nor*-HN-2. *Analytical Chemistry*, **33**, 906–910.

114 Sperry, M.L., Skanchy, D. and Marino, M.T. (1998) High-performance liquid chromatographic determination of *N*-[2-(hydroxyethyl)-*N*-(2-(7-guaninyl)ethyl] methylamine, a reaction product between nitrogen mustard and DNA and its application to biological samples. *Journal of Chromatography B*, **716**, 187–193.

115 Baumann, F. and Preiss, R. (2001) Cyclophosphamide and related anticancer drugs (review). *Journal of Chromatography B*, **764**, 173–192.

116 Mattes, W.B., Hartley, J.A. and Kohn, K.W. (1986) DNA sequence selectivity of guanine-N7 alkylation by nitrogen mustards. *Nucleic Acid Research*, **14**, 2971–2987.

117 Lemire, S.W., Ashley, D.L. and Calafat, A.M. (2003) Quantitative determination of the hydrolysis products of nitrogen mustards in human urine by liquid chromatography–electrospray ionization tandem mass spectrometry. *Journal of Analytical Toxicology*, **27**, 1–6.

118 Lemire, S.W., Barr, J.R., Ashley, D.L., Olson, C.T. and Hayes, T.L. (2004) Quantitation of biomarkers of exposure to nitrogen mustards in urine from rats dosed with nitrogen mustards and from unexposed human population. *Journal of Analytical Toxicology*, **28**, 320–326.

119 Sandí, G., Brubaker, K.L., Schneider, J.F., O'Neill, H.J. and Cannon, P.L., Jr (2001) A coulometric iodometric procedure for measuring the purity of lewisite. *Talanta*, **54**, 913–925.

120 Webb, D.R. and Carter, D.E. (1984) An improved wet digestion procedure for the analysis of total arsenic in biological samples by direct hydride atomic absorption spectrophotometry. *Journal of Analytical Toxicology*, **8**, 118–123.

121 Logan, T.P., Smith, J.R., Jakubowski, E.M. and Nielson, R.E. (1999) Verification of lewisite exposure by the analysis of 2-chlorovinyl arsonous acid in urine. *Toxicology Methods*, **9**, 275–284.

122 Fowler, W.K., Stewart, D.C., Weinberg, D.S. and Sarver, E. (1991) Gas chromatographic determination of the lewisite hydrolysate, 2-chlorovinyl arsonous acid, after derivatization with 1,2-ethanedithiol. *Journal of Chromatography*, **558**, 235–246.

References for Section 32.4

123 Di Consiglio, E., de Angelis, G., Testai, E. and Vittozzi, L. (2001) Correlation of a specific mitochondrial phospholipid–phosgene adduct with chloroform acute toxicity. *Toxicology*, **159**, 43–53.

124 Noort, D., Hulst, A.G., Fidder, A., van Gurp, R.A., de Jong, L.P.A. and Benschop, H.P. (2000) *In vitro* adduct formation of phosgene with albumin and hemoglobin in human blood. *Chemical Research in Toxicology*, **13**, 719–726.

125 Muir, B., Carrick, W.A. and Cooper, D.B. (2002) Application of central composite design in the optimization of thermal desorption parameters for the trace level determination of the chemical warfare agent chloropicrin. *Analyst*, **127**, 1198–1202.

126 Kientz, C.E. (1998) Chromatography and mass spectrometry of chemical warfare agents, toxins and related compounds: state of the art and future prospects (review). *Journal of Chromatography A*, **814**, 1–23.

127 Black, R.M. and Read, R.W. (1997) Application of liquid chromatography–atmospheric pressure chemical ionization mass spectrometry and tandem mass spectrometry to the analysis and identification of degradation products of chemical warfare agents. *Journal of Chromatography A*, **759**, 79–92.

128 Katz, S.A. and Salem, H. (2004) Synthesis and chemical analysis of riot control agents, in *Riot Control Agents* (eds E.J. Olajos and W. Stopford), CRC Press, Boca Raton, FL, pp. 25–36.

129 Szinicz, L. and Baskin, S.I. (2004) Chemische und biologische Kampfstoffe, in *Lehrbuch der Toxikologie* (eds. H. Marquardt and S. Schäfer). Wissenschaftliche Verlagsgesellschaft mbH, Stuttgart, pp. 865–895.

130 Klimmek, R., Szinicz, L. and Weger, N. (1983) *Chemische Gifte und Kampfstoffe*. Hippokrates Verlag, Stuttgart.

131 Weber, M. (2000) Zur Problematik der Entgiftung sowie der Nachweisverfahren von Methylthiophosphonsäure-O-alkyl-S-(2-N,N-dialkylaminoethyl)estern (sog. V-Stoffen). Thesis, Christian-Albrechts-Universität, Kiel.

33
Biochemical Investigations in Toxicology

W.R. Külpmann

33.1
Basic Biochemical Investigations

33.1.1
Introduction

Usually, a clinical laboratory provides 24 h services for emergencies. It may be suitable for performing simple toxicological analyses in time. In addition, these laboratories perform routinely biochemical investigations, which can assist in case of poisoning:

Exclusion of a suspected diagnosis
Supporting a suspicion of poisoning
Decision in differential diagnosis
Estimation of prognosis
Choice and monitoring of treatment.

33.1.2
Relevant Biochemical Investigations

Diagnosis
Some quantities in clinical chemistry are rather good indicators of a particular poison: (Pseudo-) Cholinesterase activity in serum (Section 33.2), carbon monoxide hemoglobin (Section 30.1), methemoglobin (Section 30.2), and thromboplastin time (Quick test) (Chapter 17). The determination of these quantities is requested when a specific suspicion arises. The measurement of the ethanol concentration in blood should be performed for all comatose patients and in all cases of doubtful symptoms because of the widespread alcohol abuse. Other biochemical quantities are helpful to find out if the patient is in a life-threatening situation and detect impaired organ function. Renal or hepatic disorders will have an impact on the elimination of poisons and on dosage of therapeutic drugs. The results of the investigations may cast suspicion on a special poison (Table 33.1), which has to be confirmed or ruled out by a dedicated toxicological analysis. The biochemical measurements are important to

Clinical Toxicological Analysis: Procedures, Results, Interpretation. Edited by Wolf-Rüdiger Külpmann

ISBN: 978-3-527-31890-2

Table 33.1 Biochemical investigations in suspected exogenous poisoning.

Basic investigations	
Electrolytes (serum)	Potassium, sodium
Enzymes (serum)	Alanine aminotransferase (ALT, previously GPT) Aspartate aminotransferase (AST, previously GOT) Creatine kinase (CK)
Substrates (serum)	Glucose, creatinine
Blood gas analysis and acid–base balance (arterial blood)	pH, pO_2, pCO_2, act. HCO_3, base excess, O_2 saturation
Hematology (blood)	Blood count, hemoglobin, hematocrit, platelet count
Urinalysis	pH, protein, glucose, hemoglobin, urobilinogen, bilirubin, nitrite, ketone bodies, leucocytes, sediment
Additional investigations	
Electrolytes (serum)	Calcium, chloride
Anion gap (serum)	
Osmolality (serum)	
Osmolal gap (serum)	
Enzymes (serum)	γ-Glutamyltransferase, butyrylcholinesterase
Substrates (serum)	Bilirubin, urea
Substrates (plasma)	Lactate
Ethanol (serum)	
Coagulation (plasma)	Prothrombin time, thrombin time, activated partial thromboplastin time

Modified from Ref. [2].

discover the possible endogenous nature of the coma (coma hepaticum, coma diabeticum).

The following biochemical investigations require special attention in poisoning (Table 33.1):

(1) Anion gap.

The anion gap (A) is calculated as follows:

$$A = Na^{+} - (Cl^{-} + HCO_3{}^{-}),$$

where Na^{+} is the sodium concentration in serum/plasma (mmol/l); Cl^{-}, the chloride concentration in serum/plasma (mmol/l); and $HCO_3{}^{-}$, the actual bicarbonate concentration in plasma (mmol/l).

Its reference interval is 8–16 mmol/l and reflects that more anions (e.g., proteins) than cations are neglected in the formula. The anion gap is often increased, because the amount of substance concentration of (low molecular) organic acids is elevated considerably. The most important settings/substances that contribute to an increased anion gap are ketone bodies, uremia, salicylate, methanol, alcohol (ethanol), lactate, and ethylene glycol, summarized by the acronym "KUSMALE." An extended anion gap may be accompanied by an increased osmolal gap (see below) [1].

During metabolic degradation of methanol, formic acid is produced and anion gap, too, is increased. The same holds true for ethylene glycol and its metabolites glycolic acid, glyoxylic acid, and oxalate.

(2) Osmolal gap [1]

Osmolal gap may be calculated as follows:

$$O_g = O_m - O_c = O_m - (1.86 \times c_{Na} + c_{Glu} + c_{Urea} + 9),$$

where O_g is the osmolal gap, O_m is the serum osmolality, measured (mmol/kg), O_c is the serum osmolality, calculated (mmol/l), c_{Na} is the serum sodium concentration (mmol/l), c_{Glu} is the serum glucose concentration (mmol/l), and c_{Urea} is the serum urea concentration (mmol/l).

Reference interval: $\leq$10 mmol/kg water.

It should be mentioned that other formulae are published for which other reference intervals may apply.

The osmolal gap is extended when other compounds apart from electrolytes, glucose, and urea are present at high amount of substance concentration.

In practice, the osmolal gap is rather often increased by elevated lactate concentration. Otherwise, exogenous compounds should be considered. Substances that contribute frequently to an increased osmolal gap are mannitol, alcohols, diatrizoate, glycerol, acetone, sorbitol, summarized by the acronym "MADGAS." However, Table 33.2 clearly indicates that some solvents do not extend the osmolal gap, even though they may be present at fatal concentrations [6a].

Usually, the oxygen saturation of blood is not measured by spectrophotometry but calculated. However, the calculated result is only valid, if dyshemoglobins (e.g., CO-hemoglobin or methemoglobin) are not present. Although oxygen saturation may be calculated as 98%, there can be a dangerous tissue hypoxia, if the dyshemoglobin fraction exceeds, for example, 50% at the same time.

Routine procedures for the determination of total hemoglobin do not exclude dyshemoglobins. A "hemoglobin concentration" of, for example, 150 g/l is misleading, if a high fraction of, for example, methemoglobin is present.

The detection of a hepatitis B or HIV infection should cast suspicion on possible drug addiction.

Table 33.2 Impact of solvents on osmolality at fatal concentration.

Solvent	Fatal concentration in serum (mg/l)	Contribution to osmolal gap (mmol/kg)
Ethanol	3500	81.7
Isopropanol	3400 (toxic)	60.8
Methanol	800	26.8
Diethyl ether	1800	26.1
Acetone	550	10.2
Trichloroethane	1000	8.1
Paraldehyde	500	4.1
Ethylene glycol	210	3.6
Chloroform	390	3.5
Toluene	10	0.1

Modified from Ref. [1].

A conspicuous color of the urine may assist in detection and identification of drugs (Section 3.8).

Tables 33.3–33.11 present drugs and chemicals that can cause pathological findings for biochemical quantities in case of poisoning. Chlorinated hydrocarbons, for instance, may be the cause of elevated activity of transaminases in serum (Tables 33.3 and 33.4).

Cadmium is a renal toxin and leads to renal failure and to an increase of creatinine and urea concentration in serum (Table 33.5).

Creatinine kinase (serum) activity is elevated, if toxins stimulate convulsions or cause muscle damage (Tables 33.6 and 33.7). Hemoglobin concentration in blood is decreased by toxins, which cause hemolysis, bleedings, or hepatic failure with concomitant coagulation disorder (Tables 33.8 and 33.9).

Often, severe intoxication is accompanied by acid–base disturbances: Insufficient oxygen delivery in CO-hemoglobinemia will lead to acidosis by increased lactate concentration [6a]. Retention acidosis will develop as result of renal failure during

Table 33.3 Liver toxins.

Kind of liver damage	Toxin(s)
Acute necrosis	Arsenic
	Borate
	Chlorinated hydrocarbons
	Chlofenotane (DDT)
	Copper
	Iron
	Paracetamol (acetaminophen)
	Paraquat
	Phosphorus
	Toxic mushrooms
	Thallium
Subacute necrosis	Chloronaphthalene
	Dimethyl nitrosamine
	Dinitrobenzene
	Polychlorinated biphenyls
	Tetrachloroethane
	Trinitrotoluene
Chronic liver damage	Aflatoxin
	Arsenic
	Ethanol
	Pyrrolizidine alkaloids
	Tetrachloromethane
	Thorotrast (thorium dioxide)
	Vinyl chloride
	Vitamin A
	Compounds listed for "subacute necrosis"

Modified from Ref. [4].

Table 33.4 Hepatotoxic chemicals.

Metals and inorganic compounds	
Antimony (acute toxic)	Manganese
Arsine (acute toxic)	Phosphorus (yellow)
Beryllium	Selenium
Bismuth	
Organic compounds	
Acetonitrile	Hydrazine
Acrylonitrile	Cresol
Benzene	Methanol
Bromoform	Methyl chloride
Chlorbutadiene	Naphthalene
Chlorinated benzenes (incl. DDT)	Nitrobenzene
Chlorinated biphenyls	Phenol
Chlorinated naphthalenes	Phenylhydrazine
Chloroform	Pyridine
1,2-Dichloropropane	Styrene
Dimethyl sulfate	Tetrabromomethane
Dinitrophenol	Tetrachloroethane
Dioxan	Tetrachloroethene
Epichlorohydrin	Tetrachloromethane
Ethanol	Toluene
Ethylbromide	Trichloroethane
Ethylsilicate	Trichloroethene
Ethylene chlorohydrin (2-chloroethanol)	

Modified from Ref. [4].

Table 33.5 Nephrotoxic compounds.

Metals and metalloorganic compounds	Antimony, arsenic, arsine, beryllium, bismuth, cadmium, chromium, iron, lead, magnesium, manganese, silver, uranium
Solvents	Carbon tetrachloride, methanol, methyl cellulose, tetrachloroethane, tetrachloroethene, trichloroethane
Glycols	Diethylene glycol, ethylene glycol, glycerol, propylene glycol, xylite
Pesticides/herbicides	3,4-Benzopyrene, chlorinated dibenzodioxins (e.g., TCDD), DDT, diquat, hexachlorobenzene, malathion, paraquat, polybrominated biphenyls (PBB), polychlorinated biphenyls (PCB)
Miscellaneous	Carbon disulfide, carbon monoxide, colchicine, oxalic acid, toxins from mushrooms (e.g., amatoxins) tartrate

Modified from Ref. [4].

mercury intoxication (Table 33.10). A synopsis of the most important biochemical findings during frequently occurring intoxications is presented in Table 33.11.

Prognosis

The momentary concentration of a poison in blood may have limited impact on the estimation of prognosis, for example, because time of drug intake is unknown.

Table 33.6 Elevated creatine kinase activity in serum from poisoning.

Amphetamines
Carbon monoxide
Clofibrate
ε-Aminocaproic acid
Ethanol
Ethylene glycol
Glutethimide
Heroin
Isopropanol
Lysergic acid diethylamide (LSD)
Methadone
p-Phenylenediamine
Phencyclidine
Phenylpropylamine
Salicylates
Strychnine
Succinylcholine
Toluene
Hornet toxins
Spider toxins
Wasp toxins

Modified from Ref. [4].

Table 33.7 Drug-induced seizures.

Anesthetics (e.g., halothane)	Lead
Antidepressant drugs	Lindane
Antihistamines	Lithium
Antidiabetics	Local anesthetics (e.g., lidocaine)
Antipsychotics	Mefenamic acid
Baclofen	Methylxanthine
Camphor	Phenytoin
Carbon monoxide	Metronidazol
Chlorambucil	Nalidixic acid
Chlorinated hydrocarbons	Opioids (e.g., fentanyl, meperidine, pentazocine, propoxyphene)
Cholinesterase inhibitors (e.g., physostigmine, organophosphates)	Organophosphates (pesticides)
Ciclosporin	Phencyclidine
Cocaine	Oxytocin
Disulfiram	Phenobarbital
Folate	Phenol
Hypo-osmolar infusion	β-Receptor-blocking agents (e.g., propranolol)
Isoniazid	Strychnine
Iodine radiopaque aids (aqueous)	Sympathomimetic drugs (e.g., amphetamines, ephedrine)
	Withdrawal

Supplemented from Ref. [5].

Table 33.8 Drugs and chemicals associated with anemia.

Symptoms	Drug(s)/chemicals
Bleeding from upper gastrointestinal tract	Alcohols Anticoagulants Glucocorticoids Hydralazine Heavy metals, for example, iron Indometacin Nonsteroidal antiinflammatory drugs Phenylbutazone Reserpine Salicylates
Hemolysis	Arsine Dichloroethane Stibine
Hemolysis in case of glucose-6-phosphate dehydrogenase deficiency	Aniline derivatives Antimalarial agents Nitrobenzene derivatives Nitrofurantoin Phenacetin

Modified from Ref. [4].

Table 33.9 Toxic hemorrhagic disorders.

Symptom	Poison(s)
Disseminated intravascular coagulation (DIC)	Iron Monoamine oxidase inhibitors Phencyclidine Toxins from mushrooms Toxins from snakes (shock)
Impaired platelet aggregation	Salicylates
Prolonged prothrombin time (from impaired liver function)	Carbon tetrachloride Paracetamol (acetaminophen) Toxins from mushrooms (e.g., amatoxins)
Prolonged prothrombin time (inhibition of vitamin K synthesis)	Anticoagulant rodenticides, coumarins

Modified from Ref. [4].

Biochemical quantities may be helpful in this setting, as they reflect the degree of actual malfunction of several organ systems.

One should consider that symptoms of poisoning could appear late for several reasons, for example, delayed absorption. An interval free from symptoms may be even characteristic for some very dangerous poisons (e.g., amanita phalloides) or some drugs (e.g., acetaminophen).

Table 33.10 Disorders of acid–base balance from poisoning.

Kind of disorder	Toxic compound(s)
Metabolic acidosis	
Lactic acidosis	Biguanides
	Carbon monoxide
	Cyanide
	Ethanol
	Ethylene glycol
	Isoniazid
	Methemoglobin producing drugs
	Methanol
	Paraldehyde
	Salicylates
	Seizures inducing drugs
Retention of acids	Acetazolamide
	Aminoglycosides
	Amphotericin B
	Analgesics
	Cadmium
	Lead
	Lithium
	6-Mercaptopurine
	Mercury
	Toluene
Metabolic alkalosis	Bumetanide
	Etacrynic acid
	Furosemide
	Thiazide diuretics
Respiratory acidosis	
via respiratory center	Anesthetics
	Opiates
	Sedative drugs
via neuromuscular blockade	Muscle relaxants
Respiratory alkalosis	
via respiratory center	Analgesics
	Catecholamines
	Salicylates
	Theophylline

Modified from Ref. [4].

Treatment

Treatment depends on the toxin and the severity of the intoxication. For the therapy of some poisonings, special antidotes are available (see Appendix D). Before final choice of treatment is made, the organ system, which would be most challenged, should be

Table 33.11 Acute poisoning: biochemical findings.

	Hypoxia, respiratory acidosis	Respiratory alkalosis	Metabolic acidosis	Hypokalemia	Hyperkalemia	Hypocalcemia	Hypoglycemia	Hyperglycemia	Hyperosmolality
Antidiabetics, oral							+		
Barbiturates	+		◇						
Benzodiazepines	+		◇						
Carbon monoxide	+		+						
Cyanide	+		+						
Digoxin					◇				
Ethanol			◇				◇		+
Ethylene glycol			+			◇	◇		+
Insulin				+			+		
Methanol			+						+
Opioids	+				◇				
Paracetamol			◇					([a]	
Salbutamol			◇	+				+	
Salicylates		◇	+				◇		◇
Theophylline			+	+		◇		+	
Tricyclic antidepressants	+		◇						

Modified from Ref. [6]. +, Frequent; ◇, less frequent.
[a]False elevated (interference of paracetamol with glucose-oxidase perid procedure).

Table 33.12 Monitoring of treatment by biochemical measurements.

Treatment	Monitoring	
Forced diuresis	Serum	Na, K, Cl, protein, osmolality
	Urine	Na, K, Cl, Ca, osmolality
Hemodialysis	Serum	Na, K, Cl, Ca, glucose, urea, creatinine osmolality
	Blood	Acid–base balance, blood gas analysis
	Plasma	Coagulation tests
Hemoperfusion	Serum	K
	Blood	Platelets
	Plasma	Coagulation tests

Modified from Ref. [2].

Table 33.13 Monitoring of antidote treatment by biochemical measurements.

Toxin	Treatment	Monitoring
Cyanide	4-DMAP[a]	Methemoglobin (blood)
Ethylene glycol	Ethanol	Ethanol (blood)
Methanol	Ethanol	Ethanol (blood)
Phenobarbital[b]	$NaHCO_3$	pH (urine)
Salicylates[b]	$NaHCO_3$	pH (urine)

Modified from Ref. [2].
[a] 4-Dimethylaminophenol.
[b] Or other "acidic" toxins. The lower their pK, the more quickly they are eliminated at alkaline pH by kidneys.

checked by appropriate biochemical measurements; forced diuresis should not be applied in case of renal failure (Table 33.12).

Monitoring is mandatory during therapy with some antidotes, which may otherwise poison the patient themselves (Table 33.13). Overdosage of 4-Dimethylaminophenol for treatment of cyanide poisoning will lead to a dangerous methemoglobinemia.

Ethanol concentration during therapy of methanol intoxication should be about 1 g/kg blood. Lower concentrations are less effective, and higher concentrations increase toxicity.

Conclusion

A close cooperation between the biochemical laboratory and the toxicological laboratory will considerably improve diagnosis and therapy of poisoning. Usually, toxicological analysis provides specific information about the drugs or chemicals that are involved. Clinical chemistry provides rather quickly, but less specific, indications of poisoning and supports differential diagnosis, therapy, and assessment of prognosis.

One should bear in mind that drugs and chemicals may interfere with biochemical procedures and may act as influencing factors [3].

33.2 Cholinesterase

P. Eyer and F. Worek

33.2.1 Introduction

The determination of cholinesterase activity is important in monitoring exposure to anticholinesterase agents, which are used as pesticides, nerve agents, and drugs as in Alzheimer's disease. The agents comprise organophosphorus compounds, carbamates, and reversible inhibitors such as huperzine A and tacrine. Studies on the toxic mechanisms of theses agents have historically focused on their interactions with serine hydrolases, particularly acetylcholinesterase (AChE, EC 3.1.1.7). In each animal species, this enzyme is encoded by a single gene, although a wide variety of molecular forms exist. While inhibition of this key enzyme explains most of the prominent symptoms of acute poisoning with anticholinesterase agents, subtle differences among the toxic actions of different inhibitors remain to be explained. In addition, other serine hydrolases may be affected, although to a different extent, including chymotrypsin and trypsin, carboxylesterases, plasma butyrylcholinesterase (BChE, EC 3.1.1.8), and neuropathy target esterase. Furthermore, processing of other signal peptides, for example, their formation from preproteins or their inactivation by hydrolytic cleavage may possibly be altered by anticholinesterase agents.

AChE is found mostly in nervous tissue, where its major function is to inactivate the neurotransmitter acetylcholine in the synaptic cleft and curiously in the membrane of red blood cells (RBCs), where its function is not fully understood.

Occurrence

The cholinergic synapse is a phylogenetically old system that is found in mammalians, birds, reptiles, amphibians, and fish as well as in arthropods, insects, molluscs, and worms [7]. AChE is found at synapses, neuromuscular, and musculotendinous junctions, cerebrospinal fluid, neuronal cell bodies, skeletal and smooth muscle cells, and in blood. The activities of AChE vary widely in different species, particularly in RBCs. Generally, primates have the highest activity, cow has 90%, horse and guinea pig only one third, rabbits one fifth, rats one tenth, and cats roughly 1% compared to humans [8]. Bird RBCs are virtually devoid of AChE [9]. Interestingly, an inverse rank order is found with platelet AChE; human platelets are virtually without any activity, while cat platelets and megakaryocytes are particularly rich in AChE [8]. The significance of these differences is unknown. Human T-lymphocytes have been shown to possess AChE while B-lymphocytes are devoid of it [10]. When determining AChE in human blood cells, the contribution of AChE in cells other than erythrocytes is negligible, because the relative amounts of these cells is several orders of magnitude less than that of RBCs [11].

AChE in blood is not restricted to the membrane-bound enzyme; true AChE is also found in plasma. While its content is low in human plasma, less than 2% [12], the relative contribution of true AChE to acetylthiocholine hydrolysis in rat plasma is about 50% [13]. AChE activity in fetal calf serum is high enough to be used as a source for purifying the enzyme [14].

BChE is similarly distributed as is AChE, but is particularly high in pancreas and liver. The latter is undoubtedly the source of the enzyme in plasma [15].

Plasma BChE activity, determined with butyrylthiocholine, is similarly high in man, dog, cat, and horse, but is somewhat lower in rat and mouse. Species differences become more marked when butyrylthiocholine is substituted by benzoylcholine, while propionylcholine is equally hydrolyzed in most mammals [11].

These species differences should be borne in mind when animals (wildlife, pets, or livestock) are used as sentinels of exposure to anticholinesterase agents. Animals can help monitor the environment, giving early warnings of environmental contamination by excessive salivation and tear production [16]. Determination of blood cholinesterase activities may then confirm exposure as has been proposed for biomonitoring of nerve agents' incorporation [17].

Several phenotypes of BChE are expressed in humans. For instance, in Caucasian populations, 95% have the usual (UU) phenotype, 1% have the atypical (AA), and about 4% have the expected heterozygote (UA) phenotype. About 0.01% of the Caucasians have no detectable BChE activity in plasma [18]. In atypical BChE, aspartate 70 at the rim of the active site gorge is replaced by glycine. Patients with this mutation respond abnormally to the muscle relaxant succinyldicholine, experiencing hours of apnea instead of the intended 3–5 min. This variant has a reduced affinity for all positively charged substrates and inhibitors (for review see Ref. [19]). Besides, the products of the BChE gene on chromosome 3 exist in multiple molecular forms in human plasma that can be separated by gel filtration and electrophoretic methods. These forms are poorly defined and there is no evidence for a definite physiological significance of this heterogeneity.

Physiological function of cholinesterases

The particular localization of AChE in the outer basal lamina of the synapse points to acetylcholine cleavage as the predominant function of AChE, enabling the hydrolysis of the transmitter in submilliseconds. This outstanding function, however, may be shared by other enzymes such as BChE. The successful development and survival of knockout mouse mutants completely lacking AChE point to alternative signaling pathways [20]. Contrary to expectation, mice without AChE activity survived up to 3 weeks after birth, though their physical development was delayed. The nullizygous mice were highly sensitive to organophosphates (OPs) and the BChE-specific inhibitor bambuterol. These findings indicate that BChE and possibly other enzymes may be capable of compensating for some functions of AChE [21]. Hence, BChE may serve as a second-line defender against excess acetylcholine, which is not essential as long as AChE is active. It explains why isolated inhibition of BChE is not associated with impaired body functions or illness.

Tissue cholinesterases may have largely unknown roles in addition to their regulation of neuronal transmission. Cholinesterases may be involved in neuronal outgrowth during development and they may retard neurodegeneration as in Alzheimer's disease [22].

The significance of the conservation of AChE in mature erythrocytes of many but not all animals remains an enigma. Nevertheless, RBC-AChE serves as a valuable surrogate marker of synaptic – and functionally important – AChE, the inhibition of which correlates favorably with RBC-AChE activity [23]. Determination of the cholinesterase status in blood is therefore widely accepted as a means of assessing risk [24], verifying exposure [25], monitoring of drug safety, such as in case of Alzheimer's disease [22], and monitoring of therapy in anticholinesterase pesticide poisoning [26].

33.2.2 Determination of Cholinesterase Activity

A variety of methods have been developed in the past decades for the sensitive and specific determination of AChE and BChE activities in blood [27–33].

At present, the colorimetric Ellman procedure [34] is generally preferred for occupational health screening [9, 35–37], and therapeutic monitoring of pesticide-poisoned patients [38, 39]. The principal reactions of the Ellman assay are shown in Scheme 33.1.

Although rather rapid, simple, and cheap, the Ellman method does not allow to exactly determine AChE in whole blood or erythrocytes due to interference with hemoglobin absorption. The absorption maximum of the colored indicator TNB^- at 412 nm [40] coincides with the Soret band of hemoglobin. Therefore, only highly diluted blood samples can be used, which reduces the assay sensitivity. In addition, reaction of the Ellman reagent DTNB (5,5′-dithio-bis-2-nitrobenzoate) with slowly reacting matrix sulfhydryl groups [41] may disturb the assay, a side reaction that is being usually disregarded [11] but may be an important factor at low AChE activities

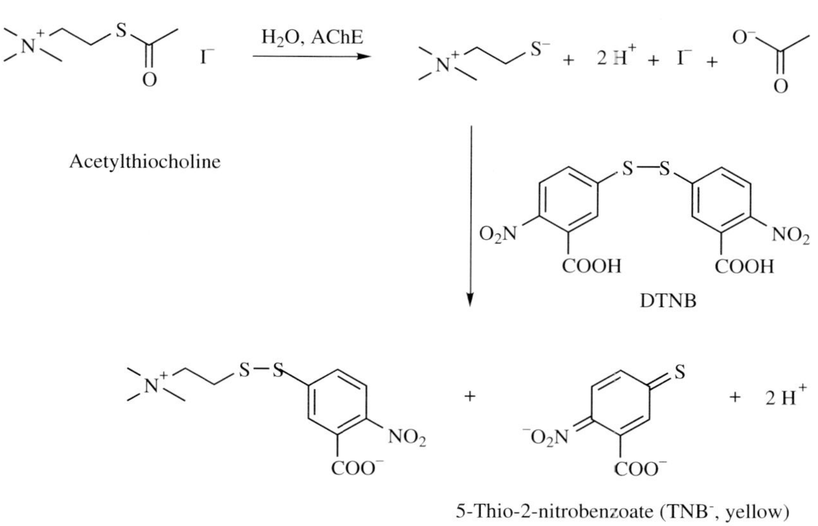

Acetylthiocholine I^- —(H_2O, AChE)→ thiocholine (S^-) + $2\,H^+$ + I^- + acetate

+ DTNB (O_2N, COOH, S–S, COOH, NO_2)

→ mixed disulfide (NO_2, COO^-) + 5-Thio-2-nitrobenzoate (^-O_2N, COO^-, S) + $2\,H^+$

5-Thio-2-nitrobenzoate (TNB⁻, yellow)

Scheme 33.1 Principal reactions of the Ellman assay.

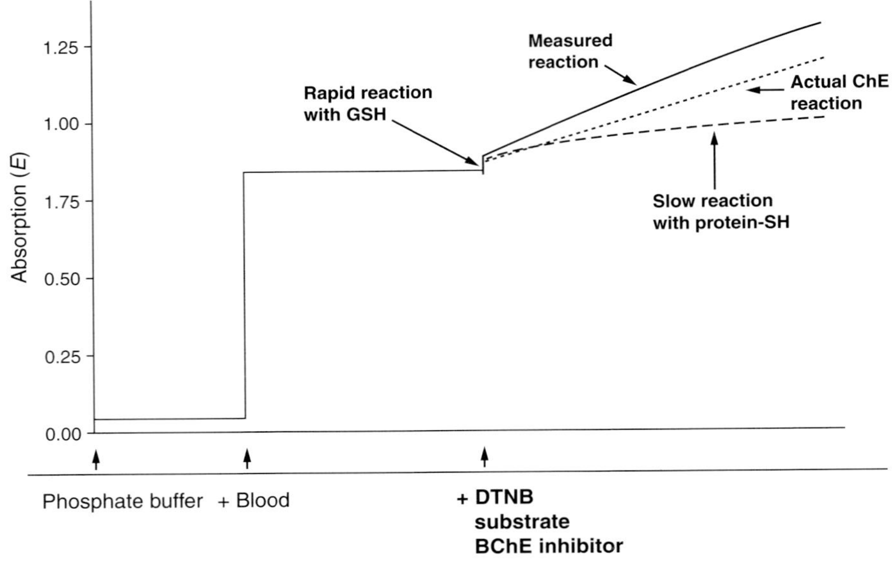

Figure 33.1 Scheme of reactions between blood, DTNB, and substrate during the Ellman assay.

(Figure 33.1). These shortcomings have led to numerous modifications of the original method. Several investigators have attempted to reduce the spectral interference of hemoglobin by using other wavelengths, dual-beam photometers or other chromogenic disulfides, for example, 4,4′-dithiopyridine [42, 43].

To measure the AChE activity in whole-blood samples, various selective BChE inhibitors such as quinidine [43, 44] and phenothiazine derivatives [42, 45] have been proposed. In another approach, two substrates were used to distinguish between AChE and BChE [46], or BChE activity is calculated from the difference of total cholinesterase activity, using 1 mmol/l acetylthiocholine, and the activity remaining after BChE inhibition with 20 µmol/l ethopropazine [47]. This procedure can be performed with diluted and frozen whole-blood samples according to the standard operating procedure (SOP) given below. (Substrate for both enzymes is 1 mmol/l acetylthiocholine; Triton X-100 as diluent has to be omitted since it inhibits BChE.)

Recently, a modification of the original Ellman procedure was presented that enables the sensitive and selective determination of AChE activity in human whole blood [48, 49]. In this modified method, the wavelength was changed to 436 nm. This reduced the indicator absorption to 80% and the hemoglobin absorption to 25%. The signal-to-noise ratio was further enhanced by decreasing the pH from 8.0 to 7.4 and by diminishing the substrate concentration from 1.0 to 0.45 mmol/l acetylthiocholine (ASCh), that is, $5 \times K_m$. These changes diminished the activity by roughly 25% but reduced markedly the spontaneous hydrolysis of ASCh, thus enabling the precise determination of 3% residual activity of human RBCs. Assays were strictly performed at constant temperature since both AChE and BChE activities depend markedly on temperature (Figure 33.2) as does the indicator

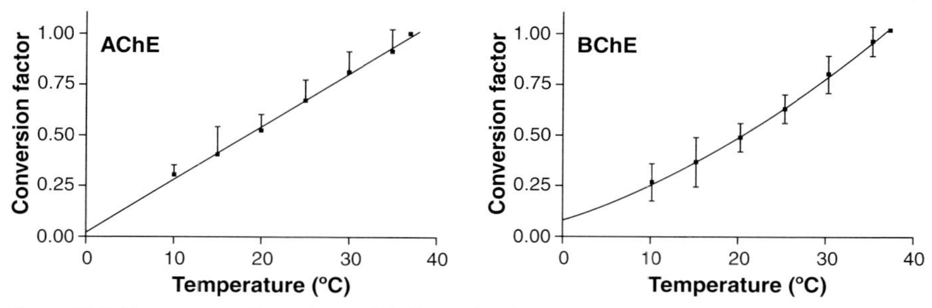

Figure 33.2 Temperature dependence of AChE and BChE activities. Mean ± SE ($n = 12$), with the conversion factor set to unity at 37 °C.

absorbance. A reaction temperature of 37 °C was chosen because most autoanalyzers used in clinical chemistry run at this temperature, and reactions with matrix sulfhydryls are completed faster.

AChE activity is determined in whole-blood samples (EDTA or heparin to prevent clotting) in the presence of the selective BChE inhibitor, ethopropazine. At 20 μmol/l ethopropazine, BChE of the most abundant UU phenotype is inhibited by 98 ± 2% while AChE is inhibited by 5% only [12, 48]. Bedside dilution of blood samples (1 : 100; ice-cold dilution reagent) slows down bimolecular secondary reactions in the presence of inhibitor (organophosphate) and reactivator (oxime) by a factor of at least 10 000 (Scheme 33.2). Normalization of the AChE activity to the hemoglobin content, determined as cyanmethemoglobin [50], compensates dilution errors.

BChE activity is determined in plasma obtained by centrifugation of whole-blood samples (EDTA or heparin) at 3000 × g for 5 min. Diluted blood samples and plasma may be frozen at −20 °C if enzyme activity determination has to be postponed.

Diluted whole-blood and plasma samples may be kept frozen at ≤−20 °C for at least 3 months without any loss of AChE or BChE activity. Recently, we analyzed duplicates of such samples (shipped from Sri Lanka in dry ice) 3 months apart and did not detect any differences [51].

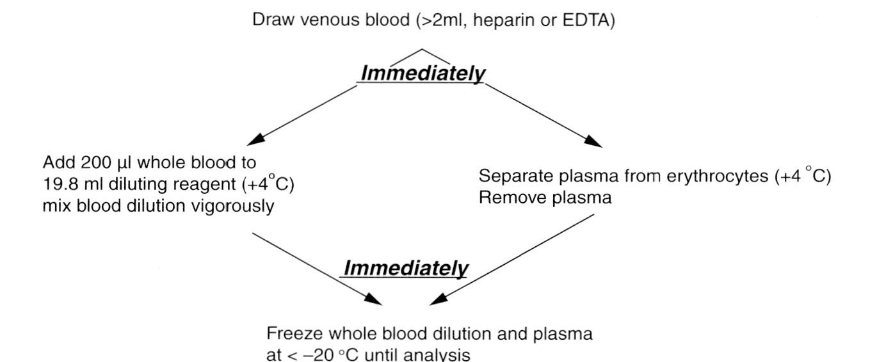

Scheme 33.2 Procedure for the handling of blood samples.

Standard operation procedure [52–54]

Diluting reagent
A 0.1 mol/l sodium phosphate buffer, pH 7.4, containing 0.03% Triton X-100, 19.8 ml in 20 ml polyethylene vials as used for scintillation counting. This reagent can be kept frozen at −20 °C until needed.

Ellman's reagent
Dissolve 396.3 mg 5,5′dithio-bis-2-nitrobenzoic acid (Sigma Chemicals) in 100 ml 0.1 mol/l sodium phosphate buffer, pH 7.4, by magnetic stirring. Store in 5 ml aliquots at −20 °C and protect from light when using it because DTNB degrades to various yellow products upon illumination [55, 56].

Ethopropazine
Dissolve 20.94 mg ethopropazine (Sigma Chemicals on request, cat no. L308765) in 10 ml of 12 mmol/l HCl (dissolves slowly) and store in 0.5 ml aliquots at −20 °C.

Transformation solution [48, 50]
Dissolve 200 mg potassium ferricyanide, 50 mg potassium cyanide, and 1000 mg sodium bicarbonate in 1000 ml of distilled water. Add 0.5 ml Triton X-100 and store the solution in an amber bottle at room temperature.

AChE determination

Mix. in polystyrene cuvettes (1 cm light path)	Final concentration
2.000 ml 0.1 mol/l sodium phosphate buffer, pH 7.4	100 mmol/l
0.100 ml 10 mmol/l DTNB	0.30 mmol/l
0.010 ml 6 mmol/l ethopropazine in 12 mmol/l HCl	0.02 mmol/l
1.000 ml hemolysate (whole blood 1 : 100)	
Equilibrate at 37 °C for 10 min, then add 0.050 ml 28.4 mmol/l acetylthiocholine	0.45 mmol/l
Record color development for 3–5 min at 436 nm (e.g., in a filter photometer at 1.0 AUFS; $E = 11.28 \times 10^3 \, l \, mol^{-1} \, cm^{-1}$)	
Corrections for spontaneous hydrolysis (substitute hemolysate by buffer)	
Typical readings	
Human hemolysate	160 mE/min
Blank	2 mE/min

Hemoglobin determination

Mix 1.500 ml of hemolysate (whole blood 1 : 100) with 1.500 ml of transformation solution and incubate the mix for 10 min at room temperature. Read absorbance at 546 nm (e.g., in a filter photometer; $E = 10.8 \times 10^3$ l mol^{-1} cm^{-1}).

Typical reading with normal blood (15 g Hb/100 ml) 0.48 E corresponding to 44 μmol/l Hb (Fe).

For calculation of the specific activity of RBC-AChE, presented as U/μmol Hb (Fe), calculate the quotient of AChE (μmol/l·min^{-1}) and μmol/l Hb obtained from the measurement in hemolysate.

Typical activity of RBC-AChE: 0.6 U/μmol Hb (Fe) or 37.3 U/g Hb.

Plasma BChE

Mix in polystyrene cuvettes (1 cm light path)	Final concentration
3.000 ml 0.1 mol/l sodium phosphate buffer, pH 7.4	100 mmol/l
0.100 ml 10 mmol/l DTNB	0.30 mmol/l
0.01 ml plasma (undiluted)	
Equilibrate at 37 °C for 10 min, then add 0.050 ml 63.2 mmol/l butyrylthiocholine	1.0 mmol/l
Record color development for 3–5 min at 436 nm (e.g., in a filter photometer at 1.0 AUFS; $E = 11.28 \times 10^3$ mol^{-1} cm^{-1})	
Corrections for spontaneous hydrolysis (substitute plasma by buffer)	
Typical readings	
Human plasma	170 mE/min
Blank	1.5 mE/min
Typical activity	3–8 kU/l plasma

Comment

Though 0.1 mol/l sodium phosphate is preferable as diluent solution, saline is also acceptable. If rapid determination of cholinesterases following venipuncture is required, ice-cold distilled water can be used to promote hemolysis without freezing. Adding 0.03% Triton X-100 fastens hemolysis and does not significantly affect AChE. If the method of Reiner *et al.* [47] is employed, Triton X-100 has to be omitted.

33.2.3 Determination of Cholinesterase Activity in the Presence of Reversible Inhibitors

Introduction

Monitoring of the effects of reversible inhibitors on blood ChEs as surrogate parameters is being used in patients with Alzheimer's disease [57]. Table 33.14 compares the half-lives of spontaneous reactivation of acetylcholinesterases inhibited

Table 33.14 Half-lives of spontaneous reactivation (min) of inhibited acetylcholinesterases.

Inhibitor	Enzyme source	pH	Temperature (°C)	Half-life (min)	Reference
Tacrine	Mouse AChE	7.0	22	0.002	[58]
Huperzine A	Human RBC-AChE	7.4	37	4.6	Eckert and Eyer, unpublished
Physostigmine	Human RBC-AChE	7.4	37	15	[60]
Pyridostigmine	Human RBC-AChE	7.4	37	31	[61]
Paraoxon-methyl	Human RBC-AChE	7.4	37	40	[62]
Fasciculin 2	Mouse AChE	7.0	22	150	[63]
Paraoxon-ethyl	Human RBC-AChE	7.4	37	1900	[64]
Rivastigmine	rHu-AChE	7.5	37	2200	[59]
Eptastigmine	Electric eel AChE	8.0	25	15 000	[59]

rHu-AChE: recombinant human AChE.

by nonacylating and acylating inhibitors. Nonacylating inhibitors, including drugs used in Alzheimer's disease such as tacrine, donepezil, and huperzine A, are primarily bound to the choline-binding site of AChE, while fasciculin 2, a mamba snake venom, binds only to the peripheral site and covers the entrance to the active site gorge [58]. In contrast, acylating inhibitors form covalent conjugates with the active site of serine hydroxyl group with widely varying stabilities. The *N*-monomethyl and *N,N*-dimethyl carbamoylated enzyme conjugates formed by physostigmine and pyridostigmine, respectively, are easily hydrolyzed while the long-chain analogues of physostigmine are quite resistant toward hydrolysis. An exception is rivastigmine that forms an *N,N*-ethylmethylcarbamylated enzyme along with the leaving moiety tightly bound to the anionic site, thereby impeding the approach of a water molecule for nucleophilic attack at the carbamylated phosphorus atom [59].

These data make clear that dilution of blood samples in case of inhibitors such as tacrine, donepezil, and huperzine A may falsify the result because of the rapid spontaneous reactivation of inhibited enzyme in the (virtual) absence of the (diluted) inhibitor. Here, undiluted samples were incubated with radioactively labeled acetylcholine after which residual substrate [57] or the product was determined.

More convenient, though possibly less accurate, is the AChE determination by the modified Ellman method in a double-beam photometer. In such a case, whole blood (EDTA or heparin) should not be diluted at the spot but brought to −20 °C as early as possible. The thawed hemolysate (10 μl) is introduced in a pair of cuvettes containing the buffer, DTNB, and ethopropazine, and the reaction in the sample cuvette is *immediately* started by addition of ASCh. The reaction of DTNB with matrix sulfhydryls proceeds in both cuvettes and is thus subtracted from the AChE reaction. When the reaction is followed for a few minutes only, inhibition of AChE is readily observed and a reversible inhibitor revealed by an accelerating reaction.

If a double-beam photometer is not available, an undiluted whole-blood sample (e.g., 50 μl) can be preincubated at 37 °C for 10 min with an equal volume of DTNB solution (10 mmol/l), sufficient to mask matrix sulfhydryls. Then, 20 μl of this mixture is introduced into the usual Ellman assay (containing the normal concentration of DTNB). Of course, these modifications require the volume compensation of phosphate buffer to give a final volume of 3.16 ml. Hemoglobin determination is brought about by diluting the thawed whole-blood sample 1 : 100 in the laboratory and proceeding the usual way. This method can also be used when intoxication by a carbamate is suspected.

If BChE should be determined in the frozen whole-blood sample, the method of Reiner *et al.* [47], as already mentioned can be applied.

Enzyme stability

EryAChE and plasma BChE are rather stable enzymes. Hemolysates can be stored for at least 3 months at −20 °C without loss of enzyme activity [46]. Lyophilized samples stored in the fridge are stable for months and can serve as standards [11]. When diluted enzyme preparations are incubated at higher temperature, for example, 37 °C for extended time periods, the inclusion of a stabilizing protein, such as 0.01% gelatine, is of advantage to preserve the activity [65].

Variability in RBC-AChE activity

In healthy adults (51 men, 50 women), RBC-AChE activity was normally distributed with 90% of the values falling between the limits 0.85 and 1.15 when corrected for hematocrit. The mean value for women tended to be slightly higher (2%) than for men, but the difference was not significant. The adult RBC-AChE activities were also found in children and older babies ($n = 18$). In contrast, full-term babies from birth to 15 days had only two-thirds of the activity of adults ($n = 7$; range 0.57–0.75) [66]. In a large study using blood specimens from blood-bank donors ($n = 800$), RBC-AChE activity as determined by the Michel method [29] showed a standard deviation of 10.6% in men and 10.9% in women ranging (98%) from 0.76 to 1.24 and 0.75 to 1.25, respectively [67].

A similar variability in RBC-AChE was found when the activity was referred to the red cell count or the hemoglobin content. Sidell and Kaminskis [68] reported for healthy humans the following AChE values (U/g Hb; 37 °C): 37.6 ± 4.9 ($n = 40$) for men and 39.3 ± 4.5 ($n = 38$) for women. The authors using a Technicon autoanalyzer found that the annual average range of AChE values (biweekly for a year, $n = 22$) varied less than hematocrit, Hb or, RBC count, 8% for men and 12% for women [68]. Other laboratories measuring under different conditions found 14.6 ± 1.2 U/g Hb ($n = 13$) in volunteers and 14.6 ± 2.6 U/g Hb ($n = 894$) in migrant residents [11]. In healthy persons, RBC-AChE activity is not generally regarded as depending on racial background [69]. These examples show that RBC-AChE – when corrected to cell count, hematocrit or hemoglobin – varies usually less than sometimes claimed and that patients who have RBC-AChE levels lower than normal due to fluid replacement would have their erythrocyte AChE levels correctly adjusted by hemoglobin, hematocrit, or cell count [70]. Moreover, the variability in RBC-AChE activity of humans reported by many groups may have a methodological basis and may not be inherent to the physiology of the subjects studied [9].

RBC-AChE activity was highest in younger cells (lower density) and lowest in older cells (highest density) [71]. Hence, it is to be expected that enhanced erythropoiesis results in higher specific activity of RBC-AChE, such as hemolysis and acute blood loss [72]. In pregnant women ($n = 259$), RBC-AChE activity was 13% higher (normalized to hematocrit) than in nonpregnant women, the effect being highest at the third trimester [69].

Patients with hereditary spherocytosis can have very high RBC-AChE activity – that may be 2–3 standard deviations above the normal mean – without clinical manifestations. Similarly in sickle-cell anemia, RBC-AChE activity is usually high. It should be noted that high concentrations of the enzyme may persist for longer times when stainable reticulum has already disappeared [73]. Patients of both gender with chronic renal failure who are maintained on chronic hemodialysis (thrice weekly) and receiving erythropoietin to target hematocrit at 36–36.5% had higher RBC-AChE activity than control subjects. It may be inferred that patients with iron deficiencies with hypochromic RBCs show near-normal RBC-AChE activity when referred to cell count, but show increased RBC-AChE activity when referred to hemoglobin.

Reduced RBC-AChE activity has been observed in paroxysmal nocturnal hemoglobinuria (PNH) in which clones of RBCs exhibit increased susceptibility to complement-mediated lysis and a deficiency of AChE activity. This loss is restricted to this clone of cells whereas apparently normal RBCs carry AChE with normal activity [72].

Variability in plasma BChE activity

Besides the genetic variations outlined above, wide intra- and interindividual variability in plasma BChE activity was observed in healthy subjects. Such variations are related to body weight, height, and sex, and may be influenced by hormone status, food, and drugs. Unexposed reference group showed an intraindividual variation (4–6 week intervals within 10 months) of 14% ($n = 131$) [18]. The interindividual variation is still higher as the following mean values (range 2.5–97.5%) were found (U/ml; 7 mmol/l butyrylthiocholine; pH 7.7; 25 °C): men ($n = 715$) 5.63 (3.20–8.53); women without oral contraceptives ($n = 473$) 4.69 (3.52–8.80); women with oral contraceptives ($n = 79$) 2.43–5.86; and pregnant woman ($n = 162$) 2.35–6.55 [74]. A similar spread was observed with acetylcholine as substrate (Michel method) with the following means (range 1–99%): men ($n = 400$) 0.963 (0.52–1.39); woman ($n = 400$) 0.817 (0.38–1.25).

Plasma BChE activity is reduced in chronic liver disease, such as in cirrhosis, is often diminished in several cancers [75], and is increased in patients with nephrotic syndrome [74].

It should be kept in mind that the reference values for plasma BChE are very wide, and individuals may have lost half of the enzyme activity and still be inside the reference range [75]. A sudden decline in the enzyme activity from a previous or baseline value can be taken to indicate exposure to an anticholinesterase agent. In the absence of a baseline value, an increase of the activity in the postexposure phase by 15–20% within some days may indicate exposure to anticholinesterase agents.

Instrumentation suitable for the modified Ellman method

Originally, the modified Ellman method was developed for use in simple filter photometers or spectrophotometers [48]. The method may be easily adapted for autoanalyzers with self-prepared reagents. The kits on the market have been criticized as being not optimal because of high substrate concentrations (5 mmol/l) [11]. However, the most common applications for the determination of AChE activity in blood include on-site occupational health screening of pesticide applicators and high-throughput screening after a potential attack of organophosphate chemical warfare agents. Owing to the size and power requirements of both conventional photometers and automated analyzers used in research and clinical laboratories, they are not suitable for the field measurement of AChE activity. To overcome these disadvantages, the Test-mate ChE Cholinesterase Test System (EQM Research, Cincinnati, Ohio – www.eqmresearch.com), a small battery-powered Ellman-based cholinesterase test system, was developed for operation without temperature regulation. The Test-mate ChE is a portable cholinesterase test system that is packaged in a watertight storage case, measures 7″ × 10″ × 11″, weighs less than 5 kg, is powered by a 9-V battery, and produces test results within 5 min from a 10 μl whole-blood finger-stick sample [44, 76]. Hereby, AChE activities are normalized to 25 °C and are referred to the individual hemoglobin concentration. Recent modifications [77], that is, shift of wavelength from 450 to 470 nm by using InGaN/SiC blue LEDs, individual correction of DTNB blank reactions with protein SH-groups and an extended temperature correction algorithm allowing operation from 10 to 50 °C (Figure 33.3), resulted in a further optimization of the system (Figure 33.4) and extended the applicability of this system for determination of AChE activity in the field.

The major shortcoming of the Test-mate ChE system, however, is the low throughput with a maximum of 12 assays per hour. Therefore, it was tempting to modify the manual Ellman assay [48] for use with automatic analyzers and microplate

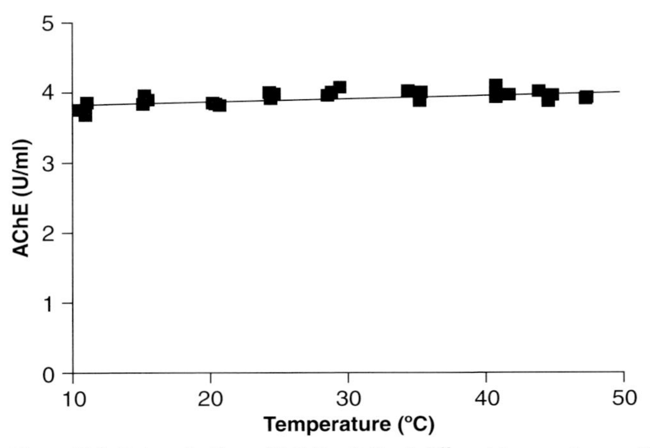

Figure 33.3 Determination of AChE activity at different temperatures with Test-mate ChE.

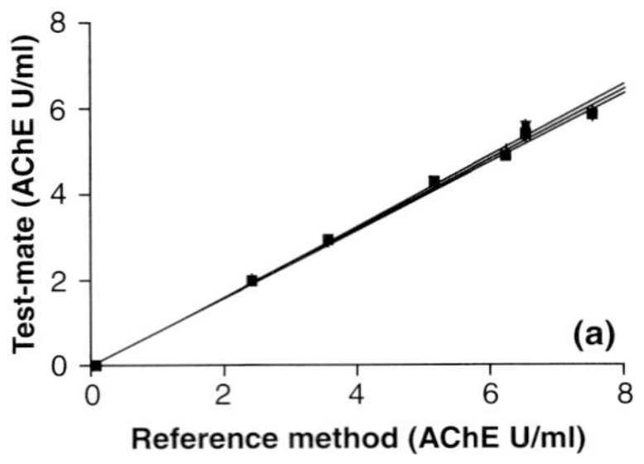

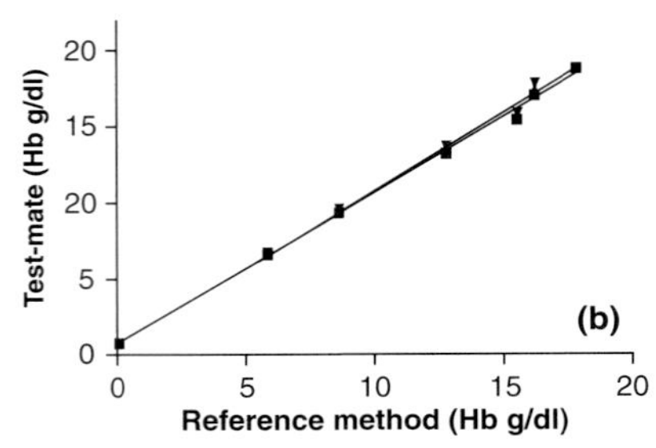

Figure 33.4 Correlation of AChE activity determined with the reference method [48] and three different Test-mate ChE devices. Reference method run at 37 °C and Test-mate ChE normalizing to 25 °C.

Table 33.15 Characteristics of the manual and microplate-based ChE assays.

Macro method	Micro method
Manual method	Automated method
Labor intensive	Robotic system
Manual pipetting	Automatic pipetting
Cuvette-based system	Microplate-based system
Manual data analysis	Automated data analysis
Large waste volumes	Small waste volumes
Less than 100 assays/day	More than 600 assays/day

readers thus enabling high-throughput analysis of blood samples. Combining a microplate-based assay (micro method) with a robotic system markedly improved the reproducibility of the measurements. Table 33.15 compares the major characteristics of the manual (macro method) and the microplate-based assays using a TECAN Genesis RMP 150 and a Sunrise apparatus (unpublished results).

33.2.4 Application of the Modified Ellman Method

33.2.4.1 Exposure Monitoring

RBC-AChE and plasma BChE determinations may indicate exposure and incorporation of anticholinesterase agents. Experience in California led to a recommendation that individuals showing a 30% or more decrease from normal plasma BChE level should discontinue participation in the work until cholinesterase levels have returned to normal [11]. Of course, such a depression does not answer questions concerning health effects. Depression of RBC-AChE to 70% of normal may be of more significance because the extent of inhibition usually correlates with inhibition in target tissues (brain, muscle) [24] at least initially after exposure. It should be considered, however, that diminished RBC-AChE recovers only slowly, 0.9% per day, while synaptic AChE recovers much faster as does BChE. For example, the recovery of BChE in workers after overexposure to an organophosphate exhibited an exponential kinetics with a half-life of 12 days, whereas the recovery of RBC-AChE was linear over time, attaining unexposed activity after about 82 days [78]. Hence, RBC-AChE depression may result from acute contamination or long-lasting exposure to very small doses that hardly affect enzyme sources with rapid turnover.

RBC-AChE reactivation assays are useful to assess the likelihood that recent significant exposure to organophosphate compounds has occurred, especially when baseline RBC-AChE levels are lacking. Hansen and Wilson proposed an assay using washed RBCs that were incubated with 2-PAM chloride, 10 mmol/l final concentration, at pH 8.0, at 25 °C for 40–60 min, followed by another RBC wash, to reduce oxime-induced acetylthiocholine hydrolysis in the subsequent Ellman assay [79].

33.2.4.2 Monitoring of Intoxicated Patients and Assessment of the Efficacy of Reactivators

The automated procedure enables the extension of the AChE assay toward therapeutic drug monitoring [39, 49]. Determination of AChE and BChE activities in whole-blood dilution and plasma, respectively, as well as reactivatability of patient's AChE, determined after incubation with high oxime concentrations, and inhibitory activity in patient's plasma, determined by incubating plasma with donor AChE, provides a meaningful tool for defining the necessity and duration of oxime treatment in OP-poisoned patients [80]. The corresponding flow chart is shown in Figure 33.5.

Using the above-depicted SOP, a reduction of RBC-AChE activity below 0.4 U/μ mol Hb and/or of BChE below 3 U/ml plasma serves as an indicator of incorporated anticholinesterase agents. RBC-AChE activity below 0.2 U/μmol Hb

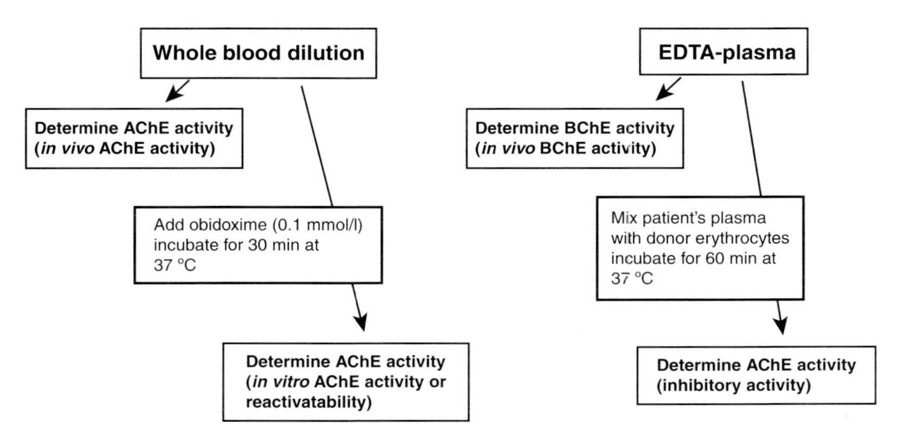

Figure 33.5 Analysis of cholinesterase status in patient's blood. Scheme of different procedures (*in vivo* AChE and BChE activity, reactivatability of inhibited AChE, and inhibitory activity of patient's plasma).

may be associated with clinically relevant cholinergic signs and symptoms. Isolated inhibition of plasma BChE is without clinical significance, but it may point to exposure to anticholinesterase agents. In serious poisoning, anticholinesterase agents may be detectable in plasma owing to their ability to inhibit added test AChE.

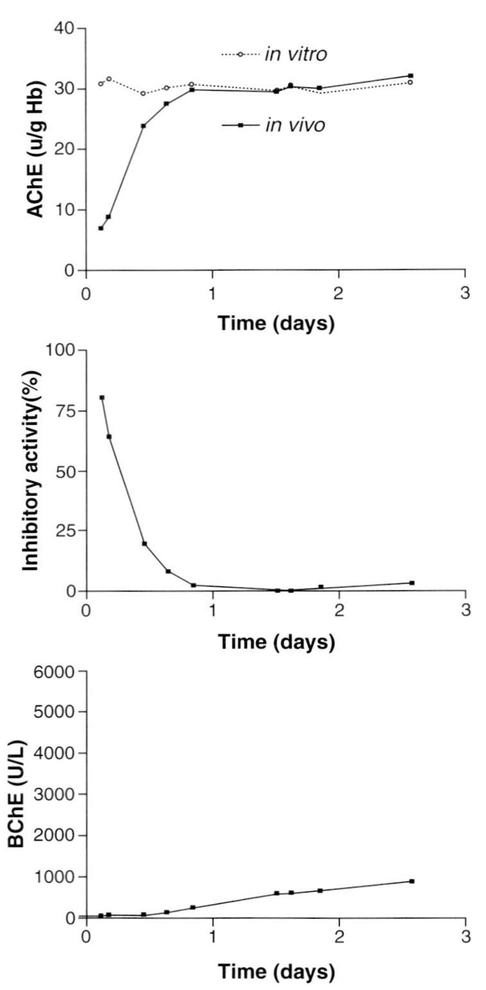

Figure 33.6 *Cholinesterase status in a patient poisoned with parathion.* About 3 h after ingestion, obidoxime was administered as an i.v. bolus (250 mg) followed by continuous infusion at 750 mg/24 h. *Upper panel*: RBC-AChE *in vivo* as measured in patient's blood and the reactivatability *in vitro* as determined after incubation of diluted patient's blood with 0.1 mmol/l obidoxime for 30 min. *Middle panel*: Inhibitory activity of patient's plasma that results in the inhibition of test RBC-AChE. *Lower panel*: BChE activity of patient's plasma. It is obvious that obidoxime is able to reactivate RBC-AChE when inhibitory material in plasma has disappeared. Then, newly synthesized BChE from the liver in plasma is no longer inhibited.

Such a test requires EDTA plasma (5–7 mmol/l EDTA) to inhibit Ca^{++}-dependent organophosphate hydrolases, such as paraoxonase.

Figure 33.6 shows the cholinesterase status of a patient poisoned by parathion and treated with obidoxime. The example reveals that reactivation of inhibited RBC-AChE comes to completion when inhibitory material is not present any longer. In this case, *de novo* synthesis of BChE leads to reappearance of activity in plasma.

In the case of mass casualties due to anticholinesterase agents (cf. the sarin attack in Tokyo and Matsumoto), determination of the cholinesterase status of the exposed patients may give first hints of the group of compounds involved. Good reactivation to near-normal values by oximes of RBC-AChE in diluted blood samples, for example, incubation of 1 ml together with 0.1 mmol/l obidoxime at 37 °C for 30 min, points to organophosphates, particularly to diethyl esters. Initial partial reactivation, which fades within 8 h, points to dimethyl esters. High rates of spontaneous reactivation seen by a gradual increase in activity during determination (accelerating curve) points to mono- and dimethyl carbamates or to reversible inhibitors such as those tried in Alzheimer's disease, for example, huperzine A. No significant reactivation by oximes points to the involvement of some nerve agents that produce a rapidly aging enzyme (soman) or an agent that forms a stable adduct not suitable for obidoxime reactivation, for example, the nerve agents cyclosarin, tabun, and choline methyl phosphonic acid derivatives [81]. A similar picture may be observed with several carbamates, which form a quite stable carbamoylated AChE, for example, rivastigmine and eptastigmine that have been tested in Alzheimer's disease [22].

References

References for Section 33.1

1 Glasser, L., Sternglanz, P.D., Combie, J. and Robinson, A. (1973) Serum osmolality and its applicability to drug overdose. *American Journal of Clinical Pathology*, **60**, 695–699.

2 Gibitz, H.J. and Külpmann, W.R. (1998) Poisonings, in *Clinical Laboratory Diagnostics* (ed. L. Thomas), TH-Books, Frankfurt/M.

3 Tryding, N. and Roos, K.A. (1992) *Drug Effects in Clinical Chemistry*, 6th edn, Apoteksbolaget AB, Stockholm.

4 Dart, R.C.(ed.) (2004) *Medical Toxicology*, 3rd edn, Lippincott Williams and Wilkins, Philadelphia, PA.

5 Messing, R.O., Closson, R.G. and Simon, R.P. (1984) Drug induced seizures: a 10 year experience. *Neurology*, **34**, 1582–1586.

6 Volans, G. and Widdop, B. (1984) ABC of poisoning. Laboratory investigations in acute poisoning. *British Medical Journal*, **289**, 426–428; (a) Külpmann, W.R., Stummvoll, H.K. and Lehmann, P. (2007) *Electrolytes, Acid-Base Balance and Blood Gases*, 2nd edn, Springer, Wien-New York.

References for Section 33.2

7 Wächtler, K. (1988) Phylogeny of the cholinergic synapse, in *The Cholinergic Synapse* (ed. V.P. Whittaker) Springer, Berlin, pp. 57–85.

8 Zajicek, J. (1957) Studies on the histogenesis of blood platelets and megakaryocytes; histochemical and gasometric investigations of acetylcholinesterase activity in the erythrocyte-erythropoietic and platelet-

megakaryocytic systems of various mammals. *Acta Physiologica Scandinavica*, **40** (Suppl 138), 1–32.

9 Wilson, B.W., McCurdy, S.A., Henderson, J.D., McCarthy, S.A. and Billitti, J.E. (1998) Cholinesterases and agriculture. Humans, laboratory animals, wildlife, in *Structure and Function of Cholinesterases and Related Proteins* (eds B.P. Doctor, P. Taylor, D.M. Quinn, R.L. Rotundo and M.K., Gentry), Plenum Press, New York and London, pp. 539–546.

10 Szelenyi, J.G., Bartha, E. and Hollan, S.R. (1982) Acetylcholinesterase activity of lymphocytes: an enzyme characteristic of T-cells. *British Journal of Haematology*, **50**, 241–245.

11 Wilson, B.W. (2001) Cholinesterases, in *Handbook of Pesticide Toxicology*, (ed. R.I. Krieger) Academic Press, San Diego, pp. 967–985.

12 Simeon-Rudolf, V., Sinko, G., Stuglin, A. and Reiner, E. (2001) Inhibition of human blood acetylcholinesterase and butyrylcholinesterase by ethopropazine. *Croatica Chemica Acta*, **74**, 173–182.

13 Traina, M.E. and Serpietri, L.A. (1984) Changes in the levels and forms of rat plasma cholinesterases during chronic diisopropylphosphorofluoridate intoxication. *Biochemical Pharmacology*, **33**, 645–653.

14 Ralston, J.S., Rush, R.S., Doctor, B.P. and Wolfe, A.D. (1985) Acetylcholinesterase from fetal bovine serum. Purification and characterization. *The Journal of Biological Chemistry*, **260**, 4312–4318.

15 Koelle, G.B. (1963) Cytological distributions and physiological functions of cholinesterases, in *Handbook of Experimental Pharmacology* (eds. O. Eichler and A. Farah), Springer, 15, pp. 187–298.

16 Oehme, F.W. and Mannala, S. (2001) Pesticide use in veterinary medicine, in *Handbook of Pesticide Toxicology*, 2nd edn (ed. R.I. Krieger), Academic Press, San Diego, pp. 263–274.

17 Halbrook, R.S., Shugart, L.R., Watson, A.P., Munro, N.B. and Linnabary, R.D. (1992) Characterizing biological variability in livestock blood cholinesterase activity for biomonitoring organophosphate nerve agent exposure. *Journal of the American Veterinary Medical Association*, **201**, 714–725.

18 Brock, A. (1991) Inter and intraindividual variations in plasma cholinesterase activity and substance concentration in employees of an organophosphorus insecticide factory. *British Journal of Industrial Medicine*, **48**, 562–567.

19 Masson, P., Legrand, P., Bartels, C.F., Froment, M.-T., Schopfer, L.M. and Lockridge, O. (1997) Role of aspartate 70 and tryptophan 82 in binding of succinyldithiocholine to human butyrylcholinesterase. *Biochemistry*, **36**, 2266–2277.

20 Li, B., Stribley, J.A., Ticu, A., Xie, W., Schopfer, L., Hammond, P., Brimijon, S., Hinrichs, S.H. and Lockridge, O. (2000) Abundant tissue butyrylcholinesterase and its possible function in the acetylcholinesterase knockout mouse. *Journal of Neurochemistry*, **75**, 1320–1331.

21 Xie, W., Stribley, J.A., Chatonnet, A., Wilder, P., Rizzino, A., McComb, R., Taylor, P., Hinrichs, S.H. and Lockridge, O. (2000) Postnatal developmental delay and supersensitivity to organophosphate in gene-targeted mice lacking acetylcholinesterase. *The Journal of Pharmacology and Experimental Therapeutics*, **293**, 896–902.

22 Sramek, J.J. and Cutler, N.R. (2000) RBC cholinesterase inhibition: a useful surrogate marker for cholinesterase inhibitor activity in Alzheimer disease therapy? *Alzheimer Disease & Associated Disorders*, **14**, 216–227.

23 Storm, J.E., Rozman, K.K. and Doull, J. (2000) Occupational exposure limits for 30 organophosphate pesticides based on inhibition of red blood cell acetylcholinesterase. *Toxicology*, **150**, 1–29.

24 Padilla, S. (1995) Regulatory and research issues related to cholinesterase inhibition. *Toxicology*, **102**, 215–220.

25 Wilson, B.W. and Henderson, J.D. (1992) Blood esterase determinations as markers of exposure. *Reviews of Environmental Contamination and Toxicology*, **128**, 55–69.

26 Eyer, P., Worek, F. and Thiermann, H. (1999) Easy laboratory tests to follow the acetylcholinesterase status during oxime therapy in organophosphate poisoning. in *XIX International Congress of the European Association of Poisons Centres and Clinical Toxicologists, 22–25 June, Dublin, p.* 110.

27 Wills, K.H. (1972) The measurement and significance of changes in the cholinesterase activities of erythrocytes and plasma in man and animals. *Critical Reviews in Toxicology*, **1**, 153–202.

28 Ammon, R. (1933) Die fermentative Spaltung des Acetylcholins. *Pflügers Archiv für die gesamte Physiologie des Menschen und der Tiere*, **233**, 486–491.

29 Michel, H.O. (1949) An electrometric method for the determination of red blood cell and plasma cholinesterase activity. *The Journal of Laboratory and Clinical Medicine*, **34**, 1564–1568.

30 St Omer, V.E.V. and Rottinghaus, G.E. (1992) Biochemical determination of cholinesterase activity in biological fluids and tissues, in *Clinical and Experimental Toxicology of Organophosphates and Carbamates* (eds B. Ballantyne and T.C. Marrs), Butterworth-Heinemann Ltd, Oxford, pp. 15–27.

31 Glick, D. (1937) Properties of cholinesterase in human serum. *The Biochemical Journal*, **31**, 521–525.

32 Nabb, D.P. and Whitfield, F. (1967) Determination of cholinesterase by an automated pH stat method. *Archives of Environmental Health*, **15**, 147–154.

33 Johnson, C.D. and Rusell, R.L. (1975) A rapid, simple radiometric assay for cholinesterase, suitable for multiple determinations. *Analytical Biochemistry*, **64**, 229–238.

34 Ellman, G.L., Courtney, K.D., Andres, V. and Featherstone, R.M. (1961) A new and rapid colorimetric determination of acetylcholinesterase activity. *Biochemical Pharmacology*, **7**, 88–95.

35 Ciesielski, S., Loomis, D.P., Mims, S.R. and Auer, A. (1994) Pesticide exposure, cholinesterase depression, and symptoms among North Carolina migrant farm workers. *American Journal of Public Health*, **84**, 446–451.

36 McCurdy, S.A., Hansen, M.E., Weisskopf, C.P., Lopez, R.L., Schneider, F., Spencer, J., Sanborn, J.R., Krieger, R.I., Wilson, B.W., Goldsmith, D.F. and Schenker, M.B. (1994) Assessment of azinphosmethyl exposure in California peach harvest workers. *Archives of Environmental Health*, **49**, 289–296.

37 London, L., Thompson, M.L., Sacks, S., Fuller, B., Bachmann, O.M. and Myers, J.E. (1995) Repeatability and validity of a field kit for estimation of cholinesterase in whole blood. *Occupational and Environmental Medicine*, **52**, 57–64.

38 Thiermann, H., Mast, U., Klimmek, R., Eyer, P., Hibler, A., Pfab, R., Felgenhauer, N. and Zilker, T. (1997) Cholinesterase status, pharmacokinetics and laboratory findings during obidoxime therapy in organophosphate poisoned patients. *Human & Experimental Toxicology*, **16**, 473–480.

39 Thiermann, H., Szinicz, L., Eyer, F., Worek, F., Eyer, P., Felgenhauer, N. and Zilker, T. (1999) Modern strategies in therapy of organophosphate poisoning. *Toxicology Letters*, **107**, 232–239.

40 Ellman, G.L. (1958) A colorimetric method for determining low concentrations of mercaptans. *Archives of Biochemistry and Biophysics*, **74**, 443–450.

41 Ellman, G.L. (1959) Tissue sulfhydryl groups. *Archives of Biochemistry and Biophysics*, **82**, 70–77.

42 Augustinsson, K.-B., Erikson, H. and Faijersson, Y. (1978) A new approach to determining cholinesterase activities in samples of whole blood. *Clinica Chimica Acta; International Journal of Clinical Chemistry*, **89**, 239–252.

43 George, P.M. and Abernethy, M.H. (1983) Improved Ellman procedure for erythrocyte cholinesterase. *Clinical Chemistry*, **29**, 365–368.

44 Magnotti, R.A., Dowling, K., Eberly, J.P. and McConnell, R.S. (1988) Field measurement of plasma and erythrocyte cholinesterases. *Clinica Chimica Acta; International Journal of Clinical Chemistry*, **176**, 315–332.

45 Gordon, J.J. (1948) N-Diethylaminoethylphenothiazine: a specific inhibitor of pseudocholinesterase. *Nature*, **162**, 146.

46 Meuling, W.J.A., Jongen, M.J.M. and van Hemmen, J.J. (1992) An automated method for the determination of acetyl- and pseudocholinesterase in hemolyzed whole blood. *American Journal of Industrial Medicine*, **22**, 231–241.

47 Reiner, E., Bosak, A. and Simeon-Rudolf, V. (2004) Activity of cholinesterases in human whole blood measured with acetylthiocholine as substrate and ethopropazine as selective inhibitor of plasma butyrylcholinesterase. *Arhiv za Higijenu Rada i Toksikologiju*, **55**, 1–4.

48 Worek, F., Mast, U., Kiderlen, D., Diepold, C. and Eyer, P. (1999) Improved determination of acetylcholinesterase activity in human whole blood. *Clinica Chimica Acta; International Journal of Clinical Chemistry*, **288**, 73–90.

49 Eyer, P. (2003) The role of oximes in the management of organophosphorus pesticide poisoning. *Toxicological Reviews*, **22**, 165–190.

50 Van Kampen, E.J. and Zijlstra, W.G. (1961) Standardization of hemoglobinometry. II. The hemiglobincyanide method. *Clinica Chimica Acta; International Journal of Clinical Chemistry*, **6**, 538–544.

51 Eddleston, M., Eyer, P., Worek, F., Mohamed, F., Senarathna, L., von Meyer, L., Juszczak, E., Hittarage, A., Azhar, S., Dissanayake, W., Sheriff, M.H.R., Szinicz, L., Dawson, A.H. and Buckley, N.A. (2005) Differences between organophosphorus insecticides in human self-poisoning: a prospective cohort study. *Lancet*, **366**, 1452–1459.

52 Szinicz, L., Eyer, P., Worek, F., Kiderlen, D. and Mast, U. (1999) Development of a standard operation procedure for determination of acetylcholinesterase (AChE) activity in blood. *The Proceedings of the Chemical and Biological Medical Treatment Symposium – Industry I, Zagreb-Dubrovnik, Croatia, 25–31 October 1998* (eds S. Bokan and Z. Orehovec), R. Price, ASA, Portland, Maine, USA, pp. 298–301.

53 Eyer, P., Worek, F., Kiderlen, D., Sinko, G., Stuglin, A., Simeon-Rudolf, V. and Reiner, E. (2003) Molar absorption coefficient for the reduced Ellman reagent: reassessment. *Analytical Biochemistry*, **312**, 224–227.

54 Worek, F., Koller, M., Thiermann, H. and Szinicz, L. (2005) Diagnostic aspects of organophosphate poisoning. *Toxicology*, **214**, 182–189.

55 Walmsley, T.A., Abernethy, M.H. and Fitzgerald, H.P. (1987) Effect of daylight on the reaction of thiols with Ellman's reagent, 5,5′-dithiobis(2-nitrobenzoic acid). *Clinical Chemistry*, **33**, 1928–1931.

56 Kiderlen, D. (2004) On the phosphoryl oxime hydrolase of plasma, an enzyme that markedly increases the efficacy of oximes in organophosphate poisoning (Doctoral Thesis), Ludwig-Maximilian-University, Munich.

57 Thomsen, T., Kewitz, H. and Pleul, O. (1988) Estimation of cholinesterase activity (EC 3.1.1.7; 3.1.1.8) in undiluted plasma and erythrocytes as a tool for measuring *in vivo* effects of reversible inhibitors. *Journal of Clinical Chemistry and Clinical Biochemistry*, **26**, 469–475.

58 Reiner, E. and Radic, Z. (2000) Mechanism of action of cholinesterase inhibitors, in *Cholinesterases and Cholinesterase Inhibitors* (ed. E. Giacobini) Martin Dunitz Ltd, London, pp. 103–119.

59 Bar-On, P., Millard, C.B., Harel, M., Dvir, H., Enz, A., Sussman, J.L. and Silman, I. (2002) Kinetic and structural studies on the interaction of cholinesterases with the anti-Alzheimer drug rivastigmine. *Biochemistry*, **41**, 3555–3564.

60 Wetherell, J.R. and French, M.C. (1991) A comparison of the decarbamoylation rates of physostigmine – inhibited plasma and red cell cholinesterases of man with other species. *Biochemical Pharmacology*, **42**, 515–520.

61 Ellin, R.I. and Kaminskis, A. (1989) Carbamy enzyme reversal as a means of predicting pyridostigmine protection against soman. *The Journal of Pharmacy and Pharmacology*, **41**, 633–635.

62 Worek, F., Diepold, C. and Eyer, P. (1999) Dimethylphosphoryl-inhibited human cholinesterases: Inhibition, reactivation, and aging kinetics. *Archives of Toxicology*, **73**, 7–14.

63 Radic, Z., Quinn, D.M., Vellom, D.C., Camp, S. and Taylor, P. (1995) Allosteric control of acetylcholinesterase catalysis by fasciculin. *The Journal of Biological Chemistry*, **270**, 20391–20399.

64 Worek, F., Bäcker, M., Thiermann, H., Szinicz, L., Mast, U., Klimmek, R. and Eyer, P. (1997) Reappraisal of indications and limitations of oxime therapy in organophosphate poisoning. *Human & Experimental Toxicology*, **16**, 466–472.

65 Doctor, B.P., Toker, L., Roth, E. and Silman, I. (1987) Microtiter assay for acetylcholinesterase. *Analytical Biochemistry*, **166**, 399–403.

66 Sabine, J.C. (1955) The clinical significance of erythrocyte cholinesterase titers. *Blood*, **10**, 1132–1138.

67 Rider, J.A., Hodges, J.L., Swader, J. and Wiggins, A.D. (1957) Plasma and red cell cholinesterase in 800 "healthy" blood donors. *The Journal of Laboratory and Clinical Medicine*, **50**, 376–383.

68 Sidell, F.R. and Kaminskis, A. (1975) Temporal intrapersonal variability of cholinesterase in human plasma and erythrocytes. *Clinical Chemistry*, **21**, 1961–1963.

69 de Peyster, A., Willis, W.O. and Liebhaber, M. (1994) Cholinesterase activity in pregnant women and newborns. *Journal of Toxicology: Clinical Toxicology*, **32**, 683–696.

70 Taylor, P.W., Lukey, B.J., Clark, C.R., Lee, R.B. and Roussel, R.R. (2003) Field verification of Test-mate ChE. *Military Medicine*, **168**, 314–319.

71 Prall, Y.G., Gambhir, K.K. and Ampy, F.R. (1998) Acetylcholinesterase: an enzymatic marker of human red blood cell aging. *Life Sciences*, **63**, 177–184.

72 Lawson, A.A. and Barr, R.D. (1987) Acetylcholinesterase in red blood cells. *American Journal of Hematology*, **26**, 101–112.

73 Sabine, J.C. (1959) Erythrocyte cholinesterase titers in hematologic disease states. *The American Journal of Medicine*, **27**, 81–96.

74 den Blaauwen, D.H., Poppe, W.A. and Tritschler, W. (1983) Cholinesterase (EC 3.1.1.8) mit Butyrylthiocholinjodid als Substrat: Referenzwerte in Abhängigkeit von Alter und Geschlecht unter Berücksichtigung hormonaler Einflüsse und Schwangerschaft. *Journal of Clinical Chemistry and Clinical Biochemistry*, **21**, 381–386.

75 McQueen, M.J. (1995) Clinical and analytical considerations in the utilization of cholinesterase measurements. *Clinica Chimica Acta; International Journal of Clinical Chemistry*, **237**, 91–105.

76 Magnotti, R.A., Jr, Eberly, J.P., Quarm, D.E.A. and McConnell, R.S. (1987) Measurement of acetylcholinesterase in erythrocytes in the field. *Clinical Chemistry*, **33**, 1731–1735.

77 Eberly, J.P., Eyer, P., Pfeiffer, B., Szinicz, L. and Worek, F. (2004) Modification of the Test-mate ChE cholinesterase test system for use over an extended temperature range. *Medical Defense Bioscience Review, Baltimore*, pp. 1–11.

78 Mason, H.J. (2000) The recovery of plasma cholinesterase and erythrocyte acetylcholinesterase activity in workers after over-exposure to dichlorvos. *Occupational Medicine*, **50**, 343–347.

79 Hansen, M.E. and Wilson, B.W. (1999) Oxime reactivation of RBC acetylcholinesterases for biomonitoring. *Archives of Environmental Contamination and Toxicology*, **37**, 283–289.

80 Eyer, P., Kiderlen, D., Meischner, V., Szinicz, L., Thiermann, H., Worek, F., Eyer, P., Felgenhauer, N., Pfab, R., Zilker, T., Eddleston, M., Senarathna, L., Sheriff, R. and Buckley, N. (2003) The current status of oximes in the treatment of OP poisoning-comparing two regimes. *Journal of Toxicology: Clinical Toxicology*, **41**, 441–443.

81 Worek, F., Thiermann, H. and Szinicz, L. (2004) Reactivation and aging kinetics of human acetylcholinesterase inhibited by organophosphonylcholines. *Archives of Toxicology*, **78**, 212–217.

34
Therapeutic Drug Monitoring

W.R. Külpmann

34.1
Introduction

The aim of therapeutic drug monitoring (TDM) is to achieve an optimal therapeutic effect of drugs by monitoring their concentration in plasma or blood to attain the desired pharmacodynamic effect as quickly as possible and to maintain it during the period of treatment. Toxic effects from overdosage and from exceeding the therapeutic range should be avoided by TDM. In general, TDM is a better guide for treatment than dosage, because the relationship between the therapeutic effect of a drug and its concentration in plasma or blood is better than that between effect and dose. For example, altered pharmacokinetics can only be detected by TDM and then be taken into account by the dosage regimen.

Therapeutic drugs for TDM

TDM is most appropriate for the following settings:

(1) The effect of the drug cannot be observed easily, for example, antiepileptic drugs for the prevention of seizures in epilepsy.
(2) A narrow therapeutic range.
(3) Alteration of pharmacokinetics, for example, from impaired absorption, distribution, or elimination.

However, there are several drugs that can be usually administered without guidance by TDM, because their effects can be easily observed, for example, antihypertensive drugs, antidiabetics, or anticoagulants, by measurement of blood pressure, glucose concentration or coagulation tests. Only if noncompliance is suspected, the determination of their concentration in plasma is indicated. Therefore, procedures for the determination of these drugs are, in general, not available in a TDM laboratory to be used for toxicological purposes.

A further requirement for TDM is a close relationship between plasma/blood concentration and the pharmacodynamic effect. This prerequisite is not fulfilled, if

Clinical Toxicological Analysis: Procedures, Results, Interpretation. Edited by Wolf-Rüdiger Külpmann

ISBN: 978-3-527-31890-2

- the effect is localized in a deep compartment;
- the effects are irreversible;
- tolerance is developing.

34.2 Pharmacokinetics

In pharmacokinetics, liberation, absorption, distribution, metabolism, and excretion (LADME) of an applied drug in the body is monitored in total. After ingestion, the drug is absorbed in the gastrointestinal tract. The velocity of absorption is dependent, for example, on the drug and its formulation. During the passage through the liver, a considerable portion of the drug may be metabolized to inactive metabolites (first-pass effect). The (absolute) bioavailability reflects the fraction of a dose of a drug reaching the systemic circulation in relation to a corresponding intravenous dose. During the α-phase, the drug is distributed in the body (Figure 34.1). The β-phase reflects the velocity of drug elimination. During the α-phase, there is no balance between the drug concentration in plasma and its concentration in the relevant compartment. Sampling before the end of the α-phase is not recommended for TDM. During intravenous infusion at constant rate or multiple dosing, the drug will accumulate depending on velocity of application and elimination. The steady-state concentration is attained after the elapse of five half-life times, if the drug is administered at a constant dose and once during every half-life time. The steady-state concentration depends on the dose, the dosing interval, and the elimination half-life. If the dosing interval is shorter as compared to the elimination half-life, there is little fluctuation of the concentration and vice versa (Figure 34.2). Therapeutic ranges are usually related to the steady-state concentration.

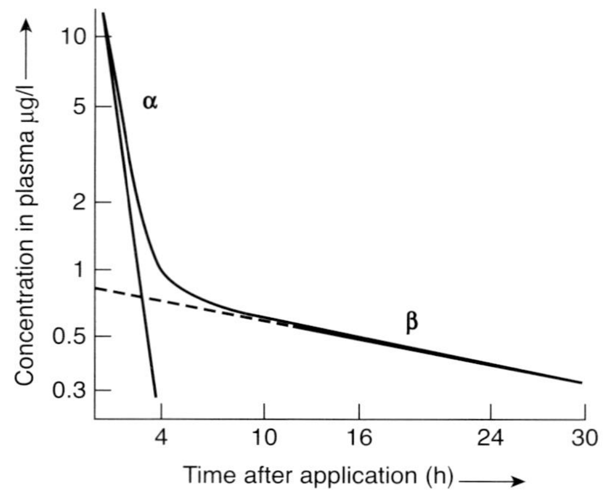

Figure 34.1 Distribution phase (α-phase) and elimination phase (β-phase) after intravenous application of a drug (open two-compartment model).

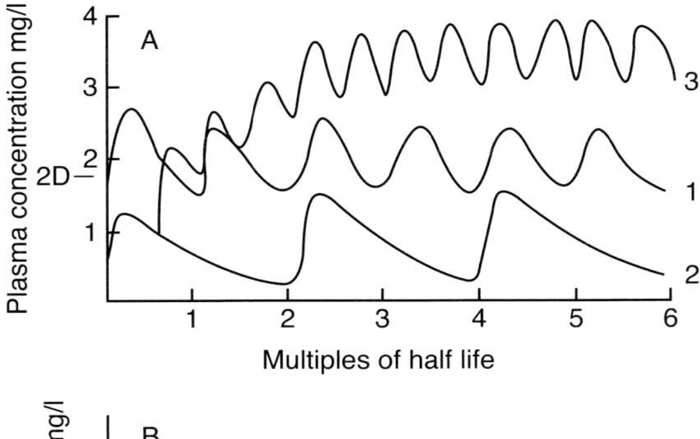

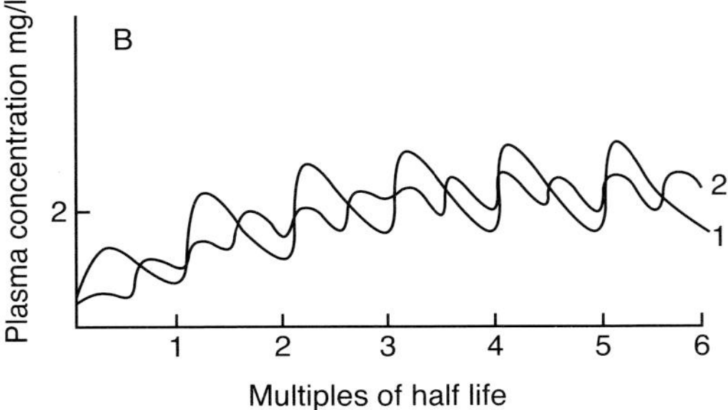

Figure 34.2 Multiple dosing. Influence of dose and dosing interval on the concentration of a drug in plasma. (A) (1) Maintenance dose administered once during one elimination half-life. Initial dose: two times maintenance dose. (2) Maintenance dose administered once during two elimination half-lives. (3) Maintenance dose administered twice during one half-life. (B) (1) Maintenance dose administered once during one elimination half-life. (2) Maintenance dose administered as two portions during one elimination half-life. D: Maintenance dose.

Sampling

In general, a sample should be taken after the end of the distribution and during the steady state. Distribution time varies considerably: it ends, for example, 30 min after intravenous application of an aminoglycoside and 6–10 h after oral application of digoxin. The peak concentration after distribution is a measure for toxic risks. The trough concentration is determined in a sample taken just before the next dosing. It shows whether the concentration is sufficiently high to maintain the pharmacodynamic effects during the whole dosing interval. Only one sample may be taken at any, but constant time after the end of distribution in case of very long half-life (e.g., phenobarbital 4 days).

34.3 Interpretation

The momentary concentration of a drug in blood depends on several influence factors (Figure 34.3)that are summarized by the mnemonic term LADME. Apart from dose, concentration in blood depends on the size of the volume of distribution.

Dose

← Completeness of absorption

← Bioavailability

← Apparent volume of distribution (body mass and composition, binding to plasma and tissue proteins, involved compartments)

← Rate of elimination (metabolism, excretion)

Plasma concentration

← Penetration into tissue

← Diffusion

← Active transport

← Plasma protein binding

Concentration at relevant receptors

← Functional condition

← Disorder

← Development of tolerance

← Competing other drugs

Pharmacodynamic effect

Figure 34.3 Relationship between dose, concentration in plasma, and concentration at relevant receptors.

Therefore, dosing is often adjusted according to body mass, nonfat body mass, or body surface area. Volume of distribution is enlarged during pregnancy or by ascites.

The absorption depends on the function of the gastrointestinal tract, on food ingested simultaneously, and on systemic and local blood circulation. The free fraction of a protein-bound drug changes with protein concentration. It is increased in hypalbuminemia, if bound to albumin; it is decreased, if bound to acid α_1-glycoprotein during acute phase reaction, when the concentration of this protein is elevated. Renal or hepatic disorders may delay the elimination of a drug or its metabolites, which may be effective as well.

Development of tolerance is associated with the decreasing effect, although concentration in blood is still within the normal therapeutic range. If different drugs are administered simultaneously, they may compete for binding with the same proteins and the free (effective) fraction may increase at constant total concentration. Electrolyte concentration may have an impact on a drug effect. Hypokalemia and hypercalcemia increase the effect of cardiac glycosides, whereas hyperkalemia and hypocalcemia decrease the effect. Therapeutic range may change depending on indication: it is lower for digoxin applied for improving cardiac contractility and higher for the treatment of atrial fibrillation. The therapeutic range is only given for orientation and nevertheless requires that the physician consider the special situation of the individual patient (Figure 34.4).

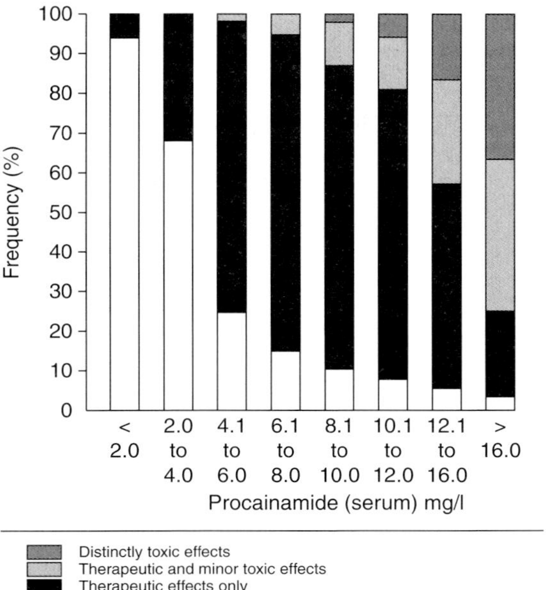

Figure 34.4 Frequency of therapeutic/toxic effects of procainamide in relation to serum concentration [7].

In case of poisoning, the determination of the concentration of a drug or toxic agent and some calculations help in answering the following questions:

How severe is the intoxication?
How long will the patient suffer from poisoning, if

(a) elimination is not enhanced?
(b) hemodialysis or hemoperfusion are applied?

The calculations are based on pharmacokinetic data, which, however, are primarily related to healthy individuals and may not be fully applicable to the patient poisoned by the drug.

The following parameters are of clinical significance:

(1) Maximum plasma concentration of the toxic agent:

$$C_{max} = \frac{m}{V_d \times M},$$

where C_{max} is the maximal plasma concentration (mg/l), m is the mass of absorbed toxic agent (mg), M is the body mass (kg), and V_d is the volume of distribution (l/kg).
(2) Mass of absorbed toxic agent:

$$m = C_{max} \times V_d \times M,$$

where C_{max} is the maximal plasma concentration (mg/l), m is the mass of absorbed toxic agent (mg), M is the body mass (kg), and V_d is the volume of distribution (kg/l).

(3) Time required to achieve nontoxic plasma concentration $C_t = C_{act}\,e^{-kt}$:

$$t = \frac{T_{1/2}}{\ln 2} \times \ln \frac{C_{act}}{C_t},$$

where C_{act} is the actual (toxic) plasma concentration (mg/l), C_t is the nontoxic plasma concentration (mg/l), k is the elimination constant (min^{-1}), t is the period of time, after which C_{act} has dropped to C_t (min), and $T_{1/2}$ is the plasma elimination half-life (min).

This equation is only applicable after completion of absorption and distribution and is valid for first-order kinetics. It is not applicable for zero-order kinetics, which implies a constant rate of metabolic degradation independent of plasma concentration, because of, for example, a saturation of the metabolizing enzyme (e.g., ethanol or salicylate at high concentrations).

(4) Time required achieving nontoxic plasma concentration if hemodialysis or hemoperfusion are applied. Formula as shown in (3) is applied. $T_{1/2}$ refers to plasma elimination half-life valid for hemodialysis or hemoperfusion. In general, prerequisites for the application of these therapeutic procedures are sufficiently high dialysance and small volume of distribution.

(5) Estimation of the completion of absorption. Blood samples are taken every 30 min. The absorption is considered as completed, if the plasma concentration is decreasing.

34.4 Analytical Methods for Drug Monitoring

Analytical toxicology has to find out whether the present clinical symptoms are caused only or partly by one or several toxic agents. After identification, the respective agent(s) are measured quantitatively. A particular agent is often identified more easily, if, as a first step, drug screening by immunoassays is carried out using a group test.

In therapeutic drug monitoring, the analyte is given and the respective immunoassay shall determine its concentration accurately and specifically. Metabolites should not interfere with the measurement, but should be determined separately, if effective. Drug screening by immunoassays used in toxicology is in general not applicable for TDM.

Cross-reactivity is very different for individual members of a group; metabolites (effective and noneffective) may contribute to the value as well as further substances of the group that may also be present in the sample. On the other hand, if poisoning by a particular drug is strongly suspected, the respective specific immunoassay may be applied for measurement.

In general, the immunochemical determination of a substance for TDM is carried out with ready-for-use reagents and with the mechanized analyzers used in clinical chemistry. The concentration of some drugs can be measured with unit-use reagents, for example, theophylline (Johnson and Johnson). For point-of-care testing, the AccuLevel system (Syva) is available: after enzyme immunochromatography, the concentration of carbamazepine, phenobarbital, phenytoin, or theophylline is read visually from a scale.

Many drugs subject to TDM are determined by chromatographic procedures such as HPLC and gas chromatography, particularly, if an immunoassay is not available. Details regarding individual agents are given in the literature [1] or can be obtained from the manufacturers of immunoassays (e.g., Abbott, Dade-Behring, Microgenics, Roche). Drugs often subject to TDM are listed in Table 34.1.

Table 34.1 Drugs subject to monitoring (TDM) (selection).

Group	Drug	Immunoassay available
CNS stimulants	Caffeine	+
	Theophylline	+
Antidysrhythmics	Amiodarone	−
	Disopyramide	+
	Lidocaine	+
	Mexiletine	− [2]
	N-Acetylprocainamide	+
	Procainamide	+
	Propafenone	− [3]
	Quinidine	+
Antibiotics	Amikacin	+
	Chloramphenicol	−
	Gentamicin	+
	Tobramycin	+
	Vancomycin	+
Anticonvulsants	Carbamazepine	+
	Clonazepam[a]	+
	Ethosuximide	+
	Felbamate	−
	Gabapentin	−
	Lamotrigine	−
	Levetiracetam	−
	Oxcarbazepine	−
	Phenobarbital	+
	Phenytoin	+
	Primidone	+
	Tiagabine	−
	Topiramate	−
	Valproic acid	+
	Vigabatrin	−
	Zonisamide	−

(*Continued*)

Table 34.1 (*Continued*)

Group	Drug	Immunoassay available
Antifungal agents	Amphotericin B	−
	Flucytosine	−
Bronchodilators	Theophylline	+
Immunosuppressants	Ciclosporin	+
Cardiac glycosides	Digitoxin	+
	Digoxin	+
Anesthetics	Etomidate	− [4]
	Methohexital	− [5]
	Midazolam[a]	+
	Thiopental	− [6]
Neuroleptics and antipsychotics	Amisulpride	−
	Amitriptyline[b]	+
	Citalopram	−
	Clobazam[a]	+
	Clomipramine[b]	+
	Clozapine	−
	Desipramine[b]	+
	Dosulepin	−
	Doxepin[b]	+
	Fluoxetine	−
	Fluphenazine	−
	Fluvoxamine	−
	Haloperidol	−
	Imipramine[b]	+
	Lithium	−
	Maprotiline	−
	Mirtazapine	−
	Norclomipramine	−
	Nordoxepin[b]	+
	Nortriptyline[b]	+
	Olanzapine	−
	Opipramol[b]	+
	Paroxetine	−
	Perphenazine	−
	Quetiapine	−
	Risperidone	−
	Sertraline	−
	Thioridazine	−
	Trimipramine	+
	Venlafaxine	−
	Zuclopenthixol	−
Antineoplastic agent	Methotrexate	+

[a, b]The detectability for a particular drug of a group depends on the antibody used in the assay and may be rather poor. For details, see package insert of the kit or contact the manufacturer. CNS: central nervous system.
[a]Immunochemical group test for benzodiazepines.
[b]Immunochemical group test for tricyclic antidepressants.

References

1 Moyer, T.P. and Boeckx, R.L. (1984) *Applied Therapeutic Drug Monitoring*, American Association of Clinical Chemistry, Washington, DC.

2 Elfving, S.M., Svens, E.H. and Leskinen, E.E.A. (1981) Gas–liquid chromatographic determination of mexiletine in human serum with nitrogen sensitive detection. *Journal of Clinical Chemistry and Clinical Biochemistry*, **19**, 1189–1191.

3 Külpmann, W.R. and Klein, H. (1986) Gas chromatographic determination of propafenone by use of "530 μm" capillary column. *Journal of Clinical Chemistry and Clinical Biochemistry*, **24**, 803–804.

4 Külpmann, W.R., Kloppenborg, A. and Kohl, B. (1984) Drug monitoring by gas chromatography. (A) Tocainide – (B) Etomidate. *Fresenius Zeitschrift fur Analytische Chemie*, **317**, 667–668.

5 Kohl, B. and Külpmann, W.R. (1982) Rapid sample preparation for the gas chromatographic determination of methohexital in serum using reversed phase cartridges. *Fresenius Zeitschrift fur Analytische Chemie*, **311**, 413–414.

6 Külpmann, W.R., Fitzlaff, R., Spring, A. and Dietz, H. (1983) Thiopental monitoring by gas chromatography. *Journal of Clinical Chemistry and Clinical Biochemistry*, **21**, 181–184.

7 Koch-Weser, J. and Klein, W. (1971) Procainamide dosage schedules, plasma concentration and effects. *Journal of the American Medical Association*, **215**, 1454–1460.

35 Poisonous Plants

M. Geldmacher-von Mallinckrodt and L. von Meyer

35.1 General Aspects

35.1.1 Frequency of Poisoning

Poisonous plants are found not only in the open countryside, for example, *Hyoscyamus niger* L. (henbane), *Datura stramonium* L. (jimsonweed, thorn apple), *Atropa belladonna* L. (deadly nightshade), and *Conium maculatum* L. (poison hemlock), in gardens, for example, *Aconitum napellus* L. (monkshood) and *Colchicum autumnale* (autumn crocus, meadow saffron), but also in pedestrian areas and on terraces and balconies, for example, *Brugmansia sanguinea* L. (angel's trumpet) and *Ricinus communis* L. (castor oil plant).

Statistics on the frequency of intoxications by plants differ considerably and also their ranking, because region, climate, and culture play an important role. Further, it is not always clearly distinguished – especially if children are involved – between ingestion and intoxication [1]:

In case of *ingestion*, a plant or parts of a plant are involved, but without poisoning (at most nausea and vomitus once), which would require medical assistance, whereas in case of *intoxication* (be it mild or severe), treatment by a physician is necessary.

In children, accidental ingestion occurs more often than intoxication, but the amount that can cause intoxication is considerably lower especially for small children than for adults. Poisoning by plants is rather rare in grown-ups. It may happen from misidentified plants, for example, use of *Laburnum anagyroides* L. (golden chain) instead of beans or peas [2], and *Conium maculatum* L. (poison hemlock) or *Cicuta virosa* L. (water hemlock) instead of parsley [3]. It has become known for so-called alternative treatment, such as sedative tea from leaves of *Nerium oleander* L. (oleander) [4].

Chains from exotic fruits may be hazardous, because the hard seed shell is pierced during manufacture and toxic compounds may be released.

Clinical Toxicological Analysis: Procedures, Results, Interpretation. Edited by Wolf-Rüdiger Külpmann

ISBN: 978-3-527-31890-2

It may be difficult to find out the real plant toxin involved in a particular intoxication: poisoning from "Pontic honey" especially of Turkish people has been reported repeatedly. In a residue from the said honey, grayanotoxin 1 was discovered, which is the most relevant toxic agent of rhododendron species, for example, of *Rhododendron ponticum*.

Rather often adults poison themselves by attempting suicide by plants or parts of them, such as needles from *Taxus baccata* L. or *Taxus brevifolia* L. (yew), roots from *Conium maculatum* L., seeds from *Datura stramonium* L., leaves from *Digitalis purpurea* L. (foxglove), and flowers or roots from *Aconitum napellus* L. Also, the abuse of hallucinogenic plants may end in poisoning, for example, from tea brewed from the leaves of *Brugmansia sanguinea* L. or from seeds of *Datura stramonium* L.

In 1992, the IPCS (International Program on Clinical Safety of WHO) distributed questionnaires to poison information centers of 55 American, African, Asian, European countries, and Australia as well to get an overview on the frequency of intoxications by plants and their ranking. The participants were asked to give relevant information on the exposure cases in 1991 [5]. The plants, which were most often reported, were (in descending order) *Dieffenbachia* (dumb cane), *Hyoscyamus niger* L., *Nerium oleander* L., *Atractylis gummifera* (white chameleon), and *Taxus baccata* L. Poisoning by plants and mushrooms occurred rather often in Japan (24% of all intoxications) and seldom in Finland (0.02%).

In the United States, plant exposures are a common category of agents responsible for calls to poison information centers [6]. They rank fourth only behind exposures involving medications, household cleaning agents, and health/beauty aids. They are an imposing category, reflected not only by the shear volume of cases, but also by the general lack of knowledge about which plants are poisonous and what should be done, if a plant exposure has occurred.

All exposures reported to the American Association of Poison Control Centers (AAPCC) for the 1985–1994 period were extracted by Krenzelok and Jacobsen [7] from an electronic file of all plant exposures in the AAPCC database over the same period of time and analyzed using a relational database. The results are based of analysis of 912 534 exposures to plants. There was considerable variability among states and regions as to the rank of the most common exposures. The number of exposures ranged from 61 200 to 1, with plant genera having in excess of 17 000 reported exposures. The top 20 most commonly reported plant exposures during that period are listed in Table 35.1. The most common plant genera represented 31% of the total exposures to plants. *Philodendron* spp. was the most common exposure, followed by *Dieffenbachia* spp., *Euphorbia* spp., *Capsicum* spp., and *Ilex* spp.

There were considerable differences between states with regard to indoor versus outdoor plants, native versus introduced varieties, and as to the rank of the most common exposures. For example, *Capsicum*, *Phytolacca*, Euphorbiaceae, and *Toxicodendron* (poison oak) were the leading plant exposures in New Mexico, North Carolina, New Jersey, and Pennsylvania, respectively [8].

The annual report of AAPCC TESS (Toxic Exposure Surveillance System) discusses plants most commonly associated with a fatal outcome, creates a perspective about plant toxicity, and examines the mechanism of toxicity. These reports covering

Table 35.1 The 20 most common plant exposures occurring over 10 years (1985–1994) in the United States [7].

Plant genera	Exposures
Unidentified species	84 593
Philodendron species	61 200
Dieffenbachia species	35 645
Euphorbia species	31 414
Capsicum species	29 461
Ilex species	23 904
Crassula species	22 295
Ficus species	20 450
Toxicodendron species	19 395
Phytolacca species	18 562
Schefflera (*Brassaia*) species	17 708
Solanum species	17 177
Spathiphyllum species	14 380
Epipremnum species	13 471
Saintpaulia species	12 238
Unidentified berry	11 384
Pyracantha species	11 227
Taxus species	11 217
Rhododendron species	9 590
Schlumbergera species	9 423
Chrysanthemum species	8 058

1983–2000 have been reviewed systematically by Krenzelok [9]. Results: over 18 years 29 172 989 exposures were reported by poison centers and 1 709 805 (5.9%) involved exposures to plants. During the same period, 10 931 fatalities occurred and 30 involved exposures to plants (0.3%). Twenty-four different botanical species or derivatives (see Table 35.2) were involved in the fatalities. Fatal outcomes were most commonly associated with the intentional abuse or misuse of botanical agents. *Cicuta* species accounted for 23.3% of the botanical related fatalities. Commonly referred to as "water hemlock" (beaver poison, children's bane, cow bane, death-of-man, false parsley, wild parsnip), four species predominate: *C. bulbifera* L., *C. douglasii*, *C. maculata* L., and *C. virosa* L. The *Cicuta* species thrive in wet habitats. The geographical distribution includes most of North America and Europe.

The annual report of the Poison Information Center in Zurich may be considered as representative for Western and Central Europe [10]. In the year 2002, in total 24 772 calls regarding possible human intoxication were registered. 11.5% were related to plants and most often children were involved. Ten moderate to series intoxications from plants were observed, including three lethal poisoning from *Colchicum autumnale* L. in children. The remaining intoxications were caused mainly by *Nerium oleander* L., *Taxus baccata* L., *Alocasia* (closely related to *Dieffenbachia*), *Laburnum anagyroides* L., *Brugmansia sanguinea* L., and *Helleborus foetidus* L. (stinking hellebore).

Vichova and Jahodar [11] present a retrospective study conducted to review the hospital admissions following acute childhood poisoning with plants in the Czech

Table 35.2 Botanicals involved in fatalities (AAPCC TESS 1983–2000) in the United States [9].

Botanical	Number of reports
Datura stramonium	5
Cicuta douglasii	3
Cicuta maculata	2
Cicuta spp ("water hemlock")	2
Pennyroyal tea	1
Pennyroyal + black cohosh	1
Conium maculatum	1
Fraxinus americana + senna/cascara	1
Phytolacca americana	1
Ginseng extract	1
Marah oreganos	1
Melia azedarach	1
Oleander extract	1
Cayenne pepper + garlic oil	1
Xanthoriza simplicissima	1
Herbal tea	1
Aloe vera gel	1
Golden seal root	1
Helenium	1
Gloriosa superba	1
Ambrosia artemisifolia + cow parsnip	1

Republic over a 6-year period from 1996 to 2001. According to the statistics of the Toxicology Information Center (operates in association with the University Hospital of the first Faculty, Charles University, Prague), plant ingestions represent about 10% of total telephone calls. Six university hospital pediatric departments and two local hospital pediatric departments were involved in the study. Information and complete data on the cases were collected on the basis of all hospital medical records and internal hospital database outcomes. A total of 174 plant exposures were analyzed to tabulate the list of top species involved in plant poisoning. The aims were to provide classification according to agent frequency, clinical presentations, severity of symptoms expressed, effected age groups, and gender of patients as well as to evaluate the treatment according to patient outcome. The most frequent ingestions were from the seeds of thorn apple (*Datura stramonium* L.) (14.9%), followed by *Dieffenbachia* exposures (11.5%) and common yew (*Taxus baccata* L.) (9.8%). Thorn apple, golden chain (Laburnum anagyroides), and raw beans (*Phaseolus vulgaris* L.) caused the most serious symptoms. There were no fatalities according to the reviewed medical reports.

35.1.2 Symptoms from Poisoning by Plants

According to Mrvos *et al.* [12], toxicological and botanical references describe a myriad of symptoms associated with the ingestion of plants. The symptoms are based

Table 35.3 The top 20 plants accounting for 66.9% of the symptoms reported in 54.2% of the plant ingestion patients (AAPCC TESS1992–1999) [12].

1. *Capsicum*	11. *Zantedeschia*
2. *Dieffenbachia*	12. *Ilex*
3. *Philodendron*	13. *Epipremnum*
4. *Caladium*	14. *Iris*
5. *Spathiphyllum*	15. *Eucalyptus*
6. *Datura*	16. *Ephedra*
7. *Euphorbia*	17. *Hedera*
8. *Ficus*	18. *Arisaema*
9. *Narcissus*	19. *Piper*
10. *Phytolacca*	20. *Begonia*

largely on a limited number of cases and anecdotal reports or the personal experience of symptoms associated with the ingestion of plants. The authors compiled the typical symptoms according to the data from AAPCC TESS 1992–1999. These data were queried electronically to identify all plant ingestions that were associated with the occurrence of symptoms. The data were analyzed using a relational database (MS Excel). The most frequently ingested plant or plant substances were identified and stratified to identify the top 20 (Table 35.3). The 10 most frequently reported symptoms associated with these plants are documented including the genus name of the plant, a common example of a species within that genus, and the percentage of symptoms within that genus that was represented by the 10 most frequent symptoms and the 10 common symptoms (Table 35.4). There were no fatalities among this set of patients. Data analyses were restricted to exposures associated with the ingestion of a single plant and no concomitant agents. Plants were grouped and analyzed at the genus level.

A total of 768 284 plant exposures were analyzed. Symptoms were reported in 53 081 (6.9%) of the patients. The top 20 plants (Table 35.3) accounted for 35 490 (66.9%) of the symptoms reported in 28 804 (54.2%) of the plant ingestion patients. The most frequent symptoms associated with these plants are organized by the genus name of the plant, a common example of a species within that genus, and the percentage of symptoms within that genus that was represented by the 10 most frequent symptoms and the 10 common symptoms (Table 35.4).

The most common symptom groups were categorized as gastrointestinal 63.7%, dermatological 16.9%, neurological 5.0%, ophthalmologic 4.0%, respiratory 3.9%, cardiovascular 2.0%, and others 4.5%.

Most plant ingestions were not associated with the development of morbidity. Gastrointestinal symptoms predominate presumably due to the presence of chemical irritants in the plants. Vomiting occurred as a frequent symptom, but its occurrence was not associated with the use of ipecac. Dermatological symptoms were common – while some were systemic, others occurred most likely as secondary contamination of the skin (e.g., perioral, hands) during the process of ingestion of the plant. Similar circumstances were probably responsible for the high incidence of ocular irritations as well.

Table 35.4 Symptoms associated with ingestion of plants (Table 35.3) according to Ref. [12].

Toxic plant	Ten most common symptoms (%)		Toxic part of the plant	Toxic compound
Arisaema spp.	Oral irritation	48.9		
Example: *Arisaema triphyllum*	Vomiting	12.6		
Top 10 = 91.1% of all symptoms	Throat irritation	10.1		
	Nausea	4.7		
	Ocular irritation	4.3		
	Dermal irritation	2.7		
	Dermal edema	2.7		
	Pain	1.7		
	Dysphagia	1.6		
	Abdominal pain	1.3		
	Cough	1.3		
Begonia spp.	Oral irritation	40.7	All parts, particularly tubers and rhizomes	Main toxins: oxalates [33, 34]
Example: *Begonia argenteo guttata*	Throat irritation	15.7		
Top 10 = 88.7% of all symptoms	Vomiting	10.6		
	Confusion	5.8		
	Nausea	3.8		
	Cough	2.6		
	Dermal irritation	2.2		
	Abdominal pain	2.1		
	Diarrhea	1.9		
	Erythema	1.6		
	Ocular irritation	1.6		

Caladium spp.	Oral irritation	65.9	All parts of the plant	Oxalic acid, calcium oxalate [33, 34]
Example: *Caladium bicolour* (angel's wings)	Throat irritation	7.1		
Top 10 = 93.9% of all symptoms	Vomiting	5.4		
	Erythema	2.8		
	Dermal irritation	2.8		
	Pruritus	2.4		
	Nausea	1.7		
	Dermal edema	1.8		
	Oral burns	1.6		
	Cough	0.9		
Capsicum spp.	Oral irritation	35.9	All parts, particularly the fruits	Capsaicin and derivatives [33, 34]
Example: *Capsicum annuum* L. (chilli pepper)	Dermal irritation	12.9		
Top 10 = 92.8% of all symptoms	Dermal edema	7.6		
	Dermal erythema	7.6		
	Ocular irritation	5.6		
	Throat irritation	5.3		
	Rash	5.3		
	Hives	5.2		
	Pruritus	5.1		
	Vomiting	3.1		
Datura spp.	Mydriasis	22.3	Whole plant	Tropane alkaloids: scopolamine, atropine, hyoscyamine [33, 34]
Example: *Datura stramonium* (jimsonweed)	Agitation	19.3		
Top 10 = 76.1% of all symptoms	Tachycardia	10.6		
	Confusion	4.3		
	Blurred vision	4.0		
	Hypertension	3.7		

(*Continued*)

Table 35.4 *(Continued)*

Toxic plant	Ten most common symptoms (%)		Toxic part of the plant	Toxic compound
	Vomiting	3.6		
	Drowsiness	3.3		
	Fever	2.9		
	Nausea	2.5		
Dieffenbachia spp.	Oral irritation	50.4	All parts of the plant	Oxalic acid, calcium oxalate, proteolytic enzymes, and other substances [33, 34]
Example: *Dieffenbachia maculata* (dump cane)	Vomiting	10.3		
Top 10 = 91% of all symptoms	Throat irritation	9.0		
	Erythema	6.1		
	Pruritus	3.2		
	Excessive secretions	2.9		
	Dermal irritation	2.7		
	Rash	2.6		
	Oral burns	2.4		
	Hives	1.4		
Ephedra spp.	Tachycardia	21.0		Many alkaloids; L-ephedrine and derivatives [23]
Example: *Ephedra sinica*	Nausea	8.7		
Top 10 = 72% of all symptoms	Oral irritation	8.6		
	Vomiting	6.5		
	Dermal irritation	6.5		
	Throat irritation	5.3		
	Agitation	5.1		
	Hypertension	4.0		
	Dizziness	3.7		
	Drowsiness	2.4		

Epipremnum spp.	Vomiting	33.6	Whole plant	Main toxin: calcium oxalate [33, 34]
Example: *Epipremnum aureum*	Oral irritation	28.4		
Top 10 = 88.9% of all symptoms	Dermal irritation	5.4		
	Nausea	4.3		
	Cough	4.3		
	Throat irritation	3.3		
	Erythema	3.2		
	Agitation	2.3		
	Abdominal pain	2.1		
	Rash	1.9		
Eucalyptus spp.	Vomiting	47.3	Leaves	Ethereal oils, main component cineol [23]
Example: *Eucalyptus globules*	Oral irritation	11.9		
Top 10 = 86.9% of all symptoms	Nausea	6.2		
	Cough	6.0		
	Rash	3.3		
	Dysphagia	3.0		
	Diarrhea	2.7		
	Drowsiness	2.3		
	Agitation	2.1		
	Abdominal pain	2.1		
Euphorbia spp.	Oral irritation	28.5	Whole plant	Diterpene esters, allergens [33, 34]
Example: *Euphorbia pulcherrima* (Christmas flower)	Vomiting	24.0		
Top 10 = 85.1% of all symptoms	Throat irritation	15.0		
	Nausea	4.4		

(Continued)

Table 35.4 (*Continued*)

Toxic plant	Ten most common symptoms (%)		Toxic part of the plant	Toxic compound
	Cough	3.6		
	Erythema	2.5		
	Anorexia	2.3		
	Dermal irritation	2.1		
	Abdominal pain	2.0		
	Headache	0.8		
Ficus spp. Example: *Ficus benjamina* (Weeping fig) Top 10 = 85.7% of all symptoms	Vomiting	38.1	Sap	Coumarins, tannic acid, allergens [33, 34]
	Cough	14.3		
	Oral irritation	13.1		
	Nausea	4.2		
	Throat irritation	4.2		
	Dermal irritation	2.7		
	Agitation	2.4		
	Fever	2.4		
	Rash	2.1		
	Diarrhea	1.4		
Hedera spp. Example: *Hedera helix* Top 10 = 85.1% of all symptoms	Vomiting	41.2	All parts, particularly leaves and fruits	Allergenic compounds including falcarinol and didehydrofalcarinol; saponins, emetine [33, 34]
	Cough	13.2		
	Oral irritation	10.1		
	Agitation	3.5		
	Nausea	3.4		
	Abdominal pain	2.9		
	Throat irritation	2.7		

	Diarrhea	2.6		
	Erythema	2.6		
	Fever	1.3		
Ilex spp.	Vomiting	26.5	Whole plant	Main toxins: saponins; *Ilex aquifolium*: also cyanogenic glycosides [33, 34]
Example: *Ilex aquifolium*	Abdominal pain	15.1		
Top 10 = 81.3% of all symptoms	Nausea	8.9		
	Oral irritation	5.8		
	Fever	5.5		
	Diarrhea	5.5		
	Drowsiness	4.4		
	Agitation	3.6		
	Cough	3.0		
	Erythema	3.0		
Iris spp.	Oral irritation	33.4	All parts, particularly the rhizomes	Main toxins unknown [33, 34]
Example: *Iris germanica*	Throat irritation	28.3		
Top 10 = 89.9% of all symptoms	Vomiting	11.1		
	Nausea	8.9		
	Abdominal pain	5.7		
	Diarrhea	3.7		
	Dysphagia	2.0		
	Ocular irritation	1.0		
	Pain	0.9		
	Cough	0.7		

(*Continued*)

Table 35.4 (*Continued*)

Toxic plant	Ten most common symptoms (%)		Toxic part of the plant	Toxic compound
Narcissus spp.	Vomiting	44.8	Whole plant, particularly the outer layer of the bulb	Many alkaloids, calcium oxalate [33, 34]
Example: *Narcissus pseudonarcissus* (daffodil)	Nausea	14.0		
Top 10 = 92.7% of all symptoms	Oral irritation	12.8		
	Abdominal pain	9.9		
	Diarrhea	3.3		
	Throat irritation	2.6		
	Dysphagia	1.8		
	Lacrimation	1.0		
	Drowsiness	1.0		
	Agitation	0.7		
Philodendron spp.	Oral irritation	30.1	All parts of the plant	Oxalic acid, calcium oxalate, allergens [33, 34]
Example: *Philodendron cordatum*	Vomiting	21.8		
Top 10 = 93.6% of all symptoms	Cough	13.9		
	Throat irritation	11.1		
	Nausea	4.1		
	Dermal irritation	3.8		
	Diarrhea	2.3		
	Erythema	1.9		
	Abdominal pain	1.8		
	Rash	1.4		
	Dysphagia	1.4		

Phytolacca spp.	Vomiting	23.9	Whole plant	Triterpenes, saponins, lectins, glycoproteins, oxalic acid [33, 34]
Example: *Phytolacca americana* (Virginian poke)	Diarrhea	11.4		
Top 10 = 77.3% of all symptoms	Oral irritation	11.3		
	Abdominal pain	9.9		
	Nausea	7.5		
	Fever	3.1		
	Dysphagia	2.5		
	Agitation	2.5		
	Drowsiness	2.3		
	Cough	2.3		
Piper spp.	Oral irritation	29.2		
Example: *Piper nigrum*	Vomiting	22.5		
Top 10 = 76.7% of all symptoms	Dermal edema	5.2		
	Ocular irritation	4.0		
	Drowsiness	3.2		
	Throat irritation	3.2		
	Abdominal pain	2.6		
	Cough	2.8		
	Agitation	2.3		
	Erythema	2.1		
Spathiphyllum spp.	Throat irritation	42.2	Whole plant	Oxalic acid, calcium oxalate [33, 34]
Example: *Spathiphyllum floribundum* (peace lily)	Oral irritation	21.7		
Top 10 = 90.4% of all symptoms	Cough	12.6		
	Nausea	3.4		
	Fever	2.3		

(Continued)

Table 35.4 (*Continued*)

Toxic plant	Ten most common symptoms (%)		Toxic part of the plant	Toxic compound
	Diarrhea	2.1		
	Abdominal pain	1.7		
	Rash	1.7		
	Agitation	1.5		
	Dysphagia	0.1		
Zantedeschia spp.	Oral irritation	60.0	Whole plant	Main toxin: calcium oxalate [33, 34]
Example: *Zantedeschia aethiopica*	Throat irritation	7.5		
Top 10 = 89.1% of all symptoms	Vomiting	6.1		
	Dermal irritation	3.8		
	Cough	3.7		
	Nausea	2.1		
	Fever	2.0		
	Abdominal pain	1.3		
	Dermal edema	1.3		
	Lacrimation	0.7		

Limitations of the analysis include the use of poison information center reports where the ingestions were not confirmed analytically. A further limitation was the grouping of all plants within a genus instead of being species specific.

The identified "toxidromes" [12] should assist clinicians in the diagnosis and assessment of patients who ingested plants.

35.1.3 Traditional Chinese Medicine

Recently, traditional Chinese medicine, in which preparations from plants play a major role, has become rather popular in Western countries. However, misjudgment or mix-up may occur easily, because the formulations are usually not known outside China. Even though the same term is used the contents of the preparations may differ considerably, because the same naming may designate different plants depending on the region. Repeatedly poisoning by heavy metals (e.g., lead or cinnabar) has been reported after administration of Chinese herbal medicine, the major part probably from pollution. Contamination with residues of pesticides and microbes (e.g., aflatoxins) has been observed. Medicines designated as phytopharmaceuticals by the manufacturer contained glucocorticoids, indometacin, or benzodiazepines without any declaration [13–15].

Plants from Africa, Arabia, or South America play a minor role in comparison.

35.1.4 Special Herbal Preparations

Areca catechu (betel nut): The nuts contain several pyridine alkaloids: arecoline is a cholinergic agent and arecaidine is considered as a carcinogen. Indeed, regular nut chewing is associated with an increased rate of oral cancer.

Catha edulis (khat): The shrub is cultivated in East Africa and on the Arabian Peninsula. The leaves that are known as khat (quat, qat, mirra) are used for chewing or are extracted for tea. Khat has stimulant effects, as it contains, for example, norpseudoephedrine and methcathinone, which are related to amphetamine (see Chapter 26).

Cycas circinalis (false sago palm): The female cones have been used as a dietary source of starch on Guam, on the Japanese Kii peninsula and in Irian Jaya (West Papua). The consumption seems to be related to the Guamanian motor neuron disease. The relevant toxin may be cycasin or less likely β-*N*-methylamino-L-alanine (BMAA) [16].

Ephedra vulgaris: A popular herbal product, particularly in China, is *Ma huang*, which is produced from *Ephedra vulgaris*. It contains L-ephedrine 4/100 to 8/100 g and may be present in products to slim down, food supplements, and energizer drinks. It is used for the production of methamphetamine.

Hedeoma pulegioides and *Mentha pulegium*: The leaves of the plants are extracted for the preparation of *pennyroyal oil*, which is used for regulation of menstrual cycle and as an abortifacient. Its toxin pulegone may cause hepatic failure from glutathione depletion of the liver. Pennyroyal oil may be combined with black cohosh (*Cimicifuga racemosa*), also an abortifacient.

Lathyrus sativus (chickling pea): Regular consumption of *L. sativus* (about 400 g/day) may lead to *lathyrism*, which is characterized by spastic paresis and later by loss of control of the bladder and rectum. The disease is caused by β-*N*-oxalylamino-L-alanine (BOAA), a neurotoxin of *L. sativus* [17].

Manihot esculenta (cassava): Cassava has been estimated as the second largest carbohydrate crop in the world. The cassava plant contains linamarin, a cyanogenic glycoside. The toxin is destroyed if the plant is processed appropriately. Otherwise cyanide is liberated in the gut. The intoxication contributes to the development of nutritional neuropathies: tropical ataxic neuropathy (Nigeria; mainly in adult males) and epidemic spastic paraparesis (Mozambique, Tanzania, Zaire; mainly in women and children).

Piper methysticum (kava): The dried root contains several kava lactones (e.g., kavain). The aqueous extract is a popular beverage in Polynesia. Its main effects are euphoria, loquacity, and hepatotoxicity.

For *caffeine* and *nicotine*, see Sections 18.2 and 31.2, respectively, for *cocaine* (*Erythroxylum coca*) Section 26.2, for *absinth* Section 27.1.1, for *opiates* (*Papaver somniferum*) Section 14.7, and for *mescaline* Section 26.7.

Favism: Some individuals consuming fava beans from *Vicia faba* (broadbean) will suffer from a fava crisis with intravascular hemolysis. These persons are deficient in glucose-6-phosphate-dehydrogenase (G6PD). The enzyme assists to restore NADPH, which is necessary to protect –SH groups and the β-chain of hemoglobin from oxidation. In the absence of sufficient G6PD activity, hemolysis develops in case of oxidative stress. About 7% of the world's population has a G6PD deficiency, with a higher frequency being reported in the Mediterranean area, Middle East, Taiwan, and South China.

35.1.5 Poisoning of Animals

Animals poisoned by plants are the subjects of an increasing number of inquiries made to poison control centers in Europe [18] as well as in many other parts of the world. Plant poisonings, together with rabies and botulism, are the main causes of death in adult cattle, for example, in Brazil and other countries in South America [19]. Estimates indicate that about one million head of cattle die annually from poisoning in Brazil. There are approximately 75 plants of practical importance to animal husbandry that have had their toxicity confirmed by experiments with the animal species affected under natural conditions. The great majority of these plants only occur in Brazil and other countries in South America.

35.2 Suspected Plant Poisoning: First Diagnostic Steps

If intoxication is suspected from plants or parts of them, a physician or a hospital should be visited as soon as possible. The respective plants should be brought at best intact. The total amount of vomitus should be kept to assist in the identification of the ingested parts of the plant and in the estimation of the quantity involved. The stuff of an (university) institute of pharmaceutical biology or botany, of a garden as well as a pharmacist, a ranger, a gardener, or a florist may be of help in identification. A photocopy of the plant or of its available parts can be faxed or a photo can be sent via e-mail to an expert, for example, in a Poison Information Center to save time [20]. The identification can always be performed faster and more easily, if the plant or at least its parts are intact.

The following items should be addressed to the patient or accompanying persons:

Intact plant:

Appearance of the plant: for example, tree, shrub, herb, and stout.

Occurrence:

For example, meadow, hedge, edge of a path, garden, and park.

Leaves:

Size, formation, and shape (e.g., almost round, oval, pointed, thorny, stemmed/not stemmed, spread of the leaf veins).

Fruits:

Kind of fruit (e.g., berry, nut)

Size, color, and consistency (e.g., juicy, fleshy, solid)

Number of kernels in the fruit

Formation regarding the plant (single, in duplicates, numerous closely packed, stemmed/not stemmed).

Patient:

Age

Parts of the plant ingested (simultaneously) (e.g., leaves, roots) and estimated quantity

Time elapsed since ingestion

Occurrence of vomitus.

35.3 Aids for Identification of Plants

35.3.1 Handbooks

Several handbooks for identification preferably with colored pictures are available, for example, Refs [1, 2]. An appropriate guide to herbal medicines is the "Physician Desk Reference" (PDR) [23].

Helpful reference books for plants in Western Europe: Frohne and Pfänder [1], Roth *et al.* [24], Hiller and Bickerich [25], Hager [26], and European Pharmacopoeia 2005/2007 [27].

Especially, the "Pharmacopoeia of Chinese Medicine" is recommended for the identification of the approximately 250 drugs of the traditional Chinese medicine, which are currently available (German version [28]). The drugs are described in alphabetical order often including drawings, which show typical microscopic details [28]. Huang has published a book on the pharmacological effects of Chinese plants [29].

African medicinal plants and arrow poisons are presented in the comprehensive work of Neuwinger [30] and in the handbook of African medicinal plants from Iwu [31]. The toxicological aspects of African medicinal plants are described by Stewart [32]. A compendium of medicinal plants from Arabia has been published by Ghazanfar [33]. The search for medicinal plants is considerably facilitated by the "Cross Name Index" [21].

35.3.2 Electronic Media

Experts of the Royal Botanic Gardens in Kew (UK) and the Medical Toxicology Unit of the Guy's & St. Thomas Hospital have compiled on a compact disc "Poisonous Plants and Fungi in Britain and Ireland" [34] (German version [35]). A computer-based drug and poison information system for veterinarians has been published by Furler [18]. It is available in English, French, German, and Italian language. It shows numerous pictures of poisonous plants, which may be helpful also in (suspected) human intoxications and is available via Internet or as compact disc [18]. A comprehensive database on Chinese Herbal Medicine Toxicology is under development [36].

Much information is documented on the Internet. Special details can be found easily by search using Google, Lycos, Altavista, or Yahoo, which, however, have to be critically appreciated as, for example, most recent publications may not yet be mentioned. One should remember that an unambiguous identification is only possible by the botanical names. The search is facilitated by the comprehensive book of Torkelson [21].

35.4 Identification of Plants

35.4.1 Macroscopic Identification

In Ref. [1], colored pictures of poisonous plants are shown accompanied by a short description. Conspicuous fruits, their formation on the branch, and the member of seeds are presented as well as the leaves, flowers, and distinctive marks. Fruits: conspicuous fruits are presented in a table ordered according to color as a key to their identification [1]. In addition are mentioned: consistency of flask (juicy, mealy, etc.), number of kernels of a fruit, number of fruits per unity, and appearance of the plant (woody, herbaceous, rambling, climbing, and epiphytic). These details are not sufficient for an unambiguous identification of an unknown plant, but the number of plants, which has to be considered, is markedly reduced. Further detailed information can be found in the comprehensive descriptions of the respective plants. Leaves: macroscopic features of the leaves are compiled and ordered according to their morphological aspects [1]. Shape and formation of leaf veins combined with other characteristics can be very helpful for identification, particularly as they are usually still present at the time of fruit bearing. The black-and-white photos (scale given for reduced size) show very markedly the leaf veins. Even though it may be difficult to discern them in thick-skinned leaves, the photos may assist in identification. However, the leaves of herbaceous plants do not only show considerable differences in size depending on age and time of insertion but also vary in shape (heterophylia). Pharmacopoeia of Chinese medicine may be helpful in the identification of Chinese medicines [28, 29] and respective compilations in identification of African and Arabic plants [30–33].

35.4.2 Microscopic Identification

Valuable clues to the identification of a plant can be obtained from the structures of fruits and leaves visible by microscopy.

Fruits: The structure of the epidermis of the wall is most important for identification. Their features are well preserved even after an extended stay in the gastrointestinal tract and are still well discernable in vomitus or stomach contents.

Preparation for microscopic examination [1].

Fruits: A rectangular piece of 5–10 mm^2 is cut by a razor blade from the fruit. It is put upside down on a cover bent and the flesh is removed from the exocarp with a bent forceps or a micro spoon spatula. In general, the epidermis of the wall is resistant to this procedure and is fixed simultaneously to the cover glass by mild pressure. Both are transferred to a slide with a drop of water. After removal, the flask can also be transferred with a drop of water for microscopic examination.

Leaves: Microscopic characteristic features of leaves, such as composition and surface structure, can be used for identification, too. Further examinations can be carried out by raster electron microscope for confirmation (e.g., in forensic medicine).

35.5 Toxic Plants' Substances

35.5.1 Toxic Compounds

The unambiguous identification of a poisonous plant should be accompanied by a proof of its supposed or actual toxicity.

The toxic compounds can be ordered in different ways, for example, according to chemical grouping:

Alkaloids: Usually, a mixture of chemically closely related alkaloids is found in a plant, one of them playing a major part (principal alkaloid). The alkaloids can be ordered according to basic structure:

Quinolizidine-related alkaloids, for example, cytosine and sparteine

Isoquinoline-related alkaloids, for example, morphine

Pyridine/piperidine-related alkaloids, for example, coniine and nicotine

Steroid-related alkaloids, for example, solanine and alkaloids of *Veratrum*

Terpene-related alkaloids, for example, aconitine

Tropane-related alkaloids, for example, atropine and hyoscyamine

Other alkaloids, for example, colchicine and taxine.

Furanocoumarins: They are found in umbelliferae and exhibit photosensitizing effects, such as erythema and blisters.

Glycosides: Anthraquinone glycosides that often have a laxative effect (e.g., *Rhamnus frangulus* L. (sacred bark tree)).

Cyanogenic glycosides , for example, amygdalin (Section 31.4) that are present in the pits of certain *Prunus* spp. (e.g., bitter almonds from *Prunus dulcis* var. *amara*, apricot pits from *Prunus armeniaca*, peach pit from *Prunus persica*). Cyanide can be released in the stomach after enzymatic cleavage.

Cardiac glycosides that are found, for example, in *Digitalis purpurea* L. (purple foxglove) or *Convallaria majalis* L. (lily of the valley).

Saponins are present in the seeds of *Aesculus hippocastanum* L. (horse chestnut).

Acids: Oxalic acid, for example, in *Oxalis acetosella* L. (wood sorrel). In *Dieffenbachia* (dumb cane), oxalate crystal needles are contained in raphides and are fired on chewing. They may cause swelling of lips, mouth, and tongue. *Further:* parasorbic acid in *Sorbus aucuparia* L. (rowan quickbeam).

Polyenes , for example, cicutoxin of *Cicuta virosa* L. (water hemlock).

Proteins: Toxalbumins (e.g., abrin from the seeds of *Abrus precatorius* L. (jequirity or precatory bean), ricin from the seeds of *Ricinus communis* (castor bean)), and lectins as phytohemagglutinins.

Terpenes:

Monoterpenes: toxicologically relevant are some monoterpenes, which are present in ethereal oils, such as thujone from *Artemisia absinthium* L. (common warmwood).

Diterpenes:toxic diterpenes are found in numerous Euphorbiaceae (*Spurge* spp) and in *Daphne mezereum* L. (*Daphne*): mezerein.

Triterpenes:toxic triterpenes are contained in *Lantana camara* L. (red or yellow sage).

Further chemical groups or groupings based on other criteria have been published [37].

35.5.2 Methods for Detection

Color reaction: In the past, color reactions have been carried out directly and the results were usually not specific. It is recommended to use color reactions only after previous extraction or thin-layer chromatography [37].

Thin-layer chromatography: In general, thin-layer chromatography (TLC) is an easy and appropriate method for a rapid screening to detect toxic compounds after extraction of a plant's part. The extraction procedure – for example, pH, solvent, purification, and accumulation steps – as well as the TLC procedure – for example, solvent for development, spray reagent – should be chosen considering the chemical characteristics of the suspected toxin and the respective part of the plant. Appropriate procedures are described in detail in a comprehensive monograph [37], as general rules are not applicable.

The investigation of mixtures of plants by TLC is very difficult, because overlapping of zones may occur impeding identification. It may be necessary to fractionate by column chromatography or other methods to separate a special group. Findings by TLC should be confirmed by another method, for example, HPLC, GC–MS, or LC–MS/MS [41].

35.6 Investigation of Biologic Materials

Stomach contents: Parts of plants, which are discernible in stomach contents or vomitus, can be investigated as described in Sections 35.4.1, 35.4.2, and 35.5.2.

Urine, blood, and plasma: The detection of several toxic compounds from plants may be difficult. The concentration of the compound may be rather low, and its distribution and metabolism unknown. However, for agents present in plants, which have been applied in the past or are administered currently in therapy or drug abuse, appropriate and validated procedures are available. Of these, numerous procedures are described in detail in this book. They are listed in Table 35.5 and should be applied in case of intoxication for confirmation and also in suspected poisoning considering the detection limit.

Table 35.5 Toxic compounds of poisonous plants [1].

Plant: systematic (trivial) name	Toxin	Analysis
Cardiac toxins		
Apocynum cannabinum (dog bane)	Strophantidin glycosides	See Chapter 20
Convallaria majalis (lily of the valley)	Cardenolides	See Chapter 20
Digitalis purpurea (foxglove)	Digitalis glycosides	See Chapter 20
Nerium oleander (oleander)	Cardiac glycoides	See Chapter 20
Strophantus spp.	Cardenolides	See Chapter 20
Convulsants		
Strychnos nux vomica	Strychnine	See Section 31.3
Cyanogenic glycosides		
Manihot esculenta (cassava)	Linamarin	See Section 31.4
Prunus spp.	Amygdalin	See Section 31.4
Cherry laurel		See Section 31.4
Plum	Amygolalin	See Section 31.4
Bitter almond	Amygolalin	See Section 31.4
Peach		See Section 31.4
Apricot		See Section 31.4
Sambucus nigra (elderberry)	Amygdalin	See Section 31.4
Sorghum spp.	Cyanogenes	See Section 31.4
Hallucinogenics		
Argyreia nervosa (Hawaiian baby wood rose)	Clavine alkaloids	See Section 26.6
Cannabis spp. (hemp)	Tetrahydrocannabinol	See Section 26.3
Ipomoea violacea (morning glory)	Clavine alkaloids, Ergoline derivatives	See Section 26.6
Nicotine		
Nicotiana glauca (tree tobacco)	Nicotine	See Section 31.2
Nicotiana tabacum (American tobacco)	Nicotine	See Section 31.2

Sometimes a dedicated immunoassay can be applied for a related compound of a plant, for example, a particular digitoxin assay may be appropriate for the detection of cardiac glycosides after ingestion of *Nerium oleander* L. [4, 38]. The cross-reactivity of antibodies from other assays may be too poor for this purpose.

References

1 Frohne, D. and Pfänder, H.J. (2005) *Poisonous Plants. A Handbook for Doctors, Pharmacists, Toxicologists, Biologists and Veterinarians,* 2nd edn, Manson Publishing Ltd, London.

2 McGrath-Hill, C.A. and Vicas, I.M. (1997) Case series of thermopsis exposures. *Clinical Toxicology,* **35**, 659–665.

3 Krenzelok, E.P., Jacobsen, T.D. and Aronis, J.M. (1996) Hemlock ingestions:

the most deadly plant exposures. *Clinical Toxicology*, **34**, 601–602.

4 Monzani, V., Rovellini, A., Schinco, G. and Rampoldi, E. (1997) Acute oleander poisoning after self-prepared tisane. *Clinical Toxicology*, **35**, 667–668.

5 IPCS/INTOX (1994) An international survey of poison centres experience of human poisoning due to plants, fungi, and their derivatives. Final Draft IPCS/INTOX/94, WHO, Geneva.

6 Litowitz, T.L., Felberg, L., White, S. and Klein-Schwartz, W. (1996) 1995 Annual Report of the American Association of Poison Control Centers Toxic Exposure Surveillance System. *American Journal of Emergency Medicine*, **14**, 487–537.

7 Krenzelok, E.P. and Jacobsen, T.D. (1997) Plant exposures. A national profile of the most common plant genera. *Veterinary and Human Toxicology*, **39**, 248–249.

8 Krenzelok, E.P., Jacobsen, T.D. and Aronis, J.M. (1996) Plant exposures: a state profile of the most common species. *Veterinary and Human Toxicology*, **38**, 289–298.

9 Krenzelok, E.P. (2002) Lethal plant exposures reported to poison centres: prevalence, characterization and mechanisms of toxicity [abstract]. *Clinical Toxicology*, **40**, 303–304.

10 Swiss Toxicological Information Centre (2002) Freie Strasse 16, CH-8028 Zurich, Annual Report.

11 Vichova, P. and Jahodar, L. (2003) Plant poisonings in children in the Czech Republic 1996–2001. *Human & Experimental Toxicology*, **22**, 467–472.

12 Mrvos, R., Krenzelok, E.P. and Jacobsen, T.D. (2001) Toxidromes associated with the most common plant ingestions. *Veterinary and Human Toxicology*, **43**, 366–369.

13 Auyeung, T.W., Chang, K.K., To, C.H., Mak, A. and Szeto, M.L. (2002) Three patients with lead poisoning following use of a China herbal pill. *Hong Kong Medical Journal*, **8**, 60–62.

14 Traub, S.J., Hoffmann, R.S. and Nelson, L.S. (2002) Lead toxicity due to use of an ayurvedic compound [abstract]. *Clinical Toxicology*, **40**, 322.

15 Cheng, S.W., Hu, W.H., Hung, D.Z. and Yang, D.Y. (2002) Anticholinergic poisoning from a large dose of *Scopolia* extract. *Veterinary and Human Toxicology*, **44**, 222–223.

16 de Wolff, F.A. *et al.* (1995) *Handbook of Clinical Neurology: Intoxication of the Nervous System, Part II* (ed. F.A. de Wolff), Elsevier, Amsterdam, pp. 21–24.

17 Spencer, P.S. (1995) *Handbook of Clinical Neurology: Intoxication of the Nervous System, Part II* (ed. F.A. de Wolff), Elsevier, Amsterdam, pp. 1–20.

18 Furler, M., Demuth, D., Althaus, F.R. and Nägeli, H. (2000) Computer-unterstütztes Giftpflanzen-Informationssystem für die Veterinärmedizin. *Schweizer Archiv für Tierheilkunde*, **142**, 323–331.

19 Hubinger Tokarnia, C., Döbereiner, J. and Vargas Peixoto, P. (2002) Poisonous plants affecting livestock in Brazil. *Toxicon*, **40**, 1635–1660.

20 McKinney, P.E., Gomez, H.F., Phillips, S. and Brent, J. (1993) The fax machine: a new method of plant identification. *Clinical Toxicology*, **31**, 663–665.

21 Torkelson, A.R. (1995) *Cross Name Index of Medicinal Plants*, CRC Press, Boca Raton, FL.

22 22. Cooper, M.R., Johnson, A.W. and Dauncey, E.A. (2003) *Poisonous Plants and Fungi. An Illustrated Guide*, 2nd edn, The Stationary Office, London.

23 Fleming, T. (ed.) (2000) *PDR (Physician's Desk Reference) for Herbal Medicines*, 2nd edn, Medical Economics Company, Montvale, NJ.

24 Roth, L., Daunderer, M. and Kormann, K. (1994) Giftpflanzen – Pflanzengifte, in *Vorkommen – Wirkung – Therapie*, 4th edn, Ecomed Verlagsges, München.

25 Hiller, K. and Bickerich, G. (1988) *Giftpflanzen*, Enke Verlag, Stuttgart.

26 Hagers Handbuch der Pharmazeutischen Praxis, Hager ROM 2006/2007, Springer Verlag, Berlin.

27 Europäisches Arzneibuch, 5. Ausgabe 2005 mit 7 Nachträgen 2007. Deutscher

Apothekerverlag, Stuttgart; Govi Verlag, Frankfurt.

28 Stöger, E. and Friedl, F.Arzneibuch der chinesischen Medizin. 2. Auflage mit Ergänzungen, zuletzt 2006, Deutscher Apothekerverlag, Stuttgart.

29 Huang, K.C. (1998) *The Pharmacology of Chinese Herbs*, 2nd edn, CRC Press, Boca Raton, FL.

30 Neuwinger, H.D. (1998) *Afrikanische Arzneipflanzen und Jagdgifte*, Wissenschaftliche Verlagsgesellschaft, Stuttgart.

31 Iwu, M.M. (1993) *CRC Handbook of African Medicinal Plants*, CRC Press, Boca Raton, FL.

32 Stewart, M.J., Steenkamp, V. and Zuckerman, M. (1998) The toxicology of African herbal remedies. *Therapeutic Drug Monitoring*, **20**, 510–516.

33 Ghazanfar, S.A. (1994) *Handbook of Arabian Medicinal Plants*, CRC Press, Boca Raton, FL.

34 Dauncey, E.A., Rayner, T.G.J., Shak-Smith, D.A. and Bates, N.S. (2000) Poisonous plants and fungi in Britain and Ireland. Interactive identification systems on CD-ROM. Royal Botanic Gardens, Kew (GB) and Medical Toxicology Unit, Guy's & St Thomas' Hospital Trust, London.

35 Dauncey, E.A., Rayner, T.G.J., Berendsohn, W.G., Hand, R., Ritter-Franke, S., Murray, V.S.G.et al. (2000) *Die Giftpflanzen Deutschlands – Ein interaktives Identifizierungssystem auf CD-ROM*, Springer Electronic Media, Heidelberg.

36 Bensoussan, A., Myers, S.P., Drew, A.K., Whyte, I.M. and Dawson, A.H. (2002) Development of a Chinese herbal medicine toxicology database. *Clinical Toxicology*, **40**, 159–167.

37 Wagner, H. and Bladt, S. (1997) *Plant Drug Analysis. A Thin-Layer Chromatographic Atlas*, 2nd edn, Springer, Berlin.

38 Dasgupta, A. and Hart, A.P. (1997) Rapid detection of oleander poisoning using fluorescence polarization, immunoassay for digitoxin. Effect of treatment with digoxin-specific Fab antibody fragment (bovine). *American Journal of Clinical Pathology*, **108**, 411–416.

39 Herfst, M.J., Edelbroek, P.M. and de Wolff, F.A. (1980) Determination of 8-methoxypsoralen in suction-blister fluid and serum by liquid chromatography. *Clinical Chemistry*, **26**, 1825–1828.

40 Meda, H.A. *et al.* (1999) Epidemic of fatal encephalopathy in preschool children in Burkina Faso and consumption of unripe ackee (*Blighia sapida*) fruit. *Lancet*, **353**, 536–540.

41 Gaillard, Y. and Pepin, G. (1999) Poisoning by plant material: review of human cases and analytical determination of main toxins by high-performance liquid chromatography–(tandem) mass spectrometry. *Journal of Chromatography B*, **733**, 181–229.

36
Poisonous Mushrooms

F. Degel

36.1
Introduction

A major part of mushrooms is inedible as they do not taste good, and only a minor part is poisonous (Germany: 20 out of 500 edible mushrooms). It should be considered that, in general, a mushroom may become toxic when it is old, decaying, harvested after exposure to night frost or kept for hours in a closed bag after collection. Edible mushrooms may cause mild to severe diseases, if not prepared appropriately (e.g., insufficient cooking) or too big helpings are eaten (mushrooms are hard to digest). Sickness may occur from ingestion of prepared mushrooms spoilt from inadequate storage or multiple warming up.

Some mushrooms are harmful only in an uncooked state. They contain indigestible or even poisonous substances that are destroyed by cooking (e.g., even tasty edible mushrooms such as *Clitocybe nuda* (wood blewit) and *Xerocomus badius* (bay boletus)). It is emphasized, however, that the toxic agents of true toadstools (e.g., amatoxins and muscarine) are not destroyed by cooking.

Some mushrooms contain a toxic agent, which is effective only in combination with alcohol consumption from 3 h to 5 days after a meal (e.g., *Coprinus atramentarius* (inky cap)) and causes a disulfiram (Antabuse) like effect.

The concentration of poisonous compounds may vary considerably between species and even within a species (e.g., amatoxins).

Disease may also develop from allergenic properties of mushrooms. After repeated ingestion of a special mushroom, the sensitized person may fall ill acutely: *Paxillus involutus* is a really edible mushroom. After sensitization, an immunohemolytic syndrome associated with acute renal failure may occur possibly requiring hemodialysis. Therefore, it is no more listed officially as an edible mushroom in Germany. Recently, the number of diseases from *Paxillus involutus* is increasing, because immigrants from Eastern Europe are not familiar with the problems associated with this otherwise tasteful mushroom. Usually, allergenic reactions to mushrooms or their toxic agents are very rare. They cause intestinal symptoms as an immediate

Clinical Toxicological Analysis: Procedures, Results, Interpretation. Edited by Wolf-Rüdiger Külpmann

ISBN: 978-3-527-31890-2

reaction shortly after ingestion. Most often so-called allergenic reactions are due to effects from spoilt mushrooms (see above).

Real poisoning by toadstools most often occurs from misidentification or ignorance (there are old mushroom hunters and bold mushroom hunters, but there are no old, bold mushroom hunters), rarely from accidental ingestion by small children or from attempting suicide.

Most often, laboratory investigations are requested in case of gastrointestinal symptoms concomitant with ingestion of mushrooms. Poisoning shall be excluded or confirmed by the detection of the toxic agents to possibly initiate the appropriate treatment including the administration of an antidote.

36.2 Classification of Poisonous Mushrooms

In the following, only poisoning by the toxic agent is considered. The toxic agents are ordered according to their main effects, which is most appropriate for toxicological purposes (Table 36.1). The variety of mushrooms is reflected in the diversity of the toxic agents. They may cause very different symptoms, which may be present shortly after ingestion, or only hours, days, or even weeks later (Table 36.1).

An anamnesis focused on possible poisoning from mushrooms and the observed symptoms assist markedly in diagnosis and in the decision to initiate an appropriate treatment. In general, more the time elapsed between ingestion and the occurrence of symptoms, more difficult is the therapy.

36.3 Frequency of Poisoning

In Central Europe, most often *Amanita phalloides* (death cap) is involved in poisoning and fatalities from mushrooms. Of the deaths from toadstools, 90% are caused by *Amanita phalloides* corresponding to 50–100 fatal outcomes yearly. Fortunately, the frequency of poisonings has decreased during the past 25 years in this area as demonstrated by the statistical data from the Swiss Poison Information Centre in Zurich. Intoxications by other mushrooms containing amatoxins but at lower concentrations than the death cap (such as *Amanita verna, Amanita virosa, Galerina marginata*, and *epiota helveola*) are rare. In Eastern Europe, particularly in Russia, 45% of poisoning from mushrooms is caused by *Gyromitra* spp. and a considerable number of cases by *Amanita muscaria* and *Amanita pantherina*. Both are used for their psychoactive properties and poisoning is caused by overdosage (symptoms see Table 36.1). In the drug scene, intoxication from overdosage of hallucinogenic psilocybe species is observed occasionally (Section 26.9).

Rather often, mild to moderate intoxications are observed after ingestion of meals containing a mixture of different mushrooms. Usually, the gastrointestinal symptoms are not attributable to a particular species, but are caused by bacterial contamination.

Table 36.1 Poisonous mushrooms: symptoms of intoxication [14–16].

Syndrome, mushroom species (examples)	Mushroom poison and effects[a]	Onset of clinical symptoms after ingestion (L: latency; C: time course)	Symptoms of poisoning
Amatoxin syndrome *Amanita phalloides* (death cap), *A. verna, A. virosa* (destroying angel)	Amatoxins, phallotoxins Amatoxins: hepatotoxic (RNA polymerase B inhibitor in cell nucleus)	L: mostly 6–24 h	Typical time course after ingestion: Phase 1 (6–24 h): symptom-free latency period
Galerina marginata, G. autumnalis (deadly *Galerina*)			Phase 2 (12–24 h): severe gastrointestinal symptoms, vomiting, colicky, bloody diarrhea, dehydration, and occasionally acute hypotension
Lepiota helveola, L. brunneoincarnata, L. josserandi (deadly *Lepiota*)			Phase 3: apparent remission, rise in catalytic concentrations of liver-associated enzymes ALT and AST, and prolongation of prothrombin time
Pholiotina (Conocybe) filaris (deadly *Conocybe*) In general, *Amanita phalloides* contains the highest concentration of amatoxins			Phase 4 (24–48 h): coma hepaticum, hepatic and renal failure [17]
Coprine syndrome (acetaldehyde, disulfiram syndrome) *Clitocybe clavipes* (fat-footed clitocybe), *Coprinus atramentarius* (inky cap)	Coprine and its metabolite 1-amino-cyclopropanol Alcohol hypersensitivity (inhibition of acetaldehyde-dehydrogenase)	L: some minutes after alcohol ingestion C: up to 72 h after mushroom ingestion	Sense of heat, sweating, metallic taste, flush, nausea, and vomiting, In severe cases: confusion, tachycardia, severe headache, chest pain, hypotension, and collapse [15]

(*Continued*)

Table 36.1 (*Continued*)

Syndrome, mushroom species (examples)	Mushroom poison and effects[a]	Onset of clinical symptoms after ingestion (L: latency; C: time course)	Symptoms of poisoning
Gyromitrin syndrome	Gyromitrin and metabolites (*N*-formylhydrazine, methylhydrazine)	L: 6–24 h	Queasiness, severe headache, pyrexia, lethargy, feeling of bloating, gastrointestinal symptoms, hepatic and renal disorder, and CNS symptoms [12]
Gyromitra esculenta (false morel)	Hepatotoxic, CNS toxicity		
Lepidella syndrome	2-Amino-3-cyclopropylbutanoic acid and/or 2-amino-4, 5-hexadienoic acid	L: 4–11 h	Vomiting, abdominal pain, and diarrhea followed by anuria and renal failure [22–24]
spp. of amanitins, *Amanita smithiana*			
Muscarine syndrome	Muscarine	L: minutes, up to 2 h, (usually 30 min)	Sweating, salivation, lacrimation, gastrointestinal symptoms, summer cholera-like symptoms, colics, constricted pupils, and in severe cases bradycardia, hypotension [20]
Clitocybe spp., *Inocybe* spp. (fiber caps)	Reaction with cholinergic receptors (antidote: atropin)	C: several hours	
Orellanine syndrome	Orellanine, orelline	L: very long onset, several days up to weeks, seldom only several hours	First: queasiness, vomiting, later: thirst, gastrointestinal symptoms, polydipsia, polyuria, myalgia, paresthesia, tinnitus, renal pain, and in severe cases renal failure [18, 19]
Cortinarius orellanus, *C. speciosissimus*, *C. gentiles*	Nephrotoxic		

Pantherina syndrome	Ibotenic acid, muscimol, muscazon, and unknown toxins	L: from some minutes up to 2 (4) h	Drunken-like symptoms, euphoria, hallucinations, gait disorders, later jerks, and convulsions
Amanita muscaria (fly agaric), *Amanita pantherina* (panther cap)	Neurotoxins and psychoactive toxins		
Psilocybin syndrome[b]	Psilocybin, psilocin	L: from some minutes up to 2 h	Euphoria, dilated pupils, perceptual alterations, dizziness, confusion, hallucinations, deliriant symptoms, delusions, tachycardia, and vomiting [21]
Gymnopilus spectabilis (big laughing gym), *Inocybe aeruginascens* (reported in Europe), *Paneolina foensecii*, *Panaeolus* spp., *Pholiotina (Conocybe) cyanopus* (psychoactive *Conocybe*), *Psilocybe* spp. (blue-staining complex, "magic mushrooms"), *Pluteus salicinus*, *Stropharia coronilla*	Hallucinogenic effects	C: Up to 12 h	
Paxillus poisoning	After sensitization: hemolysins, hemagglutinins	L: from some minutes up to 2 h	Vomiting, gastrointestinal pains and colics, and hemolysis (onset: 2 h after ingestion)
Paxillus involutus (poison paxillus)[c]	Immunohemolytic syndrome		
Tricholoma poisoning	Unknown toxins	L: from some days up to several weeks, cumulative effect	Rhabdomyolysis, renal failure, respiratory distress, and cardiac arrhythmias
Tricholoma flavovirens[c]			

(*Continued*)

Table 36.1 (*Continued*)

Syndrome, mushroom species (examples)	Mushroom poison and effects[a]	Onset of clinical symptoms after ingestion (L: latency; C: time course)	Symptoms of poisoning
Gastrointestinal irritants For example, *Agaricus placomyces* (flat-capped psalliota), *Agaricus xanthoderma* (yellow stainer), *Boletus satanas* (satanic boletus), *Dermocybe sanguinea* (blood-red cort), *Entoloma sinuatum* (leaden entoloma), *Hebeloma crustuliniforme* (ring hebeloma), *Hypholoma fasciculare* (sulfur tuft), *Lactarius helvus* (tawny lactarius), *Megacollybia platyphylla* (broad gilled collybia), *Omphalotus olearius* (jack-o-lantern), *Ramaria pallida*, *Russula emetica* (pungent russula), *Scleroderma citrinum* (poison puffball), *Tricholoma pardinum* (tiger tricholoma)	Various toxins, partly not identified	L: from some minutes up to 2 h C: up to several days	Gastrointestinal symptoms with vomiting, diarrhea, abdominal pains, and colics

[a]Concentration of poison in mushrooms can vary by season and by habitat.
[b]The syndrome is triggered by psilocybin-containing mushrooms, also known as "magic mushrooms." Accidental poisoning with other mushrooms from misidentification is rather rare. Mushrooms with hallucinogenic effects are commonly eaten deliberately for the purpose of drug abuse. Since 1998, psilocybin-containing mushrooms fall within the scope of addictive drugs prohibited by law in Germany.
[c]In Germany, these species have recently been struck off the list of edible mushrooms.

36.4 Identification of Mushrooms

Identification should be preceded by a detailed anamnesis asking for types of ingested mushrooms, quantity and time of ingestion, further people involved, and careful investigation of the symptoms. The patient should describe particular characteristics of the mushrooms (see below).

It may be followed by showing pictures of mushrooms for confirmation. The presented particular mushrooms are identified:

1. Macroscopically
2. By examination of the spores.

1. *Macroscopic identification:* It is supported by books that show color photos or drawings of the mushrooms including a detailed botanic description of characteristics. Nonexperts should call a Poison Information Center or a dedicated regional advice center.

   The following should be documented:

   Size and shape of the mushroom
   Shape, color, and diameter of the cap (pileus)
   Gills (lamellae): shape, color, attached/not attached to the stalk
   No gills
   Stipe (stalk): shape and color, ring (shape and composition) (poisonous amanita with ring of death)
   Cup (volva) (poisonous amanita with cup of death)
   Flesh: discoloring from damage or pressure, odor, and perhaps smell (caution)
   Spores: shape and color (spore print)
   Environment, where the mushrooms have been collected.

2. *Examination of the spores:* Spores can be obtained from residues of the mushrooms, (by accumulation) from vomitus, stomach contents, feces, or left over. (Therefore, it is crucial to keep all the materials that might be involved and could be helpful for identification.) It should be emphasized that the identification of mushrooms by spores including microscopic examination should be carried out only by experts and is not recommendable as a screening procedure for less experienced personnel.

Reference books

D.G. Spoerke and B.H. Rumack. Handbook of mushroom poisoning. Diagnosis and treatment. Boca Raton: CRC Press, 1994.

A. Bresinski and H. Besl. A colour atlas of poisonous fungi. A handbook for pharmacists, doctors and biologists. London: Wolfe Publishing, 1990 [15].

L. Roth, H. Frank, and K. Kormann. Giftpilze – Pilzgifte. Landsberg am Lech: Ecomed Verlag, 1990 [14].

E.A. Dauncey (ed.). Poisonous plants and fungi in Britain and Ireland. Interactive identification system on CD-ROM. Publications sales, Sir Joseph

Banks Building, Royal Botanic Gardens, Kew, Richmond, Surrey TW9 3AE, UK, 2000 [16].
M. Jordan. The encyclopedia of fungi of Britain and Europe. London: Frances Lincoln, 2004.
M. Bon, J. Wilkinson, and D. Ovenden. The mushrooms and toadstools of Britain and North-Western Europe. London: Hodder & Stoughton, 1989.
F. Fung and R.F. Clark. Health effects of mycotoxins. A toxicological overview. *Journal of Toxicology: Clinical Toxicology* 2004; 42:217–234.

Internet

Bay Area Fungi Index: Pictures and description of more than 600 mushrooms. www.fungi.ca/mushrooms.htm.
M.W. Beug (ed.) Poisonous and hallucinogenic mushrooms. http://academic.evergreen.edu/projects/mushrooms/phm/index.htm.
Other Web sites with additional links:

www.mykoweb.com/CAF
www.grzyby.pl
www.mushroomexpert.com
www.pilzewelt.de/Pilze/engl.htm

36.5 Detection of Mushroom Toxins

Very few procedures for the specific detection and/or determination of mushroom toxins are available. The identification of mushroom toxins in biological material (urine, blood, and stomach contents) is impeded by the low concentrations and the nonavailability of specific and robust procedures except for amatoxins. At the moment, the only reliable and also practicable method is ELISA for the determination of amatoxins in urine, which is presented below.

36.5.1 Amatoxins

Toxic *Amanita* species contain amatoxins comprising α-, β-, γ-, and ε-amanitin, which are cyclic octapeptides. The principal poisonous compound is α-amanitin that is composed of a central tryptophan and an isoleucine side chain (Figure 36.1).

36.5.1.1 Screening Procedure

A popular procedure for the detection of amatoxins in native (not cooked) mushrooms is the "newspaper test" (Meixner test) using hydrochloric acid. However, it is too unspecific and too unreliable to be recommendable as a screening procedure.

Amanitin

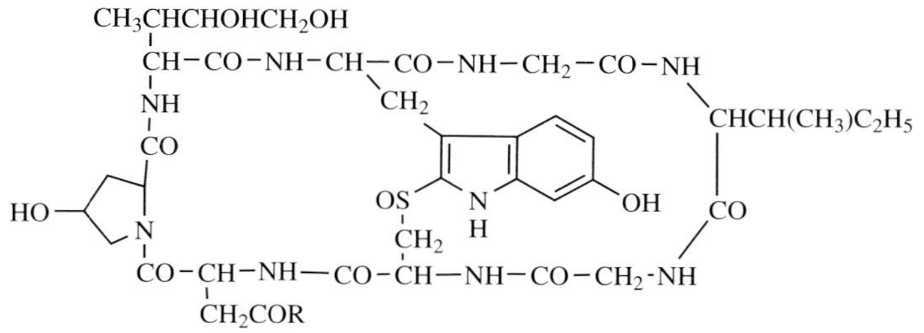

$CH_3CHCHOHCH_2OH$
$CH-CO-NH-CH-CO-NH-CH_2-CO-NH$
NH CH_2 $CHCH(CH_3)C_2H_5$
CO
HO N OS N OH CO
CH_2 H
$CO-CH-NH-CO-CH-NH-CO-CH_2-NH$
CH_2COR

$R = NH_2$ for alpha, OH for beta isomer

Figure 36.1 Amanitin.

36.5.1.2 **Immunoassay**

Introduction

An ELISA kit for the determination of amatoxins is commercially available all the year round. It has replaced the radioimmunoassay of Faulstich *et al.* labeled with tritium and the RIA from Bühlmann Lab. labeled with ^{125}I. The RIA has been available only during the mushroom season [1, 2].

Outline

The ELISA is a competitive enzyme immunoassay. The polyclonal antibodies (rabbit) are bound to the wells of the microtiter plate. During the first incubation, amatoxins from the urine sample and the calibrators compete with biotinylated amatoxins for binding to the limited number of antibodies. At the end of incubation, all nonbound antigens are removed by a wash step. During the following incubation, horseradish peroxidase labeled with streptavidin is binding to the biotin containing amatoxin–antibody complexes. Nonbound horseradish peroxidase is removed by the next wash step. During the last incubation step, a substrate (TMB) is converted to a colored complex catalyzed by (bound) horseradish peroxidase. After 15 min, the reaction is ended by addition of a stop solution. The absorbance at a wavelength of 450 nm is directly related to the enzyme activity and indirectly proportional to the amatoxin concentration in the sample.

Specimen

Urine (1 ml) without additives.

Stable for 7 days, if stored at 2–8 °C; 2 months, if stored at −20 °C.

Equipment

Devices for carrying out ELISA procedures including dedicated software.

Shaker for microtiter plates.

Washer for microtiter plates.

Chemicals
Not applicable.

Reagents
Amanitine ELISA test kit (EK-AM1, Bühlmann Lab., Allschwil, Switzerland; in Germany distributed by DPC-Biermann, Bad Nauheim), available all the year round. It contains six calibrators and two positive controls.

Negative control: urine from person not exposed to amatoxins.

Immunochemical analysis
The procedure is described in detail on the package insert of the kit. It is recommended to measure six calibrators, two control samples, and a blank in duplicate in a series.

Evaluation: The calculation of results is supported by dedicated software. Required details are provided by the manufacturer.

Analytical assessment
According to the manufacturer, the coefficient of variation for the imprecision within series varies from 2.8 to 4.5% and between series from 5.3 to 10.6% (number of measurements: 20).

The accuracy of the procedure is checked in every series with the two control samples that are present in the kit. In the range of 1.5–400 μg/l, absorbance and concentration were related consistently.

The detection limit is about 0.22 μg/l. According to the manufacturer, the coefficient of variation is below 15% for concentrations exceeding 1.5 μg/l (functional sensitivity).

A cutoff concentration for the distinction of negative from positive findings has not yet been established obligatorily. In a pilot study for the estimation of sensitivity, specificity, and predictive values, 61 urine samples from patients who suffered from gastrointestinal symptoms after ingestion of mushrooms were investigated [4]. The authors propose, from this study, a cutoff concentration of 5 μg/l for the distinction between disease from amanitin and disease from other causes.

The ELISA detects specifically α- and γ-amanitin, but not β-amanitin. However, no poisonous mushroom is known so far, which contains β-amanitin exclusively. Therefore, a false negative finding from missing cross-reaction with β-amanitin [3] is very unlikely. False positive findings from endogenous/exogenous compounds have not been observed: 25 urine samples of the routine toxicological service were investigated by the ELISA. In all samples, results were below 1.5 μg/l (mean: 0.23 μg/l; standard derivation (s): 0.09 μg/l; detection limit: 0.50 μg/l (mean + $3s$)). Similar results have been observed by Staack and Maurer [3], who investigated 100 comparable samples.

Practicability
Time of analysis is 90 min including 30 min technician time, when one patient sample, six calibrators, and one negative and two positive control samples are measured in duplicates.

36.5.1.3 High-Performance Liquid Chromatography

HPLC–UV detection
Several procedures for the determination of amatoxins have been published [5–8]. Most often plasma is used, because it is less susceptible to interference from other compounds. However, the concentration of amatoxins in plasma is only about one tenth of urine. In poisoning, plasma concentration is declining rather rapidly below the detection limit (about 10 μg/l) of the respective procedures using UV detection.

HPLC–electrochemical detection
EC detectors are superior to UV detectors regarding sensitivity and specificity. Unfortunately, they lack insufficient long-term stability of the electrodes due to fouling and specificity is considered as still unsatisfactory.

In conclusion, HPLC procedures are neither sufficiently sensitive (UV) nor specific (UV and EC) to be recommendable for detection or exclusion of amatoxin poisoning.

36.5.1.4 Liquid Chromatography–Mass Spectrometry

A LC–MS procedure has been published for the detection of amatoxins after solid-phase extraction [9] that has been improved by use of immunoaffinity extraction [10]. The method is rather demanding and problems may arise from ion suppression. Therefore, the procedure cannot yet be recommended without reservation for the detection or exclusion of amatoxin poisoning.

36.5.2 Toxins from *Russula* Species

Toxins of *Russula* species may be detectable in native material (but not in urine or plasma) using a color reaction with ferrous sulfate. The test is not specific, rather unreliable and cannot be recommended for screening.

36.6 Medical Assessment and Clinical Interpretation

Medical assessment
Poisonous *Amanita* species contain amatoxins and phallotoxins. Phallotoxins are scarcely absorbed in the gastrointestinal tract and probably play a minor role for intoxication. The amatoxins α-, β-, γ-, and ε-amanitin are considered as the crucial toxic agents for intoxication. The amanitins are cyclic octapeptides (Figure 36.1) and belong to the most potent natural toxins. 0.1–0.3 mg/kg body mass is a fatal dose for an adult. The dose may be administered by ingestion of only one *A. phalloides* mushroom (30–50 g). For small children, even lower doses are life threatening. *Amanita phalloides* contains the biggest fraction of amatoxins (200–400 mg/kg) among the *Amanita* species. The toxins are not destroyed by cooking even at

temperatures exceeding 250 °C and are inactivated *in vivo* neither by gastric juice nor by enzymes, such as proteases.

Some animals, however, such as deer, hares, or rodents are not susceptible to amanitin poisoning. In man, amatoxins penetrate rapidly the epithelial cells of the intestine and reach the blood after an extended absorption phase. They are detectable in blood only for a short while: in 70% of patients 12 h after ingestion, in 40% after 24 h, and in 0% after 48 h [11]. The severity of intoxication was not correlated to the concentration of amanitins. Of the absorbed amatoxins, 60% is excreted by the bile. α-Amanitine passes through the enterohepatic circulation that delays elimination, extends time of exposure of hepatic cells to the toxin, and augments toxicity. A minor part is eliminated by glomerular filtration through the kidneys. In urine, amanitins are detectable not before 6–12 h after ingestion. In the study (see above) [11], a positive finding had been obtained 12–24 h after ingestion for all urine samples, 48 h after ingestion still for 80%. It should be considered that concentration in plasma is only one tenth of the concentration in positive urine samples, if the samples are obtained at the same time and plasma concentration will fall more quickly below the detection limit. Therefore, urine is preferred to plasma for analysis after latency has elapsed and a concentration exceeding 5 μg/l urine is considered as an indicator of amanitin poisoning (see analytical assessment).

Clinical interpretation

After penetration, amatoxins inhibit the RNA polymerase II in eukaryotic cells and henceforth the transcription of messenger RNA followed by initiating programmed cell death after a 48 h latency. Cells with the highest replication rates and closest contact with the toxins are mainly affected, that is, cells of the liver and the tubules of the kidneys.

Clinical course of intoxication from amatoxins

Severe poisoning from amatoxins follows a sequence of four phases (time with regard to ingestion in brackets):

Phase 1: no symptoms (6–24 h)
Phase 2: severe gastroenteritis including exsiccation and possibly drop of blood pressure (12–24 h)
Phase 3: apparent improvement, but increasing activity of, for example, transaminases in plasma
Phase 4: acute hepatic failure followed by coma hepaticum, and acute renal failure (24–48 h).

The diagnosis should be established at the latest during phase 2 and be confirmed by identification of the mushroom (see above) and the detection of amatoxins in urine. Activity of transaminase ALT (GPT) and concentration of bilirubin and coagulation factors should be monitored regularly. Twenty-four hours after ingestion at the latest, pathological results are obtained. Prothrombin time is considered as the most sensitive test to reflect hepatic protein synthesis. Maximal prolongation of the prothrombin time is observed on an average 60 h after ingestion. The course of transaminase activity reflects only the dynamics of the intoxication and is considered for prognosis. Maximal activity is attained usually after 70 h. Activity of ALT (GPT) and prothrombin time reflect the severity of the intoxication.

Therapy of amanitin poisoning

Most recent recommendations for therapy are given in short for guidance. More detailed information can be obtained from the pertinent literature [13, 25–29] and from poison information centers.

1. *Decontamination:* Gastrointestinal decontamination by induced emesis and administration of charcoal. The procedure may be only effective, if applied within 4 h after ingestion of the mushrooms.
2. *Elimination enhancement:* There is an ongoing controversy regarding forced diuresis, hemodialysis, and hemoperfusion. On the whole, the procedures have been applied only with occasional success, at best, if used at an early phase of the intoxication and the presence of not yet bound toxins. Actually, the molecular adsorbent recirculation system is studied using an albumin-containing dialysate that is recycled passing a charcoal cartridge, an anion-exchanger resin absorber, and a hemodialyzer.
3. *Interruption of enterohepatic circulation:* For interruption, activated charcoal is administered orally, liquid is sucked off by a duodenal tube, and enema is given to the patient. The measures shall shorten the time of exposure of the liver cells to the amatoxins.
4. *Antidotes:* It is assumed that silibinin inhibits the hepatic uptake of amatoxins by inducing an alteration of the liver cell membranes and stimulates liver cell regeneration. There is a controversy regarding concomitant penicillin G application. A competitive inhibition of the silibinin effect is suspected and adverse effects such as cerebral seizures and allergenic shock have been observed [13, 29].
5. *Intensive care supportive measures:* These measures comprise restoration of fluid and electrolyte balance, treatment of coagulation disorders, and prevention or treatment respectively of hepatic coma according to established procedures including application of glucose and *N*-acetylcysteine.
6. *Liver transplantation:* As an ultima ratio liver transplantation is carried out.

Analytical strategy in amanitin poisoning

The determination of amatoxins should be carried out as soon as possible and should be possible even at night or at least within 12 h after admittance for confirmation of the diagnosis. It is also required for relief of the patient as early as possible. However, treatment has to be initiated in suspected poisoning as early as possible and shall not be postponed until analytical confirmation, but be continued until poisoning is excluded. Fortunately, therapy with silibinin does not cause relevant adverse effects.

Annotations

1. Mushrooms may be sprayed with pesticides that may cause the symptoms observed in the patient.

2. Not all persons eating the same meal of mushrooms must fall ill.
3. Symptoms occurring within 6 h after ingestion do not disprove reliably amanitin poisoning. A mixture of poisonous mushrooms may have been eaten.
4. A patient has to stay in the clinic, even though he does not suffer any more from gastrointestinal symptoms, when they have been observed as late as 6 h after ingestion.
5. Further poisonous mushrooms and the pertinent symptoms of intoxication are listed in Table 36.1.

References

1 Andres, R.Y. *et al.* (1986) Radioimmunoassay for amatoxins by use of rapid, ^{125}I-tracer-based system. *Clinical Chemistry*, **32**, 1751–1755.

2 Andres, R.Y. *et al.* (1987) 125I-Amatoxin and anti-amatoxin for radioimmunoassay prepared by a novel approach: chemical and structural considerations. *Toxicon*, **25**, 915–922.

3 Staack, R.F. and Maurer, H.H. (2001) New Bühlmann ELISA for determination of amanitins in urine – are there false positive results due to interferences with urine matrix, drugs or their metabolites? *Toxichem und Krimtech*, **68**, 68–71.

4 Butera, R., Locatelli, C., Coccini, T. and Manzo, L. (2004) Diagnostic accuracy of urinary amanitin analysis in suspected mushroom poisoning: a pilot study. *Journal of Toxicology: Clinical Toxicology*, **42**, 901–912.

5 Jehl, F., Jaeger, A. *et al.* (1985) Determination of α-amanitin and β-amanitin in human biological fluids by high-performance liquid chromatography. *Analytical Biochemistry*, **149**, 35–42.

6 Caccialanza, G., Gandinin, C. and Ponci, R. (1985) Direct, simultaneous determination of α-amanitin and phalloidin by high-performance liquid chromatography. *Journal of Pharmaceutical and Biomedical Analysis*, **3/2**, 179–185.

7 Rieck, W. and Platt, D. (1988) High-performance liquid chromatographic determination of α-amanitin and phalloidin in human plasma using the column-switching technique and its application in suspected cases of poisoning by the green species of *Amanita* mushroom (*Amanita phalloides*). *Journal of Chromatography*, **425**, 121–134.

8 Defendenti, C., Bonacina, E., Mauroni, M. and Gelosa, L. (1998) Validation of a high-performance liquid chromatographic method for alpha amanitin determination in urine. *Forensic Science International*, **92**, 59–68.

9 Maurer, H.H., Kraemer, T., Ledvinka, O., Schmitt, C.J. and Weber, A.A. (1997) Gas chromatography–mass spectrometry (GC–MS) and liquid chromatography–mass spectrometry (LC–MS) in toxicological analysis. Studies on the detection of clobenzorex and its metabolites within a systematic toxicological analysis procedure by GC–MS and by immunoassay and studies on the detection of alpha- and beta-amanitin in urine by atmospheric pressure ionization electrospray LC–MS. *Journal of Chromatography B*, **689**, 81–89.

10 Maurer, H.H., Schmitt, C.J., Weber, A.A. and Kraemer, T. (2000) Validated electrospray LC–MS assay for determination of the mushroom toxins alpha- and beta-amanitin in urine after immunoaffinity extraction. *Journal of Chromatography B*, **748**, 125–135.

11 Langer, M., Vesconi, S., Constantino, D. and Busi, C. (1980) Pharmacodynamics of amatoxins in human poisoning as the basis for the removal treatment, in *Amanita*

Toxins and Poisoning (eds H. Faulstich, B. Kommerell and T. Wieland), Gerhard Witzstock, Baden-Baden, pp. 90–95.

12 Baltarowich, L., Blaney, B., White, S. and Smolinske, S. (1996) Acute hepatotoxicity following ingestion of *Gyromitra esculenta* (false morel) mushrooms. 1996 North American Congress of Clinical Toxicology. *Journal of Toxicology: Clinical Toxicology*, **34**, 539–551.

13 Faulstich, H. and Zilker, T.R. (1994) Amatoxins, in *Handbook of Mushroom Poisoning. Diagnosis and Treatment* (eds D.G. Spoerke and B.H. Rumack), CRC Press Inc., Boca Raton, FL, pp. 233–247.

14 Roth, L., Frank, H. and Kormann, K. (1990) *Giftpilze – Pilzgifte: Schimmelpilze, Mykotoxine; Vorkommen, Inhaltsstoffe, Pilzallergien, Nahrungsmittelvergiftungen*, Ecomed Verlagsgesellschaft mbH, Landsberg am Lech, pp. 273–288.

15 Bresinsky, A. and Besl, H. (1985) Coprinus-Syndrom, in *Giftpilze* (eds A. Bresinsky and H. Besl), Wissenschaftliche Verlagsbuchhandlung mbH, Stuttgart, pp. 119–120.

16 Dauncey F E.A.(ed.) (1999) *Poisonous Plants and Fungi in Britain and Ireland*, Publication Sales, Royal Botanic Gardens, Kew, UK.

17 Zilker, T., Pfab, R., Kleber, J.J., Zantl, N. and Heidecke, C.D. (1999) Liver transplantation (LTX) for *Galerina marginata* poisoning [abstract]. EAPCCT XIX International congress. *Journal of Toxicology: Clinical Toxicology*, **37**, 417–418.

18 Horn, S. *et al.* (1997) End-stage renal failure from mushroom poisoning with *Cortinarius orellanus*: report of four cases and review of the literature. *American Journal of Kidney Diseases*, **30**, 282–286.

19 Rohrmoser, M., Kirchmair, M.S., Feifel, E., Valli, A., Corradini, R., Pohanka, E., Rosenkranz, A. and Pöder, R. (1997) Orellanine poisoning: rapid detection of the fungal toxin in renal biopsy material. *Journal of Toxicology: Clinical Toxicology*, **35**, 63–66.

20 de Haro, L., Prost, N., David, J.M., Arditti, J. and Valli, M. (1999) Muscarinic syndrome after mushroom ingestion (genus *Inocybe* and *Clitocybe*). Experience of the Marseille Poison Centre. *Journal of Toxicology: Clinical Toxicology*, **37**, 413–414.

21 Westberg, U. and Karlson-Stiber, C. (1999) A successful introduction of "Magic Mushrooms" on the Swedish market through the internet. *Journal of Toxicology: Clinical Toxicology*, **37**, 415–416.

22 Leathem, A.M., Pursell, R.A., Chan, V.R. and Kroeger, P.D. (1997) Renal failure caused by mushroom poisoning. *Journal of Toxicology: Clinical Toxicology*, **35**, 67–75.

23 Warden, C.R. and Benjamin, D.R. (1996) Acute renal failure associated with probable *A. smithiana* mushroom ingestions. A case series. 1996 North American Congress of Clinical Toxicology. *Journal of Toxicology: Clinical Toxicology*, **34**, 539–551.

24 Leathem, A.M. and Pursell, R.A. (1995) Suspected *Amanita smithiana* mushroom poisoning resulting in renal failure. 1995 North American Congress of Clinical Toxicology. *Journal of Toxicology: Clinical Toxicology*, **33**, 544.

25 Köppel, C. (1993) Clinical symptomatology and management of mushroom poisoning. *Toxicon*, **31**, 1513–1540.

26 Wellington, K. and Jarvis, B. (2001) Silymarin: a review of its clinical properties in the management of hepatic disorders. *BioDrugs*, **15**(7), 465–489.

27 Saller, R., Meier, R. and Brignoli, R. (2001) The use of silymarin in the treatment of liver diseases. *Drugs*, **61**(14), 2035–2063.

28 Saller, R., Brignoli, R., Melzer, J. and Meier, R. (2008) An updated systematic review with meta-analysis for the clinical evidence of silymarin. *Forschende Komplementärmedizin*, **15**(1), 9–20.

29 Ganzert, M., Felgenhauer, N., Schuster, T., Eyer, F., Gourdin, C. and Zilker, T. (2008) Amatoxin poisoning – comparison of silybinin with a combination of silibinin and penicillin. *Deutsche Medizinische Wochenschrift*, **133**(44), 2261–2267.

37
Venomous and Poisonous Animals

D. Mebs

37.1
Introduction

Two groups of animals causing envenoming or poisoning in humans can be distinguished as:

1. Venomous animals such as spiders, scorpions, bees, or snakes actively apply venom through a special venom apparatus. This device consists of venom producing tissue, usually a gland that is connected to an injecting apparatus, a stinger or a tooth. Venom is mainly of protein by nature and almost acts exclusively when applied parenterally.
2. Poisonous animals lack a specific apparatus to apply their toxic products. They make passive use of their poisons that usually consist of small molecular compounds such as alkaloids, terpenes, steroids, and so on and are mostly of nonprotein by nature. The poison is either produced in skin glands or is present in the whole body. Poisonous compounds or toxic secretions are effective when entering the body by the enteral route.

Moreover, toxic natural products may accumulate in animals, which are not typical members of the groups mentioned above, via the food chain and may cause food poisoning such as shellfish or fish poisoning.

Diagnosis of envenoming or poisoning caused by animals

For confirmed diagnosis of poisoning due to venomous or poisonous animals, identification of the animal involved is essential. This is easily achieved when venomous animals are kept in zoos or as pet animals privately. Except in the tropics, such accidents are rare in Europe (not considering stings of bees or wasps). For instance, symptoms occurring after swimming in the sea, such as painful skin reactions, are assumed to represent envenoming signs, but are rarely supported by the identification of a certain animal. In only few cases, envenoming symptoms are characteristic enough to allow the conclusion that a particular animal has been

Clinical Toxicological Analysis: Procedures, Results, Interpretation. Edited by Wolf-Rüdiger Külpmann

ISBN: 978-3-527-31890-2

involved. The occurrence of those symptoms as well as the course of envenoming or poisoning are highly variable and, of course, dose-dependent. The amount of toxic compounds injected, for example, a venom is generally unknown and in most cases evades toxicological analysis. Venomous and poisonous animals always cause acute intoxications, but in very few cases only chronic ailments.

Detection of venoms and poisons

Venoms and poisons of animal origin are rarely pure and chemically defined natural products. They are mostly mixtures of highly active toxins or other components such as enzymes. Many of these compounds represent the most toxic entities of natural origin surpassed only by bacterial toxins, such as tetanus or botulinum toxin. This implies that some of these toxins are lethal in extremely low doses. However, these properties make their detection in human body fluids, such as blood or urine, not only difficult but also even impossible. The complex composition of a venom and the structural peculiarities of the toxins require specific assay procedures such as immunochemical assays, for example, ELISA that are rarely available. In most cases, particularly when toxins of peptide or protein nature are involved, the detection or identification of the compound causing envenoming symptoms is not possible.

37.2 Venomous and Poisonous Animals

In the following chapters, venomous and poisonous animals have been selected, which can cause severe and even fatal envenoming and poisoning in humans. They are divided into two groups: marine (including seafood poisoning) and terrestrial animals. For more information, the reader is referred to books that appeared over the last 10 years [1–3]. Information on these subjects can also be retrieved from an increasing number of websites and databases.

37.2.1 Marine Animals

Cnidarians (phylum: Cnidaria)

Jellyfish, polyps, and sea anemones are able to sting upon contact due to the presence of nematocysts, a secretory product of specialized stinging cells. Following a stimulus (chemical or mechanical irritation), the cell discharges its nematocyst explosively and a stiletto-like dart is everted and penetrates the skin. With extremely high speed, an evaginated tubule covered with small spines, is ejected and injects venom. Jellyfish such as the mauve stinger (*Pelagia noctiluca*) and the lion's mane (*Cyanea capillata*) may occur in large numbers in European seas (e.g., the Mediterranean Sea, the Gulf of Biscay, around the British Isles, the North, and the Baltic Sea) at certain times of the year. The Portuguese-man-of-war (*Physalia physalis*, in the tropical Atlantic, and also worldwide) has long tentacles (several meters long), which cause extreme pain and severe skin reactions upon contact; fatalities have been reported.

Jellyfish venoms are a mixture of high-molecular proteins exhibiting cytotoxic or neurotoxic activity. Their instability and tendency to easily denature in physiological solutions prevented their thorough chemical and pharmacological characterization. The primarily local effects of these venoms on the skin are poorly understood. It has been suggested that the toxic proteins initiate autopharmacological reactions causing strong pain, massive edema, capillary spasms, urticaria, and necrosis. Venom components may also trigger immunological side effects. Treatment of these injuries is complicated and has to be symptomatic.

Systemic effects include cardiac failure and pulmonary edema, such as being caused by the Australian box-jellyfish (*Chironex fleckeri*). In this case, a specific antidote is available (in Australia only).

Venom detection: No methods or test-systems exist for the detection of cnidarian venoms or toxins in human body fluids. Eventually, after weeks or months following a jellyfish contact, antibodies specific for the jellyfish species involved can be assayed in serum using an immunochemical assay (ELISA). This can be performed only in laboratories where these tests have been developed. However, the specificity of this assay is disputed and the information achieved is mainly of academic interest [4]. Increased serum creatine kinase (CK) activity points to damage of skeletal muscles, partially due to massive edema formation. Other blood parameters are usually in normal range.

Sea urchins

Through their sharp spines, sea urchins may inflict painful injuries. The spines easily break after penetrating the skin and the calcareous fragments remain in the tissue. Surgical excision is often necessary to remove the fragments.

Fishes

Among the numerous species of fishes, only relatively few are venomous. They produce venom in glands that cover the spines of fins as an epithelial layer.

Stingrays are distributed worldwide and are present in all tropical oceans. On the dorsal surface of their tail, they have one or several barbed spines that are covered with integumentary glandular tissue producing venom. By flicking the tail, the spine eventually penetrates the skin like a dagger inflicting deep wounds. The glandular tissue remains in the lacerated wound and may cause envenoming.

Weeverfish (*Echiichthys* spp.) are found in the Mediterranean, North, and Black Sea as well as in the Atlantic. They have dorsal and opercular spines covered with venom producing glandular tissue. When the spines enter the skin, this tissue is ruptured and pressed releasing the venom into the wound inflicted and causing long-lasting and intense pain. However, chronic symptoms do not occur. Fishes with similar mechanisms for envenoming also include lionfish (*Pterois* spp.), scorpionfish (*Scorpaena* spp), and stonefish (*Synanceja* spp.).

Fish venoms are consisting of high-molecular proteins, such as the trachinin of the weeverfish (molecular mass about 340 000) that are highly thermolabile. There exist no antidotes, except for the stonefish from the Indo-Pacific. If hurt by the fin-spines, x-raying of the wound is recommended to localize residual broken spines. Treatment of the inflicted body part with hot water (60 °C) is less effective and may cause scalding instead.

Venom detection: The detection and assay of the very unstable toxins is not possible. Most blood parameters remain unchanged even during acute envenoming.

Seafood poisoning

Shellfish poisoning: After eating shellfish, such as blue mussels (*Mytilus edulis*), clams, or scallops, symptoms of poisoning may occur during certain seasons. Since mussels are farmed in maricultures, shellfish poisoning may reach epidemic proportions. Shellfish are filterfeeders and accumulate beside heavy metals, chemicals and pathogenic microorganisms, natural products, such as toxins from algae, which represent their main food source. Massive algal blooms, for example, the occurrence of toxic algae in the marine phytoplankton, induce toxicity of shellfish, because the algal toxins are stored in the mussel tissue without affecting the mussel itself.

Several algae are known to produce toxic strains, such as dinoflagellates of the genus *Alexandrium* and *Dinophyis*. The toxins they contain cause several symptoms of shellfish poisoning in humans:

(1) Paralytic shellfish poisoning (PSP), which is the "classical" type of shellfish poisoning caused by the dinoflagellate toxin saxitoxin and its congeners (gonyautoxins) inducing paralytic symptoms due to the blocking of Na^+ channels.
(2) Diarrhetic shellfish poisoning causing gastrointestinal symptoms such as nausea, vomiting, and diarrhea due to the presence of toxins like okadaic acid and dinophysis toxin.
(3) Amnesic shellfish poisoning. Rarely observed, but a highly complicated intoxication causing coma and severe neurological disorders such as amnesia. Domoic acid is the toxin involved, which is produced by diatom algae.

The incubation time of *paralytic shellfish poisoning* is short. Thirty minutes after the consumption of toxin-contaminated mussels, first symptoms of paralysis such as paresthesia of the lips, tongue, and mouth followed by numbness and general weakness may occur. The symptoms reach their maximum within the first 12 h and gradually disappear within 48 h without sequelae. In severe cases, paralysis of respiratory muscles may cause death.

Symptoms of *diarrhretic shellfish poisoning* also appear soon after the consumption of contaminated mussels. Nausea, vomiting, and diarrhea are developing usually 5–6 h after the mussel meal, persist for 2–3 days and may disappear without sequelae. *Amnesic shellfish poisoning* may result in irreversible brain damage leading to persistent deficits in short-term memory.

There exist no antidotes for the treatment of shellfish poisonings. Therefore, treatment has to be entirely symptomatic. Blood parameters are generally in the normal range, but serum electrolytes may eventually decrease due to fluid loss after vomiting and diarrhea.

Toxin detection: Toxins involved in shellfish poisoning can only be assayed by analyzing contaminated seafood. For saxitoxin and its derivatives (paralytic shellfish poisoning), a standardized bioassay, for example, toxicity testing in mice, is still mandatory. Shellfish tissue is homogenized and extracted with 0.1 mol/l HCl and neutralized with 0.1 mol/l NaOH. Various dilutions of the extract are injected intraperitoneally into mice of about 20 g body weight. One mouse unit (MU) is defined as the amount of toxin that kills a mouse in 15 min. If the mouse is not dying within 30 min, it will survive the dosage. Using a standard plot with saxitoxin, the amount of toxin in 100 g shellfish tissue can be calculated. One MU corresponds to 0.18 μg saxitoxin-HCl [5].

A bioassay for the detection of okadaic acid is also still being used. Rats are fed with mussel meat and normal food after 24 h food deprivation. Acceptance or rejection of the mussel meat and the consistency of feces is evaluated. In the usual mouse bioassay, okadaic acid is toxic, but the time until death is considerably longer than in the case of saxitoxin.

For both groups of toxins, HPLC procedures exist, which enable the quantitative assay of the toxins and of their various congeners. After separation of the compounds using various column types, the toxins are detected fluorimetrically following post-column derivatization in alkaline solutions. Toxins of diarrhretic shellfish poisoning, okadaic acid and dinophysis-toxin, are derivatized with 9-anthryldiazomethane (ADAM) before HPLC-separation and are detected fluorimetrically [6]. The analytical procedures for the assay of shellfish toxins require high technical standards and are performed in special laboratories, which also have access to the toxin standards.

The severity of symptoms occurring in paralytic shellfish poisoning corresponds to the toxin concentration in the mussels. Slight paresthesia occurs, when 2.000–10.000 MU have been ingested, paralytic symptoms are observed at 10.000–20.000 MU, the lethal dosage is considered to be in a range of 20.000–40.000 MU [7], but may be lower in children. In USA and Canada, 400 MU per 100 g shellfish meat corresponding to 80 μg saxitoxin are believed to be acceptable for human consumption. This value is also the limit in Europe [6].

Fish poisoning

Tetrodotoxic fish

Fugu, the raw meat from pufferfish (Tetraodontidae), is a delicacy in Japan. As the fish contains high concentrations of tetrodotoxin, which blocks Na^+-channels, it is carefully prepared by licensed cooks, who remove the internal organs and serve only thin slices of muscle meat. Fatal poisoning may occur, when fish is prepared by persons who are not aware of its toxic properties.

Toxin detection: Toxicity testing of fish-extracts in mice provides quick and reliable results, but is not allowing the distinction between tetrodotoxin and shellfish toxins such as saxitoxin. HPLC followed by post-column derivatization with *o*-phthaldialdehyde and fluorimetric detection is a sensitive procedure to identify tetrodotoxin and its derivatives.

Ciguatera

This is the "classical", most important form of fish poisoning, which is not of bacterial origin. It occurs in tropical oceans such as in the Indo-Pacific and the Caribbean Sea, but not in the tropical Atlantic. Sporadically and at certain times of the year, ciguatera symptoms appear after the consumption of fish that is normally safe to eat: vomiting and diarrhea followed by influenza-like symptoms such as muscle pain, arthralgia, exhaustion, but also pruritus particularly involving the palms and soles, and neurological manifestations like "hot-cold reversal". The latter is a characteristic phenomenon for the differential diagnosis of ciguatera. Ciguatoxins that cause the poisoning, originate from toxic algae (dinoflagellates, *Gambierdiscus toxicus*) and is sequestered via the food chain: herbivorous fish ingest the toxin with their food, that is sea weed where the algae have settled, store, and accumulate the toxin in their body. Preying on these herbivores by carnivorous fish further accumulate the toxin reaching considerably high levels in their body. Fish are usually not affected by ciguatoxins.

The toxins are heat-stable and are not destroyed by boiling or frying. There exist no specific antidotes to ciguatera, but the prognosis is generally good despite long-lasting complaints over weeks or even months, which may not only disappear but reappear also. Ciguatera is one of the very few chronic poisonings of animal origin. Blood parameters are in normal range.

Toxin detection: A toxin-assay in blood or urine of ciguatoxic patients is not possible. For distinguishing toxic from nontoxic fish, the mouse toxicity assay is still being used. Immunological assays have been developed, but their specificity and reliability is disputed.

Scombrotoxism

This is another important form of fish poisoning that occurs predominantly after eating canned fish like sardines or tuna. Histamine is the toxin involved, which is enzymatically produced through the decarboxylation of histidine by bacteria contaminating the fish. Red skin rash, facial flushing, pruritus, sweating, and a peppery taste of the fish are common symptoms of histamine poisoning. However, these symptoms also occur in cases of allergic reactions to seafood. Antihistamines are specific antidotes causing rapid alleviation of the symptoms.

Toxin detection: After trimethylsilylation, histamine can be assayed by gas chromatography–mass spectrometry [8]. High levels of histamine and of its metabolite *N*-methylhistamine are detected in the urine of the patient.

Mollusca

Cone snail shells (*Conus* spp.) are found mainly in the warmer regions of the Pacific Ocean. They are venomous predators and their venom consists of a very potent mixture of neurotoxic compounds, that is, peptides and proteins, the conotoxins. Envenomations occur in case of prolonged contact and are characterized by local pain, numbness, and flaccid paralysis that may lead to respiratory failure and death. Most often *C. geographus* is involved, although other *Conus* spp. may also cause envenomations.

It is interesting to note that tetrodotoxin (see tetrodotoxic fish), which is a specific inhibitor of sodium channels, is also present in various other marine animals including the *blue-ringed octopus* (*Hapalochlaena* sp.), which introduces the toxin with its saliva during a bite, as well as in several terrestrial frogs, toads, newts, and salamanders. Neither fish nor the amphibians are able to synthesize the toxin, but rather bacteria that have been identified in the skin and intestines of pufferfish and crustaceans.

37.2.2
Terrestrial Animals

Scorpions

Scorpions inject their venom, a mixture of polypeptides and proteins, with a needle-like stinger at the end of their abdomen. In Europe, there exist no scorpions of medical importance. The Mediterranean scorpion, *Euscorpius italicus*, is harmless, because it is unable to penetrate human skin with its stinger. The sting of *Buthus occitanus*, a scorpion common in South-France and Spain is painful, but exerts no stronger effects. However, stings of scorpion species (e.g., *Androctonus, Leiurus, Buthus, Parabuthus, Centruroides,* and *Tityus*) from North-Africa, Near-East, Africa, Asia, Mexico, and South-America are dangerous and produce a complex symptomatology of envenoming. The peptide toxins in their venoms cause a massive release of neurotransmitters, such as catecholamines and acetylcholine, leading to severe cardiovascular symptoms (e.g., hypotension, tachycardia) and pulmonary edema as a complication of cardiac insufficiency. Specific antivenoms exist for nonEuropean scorpion venoms, however, their application, particularly in late stages of envenoming, is a matter of controversy, because the secondary autopharmacological symptoms are not reversed with antivenom and need symptomatic intensive care treatment. Scorpion envenoming symptoms usually disappear without chronic sequelae. Blood parameters are normal in acute envenoming.

Venom Detection: Scorpion venoms represent complex mixtures of peptidic toxins exerting high affinity to nervous structures such as ion channels. The venoms can be analyzed by conventional protein biochemical methods, but no methods exist for their detection in the patient's blood or urine.

Spiders

The black widow spider (*Latrodectus tredecimguttatus*) is the only European spider that may cause severe, but not life-threatening envenoming. This spider is common not only in the Mediterranean area, but also in most tropical and subtropical parts of the world. All other European spiders, except the rare yellow sack spider (*Chiracanthium punctorium*), cannot penetrate human skin with their fangs, the chelicerae.

Few microliters of the venom, a mixture of toxic proteins, are injected through the bite of the black widow spider, causing intense pain within 10–15 min to 1 h, spreading from the bitten area over the entire body. The sudden release of neurotransmitters, particularly acetylcholine, produces a continuous stimulation of skeletal muscles and their painful contractions and cramps. This spider envenoming named latrodectism is often misdiagnosed, for instance, as acute appendicitis. For treatment, antivenom has been developed, but it is rarely available (except in Australia). After intravenous injection, dramatic improvement and disappearance of the symptoms have been reported. Analgesic drugs have only moderate or no effect in pain relief. However, pain weakens within 24 h, and, in general, prognosis of latrodectism is good even without treatment.

The bite of other spiders such as the brown or fiddleback spiders (*Loxosceles reclusa*) often causes local necrosis, but also systemic reactions including intravascular hemolysis. The Australian funnel-web spiders (*Atrax* and *Hadronyche*) are dangerous species. Their bite produces not only pain but also systemic envenoming with the risk of severe pulmonary edema. Death is now rare due to efficient antivenom therapy.

The big tarantulas (family: Mygalomorphae) that are often kept as pet are generally harmless and possess only a weak venom that may cause slight local pain, but no systemic reactions.

Venom Detection: Injected in very low quantities, the detection of spider venom is not possible in blood or urine of patients.

Insects

Statistically, bees and wasps are the most dangerous venomous creatures in Europe. Although envenoming by these stinging insects is painful, it is mostly of trivial nature. However, it may trigger severe allergic reactions including anaphylactic shock, which may lead to death within minutes in sensitized humans. The venom of these Hymenoptera insects consists of peptides and enzymes (phospholipase A_2 and hyaluronidase), which represent highly potent allergens. Envenoming following numerous stings (several hundreds) are rare and may cause intravascular hemolysis and rhabdomyolysis leading to renal failure and death. Hornet stings are less dangerous than anticipated, they cause less pain than a bee sting. Even bumblebees can apply painful stings.

Venom Detection: There exists no method for venom detection, which, on the other hand, is obsolete after a sting. In severe cases (multiple stings), the determination of parameters such as hemoglobin, hematocrit, LDH, creatine kinase, haptoglobin, clotting status, platelet count, and looking for signs of hemolysis are important to evaluate progressing envenoming and to initiate therapeutic measures such as blood transfusion, fluid, and electrolyte substitution. Allergic reactions are elicited by IgE-class antibodies. Hypersensitivity towards bee or wasp venom can be tested by assaying these antibodies in serum using the radioallergosorbent test (RAST). This test is performed in most dermatology clinics.

Snakes

The bite of European vipers such as of the sand viper (*Vipera ammodytes*), the asp viper (*Vipera aspis*) and the adder (*Vipera berus*) causes complex envenoming symptoms, depending on the amount of venom injected: painful edema, nausea, vomiting, angioneurotic edema, eventually dizziness, unconsciousness, and coma. However, death is rare. Over the past 60 years, no fatal envenoming due to an adder bite has occurred in Germany.

Snake venom is a complex mixture of protein toxins and enzymes, mostly hydrolases including proteases, and amino acid oxidases. Toxins affect the peripheral nervous system by blocking neuromuscular transmission. The enzymes produce various effects: as clotting enzymes interfering with hemostasis and finally leading to consumption coagulopathy; as hemorrhagins causing capillary damage, bleeding and local tissue necrosis; as phospholipases A_2 acting as myolysins on skeletal muscle tissue or exhibiting neurotoxic activity. The high variability in the venom composition of the snake families and species produces a variety of symptoms in envenoming. Several symptom complexes can be distinguished in snakebite:

(1) Neurotoxic, that is, paralytic symptoms occur after the bite of cobras (*Naja, Hemachatus, Ophiophagus* spp.), kraits (*Bungarus* spp.), mambas (*Dendroaspis* spp.), coral snakes (*Micrurus* spp.) and certain sea snakes (*Laticauda* spp.) as well as of some vipers and rattlesnakes (*Crotalus* spp.) due to the blockage of neuromuscular transmission at motor endplates leading to the paralysis of facial and respiratory muscles. Death is caused by respiratory failure and arrest.

(2) Damage of skeletal muscles follows bites of numerous snake species including sea snakes, vipers, and rattlesnakes. Myolysis is due to the action of venom phospholipase A_2 causing a dramatic increase of creatine kinase activity in serum and myoglobinuria, eventually leading to renal failure.

(3) Hemostatic disturbances are characteristic features of bites by many vipers and rattlesnakes, but also by some rear-fanged (opisthoglyphous) colubrid snakes (*Thelotornis* spp., *Dispholidus typus*). Complex interactions with the blood clotting system such as activation of certain factors (Factor X, prothrombin, etc.) or direct hydrolysis of fibrinogen lead to severe coagulopathy, for example, incoagulable blood and bleeding from damaged vessels.

(4) Swelling, edema, hemorrhage, and necrosis are local symptoms at the bite site occurring in many snakebites, particularly of vipers and rattlesnakes.

(5) Cardiovascular symptoms, for example, hypotension are less dangerous in snakebite, shock symptoms are mainly due to loss of fluid in bleeding or massive edema formation.

The occurrence of only a single characteristic symptom is rare in case of a snakebite; the development of a complex envenoming symptomatology is more common. For details, see Refs [1–3].

Specific treatment of snakebite is performed by intravenous injection of antivenom. Supportive and symptomatic treatment is recommended, particularly when antivenom is not available. This includes fluid replacement, infusion of clotting factors, artificial ventilation, and hemodialysis. Antivenom application may cause allergic reactions and even anaphylaxis. Late antivenom reactions, such as serum sickness may occur about a week later and are treated with corticosteroids. Snakebite has no chronic sequelae, except tissue loss due to hemorrhage and necrosis.

Venom Detection: A venom detection kit is available only for Australian snakes. It enables the immunological assay of venom antigens in swabs taken from the bite site, and also in urine from the patient. This test has been developed for the identification of the snake involved to allow the use of a specific antivenom. Experimental immunoassays have been developed for estimating the venom concentration in blood and urine, but these test kits are not available in other parts of the world. The description of the snake by the patient is often misleading and one has to rely on the symptoms developing in the course of envenoming. The correct identification of the snake is not important in Europe, because bites of all species cause similar envenoming symptoms. People keeping venomous snakes as pets usually know the scientific species names. Blood clotting tests should be performed when hemostatic disturbances are to be expected. Practically, this is also a test, whether venom had been injected (in about half of the snakebite cases, no venom was injected by the snake; so called "dry bites"). Control of the coagulation parameters also allows testing of the efficacy of the antivenom therapy (returning to normal clotting time).

37.3 Conclusion

Envenoming and poisoning by animals are often dramatic events. They are relatively rare in Europe, but are common in the tropics. The venom dosage is always unknown, therefore, the severity of envenoming or poisoning is difficult to estimate, but must be evaluated by carefully observing the developing symptomatology. Clinical tests provide helpful data only under exceptional circumstances.

As mentioned in all the chapters, toxins from animals are active in extremely low doses. Venoms and toxins of animal origin can be analyzed by biochemical methods and procedures. However, attempts to measure their concentrations in blood or other body fluids, so as to obtain pharmacokinetic data, were successful in few cases only, mostly under experimental conditions. Since such intoxications are always acute and chronic implications are rare, treatment and therapy should not be delayed and should be initiated with the occurrence of first signs of poisoning or envenoming.

References

1 Mebs, D. (2002) *Venomous and Poisonous Animals*, Medpharm Sci Publishers, Stuttgart.

2 Meier, J. and White, J.(eds) (1995) *Handbook of Clinical Toxicology of Animal Venoms and Poisons*, CRC Press, Boca Raton.

3 Junghans, T. and Bodio, M. (1996) *Notfall-Handbuch Gifttiere*, G. Thieme, Stuttgart.

4 Burnett, J.W., Calton, G.J., Fenner, P.J. and Williamson, J.A. (1988) Serological diagnosis of jellyfish envenomations. *Comparative Biochemistry and Physiology*, **91**, 79–85.

5 Williams, S.(ed.) (1984) *AOAC: Official Methods of Analysis of the Association of Official Analytical Chemists*, 14th edn, AOAC Inc., Arlington, pp. 344–345.

6 Luckas, B. (1992) Phycotoxins in seafood – toxicological and chromatographic aspects. *Journal of Chromatography*, **624**, 439–456.

7 Edwards, H.I. (1956) The etiology and epidemiology of paralytic shellfish poisoning. *Journal of Milk and Food Technology*, **19**, 331–337.

8 Henion, J.D., Nosanchuck, J.S. and Bilder, B.M. (1981) Capillary gas chromatographic mass spectrometric determination of histamine in tuna fish causing scombroid poisoning. *Journal of Chromatography*, **213**, 475–485.

Appendix A

Abbreviations

A	Mobile phase A in a gradient elution system (HPLC)
AAS	Atomic absorption spectrometry
AC	Acetate
ACE	Angiotensin-converting enzyme
AChE	Acetylcholinesterase
ADH	Alcohol dehydrogenase
ADI	Acceptable daily intake
AEGL	Acute exposure guideline levels
ALDH	Aldehyde dehydrogenase
ALT	Alanine aminotransferase
AMDIS	Automated Mass Spectra Deconvolution and Identification System
APCI	Atmospheric pressure chemical ionization
API	LC–MS instrument (trademark)
API	Atmospheric pressure ionization
AST	Aspartate aminotransferase
B	Mobile phase B in a gradient elution system (HPLC)
BA	Bromoacetone
BAL	British anti-lewisite (antidote)
BAN	British Approved Name
BAT	Biologically acceptable concentration
BChE	Butyrylcholinesterase
BDB	1-(3,4-Methylenedioxyphenyl)-2-butanamine (= 1-(1,3-benzodioxol-5-yl)-2-butanamine)
BDMSC	Butyldimethylsilyl chloride
BIPM	Bureau International des Poids et Mesures
BondElut	Solid-phase (trademark)
BSTFA	*N,O*-Bis-(trimethylsilyl) trifluoroacetamide
BTEX	Benzene, toluene, ethylbenzene, xylenes
BuChE	Butyrylcholinesterase
BZ	3-Quinuclidinyl benzilate

Clinical Toxicological Analysis: Procedures, Results, Interpretation. Edited by Wolf-Rüdiger Külpmann

ISBN: 978-3-527-31890-2

c	Concentration
C18	Octadecyl solid phase
C2	Ethyl solid phase
C8	Octyl solid phase
CE	Capillary electrophoresis
CEDIA	Cloned enzyme donor immunoassay
CEN	European Committee for Standardization (Comité Européen de Normalisation)
ChE	Cholinesterase
ChemElut	Solid phase (trademark)
CHMPA (CMPA)	Cyclohexylmethylphosphonic acid
CI	Chemical ionization
CID	Collision-induced dissociation
CIEIA	Competitive inhibition enzyme immunoassay
CIS	Cold injection system
CIPM	International Committee of Weights and Measures
CK	Creatine kinase
ClinElut	Solid phase (trademark)
CMP	Computer Monitoring Program
CN	Chloroacetophenone
CP-SIL 5 CB	Stationary phase (trademark)
CS	2-Chlorobenzylidene malononitrile
CSF	Cerebrospinal fluid
CVAA	2-Chlorovinyl-arsonous acid
CVX	Chinese VX
CWA	Chemical warfare agents
CWC	Chemical weapons convention
DAD	Diode array detector
DB 1	Stationary phase (trademark)
DB 5	Stationary phase (trademark)
DFP	Diisopropyl fluorophosphate
DGKL	German United Society for Clinical Chemistry and Laboratory Medicine
DHHS	US Department of Health and Human Services
DIN	German Institute for Standardization (Deutsches Institut für Normung)
DNA	Deoxyribonucleic acid
DPMPH	*S*-(2-Diisopropylaminoethyl)-methylphosphonohydroxide (VX metabolite)
EC	Enzyme Commission
ECD	Electrochemical detector (HPLC)
ECD	Electron capture detector (GC)
EDAPS	Ethyl-dimethylamidophosphoric acid (hydroxytabun)
EDDP	2-Ethylidene-1,5-dimethyl-3,3-diphenylpyrrolidine (methadone metabolite)

EDEA	*N*-Ethyldiethanolamine
EDT	1,2-Ethanethiol
EDTA	Ethylenediaminetetraacetic acid
EHMPA	(2-Ethyl)-hexylmethylphosphonic acid
EI	Electron-impact ionization
EI	Enzyme immunoassay
EIA	Enzyme immuno assay
ELISA	Enzyme-linked immunosorbent assay
EM	Emission
EMIT	Enzyme-multiplied immunoassay technique
EMPA	Ethylmethylphosphonic acid
EMTPA	*O*-Ethylmethylthiophosphoric acid
ENV+	Solid phase (trademark)
EPA	Ethylphosphonic acid
ES	Electrospray
ESI	Electrospray ionization
EU	European Union
EX	Excitation
EXME	Extractive methylation
FIA	Flow injection analysis
FID	Flame ionization detector
FPD	Flame photometric detector
FPIA	Fluorescence polarization immunoassay
FPLC	Fast protein liquid chromatography
FSM	Full scan mode
g	Gravitational constant
G-agents	Collective term for GA, GB, GD, and GF
GA	Tabun
GABA	γ-Aminobutyric acid
GB	Sarin
GBL	γ-Butyrolactone
GC	Gas chromatography or gas chromatograph
GC–MS	Gas chromatograph(y)–mass spectrometer(try)
GC–MS–MS	Gas chromatograph–triple quadrupole
GD	Soman
GF	Cyclosarin
GHB	γ-Hydroxybutyrate
GLP	Good Laboratory Practice
γ-GT	γ-Glutamyltransferase
HD	Sulfur mustard
HETEG	*N*7-(2-Hydroxyethyric-thioethyl)-guanine
HF	Hydrofluoric acid
HFBA	Heptafluorobutyric anhydride
HFBI	Heptafluorobutyrylimidazol
HHLB	Solid phase (trademark)

HN-1	*N*-Ethyl-2,2′-dichloroethylamine (nitrogen mustard)
HN-2	*N*-Methyl-2,2′-dichloroethylamine (nitrogen mustard)
HN-3	Tris-(2-chloroethyl) amine (nitrogen mustard)
HPLC	High-performance liquid chromatography
HPTLC	High-performance thin-layer chromatography
HY	Hydrolysis
IA	Immunoassay
IAE	Immunoaffinity extraction
ICP	Inductively coupled plasma
ID	Internal diameter
iBMPA	*iso*-Butylmethylphosphonic acid
IFCC	International Federation of Clinical Chemistry and Laboratory Medicine
ILAC	International Laboratory Accreditation Cooperation
IMPA	*iso*-Propylmethylphosphonic acid
INN	International Nonproprietary Name
IPCS	International Programme on Chemical Safety
IS	Internal standard
ISE	Ion-selective electrode
ISO	International Organization for Standardization
IST	Manufacturer (solid phase)
IUBMB	International Union of Biochemistry and Molecular Biology
IUPAC	International Union of Pure and Applied Chemistry
KAS	Cold on-column injection
kDa	Kilodalton
JCTLM	Joint Committee for Traceability in Laboratory Medicine
LAMPA	Lysergic acid methylpropylamide
LC	Liquid chromatograph(y)
LC–MS	Liquid chromatograph–quadrupole mass spectrometer
LC–MS–MS	Liquid chromatograph–triple quadrupole mass spectrometer
LDH	Lactate dehydrogenase
LiChrolut	Solid phase (trademark)
LSD	Lysergic acid diethylamide
MAK	Maximum allowable concentration at the workplace
MBDB	*N*-Methyl-1-(3,4-methylenedioxyphenyl)-2-butanamine (= *N*-methyl-1-(1,3-benzodioxol-5-yl)-2-butanamine)
MDA	3,4-Methylenedioxyamphetamine
MDE	3,4-Methylenedioxyethylamphetamine
MDEA	3,4-Methylenedioxyethylamphetamine
MDEAN	*N*-Methyldiethanolamine
MDMA	3,4-Methylenedioxymethamphetamine
$[M-H]^+$	Molecular ion with a positive charge from ionization
MIM	Multiple ion monitoring
min	Minute(s)
MPPH	5-*p*-Methylphenyl-5-phenylhydantoin

MS	Mass spectrometry(er)
MSD	Mass-selective detector
MSTFA	*N*-Methyl-*N*-trimethylsilyl-trifluoroacetamide
MTBSTFA	*N*-Methyl-*N*-(*tert*-butyldimethylsilyl)trifuoroacetamide
MW	Relative molecular mass
m/z	Mass/charge
N-7-G	*N*-[2-(Hydroxyethyl)-*N*-(2-(7-guaninyl)ethyl]methylamine
NAPA	*N*-Acetylprocainamide
*n*BMPA	*n*-Butylmethylphosphonic acid
NICI	Negative ion chemical ionization
NIDA	National Institute on Drug Abuse (USA)
NKS	Nerve agents
NIST	National Institute of Standards and Technology (USA)
NMD	*N*-Methyldiethanolamine
NPD	Nitrogen–phosphorus detector
OPC	Organophosphorus compound
OPCW	Organization for the Prohibition of Chemical Weapons
PAB	Phenylarsine–BAL complex
PCI	Positive ion chemical ionization
PCP	Phencyclidine
PBS	Phosphate buffer system
PD-10	Solid phase (trademark)
PEMA	Phenylethylmalonamide
PepMap	Separation column for peptides (trademark)
PFB	Perfluorobutyrate
PFPITC	Pentafluorophenyl-iosothiocyanate
PhPA	Phenylphosphonic acid
PIC	Poison Information Center
PIC	Prior informed consent
PMPA	Pinacolylmethylphosphonic acid
POCT	Point-of-care testing
POP	Persistent organic pollutant
PP	Polypropylene
PR_{18}	Separation column (trademark)
PTFE	Teflon (polytetrafluoroethylene)
PTT	Activated partial thromboplastin time
Q-TOF	Quadrupole/time-of-flight mass spectrometer
RAID-M	Rapid alarm and identification device – military
RBC	Erythrocyte (concentrate)
RCA	Riot control agents
ρ (rho)	Density
RI	Retention index
RIA	Radioimmunoassay
RP	Reversed phase
rpm	Revolutions per minute

RPR-1	Ion exchanger (trademark)
RR	Relative response
RRT	Relative retention time
RT	Retention time
RVC	Evaporation centrifuge (trademark)
S	Serum
SAMHSA	Substance Abuse and Mental Health Services Administration (USA)
SBMTE	1,1′-Sulfonyl-bis-[2-(methylthio)ethane]
SepPak	Solid phase (trademark)
SIM	Selected ion monitoring
SOP	Standard operating procedure
SPE	Solid-phase extraction
SPME	Solid phase microextraction
STA	Systematic toxicological analysis
t_o	Dead time
TBDMS	*tert*-Butyldimethylsilane/-silyl ester
tBDMSC	*tert*-Butyldimethylsilyl chloride
TDG	Thiodiglycol
TDG–PFB	Thiodiglycol–pentafluorobenzylate
TDM	Therapeutic drug monitoring
TEA	Triethanolamine
TFE	Trifluoroacetic acid
THC	Δ^9-Tetrahydrocannabinol
TIC	Total ion chromatogram (total ion current)
TLC	Thin-layer chromatography
TMCS	Trimethylchlorosilane/-silyl ester
TMS	Trimethylsilane/-silyl ester
TSQ	LC–MS instrument (trademark)
U	Urine
UF-2	Filter (trademark)
UKNEQAS	United Kingdom National External Quality Assessment Scheme
USAN	United States Adopted Name
UV	Ultraviolet (light)
V	Volt
v	Volume
V-agents	Collective term for VX, VR, CVX
VHHC	Volatile halogenated hydrocarbons
VIM	International Vocabulary of Metrology
VIS	Visible (light)
VR	Russian VX
v/v	Volume ratio
WHO	World Health Organization
Wz	Trademark

Appendix B

Therapeutic, toxic, and comatose-fatal blood plasma/serum concentrations of drugs and xenobiotics and their half-lives in plasma [1, 2].

Substance	Plasma/serum concentration (mg/l)			Elimination half-life (h)
	Therapeutic	Toxic (from)	Comatose-fatal (from)	
Acetaminophen	10–20 (2.5–25)	100–150[f] 75–100[g]	160	2–4
Acetyldigoxin[a]	0.0005–0.0008	0.0025–0.003	0.005	40–70
Acetylsalicylic acid[b]	20–300[c]	400–500 Children 300	500–900	0.3
Ajmaline	0.2–1		5.5[d]	About 5–6
Alfentanil	0.1–0.4[e]			0.6–2.3
Allopurinol	1–5			0.5–3
Alprazolam	0.005–0.05 (−0.1)	0.1–0.4		6–20
Amikacin	15–25 (−30)[f]	30[f]		2–3
5-Aminosalicylic acid	0.1–1.8			0.5–2.4
Amiodarone	1.0–2.5 0.5–2.0[g]	3.0		720–2880
N-Desethylamiodarone[#]	1.0–2.5			1368–1536
Amitriptyline	0.5–0.3	0.5–0.6	1.5–2	30–50
plus nortriptyline[#]	0.1–0.2	0.5	1.5–2.0	31–45
Amlodipine	0.006–0.018	0.088	0.1–0.2	34–50
Amobarbital	2–12	9	13–96	15–30
Amphetamine	0.05–0.15	0.2	0.5–1.0	4–8
Amphotericin B	1.5–3.5[f] 0.025–1.0[g]	(3–) 5–10		24–48[i]
Ampicillin	2–20[f] 0.02–1.0[g]			1
Amrinone	1–2 (−4)			3–12
Antipyrine	5–25	50–100		10–12
Aprobarbital	10–40	40	50	14–34
Aripiprazole	0.12–0.45	1.87[d]		75
Atenolol	0.1–0.6 (−1.0)	2	27[d]	4–14

Clinical Toxicological Analysis: Procedures, Results, Interpretation. Edited by Wolf-Rüdiger Külpmann

ISBN: 978-3-527-31890-2

Substance	Plasma/serum concentration (mg/l)			
	Therapeutic	Toxic (from)	Comatose-fatal (from)	Elimination half-life (h)
Atropine	0.002–0.025	0.03–0.1	0.2	2–6.5; 13–38
Azathioprine	0.05–0.3[j]			0.2
6-Mercaptopurine[#]	0.04–0.3	1–2		1–1.5
Baclofen	0.08–0.6	1.1–3.5	6–9.6	6.1–7.5
Bambuterol	0.001–0.006 (−0.01)		0.04	16–20
Barbital	5–30	20	90	57–120
Benzbromarone	2–10			2–4
Benzoylecgonine	See cocaine			
Betaxolol	0.005–0.05		36[d]	14–22
Biperiden	0.05–0.1		0.25[d]	18–24
Bisoprolol	0.01–0.06			10–12
Brallobarbital	4–8	8–10	15	20–40
Bromazepam	0.08–0.17	0.3–0.4	(1–) 2	8–22
Bromide	10–50	500–1000 (−1500)	2000	288–864
Bromisoval	10–20	30–40		Approximately 4
Brotizolam	0.001–0.02		10[d]	4–10
Bunitrolol	0.001–0.015			2–6
Bupivacaine	1–4[f]; 0.25–0.75[k]	1.5[k]; 4–5		0.5–3
Buprenorphine	0.0005–0.005	0.2	1.1[d]; 4–13	3–5
Butabarbital	See secbutabarbital			
Butalbital	1–10	10–15	15–30	20–30
Butriptyline	0.07–0.15	0.4–0.5		
Caffeine	8–20	30–50	80–100	2–10
Camazepam	0.1–0.6	2		20–24
Camphor		0.3–0.4	1.7	
Canrenone	See spironolactone			
Captopril	0.05–0.5 (−1.0)	5–6	60	1–2
Carazolol	−0.015			9
Carbamazepine	4–9 (−12)	12–15	25	12–60; 7–35[l]
Carbamazepine-10, 11-epoxide	0.5–6	15		
Carbenoxolone	About 5–30	50		8–20
Carboplatin	Maximum 10–25			2.5–6
Carbromal	2–10	15–20	40	7–15
Carteolol	0.01–0.1			3–7
Chloral hydrate as trichloroethanol[#]	(1.5) 5–15	40	60	8–30
Chloramphenicol[c]	5–15 10–20 (−25)[f] 5–10[g]	25 10[g]		2–6
Chlordiazepoxide	0.4–4.0	3–10	20	6–24
Chloroquine	0.02–0.2	0.6–1	3	72–1440[m]
Chlorpheniramine	0.01–0.017		1.1[d]	15–25
Chlorpromazine	0.03–0.5	0.5–2	2	10–30

Substance	Plasma/serum concentration (mg/l)			
	Therapeutic	Toxic (from)	Comatose-fatal (from)	Elimination half-life (h)
Chlorprothixene	0.03–0.3	0.4–0.7	0.8	10–30
Chlorthalidone	0.14–1.4	Approximately 2		44–48 (35–70)
Ciclosporin	0.1–0.4 (EDTA–blood)	0.35–0.4		10–27
Cimetidine	0.5–1.5	1.3[g]	110[d]	1.5–4
Citalopram	0.02–0.2		0.5	Approximately 33
Clenbuterol	0.0003–0.0006	0.003[d]		34–35
Clindamycin	1.5–9.5[f] 0.05–2.0[g]			2–3
Clobazam	0.1–0.4			10–32
Clofibrate	50–250			10–18
Clomethiazole	0.1–2.8	(1.6) 13–26	(8–) 50–70	3–7
Clomipramine	(0.02–) 0.09–0.25	0.4–0.6	1–2	20–26
Clonazepam	0.02–0.07	0.1		20–60
Clorazepate	See nordiazepam[#]			
Clonidine	0.0002–0.002	0.025–0.06	0.23[d]	8–25
Clotiazepam	0.1–0.7			3–15
Cloxacillin	5–30; 85[f]			0.5–1
Clozapine	0.1–0.6 (–0.8)	0.8–1.3	3	6–14
Norclozapine[#]	0.1–0.6			
Cocaine	0.05–0.3	0.25–5	1–20	0.5–1
Benzoylecgonine				5–6
Codeine	0.05–0.25[f] 0.01–0.05	0.3–1.0	1.6	3–4
Colchicine	0.0003–0.0024	0.005	0.024[d]	11–32
Cyanide	0.001–0.006 0.005–0.012[γ]	0.5	(1–) 4–5	19[d]
Cyclobarbital	2–10	10–15	20	8–17
Cyclophosphamide	10–25			4–8
Dantrolene	1.0–3.0[f] 0.3–1.4[g]			4–12
Dapsone	0.5–5	10–20	18[d]	25–31
Deferoxamine	3–15			4–6
Demoxepam	0.5–0.74	1	2.7	
Desipramine	0.01–0.5	0.5–1.0	3	15–25
Desmethyldiazepam	See nordiazepam[#]			
Dextromethorphan	0.01–0.04	0.1	3	2–4
Dextropropoxyphene	0.05–0.75	1	2	10–30
Norpropoxyphene[#]	0.1–0.15	2		
Diazepam		1.5	5	24–48
anxiolytic	0.125–0.25			
antiepileptic	0.25–0.5			
Diazoxide	10–50	50–100		20–36 (–48)
Dibenzepin	0.1–0.5[f] 0.025–0.15[g]	3	18	3.5–5
Diclofenac	0.1–2.2[f] 0.05–0.5[g]	50; 60[d]		1–2

Substance	Plasma/serum concentration (mg/l)			Elimination half-life (h)
	Therapeutic	Toxic (from)	Comatose-fatal (from)	
Dicoumarol	8–30 (−50)	50–70		24–96
Diethylpentenamide	2–10	20	45	6–7
Digitoxin	0.01–0.03	0.03	0.04–0.1	140–200
Digoxin	0.0005–0.001[n]	0.0025–0.007[n]	0.01–0.03[n]	40–70
Dihydrocodeine	0.03–0.25	0.5–1	2	3–4
Dihydroergotamine	0.001–0.01			7–9
Diltiazem	0.05–0.4	0.8	2–6	2–6
Diphenhydramine	0.1–1.0	1.0	5	4–10; 20–60
Dipyrone	10[p]	20[p]		6–8
Disopyramide	2–7	8		5–8
Disulfiram	0.05–0.4	0.5–5	8	Approximately 5–7
Domperidone	0.005–0.025 (−0.04)			12–16
Dosulepin	0.02–0.15 (−0.4)	0.8	(1–) 5–19	11–40
Dothiepin	See dosulepin			
Doxepin	0.02–0.15	0.1	1–18	8–25
plus nordoxepin[#]	0.2–0.35	0.5–1.0	2–4	
Doxylamine	0.05–0.2	1–2	5	9–11
Dronabinol	0.01–0.2			100
Enalapril as enalaprilat	0.01–0.05 (−0.1)			8–11
Enoxacin	1–4			3–6
Ephedrine	0.02–0.2	1	5[d]	3–11
Ethanol		1000–2000	3500–4000	0.15 ± 0.05 g/(kg × h)[z]
Ethinamate	5–10	50–100	100	Approximately 2
Ethosuximide	40–100	(100–) 150–200	250	30–60
Etomidate	0.1–0.5 (−1)			2.8–5.0
Famotidine	0.02–0.06 (−0.2)	0.42[d]		2–4.5
Felbamate	50–100 (−110)	120–200		20–24
Felodipine	0.001–0.008 (−0.012)	0.01–0.015		22–27
Fenfluramine	0.05–0.15	0.5–0.7	6	1–2; 18–25
Fentanyl	0.001–0.002 0.003–0.3[e]	0.002–0.02	0.003–0.02[d,o]	1–3.5
Fexofenadine	0.3–0.6			14–18
FK-506	See tacrolimus			
Flecainide	0.75–1.25[f] 0.45–0.90[g]	1.5–3	2.6[d]; 13[d]	10–20
Flucloxacillin	3–30			0.7–1.5
Flucytosine	50–100[f] 25–50[g]	100		3–5
Flumazenil	0.01–0.05 0.2–0.3[f]	0.5		1–2
Flunitrazepam	0.005–0.015	0.05		10–20
5-Fluorouracil	0.05–0.3	0.4–0.6		<0.5
Fluoxetine	Approximately 0.1–0.5	2	1.3–6.8	48–96
Norfluoxetine[#]	0.1–0.5	0.4	0.9–5.0	168–216
Fluphenazine	(0.0002–) 0.001–0.017	0.05–0.1		10–18

Substance	Plasma/serum concentration (mg/l) Therapeutic	Toxic (from)	Comatose-fatal (from)	Elimination half-life (h)
Flurazepam	0.0005–0.03 (–0.1)	0.15–0.20	0.5–17	Approximately 2
Desalkylflurazepam#	0.04–0.15	0.2–0.5 (plus flurazepam)	50–98	
Fluvoxamine	0.05–0.25	0.65	2.8[d]	15–22
Furosemide	2–5 (–10)	25–30		1–3
Gabapentin	5.9–21	85[d]		5–8
Gallopamil	0.02–0.1		8[d]	3–8
Ganciclovir	5–12.5[f]	20[f]		2–4
	0.2–1.0[g]		3–5[g]	
Gentamicin	4–15[f]	2[g]		1.5–6
	0.05–2[g]			
Glibenclamide	0.03–0.35	0.6		10
Glutethimide	2–12	12–20	30	5–20
Glyceryl trinitrate	Approximately 0.015			0.3–0.5
Griseofulvin	0.3–1.3 (–2.5)			22
Halazepam	See nordiazepam#			
Haloperidol	0.005–0.015 (–0.04)	0.05–0.1	0.5	10–35
Heptabarbital	1–4	8–15	20	6–11
Hexobarbital	1–5	(8–) 10–20	50	4–6
Hydralazine	(0.05–) 0.2–0.9			2–6
Hydrochlorothiazide	0.07–0.45			10–12
Hydrocodone	0.002–0.024 (–0.05)	0.1	0.1 (0.2)	Approximately 4
Hydromorphone	0.008–0.032	0.1	>0.1	2–3
γ-Hydroxybutyrate	50–120[e]	80[o]	250–280[o]	0.3–0.5 (–1.0)
Ibuprofen	15–30 (5–50)	100		2–3
Imipramine	0.045–0.15	0.4–0.5	2	6–20
plus desipramine#	0.15–0.3	0.5	0.8	
Indometacin	0.5–3	4–6		3–11
INH	See isoniazid			
Isoniazid	3–10[f]	20	(30–) 100	1–3
	0.2–1.0[g]			
Itraconazole	>0.25[g]			24–36
Ketamine	0.5–6.5	7[o]	7[o]	1–3
Ketanserin	0.015–0.2			10–22
	0.08–1.0[f]			
Ketazolam	0.001–0.02			1–3
Ketobemidone	0.025–0.030		0.6	2–2.5
Ketoconazole	3–10 (–20)[f]			6–10
	0.3–0.5[g]			
Ketoprofen	1–5		1100[d]	1.5–2 (–4)
	5–15 (–20)[f]			
Labetalol	0.025–0.2	0.5–1.0		3–10
Lamotrigine	11–19	25–30	50[d]	24–36
Levodopa	0.3–1.6		650[d]	1–3
Levomepromazine	0.02–0.15	0.4	0.5	15–30
Levomethadone	0.04–0.3	1.0	0.2	10–40

Substance	Plasma/serum concentration (mg/l)			Elimination half-life (h)
	Therapeutic	Toxic (from)	Comatose-fatal (from)	
Levorphanol	0.007–0.02	0.1	0.8[d] (2.7[d])	11–30
Lidocaine	(1.0–) 1.5–5.0	6–10	10–25	1–4
Lisinopril	(0.005–) 0.02–0.07	0.5		12
Lithium	0.6–0.8 mmol/l	1.5 (−2) mmol/l		8–50
Lofepramine	0.003–0.03			10–20
Loprazolam	0.03–0.01			11–20
Lorazepam	0.02–0.25	0.3–0.6		10–40
Lormetazepam	0.001–0.02			10–15
LSD	0.0005–0.005	0.001	0.002–0.005	Approximately 2–5
Lysergide	See LSD			
Maprotiline	0.075–0.25	0.3–0.8	1–5	20–60
plus desmethylmaprotiline[#]	0.1–0.4	0.75–1.0		
MDA	See methylenedioxy-amphetamine			
MDMA	See Methylenedioxy-methylamphetamine			
Mebendazole	>0.1	0.5		2.8–9
Medazepam	0.01–0.15 0.1–0.5[f]	0.6 (−1)		2–5
Melitracen	0.01–0.1			12–23
Melperone	<0.2		17.1[d]	4–8
Meperidine	0.1–0.8	(1–) 2	5	3–6 (−10)
Mepivacaine	2–5.5	6–10	50	1–3
Meprobamate	5–25	50	70	6–17
6-Mercaptopurin	0.03–0.08	1–2		
Mesalazine	0.1–1.8			0.5–2.4
Mesuximide	0.04–0.08			1–2
N-Normesuximide[#]	10–30 (−40)	40–50		20–40
Metamizole	10[p]	20[p]		6–8
Metformin	1–4	5–10	64[d]; 85[d]; 91[d]; 166[d]	2–4 (−10)
D,L-Methadone	0.05–0.5 (−1) 0.2–0.75[q]	0.2[r] 0.75[q]	0.2–1.0[s]	23–25 (13–55)
Methamphetamine	0.01–0.05	0.2–1	10–40	6–9
Methaqualone	0.4–5.0	2	8	10–40
Methimazole	0.5–2 (−3)			2–28
Methohexital	(0.5–) 1–6			1–3
Methotrimeprazine	0.02–0.15	0.4	0.5	15–30
Methyldopa	1–5	7–10		1.5–3
Methylenedioxy-amphetamine	−0.4	1 (−1.5)	2	
Methylenedioxy-methylamphetamine	0.1–0.35	0.35–0.5	0.4–0.8	9–10
Methylphenidate	0.005–0.06	(0.5–) 0.8	2.3	2–7
Methyprylone	10–20	12–75 (−128)	50 (−100)	3–6; 9–11

Substance	Plasma/serum concentration (mg/l) Therapeutic	Toxic (from)	Comatose-fatal (from)	Elimination half-life (h)
Metildigoxin as digoxin	0.0005–0.0008	0.0025–0.003	0.005	40–70
Metipranolol as desacetylmetipranolol	0.02–0.08			2–3.5
Metoclopramide	0.04–0.15	0.1–0.2	4.4[d]	3–6
Metoprolol	0.1–0.6 0.02–0.34[g]	(0.65[d]–) 1.0	(4.7[d]) 12–18	3–6
Metronidazole	(3–) 10–30	150 (200[d])		6–10 (−14)
Mexiletine	(0.5–)0.7–2	2.5	35[d]	5–26
Mianserin	0.015–0.07 (−0.14)	0.5–5		8–19
Midazolam	0.08–0.25	1–1.5		1.5–3; 8–22[h]
Mirtazapine	0.02–0.1 (−0.3)	1–2		20–40
Moclobemide	1.5–4.0[f] 0.4–1[g]	5–8	16	1–3
Modafimil	Approximately 2–3[g]			12–15
Molsidomine	0.001–0.02 (−0.2)			1–2.5
Morphine	0.01–0.12	0.15–0.5	0.05–4	1–4
Naloxone	0.01–0.03			1–2
Neostigmine	0.001–0.01			0.4–1.3
Netilmicin	7–15 (−18)[f] 0.5–2 (−3)[g]	4[g]		2–3 (31–43 terminal)
Nicotine	0.005–0.02 (−0.03)	0.4 (−1.0)	5; 13.6[d]	1–4
plus cotinine[#]	0.025–0.35[f]	0.3–1	5	16–20
Nicotinic acid	3–19			0.3–1
Nifedipine	0.02–0.1 (−0.15)	0.15–0.2	5.4[d]	2–5
Niflumic acid	2–35			2–3
Nitrazepam	0.03–0.12	0.2–0.5		20–30
Nitrofurantoin	0.5–2 (−3)	3–4		0.7–1.3
Nitroglycerin	Approximately −0.015			0.3–0.5
Nitroprusside	See thiocyanate			
Nomifensine (unbound)	0.2–0.6[f] 0.02–0.06[g]	0.8–1		2–5
Nordazepam	See nordiazepam			
Nordiazepam	0.2–0.8 (−1.8)	1.5–2.0		40–80
Nortriptyline	(0.05–) 0.075–0.25	0.25	1–3	18–56
Olanzapine	0.01–0.05 (−0.1)	0.2	1[d]	33 (21–54)
Opipramol	0.05–0.2 (−0.5)	0.5–2 (−3)	7–10	6–12
Oxazepam	(0.15–) 0.5–2.0	2	3–5	6–20
Oxcarbazepine				
10-Hydroxycarbazepine[#]	12–30 (−40)	45		1–2.5
Oxprenolol	0.05–0.3 (−1.0)	2–3	10	1–4
Oxycodone	(0.005–) 0.02–0.05	0.2	0.6	2–5
Pancuronium	0.1–0.6	0.4[d]	1.6	1.5–2.5
Papaverine	0.2–0.6 (−2)			1–2; 6–7
Paracetamol	10–20 (2.5–25) 75–100[g]	100–150[f]	160	2–4

Substance	Plasma/serum concentration (mg/l)			Elimination half-life (h)
	Therapeutic	Toxic (from)	Comatose-fatal (from)	
Paraldehyde	30–100[t] (blood) 100–300[u] blood	200–400 (blood)	500 (blood)	4–10
Paraoxon		0.005		
Parathion		0.01–0.05	0.05–0.08	
Paroxetine	0.01–0.075 0.015–0.15 (−0.25[f])	0.35–0.4		16–24
PCP	See phencyclidine			
Pentazocine	0.01–0.2 (−0.5)	1–2	3	2–5
Pentobarbital	1–10 (25–40)	(5–) 8–10	(8–) 15–25	20–40
Pentoxifylline	Approximately 0.5–2			0.5–2 (4–6)
Perazine	0.02–0.35	0.5		8–16 (−35)
Pethidine	0.1–0.8	(1–) 2	5	3–6 (−10)
Norpethidine[#]	0.3	0.5		
Phenacetin	5–20	50		Approximately 1
Phenazone	5–25	50–100		10–12
Phencyclidine		0.007–0.24	(0.3–) 1–5	1–12
Phenformin	0.03–0.1 (−0.3)	0.6	3	4–13
Pheniramine	0.01–0.27		1.9[d]	16–19
Phenobarbital	2–30 (−40)	30–40	45–120	60–130
Phenprocoumon	1–3	5		100–160
Phentermine	0.03–0.1		7.6[d]	Approximately 20
Phenylbutazone	50–100	120–200	400–500	30–175[m]
Phenylephrine	0.03–0.1 (−0.3)			2–3
Phenytoin	10–20	20–40	70	10–60[m]
Physostigmine	<0.001–0.005			0.4–1.0
Pindolol	0.02–0.08 (−0.15)	0.7		2–5
Pipamperone	0.1–0.4	0.5–0.6		<4
Piperazine	0.02–0.1	0.5		
Pirenzepine	0.03–0.45			8–20
Piroxicam	5–10 (−20)	14[d]		30–70
Prajmalium	0.06–0.5			5–7
Prazepam	0.01–0.04	1		1–3
Prazosin	0.001–0.075	0.9		2.1–3.7
Prednisolone	0.5–1.0			2–6
Primaquine	Approximately 0.1–0.2			4–7
Primidone	5–12[f]	10 (15–20)	50	4–12; 9–22
Probenecid	40–60; 100–200			3–17[m]
Procainamide	4–10	10–15	20	2–5
N-Acetylprocainamide[#]	2–12	40[v]		3–7
Procaine	2.5–10	15–20	20	−0.5
Promethazine	(0.05–) 0.1–0.4	1	2.4[d]	8–15 (−20)
Propafenone	0.4–1.1 (−1.6)	1.1–3.0	7.7[d]	5–8; 2–32
plus norpropafenone[#]		2–3		
Propallylonal	0.3–10	>10		Approximately 3

Substance	Plasma/serum concentration (mg/l)			
	Therapeutic	Toxic (from)	Comatose-fatal (from)	Elimination half-life (h)
Propofol	2–4 (−8)[w] blood			3–8 (0.5–1 β-phase)
Propoxyphene	See dextropro-poxyphene			
Propranolol	0.1–0.3[f] 0.05–0.15[g]	1–2	4–10	2–6
Propyphenazone	3–12			1–1.5 (−3)
Prothipendyl	0.05–0.2	0.5 (−1)		2–3
Protriptyline	0.07–0.17 (−0.38)	0.5	1	50–200
Pseudoephedrine	0.5–0.8		19	9–16
Pyridostigmine	0.05–0.1 (−0.2)			1–2.5
Pyrithyldione	1–10			11–20
Quetiapine	<1[za]	1.8[d]	12.7[d]	5–7
Quinidine	(1–) 2–6	6–10	30	4–12
Ranitidine	(0.05–) 0.15–0.5			2–4
Risperidone	0.01–0.09		1.8[d]	2–4
plus 9-hydroxyris-peridone#	0.01–0.1	0.08		22–24
Salicylic acid		400–500	500–900	3–20
for rheumatism therapy	200–300			
as anticoagulant	50–125			
as prostaglandin synthetase inhibitor	50–150			
Scopolamine	0.0001–0.0003 (−0.001)			Approximately 3
Secbutabarbital	5–15	10	30	34–42
Secobarbital	2–10	8	(4–) 10–50	15–30
Sertraline	0.05–0.25 (−0.5)	0.29[d]	1.6[d]; 3[d]	24–28
Sotalol	0.5–3 (−5)	5–10	40[d]; 43[d]	5–13 (−17)
Spironolactone	0.1–0.5			13–24
Canrenone#	0.05–0.25 (−0.5)			
Streptomycin	15–40[f] 1–5[g]	40–50[f] 5[g]		2–4
Strychnine		0.075–0.1	0.2–2	Approximately 10–15
Sufentanil	0.0005–0.005 0.01–0.02[f]		0.001–0.007[d]	2–5
Sulindac	0.5–5			Approximately 7
plus sulindac sulfide#	1–5 (+ sulfone)			15–18
Sulpiride	0.04–0.6 0.15–0.75[f]		3.8[d]	4–7
Sultiame	0.5–12.5	12–15	20–25	3–30
Suramin	150–250	300		44–54
Tacrolimus	0.01–0.025[f]	0.012–0.015		9–16

Substance	Plasma/serum concentration (mg/l)			Elimination half-life (h)
	Therapeutic	Toxic (from)	Comatose-fatal (from)	
	0.003–0.015[g]			
Talinolol	0.04–0.15		5[d]; 20[d]	10–14
Teicoplanin	5–40[g]	200		10–15
Temazepam	0.3–0.9	1	8.2[d]	6–25
	0.02–0.15[g]			
Tenoxicam	5–10			(50–) 70–90
Terbutaline	0.001–0.006 (−0.01)		0.04	16–20
Terfenadine	0.0015–0.0045	0.06	0.4[d]	15–22
Tetracycline	5–10	30		6–10
	1–5[g]			
Δ⁹-Tetrahydro-cannabinol	See Dronabinol			
Tetrazepam	0.05–0.6 (−1)			10–26
Thalidomide	0.5–1.5 (−8)			5–9
THC	See Dronabinol			
Theobromine	10–15	20		6–10
Theophylline	8–20	25–30	50–250	6–9 (3–6[y])
Thiocyanate	1–4[x]	35–50	200	72–96
	3–12[y]			
from nitroprusside	(5–) 6–30	50–100		
Thiopental	1–5	10 (40–50)	10–100	3–8
Pentobarbital[#]	5–10	10–15	15–25	20–40
Thioridazine	0.2–1	2 (−5)	3–10	7–13 (−36)
Mesoridazine[#]	0.2–1.6			10–14
Sulforidazine[#]	<0.6			10–16
plus sulforidazine[#]	0.75–1.5	3		
Tiapride	1–2[f]			Approximately 3–4
Tilidine	0.05–0.12		1.7[d]	Approximately 3
Nortilidine[#]	0.2		4.4[d]	
Timolol	0.005–0.05			2–6
	0.02–0.1[f]			
Tobramycin	8–15[f]	2[g,l]		2–3
	0.5–1.5[g]			
Tocainide	4–12	13–15 (20[d])	74[d]	8–25
Tolbutamide	45–100	400–500	640[d]	4–12
Tramadol	0.1–0.8 (−1.0) (blood)	1 (blood)	2 (blood)	5–10
Triamterene	0.05–0.2[f]			1.5–4
	0.01–0.1[g]			
Triazolam	0.002–0.02	0.04		2–5
Trifluoperazine	(0.001–) 0.005–0.05	0.1–0.2		7–18
Triflupromazine	0.03–0.1	0.3–0.5		Approximately 6
Trimipramine	0.01–0.3	0.5	8.7[d]	10–20 (−40)
Valproic acid	40–100 (−150)	150–200	720[d]	10–20 (7–17)
Vancomycin	20–40[f]	(30–) 50		4–11

Substance	Plasma/serum concentration (mg/l) Therapeutic	Toxic (from)	Comatose-fatal (from)	Elimination half-life (h)
	8–15[g]			
Venlafaxine				3–5
plus *O*-desmethyl-venlafaxine[#]	0.25–0.75	1–1.5	6.6[d]	10–11
Verapamil	0.02–0.35	0.9	2.5–4	6–14
plus norverapamil[#]	0.1–0.6	1		
Vigabatrin	5–15[g]			5–8
Vinylbital	1–4	5	8	18–33
Warfarin	5–10[f]	10–12	100	37–50
	0.3–3[g]			
Yohimbine	0.05–0.3			1–3
Zidovudine	1–1.5[f]	0.5–3		1–1.5
	0.1–0.3[g]			
Zolpidem	0.08–0.15 (−0.2)	0.5	2–4	2–5
Zopiclone	0.01–0.05	0.05	0.6[d]	3.5–8
	0.04–0.07[f]			
Zotepine	0.01–0.15	0.15–0.2	14–16	

[#]Active metabolite.
[a]As digoxin.
[b]As salicylic acid.
[c]For rheumatism 200–300 mg/l, as anticoagulant 50–125 mg/l, as prostaglandin synthetase inhibitor 50–150 mg/l.
[d]Case report.
[e]During mechanical ventilation.
[f]Peak concentration.
[g]Trough concentration.
[h]Intensive care patients in some cases.
[i]In some cases of intoxication and in children dramatically increased.
[j]+ Dithiothreitol 0.1 mol/l.
[k]Epidural anesthesia.
[l]Chronic administration.
[m]Dose dependent.
[n]Depending on potassium concentration in plasma.
[o]Abuse.
[p]Sum of active metabolites.
[q]Addicts.
[r]Native.
[s]Acute.
[t]Sedative.
[u]Hypnotic.
[v]Procainamide + *N*-acetylprocainamide.
[w]Narcosis.
[x]Nonsmokers.
[y]Smokers.
[z]kg body mass.
[za]Dosage: 250 mg/8 h.

It has to be taken into account that these values will never be static and might change with advancing knowledge or with other (therapeutic) use of the compounds. Using these data is for the user's own liability. Use of drug data without sufficient knowledge about the patient or victim, the case, and about pharmacokinetics, toxicokinetics, and pharmacodynamics might give a wrong interpretation.

References

1 Schulz, M. and Schmoldt, A. (2003) Therapeutic and toxic blood concentrations of more than 800 drugs and other xenobiotics. *Pharmazie*, **58**, 447–474.

2 TIAFT reference blood level list of therapeutic and toxic substances. Update 2005-03-03.

Appendix C

Biological Tolerance (BAT) Values [1]

In the following table, chemicals that are defined as hazardous at the workplace are listed. For these chemicals, maximal concentrations in blood or urine are known, which are considered harmless, even if they are present regularly due to exposure (BAT values). The data help assess the toxicological significance of measurements of noncommon compounds in particular.

Chemical	Compound indicating exposure	BAT value	Specimen
Acetone	Acetone	80 mg/l	U^{b}
Acetylcholinesterase inhibitor	Acetylcholinesterase	30% decrease of the activity before exposure	$E^{b,c}$
Aluminum	Aluminum	200 μg/l	U^{b}
Aniline	Aniline (free)	1 mg/l	$U^{b,c}$
	Aniline (released from aniline hemoglobin conjugate)	100 μg/l	$B^{b,c}$
1-Butanol	1-Butanol	10 mg/g creatinine	U^{b}
	1-Butanol	2 mg/g creatinine	U^{d}
2-Butanone (methyl ethyl ketone)	2-Butanone	5 mg/l	U^{b}
p-tert-Butylphenol	*p-tert*-Butylphenol	2 mg/l	U^{b}
Carbon disulfide	2-Thio-4-thiazolidinecarboxylic acid (TTCA)	4 mg/g creatinine	U^{b}
Carbon monoxide	CO-hemoglobin	5% of total hemoglobin	B^{b}
Carbon tetrachloride	Carbon tetrachloride	3.5 μg/l	$B^{b,c}$
Chlorobenzene	4-Chlorocatechol (total)	175 mg/g creatinine	U^{b}
	4-Chlorocatechol (total)	35 mg/g creatinine	U^{d}
Cyclohexane	1,2-Cyclohexanediol (total)	170 mg/g creatinine	$U^{b,c}$
1,2-Dichlorobenzene	1,2-Dichlorobenzene	0.14 mg/l	B^{b}
	3,4-Dichlorocatechol + 4,5-dichlorocatechol	150 mg/g creatinine	U^{b}
N,N-Dimethylacetamide	*N*-Methylacetamide	30 mg/g creatinine	$U^{b,c}$
Dimethylformamide	*N*-Methylformamide	35 mg/l	U^{b}

Clinical Toxicological Analysis: Procedures, Results, Interpretation. Edited by Wolf-Rüdiger Külpmann

ISBN: 978-3-527-31890-2

Chemical	Compound indicating exposure	BAT value	Specimen
Ethylene glycol dinitrate	Ethylene glycol dinitrate	0.3 µg/l	B[b]
Halothane	Trifluoroacetic acid	2.5 mg/l	B[b,c]
Hexachlorobenzene	Hexachlorobenzene	150 µg/l	P[a]S[a]
n-Hexane	2,5-Hexanedione + 4,5-dihydroxy-2-hexanone	5 mg/l	U[b]
2-Hexanone	2,5-Hexanedione + 4,5-dihydroxy-2-hexanone	5 mg/l	U[b]
Hydrogen fluoride and fluorides	Fluoride	7 mg/g creatinine	U[b]
		4 mg/g creatinine	U[d]
Isopropanol	Acetone	50 mg/l	B[b], U[b]
Lindane (γ-hexachlorocyclohexane)	Lindane	25 µg/l	P[b]S[b]
Mercury and inorganic mercury compounds	Mercury	25 µg/g creatinine	U[a]
Methanol	Methanol	30 mg/l	U[b,c]
4-Methyl-2-pentanone (methyl isobutyl ketone)	4-Methyl-2-pentanone	3.5 mg/l	U[b]
Nitrobenzene	Aniline (released from aniline–hemoglobin conjugate)	100 µg/l	B[c]
Parathion	*p*-Nitrophenol	500 µg/l	U[c]
	Acetylcholinesterase	30% decrease of the activity before exposure	E[c]
Styrene	Mandelic acid + phenylglyoxylic acid	600 mg/g creatinine	U[b,c]
Tetrachloromethane	Tetrachloromethane	3.5 µg/l	B[b,c]
Tetraethyl lead	Diethyl lead	25 µg/l[f]	U[b]
	Lead (total)[e]	50 µg/l	U[b]
Tetrahydrofuran	Tetrahydrofuran	2 mg/l	U[b]
Tetramethyl lead	See Tetraethyl lead		
Toluene	Toluene	1.0 mg/l	B[b]
	o-Cresol	3.0 mg/l	U[b,c]
1,1,1-Trichloroethane	1,1,1-Trichloroethane	550 µg/l	B[c,d]
Xylene (all isomers)	Xylene	1.5 mg/l	B[b]
	Methylhippuric acid	2000 mg/l	U[b]

B: blood; E: erythrocytes; P: plasma; S: serum; U: urine.
[a]Sampling at any time.
[b]Sampling at the end of exposure or at the end of shift.
[c]Sampling in the case of long-term exposure: after several preceding shifts.
[d]Sampling at the beginning of the next shift.
[e]Valid also for mixtures of tetraethyl lead with tetramethyl lead.
[f]Concentration related to lead.

References

1 Deutsche Forschungsgemeinschaft (2007) *List of MAK and BAT Values 2007.* Commission for the Investigation of Health Hazards of Chemical Compounds. Report 43. Wiley-VCH Verlag GmbH, Weinheim.

Appendix D

Toxic Agent/Toxic Symptom and Proposed Antidote[a]

Toxic agent/toxic symptom	Antidote
Acetaminophen	*N*-Acetylcysteine
Acidosis (e.g., methanol, ethylene glycol)	Sodium bicarbonate
Amanita phalloides	Silibinin
Amatoxins	Silibinin
Antidepressants, tricyclic (anticholinergic)	Physostigmine sulfate
Antidepressants, tricyclic sodium bicarbonate	Sodium bicarbonate
Antihistamines	Physostigmine sulfate
Antiparkinsonian drugs	Physostigmine sulfate
Arsenic	Succimer (*meso*-2,3-dimercaptosuccinic acid (DMSA))
Arsenic (acute, chronic p.)	Dimercaptopropanesulfonic acid (DMPS)
Atropa belladonna	Physostigmine sulfate
Benzodiazepines	Flumazenil
Bradydysrhythmia	Atropine sulfate
Brugmansia suavolens	Physostigmine sulfate
Cadmium	Tiopronine
Calcium channel blockers	Glucagon
Carbachol	Atropine sulfate
Carbamates (pesticides)	Atropine sulfate
Carbon monoxide	Oxygen, hyperbaric
Cardiac glycosides	Digoxin Immune Fab
Cesium-137	Ferric(III) hexacyanoferrate(II) (Prussian blue)
Chloroquine	Diazepam
Chromate(VI)	Ascorbic acid
Copper	D-Penicillamine
Copper	Tiopronine
Coumarins (anticoagulants)	Cholestyramine
Coumarins (anticoagulants)	Phyto(me)nadione (vitamin K_1)
Curare	Neostigmine
Cyanide	4-Dimethylaminophenol (4-DMAP)
Cyanide	Hydroxocobolamin
Cyanide	Sodium thiosulfate (following 4-DMAP or hydroxocobalamin)

Clinical Toxicological Analysis: Procedures, Results, Interpretation. Edited by Wolf-Rüdiger Külpmann

ISBN: 978-3-527-31890-2

Toxic agent/toxic symptom	Antidote
Cyanide	Amyl nitrite (US)
Datura stramonium	Physostigmine sulfate
Ethylene glycol	Ethanol
Ethylene glycol	Fomepizole (4-methylpyrazol)
Fluoride	Calcium gluconate
Formic acid	Folinic acid/folic acid
Galerina marginata	Silibinin
Gases, irritant	Beclometasone dipropionate
Gases, irritant	Theophylline
Gases, irritant (bronchospasms)	Adrenaline (inhalation aerosol)
Gases, irritant (bronchospasms)	Salbutamol
Gyromitra esculenta	Pyridoxine hydrochloride (vitamin B_6)
Heparin	Protamine sulfate
Hydrazine	Pyridoxine hydrochloride (vitamin B_6)
Hydrofluoric acid (burns)	Calcium gluconate gel (topical)
Hyoscyamus niger	Physostigmine sulfate
Hyperthermia (from succinylcholine)	Dantrolene
Hyperthermia (from volatile inhalational anesthetics)	Dantrolene
Iron	Deferoxamine
Iron	Tiopronine
Isoniazid	Pyridoxine hydrochloride (vitamin B_6)
Iodine (radioactive)	Potassium iodide
Lead	Disodium calcium edetate
Lead	Succimer (DMSA)
Lead (chronic p.)	DMPS
Lipophilic compounds (skin)	Macrogol
Mercury	Succimer (DMSA)
Mercury	Tiopronine
Mercury (acute, chronic p.)	DMPS
Methanol	Ethanol
Methanol	Folinic acid/folic acid
Methanol	Fomepizole (4-methylpyrazol)
Methemoglobinemia	Ascorbic acid
Methemoglobinemia	Methylene blue
Methotrexate	Folinic acid (leucovorin)
Neostigmine	Atropine sulfate
Nitriles	4-DMAP
Nitrogen mustard	Sodium thiosulfate
Opiates (heroin, morphine, codeine, dihydrocodeine)	Naloxone
Opioids (meperidine, methadone, dextropropoxyphene, pentazocine, tilidine, tramadol)	Naloxone
Organophosphorus compounds (pesticides)	Atropine sulfate
Organophosphorus pesticides	Obidoxime
Oxalate	Calcium gluconate
Paracetamol	*N*-Acetylcysteine
Parkinsonian symptoms (phenothiazines, butyrophenones, metoclopramide)	Biperiden

Toxic agent/toxic symptom	Antidote
Phenobarbital	Sodium bicarbonate
Physostigmine	Atropine sulfate
Plutonium	Diethylenetriaminepentaacetate (DTPA)
Polonium	Tiopronine
Pulmonary edema (from irritants)	Prednisolone
Pyridostigmine	Atropine sulfate
β-Receptor blocking agent	Adrenaline
β-Receptor blocking agent	Glucagon
β-Receptor blocking agent	Atropine sulfate
Rodenticides (anticoagulants)	Phyto(me)nadione (vitamin K_1)
Salicylate	Sodium bicarbonate
Sarin	Atropine sulfate
Sarin	Obidoxime
Soman	Atropine sulfate
Spasmolytics (anticholinergic)	Physostigmine sulfate
Stibine (SbH_3) (acute p.)	DMPS
Sulfur mustard	Sodium thiosulfate
Surfactants (anionic, nonionic)	Simeticon
Tabun	Atropine sulfate
Tabun	Obidoxime
Thallium	Ferric(III) hexacyanoferrate(II) (Prussian blue)
VX	Atropine sulfate
VX	Obidoxime
Warfarin (anticoagulant)	Phyto(me)nadione (vitamin K_1)
Zinc	Tiopronine

[a]Not all antidotes, which can be applied in the treatment of the intoxications, are listed.
p.: poisoning; US: the United States of America.
4-DMAP: 4-Dimethylaminophenol
DMPS: Dimercaptopropanesulfonic acid
DMSA: *meso*-2,3-Dimercaptosuccinic acid (succimer)
DTPA: Diethylenetriaminepentaacetate

Ipecacuanha: inducing vomiting.
Activated charcoal: inhibition of gastrointestinal absorption.
Alkaline urine: enhancing urinary excretion of acidic compounds.
Acidic urine: enhancing urinary excretion of basic compounds.

The respective antidotes have been used for the treatment of poisoning or of toxic symptoms. The physician is responsible for the choice of the appropriate antidote, its dosage, and route of application considering the individual patient and local regulations.

Appendix E

Poison Information Centers

Australia
Westmead NSW
Emergency telephone: 61-2-9845-3111

Austria
Vienna
Emergency telephone: available at www.akh-wien.ac.at/viz/

Belgium
Brussels
Emergency telephone: available at www.poisoncentre.be

Brazil
Porto Alegre
Emergency telephone: 800-780-200

Bulgaria
Sofia
Emergency telephone: (0 03 59) (2) 5 15 33 46

Canada
Calgary
Emergency telephone: (800) 332-1414 (Alberta only) or (403) 670-1414

Vancouver
Emergency telephone: (800) 657-8911 or (604) 682-5050

Croatia
Zagreb
Emergency telephone: (0 03 85) (1) 2 34 83 42 or (0 03 85) 2 34 78 84

Czech Republic
Prague
Emergency telephone: (00 42) (02) 24 91 92 93 or (00 42) (02) 24 91 54 02

Clinical Toxicological Analysis: Procedures, Results, Interpretation. Edited by Wolf-Rüdiger Külpmann

ISBN: 978-3-527-31890-2

Denmark
Copenhagen
Emergency telephone: available at www.giftinformationen.dk

Eire
Dublin
Emergency telephone: (0 03 53) (1) 8 09 25 66 or (0 03 53) (1) 8 37 99 66

Finland
Helsinki
Emergency telephone: (0 03 58) (9) 47 19 77 or (0 03 58) (9) 47 11

France
Marseille
Emergency telephone: (0033) (4) 91 75 25 25

Paris
Emergency telephone: available at www.centres-antipoison.net

Strasbourg
Emergency telephone: (0033) (3) 88 37 37 37

Germany
Berlin
Emergency telephone: available at www.charite.de or www.giftnotruf.de

Bonn
Emergency telephone: available at www.meb.uni-bonn.de/giftzentrale

Erfurt
Emergency telephone: available at www.ggiz-erfurt.de

Freiburg
Emergency telephone: available at www.giftberatung.de

Göttingen
Emergency telephone: available at www.giz-nord.de

Homburg/Saar
Emergency telephone: available at www.med-rz.uni-sb.de/med_fak/kinderklinik/Vergiftungszentrale/body-vergiftungszentrale.html

Mainz
Emergency telephone: available at www.giftinfo.uni-mainz.de

München (Munich)
Emergency telephone: available at www.toxinfo.org

Nürnberg (Nuremberg)
Emergency telephone: available at www.giftinformation.de

Greece
Athens
Emergency telephone: available at www.uoa.gr/health/poisonic

Hungary
Budapest
Emergency telephone: available at www.antsz.hu/okk/okbi/

Israel
Haifa
Emergency telephone: (0 09 72) (4) 8 54 19 00

Italy
Milano
Emergency telephone: available at www.ospedale-niguarda.it

Roma
Emergency telephone: available at www.uniroma1.it/cav or www.toxit.it

Lithuania
Vilnius
Emergency telephone: (0 03 70) (2) 36 20 92

The Netherlands
Bilthoven
Emergency telephone: (00 31) (30) 2 74 88 88

New Zealand
Dunedin
Emergency telephone: 0 800 764 766

Norway
Oslo
Emergency telephone: see Internet www.shdir.no/giftinfo

Poland
Warsaw
Emergency telephone: (00 48) (22) 6 19 08 97 or (0048) (22) 6 19 66 54

Portugal
Lisbon
Emergency telephone: available at www.inem.pt

Romania
Emergency telephone: (00 40) (1) 6 34 38 90

Russia
Moscow
Emergency telephone: (007) (95) 9 21 68 85 or (007) (95) 9 28 16 87

Spain
Madrid
Emergency telephone: available at www.mju.es/toxicologia/intframe.html

Sweden
Stockholm
Emergency telephone: available at www.giftinformation.se

Switzerland
Zürich (Zurich)
Emergency telephone: available at www.toxi.ch

Slovakia
Bratislava
Emergency telephone: (0042) (12) 54 77 41 66

Slovenia
Ljubljana
Emergency telephone: (0 03 86) (1) 2 30 24 57

Turkey
Ankara
Emergency telephone: (0090) (312) 4 33 70 01

United Kingdom
Birmingham
Emergency telephone: available at www.npis.org

Edinburgh
Emergency telephone: available at www.show.scot.nhs.uk/spib/ or www.spib.axl.co.uk

London
Emergency telephone: available at www.medtox.org

Newcastle upon Tyne
Emergency telephone: available at www.nyrdtc.nhs.uk

United States (USA)
Albuquerque, NM
Emergency telephone: (800) 432-6866 or (505) 272-2222

Atlanta, GA
Emergency telephone: (800) 222-1222 or (404) 616-9000

Boston, MA
Emergency telephone: (800) 682-9211(MA and RI only) or (617) 232-2120

Chicago, IL
Emergency telephone: (800) 222-1222 or (312) 906-6186

Cleveland, OH
Emergency telephone: (888) 231-4455 (OH only) or (216) 231-4455

Dallas, TX
Emergency telephone: (800) 222-1222
Detroit, MI
Emergency telephone: (800) 764-7661 (MI only) or (313) 745-5711

Indianapolis, IN
Emergency telephone: (800) 382-9097 (IN only) or (317) 962-2323

Memphis, TN
Emergency telephone: (800) 288-9999 (TN only) or (901) 528-6048

Miami, FL
Emergency telephone: (800) 282-3171 (FL only) or (305) 585-5253 or (305) 585-8417

Milwaukee, WI
Emergency telephone: (800) 815-8855 (WI only) or (414) 266-2222

Minneapolis, MN
Emergency telephone: (800) 222-1222 or (612) 347-3141

New York, NY
Emergency telephone: (800) 210-3985, (212) 340-4494, (212) POI-SONS or (212) VEN-ENOS

Oklahoma, OK
Emergency telephone: (800) 764-7661 (OK only) or (405) 271-5454

Philadelphia, PA
Emergency telephone: (800) 722-7112 or (215) 590-2100

Pittsburgh, PA
Emergency telephone: (800) 222-1222 or (412) 681-6669

San Francisco, CA
Emergency telephone: (800) 876-4766 (CA only)

Seattle, WA
Emergency telephone: (800) 732-6985 (WA only) or (206) 526-2121

Washington, DC
Emergency telephone: (800) 222-1222 or (202) 625-3333

Further information on European Poison Information Centers: Rote Liste
Update every year
Editor and publisher: Rote Liste Service GmbH, Frankfurt/Main

Appendix F

List of Narcotic Drugs

(According to German law valid from 2008-03-01)
Narcotic drugs: non-negotiable and not permitted to prescribe.[a]

INN	Non registered or trivial name	Abbreviation	IUPAC
Acetorphine			
	Acetyldihydrocodeine		
Acetylmethadol			
	Acetyl-α-methylfentanyl		
			4-Allyloxy-3, 5-dimethoxyphenethylazane
Allylprodine			
Alphacetylmethadol			
Alphameprodine			
Alphamethadol			
Alphaprodine			
Anileridine			
	1-(3,4-Methylenedioxyphenyl)-2-butanamine	BDB	1-(1,3-Benzodioxol-5-yl) butan-2-ylazane
Benzethidine			
Benzfetamine	Benzphetamine		
			1-(1,3-Benzodioxol-5-yl)-2-(pyrrolidine-1-yl)propane-1-on
	Benzylfentanyl		
	Benzylmorphine		
Betacetylmethadol			
Betameprodine			
Betamethadol			
Betaprodine			
Bezitramide			
Brolamfetamine	2,5-Dimethoxy-4-bromoamfetamine	DOB	

Clinical Toxicological Analysis: Procedures, Results, Interpretation. Edited by Wolf-Rüdiger Külpmann

ISBN: 978-3-527-31890-2

INN	Non registered or trivial name	Abbreviation	IUPAC
	4-Bromo-2,5-dimethoxyphenylethylamine	BDMPEA, 2CB	
	Cannabis (1)		
Carfentanil			
Cathinone			
		2CI	4-Iodo-2,5-dimethoxyphenethylazane
		6-Cl-MDMA	[1-(6-Chloro-1, 3-benzodioxol-5-yl)-propane-2-yl](methyl)azane
Clonitazen			
	Codeine-*N*-oxide		
		2C-T-2	4-Ethylsulfanyl-2,5-dimethoxyphenethylazane
		2C-T-7	2,5-Dimethoxy-4-(propylsulfanyl) phenethylazane
Codoxim			
Desomorphine	Dihydrodesoxymorphine		
Diampromide			
	Diethoxybromamfetamine		
Diethylthiambutene			
	N,N-Diethyltryptamine	DET	
	Dihydroetorphine		
Dimenoxadol			
Dimepheptanol	Methadol		
	Dimethoxyamfetamine	DMA	
	Dimethoxyethylamfetamine	DOET	
	Dimethoxymethylamfetamine	DOM, STP	
	Dimethylheptyltetrahy-drocannabinol	DMHP	
Dimethylthiambutene			
	N,N-Dimethyltryptamine	DMT	
Dioxaphetyl butyrate			
Dipipanone			
	2,5-Dimethoxy-4-chloroamfetamine	DOC	1-(4-Chloro-2,5-dimethoxyphenyl)-propane-2-yl-azane
Drotebanol			
Ethylmethylthiambutene			
	Ethylpiperidyl benzilate		
Eticyclidine		PCE	
Etonitazene			
Etoxeridine			
Etryptamine	α-Ethyltryptamine		
	N-Hydroxy-MDMA	FLEA	*N*-[1-(1,3-Benzodioxol-5-yl)propane-2-yl]-*N*-methylhydroxylamine

INN	Non registered or trivial name	Abbreviation	IUPAC
Furethidine	*p*-Fluorofentanyl		
	Hashish (1)		
	Heroin (diacetylmorphine, diamorphine)		
Hydromorphinol	14-Hydroxydihydromorphine		
	N-Hydroxyamfetamine	NOHA	
	β-Hydroxyfentanyl		
	Hydroxymethylenedioxyamfetamine	MDOH	
	Hydroxymethylenedioxyamfetamine	*N*-Hydroxy-MDA	
	β-Hydroxy-3-methylfentanyl (Ohmefentanyl)		
Hydroxypethidine			
Lefetamine		SPA	
Levomethorphan			
Levophenacylmorphan			
Lofentanil			
Lysergide	*N,N*-Diethyl-D-lysergamide	LSD, LSD-25	
		MAL	3,5-Dimethoxy-4-(2-methylallyloxy) phenethylazane
	N-Methyl-1-(3,4-methylenedioxyphenyl)-2-butanamine	MBDB	
	Mebroqualone		
Mecloqualone			
	Mescaline		
Metazocine			
	Methcathinone (ephedrone)		
	Methoxyamfetamine (*p*-Methoxyamfetamine)	PMA	
	5-Methoxy-*N*, *N*-diisopropyltryptamine	5-MeO-DIPT	
	5-Methoxy-DMT	5-MeO-DMT	
			(2-Methoxyethyl) (1-phenylcyclohexyl)azane
	Methoxymetamfetamine (*p*-Methoxymetamfetamine)	PMMA	
	Methoxymethylene-dioxyamfetamine	MMDA	
			(3-Methoxypropyl) (1-phenylcyclohexyl)azane
	Methylaminorex (4-methylaminorex)	4-MAR	
Methyldesorphine			
Methyldihydromorphine			
	Methylenedioxyethylamfetamine	*N*-Ethyl-MDA	
	Methylenedioxyethylamfetamine	MDE, MDEA	
	Methylenedioxymetamfetamine	MDMA	

INN	Non registered or trivial name	Abbreviation	IUPAC
	α-Methylfentanyl		
	3-Methylfentanyl (mefentanyl)		
	Methylmethaqualone		
	Methylphenylpropionoxypiperidine	MPPP	
	Methyl-3-phenylpropylamine	1M-3PP	
	Methylphenyltetrahydropyridine	MPTP	
	Methylpiperidyl benzilate		
	4-Methylthioamfetamine	4-MTA	
	α-Methylthiofentanyl		
	3-Methylthiofentanyl		
	α-Methyltryptamine	α-MT	
Metopon	5-Methyldihydromorphinone		
Morpheridine			
	Morphine-*N*-oxide		
Myrophine	Myristylbenzylmorphine		
Nicomorphine	3,6-Dinicotinoylmorphine		
Noracymethadol			
Norcodeine	*N*-Desmethylcodeine		
Norlevorphanol	(−)-3-Hydroxymorphinan		
Normorphine	Desmethylmorphine		
Norpipanone			
	Parahexyl		3-Hexyl-6,6, 9-trimethyl-7,8,9, 10-tetrahydro- 6*H*-benzo[*c*]chromen-1-ol
		PCPr	(1-Phenylcyclohexyl) (propyl)azane
Phenadoxone			
Phenampromide			
Phenazocine			
Phencyclidine		PCP	
	Phenethylphenylacetoxypiperidine	PEPAP	
	Phenethylphenyltetrahydropyridine	PEPTP	
Phenpromethamine	1-Methylamino-2-phenylpropane	PPMA	
Phenomorphan			
Phenoperidine			
Piminodine			
		PPP	1-Phenyl- 2-(pyrrolidine-1-yl) propane-1-on
Proheptazine			
Properidine			
	Psilocin		
	Psilocin-(eth)		
Psilocybine			
	Psilocybine-(eth)		
			2-(Pyrrolidine-1-yl)- 1-(*p*-tolyl)propane-1-on

INN	Non registered or trivial name	Abbreviation	IUPAC
Racemethorphan			
Rolicyclidine		PHP, PCPy	
	Salvia divinorum (plants and parts of plants)		
Tenamfetamine	Methylenedioxyamfetamine	MDA	
Tenocyclidine		TCP	
	Tetrahydrocannabinols and following isomers (2):		
	Δ6a(10a)-Tetrahydrocannabinol	Δ6a(10a)-THC	
	Δ6a-Tetrahydrocannabinol	Δ6a-THC	
	Δ7-Tetrahydrocannabinol	Δ7-THC	
	Δ8-Tetrahydrocannabinol	Δ8-THC	
	Δ10-Tetrahydrocannabinol	Δ10-THC	
	Δ9(11)-Tetrahydrocannabinol	Δ9(11)-THC	
	Thenylfentanyl		
	Thiofentanyl		
Trimeperidine			
	Trimethoxyamfetamine	TMA	
	2,4,5-Trimethoxyamfetamine	TMA-2	

INN: International Nonproprietary Name; IUPAC: International Union of Pure and Applied Chemistry.
[a]The restrictions pertaining to these substances are also valid for the derivatives of the compounds, their stereoisomers, their salts, and respective preparations (apart from stated exceptions). (1) Marijuana, plants and parts of plants of genus *Cannabis* (apart from stated exceptions). (2) Including their stereoisomers.

Narcotic drugs: negotiable, but not permitted to prescribe.[a]

INN	Non registered or trivial name	Abbreviation
Amfetaminil		
Amineptin		
Aminorex		
	Benzylpiperazine	BZP
Butalbital		
	Butobarbital	
Cetobemidone	Ketobemidone	
	m-Chlorophenylpiperazine	*m*-CPP
	D-Cocaine	
Cyclobarbital		
	Dextromethadone	
Dextromoramide		
Dextropropoxyphene		
Difenoxin (1)		
	Dihydromorphine	

INN	Non registered or trivial name	Abbreviation
	Dihydrothebain	
Diphenoxylate (1)		
	Ecgonine	
	Erythroxylum coca (plants and parts of plants) (2)	
Ethchlorvynol		
Ethinamate		
	3-*O*-Ethylmorphine (ethylmorphine) (1)	
Etilamfetamine	*N*-Ethylamphetamine	
Fencamfamine		
Glutethimide		
	Isocodeine	
Isomethadone		
Levamfetamine	Levamphetamine	
	Levmetamfetamine (levometamfetamine)	
Levomoramide		
Levorphanol		
Mazindol		
Mefenorex		
Meprobamate		
Mesocarb		
(*RS*)-Metamfetamine	Metamfetamine racemate	
Metamfetamine	Methamphetamine	
	Methadone intermediate (premethadone)	
Methaqualone		
(*RS*,*SR*)-Methylphenidate		
Methyprylone		
	Moramide intermediate (premoramide)	
Nicocodine	6-Nicotinoylcodeine	
Nicodicodine	6-Nicotinoyldihydrocodeine	
	Oripavine	
Oxymorphone	14-Hydroxydihydromorphinone	
	Papaver bracteatum (plants and parts of plants) (3)	
	Pethidine intermediate A (prepethidine)	
	Pethidine intermediate B (norpethidine)	
	Pethidine intermediate C (pethidinic acid)	
Phendimetrazine		
Phenmetrazine		
Pholcodine	Morpholinylethylmorphine (1)	
	Poppy straw from *Papaver somniferum* (4)	
Propiram		

INN	Non registered or trivial name	Abbreviation
Pyrovaleron		
Racemoramide		
Racemorphan		
Secbutabarbital	Butabarbital	
	Δ9-Tetrahydrocannabinol	Δ9-THC
	Tetrahydrothebain	
Thebacon	Acetyldihydrocodeinone	
	Thebain	
cis-Tilidine		
Vinylbital		
Zipeprol		

INN: International Nonproprietary Name; IUPAC: International Union of Pure and Applied Chemistry.
[a]The restrictions pertaining to these substances are also valid for their derivatives, their salts and respective preparations as well as the derivatives and salts of the compounds, which are listed in table below (except for γ-hydroxybutyric acid (GHB)), if not in medical use (apart from stated exceptions). (1) Except for low concentrated preparations. (2) Including plants and parts of plants from *Erythroxylum bolivianum, E. spruceanum, E. novogranatense.* (3) Except for seeds and use for decoration. (4) Obtained from processing plants or part of plants of *Papaver somniferum* for the concentration of the alkaloids.

Narcotic drugs: negotiable and permitted to prescribe.[a]

INN	Non registered or trivial name	Abbreviation
Alfentanil		
Allobarbital		
Alprazolam (1)		
Amfepramone (1)		
Amfetamine	Amphetamine	
Amobarbital		
Barbital (1)		
Bromazepam (1)		
Brotizolam (1)		
Buprenorphine		
Camazepam		
Cathine (1)	(+)-Norpseudoephedrine (D-norpseudoephedrine)	
Chlordiazepoxide (1)		
Clobazam (1)		
Clonazepam (1)		
Clorazepate (1)		
Clotiazepam (1)		
Cloxazolam		
	Cocaine (benzoylecgonine methyl ester)	
	Codeine (3-methylmorphine) (1)	
Dexamfetamine	Dexamphetamine	
Delorazepam		

INN	Non registered or trivial name	Abbreviation
Dexmethylphenidate		
Diazepam (1)		
Dihydrocodeine (1)		
Dronabinol		
Estazolam (1)		
Ethylloflazepate		
Ethorphine		
Fenetylline		
Fenproporex (1)		
Fentanyl		
Fludiazepam		
Flunitrazepam (1)		
Flurazepam (1)		
Halazepam (1)		
Haloxazolam		
Hydrocodone	Dihydrocodeinone	
Hydromorphone	Dihydromorphinone	
	γ-Hydroxybutyric acid (1)	GHB
Ketazolam (1)		
Levacetylmethadol	Levomethadyl acetate	LAAM
Levomethadone		
Loprazolam (1)		
Lorazepam (1)		
Lormetazepam (1)		
Medazepam (1)		
Methadone		
Methylphenidate		
Methylhenobarbital (1)	Mephobarbital	
Midazolam (1)		
	Morphine	
Nabilone		
Nimetazepam		
Nitrazepam (1)		
Nordazepam (1)	Nordiazepam	
Normethadone		
	Opium (sap from *Papaver somniferum* sp.) (1)	
Oxazepam (1)		
Oxazolam (1)		
Oxycodone	14-Hydroxydihydrocodeinone	
	Papaver somniferum (plants and parts of plants) (2)	
Pemoline (1)		
Pentazocine		
Pentobarbital		
Pethidine		
Phenobarbital (1)		
Phentermine (1)		
Pinazepam		
Pipradrol		

INN	Non registered or trivial name	Abbreviation
Piritramide		
Prazepam (1)		
Remifentanil		
Secobarbital		
Sufentanil		
Temazepam (1)		
Tetrazepam (1)		
Tilidine (1)	*trans*-Tilidine	
Triazolam (1)		
Zolpidem (1)		

INN: International Nonproprietary Name; IUPAC: International Union of Pure and Applied Chemistry.

[a]The restrictions pertaining to these substances are also valid for their salts and respective preparations if in medical use (apart from stated exceptions). (1) Except for low concentrated preparations.

Index

a

Abrus precatorius L. (jequirity bean) 804
absinth 516
absorption spectrophotometry (AS) 652–656, 668–671
ACE (angiotensin converting enzyme) inhibitors
– GC-MS 161
acebutolol
– HPLC 457
acenocoumarol 303
– HPLC 305
acephate
– ADI value 596
– MAK value 596
– TLC 594
aceprometazine
– TBPE test 179
acetaldehyde (ethanal)
– headspace-GC 171, 173
acetamidoflunitrazepam
– HPLC 355
acetaminophen, *see also* paracetamol
– antidote 857
– clinical interpretation 205
– elimination half-life 204, 841
– extraction 190
– GC 198
– GC–MS 109, 112, 116
– HPLC 109, 112, 116, 193
– immunoassay 203
– intoxication 144
– medical assessment 204
– metabolism 206
– plasma/serum concentration 841
– serum concentration 204
– toxic hemorrhagic disorder 751
acetazolamide
– acid-base balance 752
acetone
– BAT value 520, 853
– blood concentration 520
– clinical interpretation 522
– headspace-GC 171, 173, 519
– medical assessment 520
– toxicity 522
acetonitrile
– headspace-GC 171, 173
– hepatotoxicity 749
– protein precipitation 119
acetorphine 867
acetylcholinesterase (AChE)
– BAT value 853
– determination 760
– enzyme stability 763
– half-life of inhibited AChE 762
– metabolite binding 702
acetylcholinesterase (AChE) inhibitor
– BAT value 853
acetyldigoxin 336
– elimination half-life 841
– plasma/serum concentration 841
acetylmethadol 867
6-acetylmorphine
– elimination half-life 247
– GC–MS 247
– pharmacokinetics 247
N-acetylprocainamide
– clinical interpretation 281
– elimination half-life 281
– immunoassay 781
– medical assessment 281
– metabolism 281
– serum concentration 281
acetylsalicylic acid 209
– clinical interpretation 213
– elimination half-life 841
– extraction 190

Clinical Toxicological Analysis: Procedures, Results, Interpretation. Edited by Wolf-Rüdiger Külpmann

ISBN: 978-3-527-31890-2

– GC 198
– HPLC 193
– intoxication 144
– medical assessment 212
– metabolism 209
– plasma/serum concentration 841
– serum concentration 212
acids 804
aconitine
– TBPE test 180
Aconitum napellus L. (monkshood) 785–786
acrolein
– headspace-GC 171, 173
acrylonitrile
– hepatotoxicity 749
Adamsite 734
ADI (acceptable daily intake) value 561
Aesculus hippocastanum L. (horse chestnut) 804
aflatoxin 748
ajmaline 177
– elimination half-life 841
– GC 274
– HPLC 273
– plasma/serum concentration 841
– TBPE test 179
alcohols
– anemia 751
– clinical interpretation 522
– headspace-GC 517
– highly volatile 511–523
– medical assessment 520
aldicarb 564
– ADI value 574
– blood/serum concentration 574
– clinical interpretation 574
– LD_{50} value 573
– MAK value 574
– medical assessment 573
– toxicity 574
aldrin 578
– ADI value 580
– blood/plasma/serum concentration 581
– clinical interpretation 581
– MAK value 580
– medical assessment 580
– poisoning 581
– TLC 579
– toxicity 580
– toxicokinetics 580
alfentanil 873
– elimination half-life 841
– plasma/serum concentration 841
algae
– toxic 830
– toxin detection 830
alimemazine
– TBPE test 179
alkaloid 804
– clavinet 806
– plant 158
– pyrrolizidine 748
allobarbital 873
– immunoassay 346
allopurinol
– elimination half-life 841
– plasma/serum concentration 841
allylprodine 867
alphacetylmethadol 867
alphameprodine 867
alphamethadol 867
alphaprodine 867
alprazolam 873
– elimination half-life 841
– GC–MS 408
– metabolism 359
– plasma/serum concentration 841
alprenolol
– HPLC 457
– TBPE test 180
aluminium
– BAT value 853
Amanita pantherina 813
Amanita phalloides 751, 811, 819
– antidote 857
α-amanitin
– ELISA 159
amanitin 812, 817
– analytical strategy 821
– therapy 821
amatoxin 811, 816
– antidote 857
– clinical interpretation 820
– HPLC 819
– LC–MS 819
– immunoassay 817
– intoxication 820
– medical assessment 819
– nephrotoxicity 749
– screening procedure 816
– syndrome 811
– toxic hemorrhagic disorder 751
Ambrosia artemisifolia 788
ambroxol
– HPLC 136
amezine
– TBPE test 180
amfepramone 873

amfetamine 873
amfetaminil 871
amikacin
– elimination half-life 841
– immunoassay 781
– plasma/serum concentration 841
amineptin 871
ε-aminocaproic acid
– creatine kinase activity 750
aminocarb
– TLC 572
7-aminoflunitrazepam
– HPLC 355
aminoglycosides
– acid-base balance 752
aminophenazone
– extraction 190
– GC 198
– HPLC 193
aminorex 871
5-aminosalicylic acid
– elimination half-life 841
– plasma/serum concentration 841
aminothiol 699
– GC–MS 700
amiodarone
– clinical interpretation 275
– elimination half-life 275, 841
– GC 274
– GC–MS 109, 112, 116
– HPLC 109, 112, 116, 273
– medical assessment 275
– metabolism 276
– plasma/serum concentration 841
– serum concentration 275
– TBPE test 179
amisulpride
– GC–MS 408
amitriptyline 414, 432–434
– clinical interpretation 432–434
– elimination half-life 841
– GC 405
– GC–MS 109, 112, 116, 408
– HPLC 109, 112, 116, 136, 397, 401
– immunoassay 781
– medical assessment 432–434
– metabolism 433
– overdose 434
– plasma/serum concentration 841
– serum concentration 414, 435
– TBPE test 179
amitriptyline-*N*-oxide
– GC–MS 408
– TBPE test 179
amlodipine
– elimination half-life 841
– plasma/serum concentration 841
amobarbital 873
– elimination half-life 841
– immunoassay 346
– plasma/serum concentration 841
amphetamine 463–470, 873
– clinical interpretation 469
– creatine kinase activity 750
– drug-induced seizure 750
– elimination half-life 841
– GC-MS 109, 112, 116, 158, 467
– HPLC 109, 112, 116, 467–469
– immunoassay 466
– medical assessment 469
– metabolism 464
– minimal concentration to be detected 75
– plasma/serum concentration 841
– TBPE test 179
amphotericin B
– acid-base balance 752
– elimination half-life 841
– plasma/serum concentration 841
ampicillin
– elimination half-life 841
– plasma/serum concentration 841
amrinone
– elimination half-life 841
– plasma/serum concentration 841
amygdalin 806
iso-amyl acetate (*iso*-pentyl acetate)
– headspace-GC 172, 174
n-amyl acetate (pentyl acetate)
– headspace-GC 172, 174
analgesics 215–265
– acid-base balance 752
– group immunoassay 216
analgesics, nonopioid 189–214
– GC-MS 158
analysis, toxicological
– chemicals 8
– devices 5
– impact 1, 96
– management 7
– methods 6
– personnel 8
– technical requirements 7
analytical assessment 77
analytical result
– assessment 77–79
analytical toxicological report 81–86
– delivery 86
anemia

– associated drug 751
anesthetics
– acid-base balance 752
– drug-induced seizure 750
– local 158
angel's trumpet (*Brugmansia sanguinea* L.) 785–786
angiotensin receptor blockers
– GC-MS 161
anileridine 867
aniline
– BAT value 853
– headspace-GC 171, 174
aniline derivative
– anemia 751
animal
– diagnosis of poisoning caused by animal 825
– marine 826
– terrestrial 831
– venomous and poisonous 825–834
antazoline
– TBPE test 179
anticoagulants 301–311
– anemia 751
– GC-MS 161
– HPLC 303
– mechanism of action 302
– screening method 302
– toxic hemorrhagic disorder 751
anticonvulsants 287–300
– GC-MS 158
– HPLC 289
– immunoassay 288
antidepressants 393–452
– color test 394
– drug-induced seizure 750
– GC-MS 158
– group assay 393
– HPLC 396
– immunoassay 393
– intoxication 139
– tricyclic
  – antidote 433, 857
  – intoxication 432
  – minimal concentration to be detected 75
  – poisoning 753
– tetracyclic
  – minimal concentration to be detected 75
antidiabetics 613–621
– drug-induced seizure 750
– GC–MS 161
– HPLC–MS 615
– LC–MS 618–619
– oral 615–619
– poisoning 753
antidote 857–859
– monitoring 754
antidysrhythmic agent 271–284
– GC–MS 158
– HPLC 273
– immunoassay 272
antihistamines (H1 antagonists)
– antidote 857
– drug-induced seizure 750
– GC-MS 158
anti-inflammatory drug
– nonsteroidal 751
antimalarial agent
– anemia 751
antimony
– hepatotoxicity 749
– nephrotoxicity 749
antineoplastic drugs
– IA, HPLC, LC–MS 159
antiparkinsonian agents
– antidote 857
– GC-MS 158
antipsychotics
– drug-induced seizure 750
antipyrine, *see also* phenazone
– clinical interpretation 208
– elimination half-life 207, 841
– GC 198
– GC–MS 113
– HPLC 113, 193
– medical assessment 207
– metabolism 208
– plasma/serum concentration 841
– serum concentration 207
antirheumatics 189–214
antirheumatics, nonsteroidal
– GC-MS 161
Apocynum cannabinum (dog bane) 806
aprobarbital
– elimination half-life 841
– immunoassay 346
– plasma/serum concentration 841
Areca catechu (betel nut) 799
Argyreia nervosa (Hawaiian baby wood rose) 806
Arisaema spp. 789–790
aripiprazole
– elimination half-life 841
– GC–MS 408
– plasma/serum concentration 841
aromatics (BTEX) 523–531
– GC 523

– headspace-GC 526
arsenic
– antidote 857
– liver toxin 748
– nephrotoxicity 749
arsine
– anemia 751
– hepatotoxicity 749
– nephrotoxicity 749
assessment of analytical results 77
– analytical 77
– biological 78
– nosological 79
Artemisia absinthium L. 804
astemizole
– TBPE test 179
atenolol 458
– elimination half-life 841
– GC–MS 109, 112, 116
– HPLC 109, 112, 116, 457
– metabolism 458
– plasma half-life 458
– plasma/serum concentration 458, 841
atomic absorption spectrometry (AAS) 664
Atropa belladonna L. (deadly nightshade) 785
– antidote 857
atropine 791
– elimination half-life 842
– plasma/serum concentration 842
autumn crocus (*Colchicum autumnale*) 785
azathioprine
– elimination half-life 842
– plasma/serum concentration 842
azinphos-ethyl/ethylguthion
– LD_{50} value 596
– TLC 594
azinphos-methyl
– ADI value 596
– LD_{50} value 596
– MAK value 596
– TLC 594

b
baclofen
– drug-induced seizure 750
– elimination half-life 842
– plasma/serum concentration 842
bambuterol
– elimination half-life 842
– plasma/serum concentration 842
barbital 873
– elimination half-life 842
– immunoassay 346
– plasma/serum concentration 842
barbiturates 339–350
– GC-MS 158, 161
– minimal concentration to be detected 75
– poisoning 753
BDB (benzodioxazolylbutanamine) 463–464
Begonia spp. 789–790
bendiocarb
– ADI value 574
– LD_{50} value 573
– MAK value 574
benomyl
– ADI value 574
– MAK value 574
benperidol
– GC 405
– GC–MS 408
– HPLC 397
– TBPE test 180
benzaldehyde
– headspace-GC 171, 174
benzalkonium chloride
– TBPE test 179
benzbromarone
– elimination half-life 842
– HPLC 305
– plasma/serum concentration 842
benzene
– BAT value 527
– blood concentration 530
– chlorinated 749
– clinical interpretation 530
– headspace-GC 171, 173, 526
– hepatotoxicity 749
– intoxication 530
– medical assessment 527
– metabolism 528
benzethidine 867
benzfetamine 867
benzodiazepines 351–364, 415
– antidote 857
– clinical interpretation 356
– GC 356
– GC-MS 158
– HPLC 352
– immunoassay 351
– medical assessment 356
– metabolism of 1,4-benzodiazepine 360
– minimal concentration to be detected 75
– poisoning 753
benzodioxazolylbutanamine, *see* BDB
3,4-benzopyrene
– nephrotoxicity 749
benzoylecgonine
– elimination half-life 843

– GC–MS 485
benzyl alcohol
– headspace-GC 171, 174
1-benzylpiperazine (BZP, A2, Frenzy, Nemesis) 465
beryllium
– hepatotoxicity 749
– nephrotoxicity 749
beta-receptor blocking agents 455–462
– antidote 859
– clinical interpretation 457
– drug-induced seizure 750
– GC 457
– GC-MS 158
– HPLC 455
– medical assessment 457
betacetylmethadol 867
betameprodine 867
betamethadol 867
betaprodine 867
betaxolol
– elimination half-life 842
– HPLC 457
– plasma/serum concentration 842
betel nut (*Areca catechu*) 799
bevoniummetil sulphate
– TBPE test 180
bezitramide 867
biguanides
– acid-base balance 752
biochemical investigation 104, 745–769
biological exposure indices (BEI) value 563
biological tolerance (BAT) value 563
biperiden
– elimination half-life 842
– plasma/serum concentration 842
– TBPE test 179
biphenyls
– chlorinated 749
– polybrominated (PBB) 749
– polychlorinated (PCB) 748–749
3,3′-bis-chloropropyl sulfide 703
bismuth
– hepatotoxicity 749
– nephrotoxicity 749
bisnortilidine
– GC–MS 261
bisoprolol
– elimination half-life 842
– HPLC 457
– plasma/serum concentration 842
N,O-bis-(trimethylsilyl) trifluoro acetamide 695
blister(ing) agent 703–729
blood agent 732
blue-ringed octopus 830
borate
– liver toxin 748
bradydysrhythmia
– antidote 857
brain death 104
brallobarbital
– elimination half-life 842
– plasma/serum concentration 842
broadbean 800
brodifacoum 308
– half-life 310
– pharmacokinetics 310
– serum concentration 311
– toxicity 310
brolamfetamine 867
bromadiolone 308
– serum concentration 311
– toxicity 310
bromazepam 873
– elimination half-life 842
– GC–MS 408
– HPLC 136
– metabolism 359
– plasma/serum concentration 842
bromide
– clinical interpretation 380
– elimination half-life 842
– ISE 159
– medical assessment 380
– plasma/serum concentration 842
– serum concentration 380
bromisoval
– extractability 369
– elimination half-life 842
bromoacetone 734
bromodichloromethane
– headspace-GC 548
bromoform, *see* tribromomethane
4-bromophenacyl bromide 695
bromophos
– ADI value 596
– blood/plasma/serum concentration 597
– MAK value 596
– TLC 594
– toxicity 597
bromophos-ethyl 592
– ADI value 596
– MAK value 596
– TLC 594
bromperidol
– GC 405
– GC–MS 408

– HPLC 397
bronchodilator 313–316
– group assay 313
brotizolam 873
– elimination half-life 842
– plasma/serum concentration 842
brucine
– TBPE test 179
Brugmansia sanguinea L. (angel's trumpet) 785–786
Brugmansia suavolens
– antidote 857
BSTFA 695
BTEX, *see* aromatics
bumetanide
– acid-base balance 752
bunitrolol
– elimination half-life 842
– HPLC 457
– plasma/serum concentration 842
bupivacaine
– elimination half-life 842
– plasma/serum concentration 842
– TBPE test 179
bupranolol
– TBPE test 179
buprenorphine 216, 873
– clinical interpretation 218
– elimination 217
– elimination half-life 842
– GC–MS 109, 112, 116, 217
– HPLC 109, 112, 116
– HPLC-DAD 217
– immunoassay 216
– medical assessment 217
– plasma/serum concentration 842
– serum concentration 217
buspirone
– GC–MS 408
butabarbital
– elimination half-life 842
– immunoassay 346
– plasma/serum concentration 842
butalbital 871
– elimination half-life 842
– immunoassay 346
– plasma/serum concentration 842
1,4-butandiol
– GC 534
1-butanol
– BAT value 853
– headspace-GC 171, 173
iso-butanol
– headspace-GC 171, 173
tert-butanol
– headspace-GC 171, 173, 519
2-butanone
– BAT value 853
butocarboxim
– TLC 568
butocarboxim sulfoxide
– TLC 568
butoxycarboxim
– TLC 568
butriptyline
– elimination half-life 842
– plasma/serum concentration 842
n-butyl acetate
– headspace-GC 171, 173
tert-butyl-dimethylsilyl chloride (*t*BDMSC) 695
i-butyl methylphosphonic acid (*i*BMPA) 689
– GC–MS/MS 698
– LC–MS 694
n-butyl methylphosphonic acid (*n*BMPA) 689
– LC–MS 694
p-*tert*-butylphenol
– BAT value 853
butyrophenones neuroleptics 415, 434
– antidote 858
– clinical interpretation 434
– GC 405
– GC-MS 112, 158
– HPLC 112, 397
– medical assessment 434
– serum concentration 415
butyrylcholinesterase (BChE) 755
– enzyme stability 763
– plasma 755, 761, 764
– variability 763
BZ, *see* 3-quinuclidinyl benzilate
BZP, *see* 1-benzylpiperazine

c

C-peptide 613
– immunoassay 613–614
cadmium
– acid-base balance 752
– antidote 857
– nephrotoxicity 749
caffeine
– clinical interpretation 314
– elimination half-life 842
– GC–MS 109, 112, 116, 314
– HPLC 109, 112, 116, 313
– immunoassay 313
– medical assessment 314
– plasma/serum concentration 842

Caladium spp. 789–791
calcium channel blockers 317–325
– antidote 857
– GC-MS 161
calcium oxalate 791–798
camazepam 873
– elimination half-life 842
– metabolism 359
– plasma/serum concentration 842
camphechlor 578
camphor
– drug-induced seizure 750
– elimination half-life 842
– plasma/serum concentration 842
cannabinoids 470–479
– clinical interpretation 478
– GC–MS 109, 112, 475
– HPLC 109, 112, 475
– immunoassay 470
– medical assessment 478
– minimal concentration to be detected 75
– THC (Δ^9-tetrahydrocannabinol) metabolism 472
– TLC 471
Cannabis spp. (hemp) 806
canrenone
– elimination half-life 842
– plasma/serum concentration 842
capillary electrophoresis (CE) 54
– injection 54
– UV 694
capsaicin 791
Capsicum spp. 787–791
captopril
– elimination half-life 842
– plasma/serum concentration 842
carazolol
– elimination half-life 842
– HPLC 457
– plasma/serum concentration 842
– TBPE test 180
carbachol
– antidote 857
carbamazepine
– clinical interpretation 294
– elimination half-life 842
– GC–MS 109, 112, 116, 292–294
– HPLC 109, 112, 116, 136, 289
– immunoassay 288
– medical assessment 294
– metabolism 295
– plasma/serum concentration 842
– serum concentration 295
carbamazepine-10,11-epoxide
– HPLC 289
– GC–MS 292–294
– metabolism 295
– serum concentration 295
carbamates (pesticides) 564–576
– antidote 575, 857
– clinical interpretation 574–576
– elimination half-life 842
– GC–MS 112
– general unknown analysis 565
– HPLC 112
– medical assessment 573
– plasma/serum concentration 842
– poisoning 575
– TLC 565–573
carbaryl 564
– ADI value 574
– blood/plasma/serum concentration 574
– clinical interpretation 574
– LD_{50} value 573
– MAK value 574
– medical assessment 573
– poisoning 574
– TLC 568–572
carbendazim
– ADI value 574
– MAK value 574
carbenoxolone
– elimination half-life 842
– plasma/serum concentration 842
carbofuran 564
– ADI value 574
– blood concentration 575
– clinical interpretation 575
– intoxication 575
– LD_{50} value 573
– MAK value 574
– medical assessment 573
– TLC 568–572
carbon disulfide
– BAT value 853
– nephrotoxicity 749
carbon monoxide 623
– acid-base balance 752
– antidote 857
– AS 159
– BAT value 853
– creatine kinase activity 750
– drug-induced seizure 750
– clinical interpretation 626
– intoxication 624, 627
– nephrotoxicity 749
– poisoning 753
carbon tetrabromide, *see* tetrabromomethane

carbon tetrachloride, *see* tetrachloromethane
carbophenothion
– ADI value 596
– MAK value 596
carboplatin
– elimination half-life 842
– plasma/serum concentration 842
carbosulfan
– TLC 568
carboxyhemoglobin 623
– BAT value 626
– clinical interpretation 626
– elimination half-life 627
– medical assessment 626
– oximetry 624
– spectrophotometry 625
carbromal 380
– clinical interpretation 380
– elimination half-life 380, 842
– extractability 369
– GC 375
– GC–MS 109, 112, 116
– HPLC 109, 112, 116, 371
– medical assessment 380
– metabolism 380
– plasma/serum concentration 842
– quality assurance 64
– serum concentration 380
cardenolides 806
cardiac glycosides 159, 327–336, 804, 806
– antidote 857
– IA, LC-MS 159
carfentanil 868
carteolol
– elimination half-life 842
– plasma/serum concentration 842
cassava 800, 806
castor bean 806
castor oil plant (*Ricinus communis* L.) 785
catecholamines
– acid-base balance 752
– HPLC determination 53
Catha edulis (khat) 466, 799
cathine 873
cathinone 868
celiprolol
– HPLC 457
cesium-137
– antidote 857
cetobemidone 871
cetylpyridinium chloride
– TBPE test 179
chemical warfare agents (CWA) 679–734
– classification 679–680
– toxicity 679, 681
chill-X 479
chinese VX (CVX) 683
Chiracanthium punctorium (yellow sack spider) 831
Chironex fleckeri 827
chlocyclizine
– TBPE test 179
chlofenotane, *see* DDT
chloral hydrate 381
– clinical interpretation 381
– elimination half-life 381, 842
– Fujiwara reaction 542–543
– GC–MS 109, 112
– HPLC 109, 112
– medical assessment 381
– metabolism 381
– plasma/serum concentration 842
– serum concentration 381
chlorambucil
– drug-induced seizure 750
chloramphenicol
– elimination half-life 842
– plasma/serum concentration 842
chlorbutadiene
– hepatotoxicity 749
chlordane 578
– ADI value 580
– blood/plasma/serum concentration 581
– clinical interpretation 582
– MAK value 580
– medical assessment 580
– poisoning 581
– toxicity 580
– toxicodynamics 581
– toxicokinetics 580
chlordiazepoxide 136, 873
– elimination half-life 842
– GC–MS 408
– metabolism 359
– plasma/serum concentration 842
chlorfenvinphos
– ADI value 596
– blood/plasma/serum concentration 597
– LD_{50} value 596
– MAK value 596
– TLC 594
– toxicity 597
chlorinated hydrocarbons 576–582
– clinical interpretation 581
– cyclic 576
– drug-induced seizure 750
– general unknown analysis 579
– liver toxin 748

– medical assessment 580
– TLC 579
chlormephos
– TLC 594
chlormethiazole, *see* clomethiazole
chloroacetophenone (CN) 734
chlorobenzene
– BAT value 853
– headspace-GC 171, 173
o-chlorobenzylidene malonodinitrile 734
2-chloroethanol, *see* ethylene chlorohydrin
chloroform, *see* trichloromethane
chloronaphthalene
– liver toxin 748
chlorophacinone 308
– HPLC 305
– serum concentration 311
– toxicity 310
chloropicrin 730, 732
chloroquine
– antidote 857
– clinical interpretation 638
– elimination half-life 842
– GC–MS 109, 112, 116
– HPLC 109, 112, 116, 136
– intoxication 639
– medical assessment 638
– metabolism 639
– plasma elimination half-life 638
– plasma/serum concentration 842
– serum concentration 638
– TBPE test 180
– TLC 635
– toxicity 638
2-chlorovinylarsonous acid (CVAA) 726
– GC–MS 727
chlorpheniramine
– elimination half-life 842
– plasma/serum concentration 842
chlorpromazine
– elimination half-life 842
– GC–MS 408
– plasma/serum concentration 842
– TBPE test 179
chlorpropamide 616
– blood/plasma/serum concentration 620
– LC–MS 619
– metabolism 616
chlorprothixene 416, 434
– clinical interpretation 437
– elimination half-life 843
– GC 405
– GC–MS 109, 112, 116, 408
– HPLC 109, 112, 116, 136, 397
– intoxication 437
– medical assessment 434–437
– metabolism 436
– plasma/serum concentration 843
– serum concentration 416, 435
– TBPE test 179
chlorpyrifos
– blood/plasma/serum concentration 597
– LD_{50} value 596
– TLC 594
– toxicity 597
chlorpyrifos-methyl
– ADI value 596
– MAK value 596
– TLC 594
chlorthalidone
– elimination half-life 843
– plasma/serum concentration 843
chlorthion
– TLC 594
chlorthiophos
– TLC 594
CHMPA, *see* cyclohexyl methylphosphonic acid
choking agent, *see* lung-damaging agent
cholinesterase 755
– determination 757
– Ellman assay 757
– physiological function 756
– presence of reversible inhibitor 761
cholinesterase inhibitors 761
– drug-induced seizure 750
chromate (VI)
– antidote 857
chromium
– nephrotoxicity 749
Chrysanthemum spp. 787
ciclonium bromide
– TBPE test 179
ciclosporin
– drug-induced seizure 750
– elimination half-life 843
– immunoassay 781
– plasma/serum concentration 843
Cicuta douglasii 788
Cicuta maculate 788
Cicuta spp. (water hemlock) 787–788
Cicuta virosa L. 804
cicutoxin 804
ciguatera 829
cimetidine
– elimination half-life 843
– plasma/serum concentration 843
Cimicifuga racemosa 799
cineol 793

ciprofloxacin
– HPLC 136
citalopram 416, 437
– clinical interpretation 437
– elimination half-life 437, 843
– GC–MS 408
– intoxication 437
– medical assessment 437
– plasma/serum concentration 843
– serum concentration 416, 435
Clark I, *see* diphenylarsine chloride
Clark II, *see* diphenylarsine cyanide
clenbuterol
– elimination half-life 843
– plasma/serum concentration 843
– TBPE test 180
clindamycin
– elimination half-life 843
– plasma/serum concentration 843
clinical toxicological analysis
– practicability 59–62
clinical toxicological investigation
– strategy 95–105
clinical toxicological requirements 102
Clitocybe clavipes 811
clobazam 873
– elimination half-life 843
– GC–MS 109, 112, 116, 408
– HPLC 109, 112, 116
– immunoassay 781
– plasma/serum concentration 843
clobutinol
– TBPE test 180
clofencyclane
– TBPE test 179
clofibrate
– creatine kinase activity 750
– elimination half-life 843
– plasma/serum concentration 843
clomethiazole (chlormethiazole) 416
– clinical interpretation 382
– elimination half-life 382, 843
– extractability 369
– GC 375
– GC-MS 109, 112, 116, 158, 408
– HPLC 109, 112, 116, 136, 371
– medical assessment 382
– metabolism 383
– plasma/serum concentration 843
– serum concentration 382
clomipramine 416, 437
– acute poisoning 438
– clinical interpretation 438
– elimination half-life 843
– GC 405
– GC–MS 109, 112, 116, 408
– HPLC 109, 112, 116, 136, 397, 401
– immunoassay 781
– medical assessment 437
– plasma half-life 438
– plasma/serum concentration 843
– serum concentration 416, 435
– TBPE test 179
clonazepam 873
– antidote 357
– clinical interpretation 357
– elimination half-life 357, 843
– GC–MS 109, 112, 116
– HPLC 109, 112, 116, 136
– immunoassay 781
– medical assessment 357
– metabolism 358–359
– plasma/serum concentration 357, 843
clonidine
– elimination half-life 843
– plasma/serum concentration 843
– TBPE test 180
clonitazen 868
clopenthixol
– TBPE test 179
clorazepate 873
– elimination half-life 843
– metabolism 359
– plasma/serum concentration 843
clotiazepam 873
– elimination half-life 843
– GC–MS 408
– plasma/serum concentration 843
cloxacillin
– elimination half-life 843
– plasma/serum concentration 843
cloxazolam 873
clozapine 417, 438
– clinical interpretation 438
– elimination half-life 843
– GC 405, 407
– GC–MS 109, 112, 116, 408
– HPLC 109, 112, 116, 136, 397, 401
– HPLC–MS 432
– intoxication 438
– medical assessment 438
– metabolism 439
– plasma half-life 438
– plasma/serum concentration 843
– serum concentration 417, 435
– TBPE test 180
CMPA, *see* cyclohexyl methylphosphonic acid
Cnidaria 826

cobra 833
cocaine 480–488
– clinical interpretation 486
– elimination half-life 843
– drug-induced seizure 750
– GC–MS 109, 112, 116, 483
– HPLC 109, 112, 116, 482
– immunoassay 481
– medical assessment 486
– metabolism 481
– minimal concentration to be detected 75
– plasma/serum concentration 843
codeine 177, 241
– antidote 858
– elimination half-life 247, 843
– GC–MS 109, 112, 116, 247
– HPLC 109, 112, 116
– metabolism 242
– pharmacokinetics 247
– plasma/serum concentration 843
– TBPE test 179
D_3-codeine
– GC–MS 247
codoxime 868
colchicine
– elimination half-life 843
– nephrotoxicity 749
– plasma/serum concentration 843
Colchicum autumnale (autumn crocus, meadow saffron) 785
color test 175
cone snail shell 830
Conium maculatum L. (poison hemlock) 785–788
Conus spp. 830
Convallaria majalis L. 804, 806
copper
– antidote 857
– liver toxin 748
coprine syndrome 811
Coprinus atramentarius 811
coral snake 833
Cortinarius orellanus 812
cotinine 640
– clinical interpretation 641
– GC–MS 195, 640
– HPLC 190–194, 640
– immunoassay 640
– medical assessment 641
– metabolism 641
– plasma elimination half-life 641
– serum concentration 641
– toxicity 641
coumaphos
– TLC 594
coumarins 302, 309, 794
– antidote 857
– HPLC 305
– toxic hemorrhagic disorder 751
coumatetralyl 308
– HPLC 305
– toxicity 310
Crassula spp. 787
cresol
– hepatotoxicity 749
Crotalus spp. (rattlesnake) 833
curare
– antidote 857
Cyanea capillata 826
cyanide 630, 646
– absorption spectrophotometry 652–656
– acid-base balance 752
– antidote 660, 754, 857–858
– blood concentration 660
– clinical interpretation 660–661
– elimination half-life 843
– indicator tube method 647–650
– intoxication 660–661
– medical assessment 659
– paper strip 650–652
– plasma/serum concentration 843
– poisoning 753
– potentiometry 656–659
Cycas circinalis (false sago palm) 799
cyclobarbital 871
– elimination half-life 843
– plasma/serum concentration 843
cyclohexane
– BAT value 853
– headspace-GC 171, 173
cyclohexanol
– headspace-GC 171, 174
cyclohexanone
– headspace-GC 171, 174
cyclohexyl methylphosphonic acid (CHMPA, CMPA) 689
– CE – UV 694
– GC – MS 697
– GC – MS/MS 698
– LC – MS 694
– LC – MS/MS 694
cyclophosphamide
– elimination half-life 843
– plasma/serum concentration 843
cyclosarin (GF) 683
– GC–MS 687
cyfluthrin 600
– ADI value 602

– LD_{50} value 602
cyhalothrin
– ADI value 602
– LD_{50} value 602
λ-cyhalothrin 600
cypermethrin 600
– ADI value 602
cyproheptadine
– TBPE test 179

d

dantrolene
– elimination half-life 843
– plasma/serum concentration 843
Daphne mezereum L. 805
dapsone
– elimination half-life 843
– plasma/serum concentration 843
Datura stramonium L. (jimsonweed, thorn apple) 785–789, 791
– antidote 858
DDT (dichlorodiphenyltrichloroethane, chlofenotane) 576, 578
– ADI value 580
– blood/plasma/serum concentration 581
– clinical interpretation 582
– hepatotoxicity 749
– liver toxin 748
– MAK value 580
– medical assessment 580
– nephrotoxicity 749
– poisoning 581
– TLC 579
– toxicity 580
– toxicodynamics 581
– toxicokinetics 580
DDVP, *see* dichlorvos
deadly nightshade (*Atropa belladonna* L.) 785
deanol
– GC–MS 408
decane
– headspace-GC 171, 174
deferoxamine
– elimination half-life 843
– plasma/serum concentration 843
delorazepam 873
deltamethrin 600
– ADI value 602
– LD_{50} value 602
demethyldiazepam, *see* nordazepam
demethyldoxepin
– HPLC-DAD 143
demeton
– TLC 594
demeton-*S*-methyl
– LD_{50} value 596
demeton-*S*-methylsulfon
– TLC 594
demeton sulfoxide
– TLC 594
demoxepam
– elimination half-life 843
– metabolism 359
– plasma/serum concentration 843
desalkylflurazepam
– plasma/serum concentration 845
N-desethylamiodarone
– elimination half-life 841
– plasma/serum concentration 841
designer drug 463
designer drugs, amphetamine type
– GC-MS 158
designer drugs, phencyclidine type
– GC-MS 158
designer drugs, phenethylamine type
– GC-MS 158
designer drugs, piperazine type
– GC-MS 158
desipramine, *see also* imipramine 418, 440
– clinical interpretation 432, 440
– elimination half-life 843
– GC 405
– GC–MS 109, 112, 116, 408
– HPLC 109, 112, 116, 397, 401
– HPLC-DAD 143
– immunoassay 781
– medical assessment 432, 440
– metabolism 441
– plasma half-life 440
– plasma/serum concentration 843
– serum concentration 435
– TBPE test 179
desmedipham
– TLC 572
desmethyldiazepam, *see* nordazepam
desomorphine 868
detajmium
– GC 274
– HPLC 273
dexamfetamine 873
dexmethylphenidate 874
dextromethorphan
– elimination half-life 843
– GC–MS 109, 112, 116
– HPLC 109, 112, 116
– plasma/serum concentration 843
– TBPE test 179
dextromoramide 871

dextropropoxyphene, *see* also propoxyphene 218, 871
– antidote 858
– clinical interpretation 222
– elimination half-life 843
– GC–MS 109, 112, 116, 221
– HPLC 109, 112, 116, 219
– immunoassay 219
– medical assessment 221
– metabolism 218
– plasma/serum concentration 843
– serum concentration 221
– TBPE test 179
diamorphine
– elimination half-life 247
– metabolism 242
– pharmacokinetics 247
diampromide 868
diazepam 874
– antidote 358
– clinical interpretation 358
– elimination half-life 358
– GC–MS 109, 112, 116, 408
– HPLC 109, 112, 116, 136, 355
– HPLC-DAD 143
– intoxication 139
– medical assessment 358
– metabolism 359
– plasma/serum concentration 358, 843
diazinon (dimpylate) 592
– ADI value 596
– blood/plasma/serum concentration 597
– LD_{50} value 596
– MAK value 596
– TLC 594
– toxicity 597
diazomethane 695
diazoxide
– elimination half-life 843
– plasma/serum concentration 843
dibenzepin
– elimination half-life 843
– GC–MS 408
– HPLC 136
– plasma/serum concentration 843
– TBPE test 179
dibenzodioxin
– chlorinated 749
dibromochloromethane
– headspace-GC 548
dibromomethane
– Fujiwara reaction 543
dichloroacetic acid
– Fujiwara reaction 542
1,2-dichlorobenzene
– BAT value 853
– headspace-GC 171, 174
1,2-dichloroethane
– clinical interpretation 551
– headspace-GC 171, 173
dichloromethane (methylene chloride) 548
– BAT value 549
– clinical interpretation 551
– headspace-GC 171, 173, 548
– MAK value 549
– medical assessment 548–549
1,2-dichloropropane
– hepatotoxicity 749
α,α-dichlorotoluene
– Fujiwara reaction 542
dichlorvos/DDVP
– ADI value 596
– blood/plasma/serum concentration 597
– LD_{50} value 596
– MAK value 596
– TLC 594
– toxicity 597
diclofenac 199
– clinical interpretation 200
– elimination half-life 199, 843
– extraction 190
– GC 198
– GC–MS 109, 112, 116
– HPLC 109, 112, 116, 193
– medical assessment 199
– metabolism 200
– plasma/serum concentration 843
– serum concentration 199
dicoumarol
– elimination half-life 844
– HPLC 305
– plasma/serum concentration 844
dicycloverine
– TBPE test 179
didehydrofalcarinol 794
Dieffenbachia (dumb cane) 786–789, 792
dieldrin 578
– ADI value 580
– blood/plasma/serum concentration 581
– clinical interpretation 581
– MAK value 580
– medical assessment 580
– poisoning 581
– TLC 579
– toxicity 580
– toxicokinetics 580
diethyl ether
– headspace-GC 171, 173

diethylene glycol 531
– clinical interpretation 538
– GC 534
– intoxication 538
– medical assessment 535
– metabolism 536
– nephrotoxicity 749
diethylene glycol monobutyl ether
– GC 534
diethylene glycol monoethyl ether
– GC 534
diethylene glycol monomethyl ether
– GC 534
diethylpentenamide
– elimination half-life 844
– plasma/serum concentration 844
diethylthiambutene 868
difenacoum 308
– serum concentration 311
– toxicity 310
difenoxin 871
differential diagnosis 90
Digitalis purpurea L. (foxglove) 786, 804, 806
digitoxin 327
– clinical interpretation 329
– elimination half-life 844
– GC–MS 109, 112
– HPLC 109, 112
– immunoassay 328, 782
– medical assessment 329
– plasma/serum concentration 844
– quality assurance 64
– serum concentration 329
digoxin 331
– clinical interpretation 333
– elimination half-life 844
– GC–MS 109, 112
– HPLC 109, 112
– immunoassay 332, 782
– medical assessment 333
– plasma/serum concentration 844
– poisoning 753
– quality assurance 64
– serum concentration 333
dihydrocodeine 241, 874
– antidote 858
– elimination half-life 247, 844
– GC–MS 109, 112, 116, 247
– HPLC 109, 112, 116
– pharmacokinetics 247
– plasma/serum concentration 844
– TBPE test 179
dihydroergotamine
– elimination half-life 844
– plasma/serum concentration 844
– TBPE test 179
dihydromorphine
– GC–MS 247
dihydropyridines
– GC-MS 161
diisopropyl ether
– headspace-GC 171, 173
diisopropyl fluorophosphates (DFP) 683
diltiazem
– elimination half-life 844
– GC 274
– HPLC 273
– plasma/serum concentration 844
– TBPE test 180
dimenoxadol 868
dimepheptanol 868
dimetan
– TLC 569
dimethamphetamine
– TBPE test 179
dimethoate
– ADI value 596
– LD_{50} value 596
– MAK value 596
– TLC 594
N,N-dimethylacetamide
– BAT value 853
dimethylformamide
– BAT value 853
dimethyl nitrosamine
– liver toxin 748
dimethyl sulphate
– hepatotoxicity 749
dimethylthiambutene 868
dimetilan
– TLC 569–572
dimpylate, *see* diazinon
dinitrobenzene
– liver toxin 748
dinitrophenol
– hepatotoxicity 749
dioxacarb
– TLC 572
1,4-dioxane
– headspace-GC 171, 173
dioxaphethyl butyrate 868
dioxins
– hepatotoxicity 749
diphenhydramine 177, 383
– clinical interpretation 384
– elimination half-life 383, 844
– extractability 369

– GC 375
– GC–MS 109, 112, 116
– HPLC 109, 112, 116, 136, 371
– intoxication 138
– medical assessment 384
– metabolism 383
– plasma/serum concentration 844
– serum concentration 383
– TBPE test 179
diphenol laxatives
– GC-MS 158
diphenoxylate 872
diphenylarsine chloride (Clark I) 734
diphenylarsine cyanide (Clark II) 734
dipipanone 868
dipyrone, *see also* metamizole
– clinical interpretation 203
– elimination half-life 202, 844
– extraction 190
– GC 198
– GC–MS 109, 112, 116
– HPLC 109, 112, 116, 193
– medical assessment 202
– metabolism 203
– plasma/serum concentration 844
– serum concentration 202
diquat 583
– nephrotoxicity 749
disclaimer XXXI, 86
disopyramide
– clinical interpretation 277
– elimination half-life 277, 844
– GC 109, 112, 116, 274
– HPLC 109, 112, 116, 273
– immunoassay 781
– medical assessment 277
– metabolism 277
– plasma/serum concentration 844
– serum concentration 277
disulfiram
– drug-induced seizure 750
– elimination half-life 844
– plasma/serum concentration 844
disulfoton
– ADI value 596
– LD_{50} value 596
– MAK value 596
– TLC 594
diterpene ester 793
diuretics
– GC-MS 161
dixyrazine
– TBPE test 180
dodecane
– headspace-GC 171, 174
domperidone
– elimination half-life 844
– plasma/serum concentration 844
dosulepin
– elimination half-life 844
– GC–MS 408
– plasma/serum concentration 844
– TBPE test 179
dothiepin, *see* dosulepin
doxepin 417, 439
– clinical interpretation 439
– elimination half-life 844
– GC 405
– GC–MS 109, 112, 116, 408
– HPLC 109, 112, 116, 136, 397, 401
– HPLC-DAD 143
– immunoassay 781
– medical assessment 439
– metabolism 439
– overdosage 440
– plasma half-life 435, 440
– plasma/serum concentration 844
– serum concentration 417
– TBPE test 179
doxepin metabolite
– HPLC-DAD 143
doxylamine 384
– clinical interpretation 384
– elimination half-life 384, 844
– extractability 369
– GC 375
– GC–MS 109, 112, 116
– HPLC 109, 112, 116, 371
– medical assessment 384
– metabolism 384
– plasma/serum concentration 844
– serum concentration 384
– TBPE test 179
drofenine
– TBPE test 179
dronabinol 874
– elimination half-life 844
– plasma/serum concentration 844
drotebanol 868
drugs of abuse 463–507
duloxetin
– GC–MS 408
dumb cane (*Dieffenbachia*) 786–789, 792
dyshemoglobin 623–633

e

Echiichthys spp. (weeverfish) 827
ecstasy 463

EDAPA
– GC–MS/MS 698
elderberry 806
electrochemical method 46–53
– calibration 48
– detection for HPLC 50
– inverse voltammetry 49
Ellman method 757
– application 766
– instrumentation 765
emetine 794
– TBPE test 179
enalapril
– elimination half-life 844
– plasma/serum concentration 844
endrin 578
– ADI value 580
– MAK value 580
– medical assessment 580
– TLC 579
– toxicity 580
enoxacin
– elimination half-life 844
– plasma/serum concentration 844
Ephedra spp. 466, 789, 792
Ephedra vulgaris 799
ephedrine 792
– drug-induced seizure 750
– elimination half-life 844
– plasma/serum concentration 844
– TBPE test 180
epichlorohydrin
– hepatotoxicity 749
Epipremnum spp. 787–789, 793
eprazinone
– TBPE test 179
ergoline derivative 806
Erythroxylum coca 800
Eschscholtzia californica ingredients
– GC-MS 158
estazolam 874
etacrynic acid
– acid-base balance 752
ethanal (acetaldehyde)
– headspace-GC 171, 173
ethanol
– acid-base balance 752
– BAT value 520
– blood concentration 514–515
– clinical interpretation 515
– creatine kinase activity 750
– elimination half-life 844
– enzymatic determination 512
– headspace-GC 171, 173, 514, 519
– hepatotoxicity 749
– intoxication 515
– liver toxin 748
– medical assessment 514, 520
– plasma/serum concentration 844
– poisoning 753
– quality assurance 64
– serum concentration 514–516
ethchlorvynol 872
ethinamate 872
– elimination half-life 844
– plasma/serum concentration 844
ethiofencarb
– ADI value 574
– MAK value 574
– medical assessment 573
– TLC 568–572
ethion
– ADI value 596
– LD_{50} value 596
– MAK value 596
ethoprophos/ethoprop
– TLC 594
ethorphine 874
ethosuximide
– clinical interpretation 295
– elimination half-life 295, 844
– GC–MS 109, 112, 116
– HPLC 109, 112, 116, 289
– immunoassay 288
– medical assessment 295
– metabolism 296
– plasma/serum concentration 844
– serum concentration 295
2-ethoxyacetic acid
– elimation half-life 537
ethyl acetate
– headspace-GC 171, 173
ethylbenzene
– BAT value 527
– clinical interpretation 530
– headspace-GC 171, 173, 526
– medical assessment 527
– metabolism 528
– poisoning 530
ethylbromide
– hepatotoxicity 749
ethyl diethanolamine (EDEA) 724
ethylene chlorohydrin (2-chloroethanol)
– hepatotoxicity 749
ethylene glycol 531
– acid-base balance 752
– antidote 536, 539, 754, 858
– BAT value 537

– blood concentration 536
– clinical interpretation 538
– creatine kinase activity 750
– GC 534
– intoxication 538
– medical assessment 536
– metabolism 536
– nephrotoxicity 749
– plasma half-life 536
– poisoning 753
ethylene glycol dinitrate
– BAT value 854
ethylene glycol monobutyl ether
– BAT value 537
ethylene glycol monoethyl ether
– BAT value 537
(2-ethylhexyl) methylphosphonic acid (EHMPA) 689
ethylloflazepate 874
ethyl methyl ketone
– headspace-GC 171, 173
ethyl methylphosphonic acid (EMPA) 689
– CE-UV 694
– GC–FPD 697
– GC–MS 697
– GC–MS/MS 697–698
– GC–NICI-MS 697
– GC–NICI-MS/MS 697
– ion chromatography 694
– LC–MS 694
– LC–MS/MS 694
ethylmethylthiambutene 868
ethylmorphine
– TBPE test 179
ethylsilicate
– hepatotoxicity 749
eticyclidine 868
etilamfetamine 872
etodroxizine
– TBPE test 179
etomidate
– elimination half-life 844
– plasma/serum concentration 844
etonitazene 868
etoxeridine 868
etrimfos
– TLC 594
etryptamine 868
Eucalyptus spp. 789, 793
Euphorbia spp. 786–789, 793
eye irritant 733–734
exposure monitoring 766
extrelut extraction 119

f
falcarinol 794
famotidine
– elimination half-life 844
– plasma/serum concentration 844
felbamate
– elimination half-life 844
– plasma/serum concentration 844
felodipine
– elimination half-life 844
– plasma/serum concentration 844
fencamfamine 872
– TBPE test 179
fenetylline 874
fenfluramine
– elimination half-life 844
– plasma/serum concentration 844
fenitrothion
– ADI value 596
– blood/plasma/serum concentration 597
– LD_{50} value 596
– MAK value 596
– TLC 594
– toxicity 597
fenpropathrin 600
– ADI value 602
– LD_{50} value 602
fenproporex 874
fentanyl 222, 874
– clinical interpretation 227
– drug-induced seizure 750
– elimination half-life 844
– GC–MS 109, 112, 116, 224
– HPLC 109, 112, 116, 224
– immunoassay 223
– medical assessment 227
– metabolism 222
– plasma/serum concentration 844
– serum concentration 227
fenthion
– ADI value 596
– MAK value 596
– TLC 594
fenvalerate 600
– ADI value 602
– LD_{50} value 602
fexofenadine
– elimination half-life 844
– plasma/serum concentration 844
Ficus benjamina (weeping fig) 787–789, 794
fish 827
– poisoning 829
– tetrodotoxic 829
– toxin detection 829

FK-506
– elimination half-life 844
– plasma/serum concentration 844
flecainide
– clinical interpretation 278
– elimination half-life 278, 844
– GC 274
– GC–MS 109, 112, 116
– HPLC 109, 112, 116, 273
– medical assessment 278
– plasma/serum concentration 844
– serum concentration 278
– TBPE test 179
flocoumafen 308
– toxicity 310
fluanisone
– TBPE test 179
flucloxacillin
– elimination half-life 844
– plasma/serum concentration 844
flucytosine
– elimination half-life 844
– plasma/serum concentration 844
fludiazepam 874
flumazenil
– elimination half-life 844
– plasma/serum concentration 844
flunarizine
– TBPE test 179
flunitrazepam 874
– antidote 360
– clinical interpretation 358
– elimination half-life 358, 844
– GC–MS 109, 112, 116
– HPLC 109, 112, 116, 136, 355
– medical assessment 358
– metabolism 359, 362
– plasma/serum concentration 360, 844
fluoride 661, 699
– antidote 858
– BAT value 854
– clinical interpretation 664
– medical assessment 663
– plasma elimination half-life 663
– plasma/serum concentration 663
– poisoning 663–664
– potentiometry 661–663
– toxicity 663
fluoroorganophosphates 700
– GC–MS 702
5-fluorouracil
– elimination half-life 844
– plasma/serum concentration 844
fluoxetine
– elimination half-life 844
– GC–MS 109, 112, 116, 409
– HPLC 109, 112, 116, 136
– HPLC-DAD 146
– plasma/serum concentration 844
– serum concentration 435
flupentixol
– GC–MS 409
– TBPE test 179
fluphenazine
– elimination half-life 844
– GC–MS 409
– plasma/serum concentration 844
flurazepam 177, 874
– elimination half-life 845
– metabolism 359
– plasma/serum concentration 845
– TBPE test 179
fluspirilene
– GC–MS 409
fluvoxamine
– elimination half-life 845
– GC–MS 409
– plasma/serum concentration 845
folate
– drug-induced seizure 750
fomepizole (4-methylpyrazole) 523, 536, 539
fonofos
– TLC 594
forensic aspects 93f.
formetanate
– TLC 568–572
formic acid
– antidote 858
foxglove (*Digitalis purpurea* L.) 786
furanocoumarins 804
furathiocarb
– TLC 572
furethidine 869
furfural
– headspace-GC 171, 174
furosemide
– acid-base balance 752
– elimination half-life 845
– plasma/serum concentration 845

g

gabapentin 296
– elimination half-life 296, 745, 845
– GC–MS 109, 112
– HPLC 109, 112
– plasma/serum concentration 845
Galerina marginata 811

– antidote 858
gallopamil
– elimination half-life 845
– plasma/serum concentration 845
ganciclovir
– elimination half-life 845
– plasma/serum concentration 845
gas chromatography 35–38, 151
– headspace analysis, *see* headspace-gas chromatography
– injection 35
– detection 36
gas chromatography-mass spectrometry (GC-MS) 38–39, 152
– data evaluation 153
– full scan mode 39
– instrumentation 38
– ionization 38
– procedure 153
– quality assurance 154
– screening 195
gas chromatography–flame photometric detection
– (GC – FPD) 688
gas chromatography–N–P detector (GC–NPD) 688
gas chromatography–negative ion chemical ionization (GC-NICI-MS) 697
gases, irritant
– antidote 858
`general unknown' analysis 6, 107–182
– color test 175
– gas chromatography 151
– GC headspace analysis 165
– GC – mass spectrometry 152
– HPLC 108
gentamicin
– elimination half-life 845
– immunoassay 781
– plasma/serum concentration 845
GHB (γ-hydroxybutyrate) 488–492
– clinical interpretation 491
– elimination half-life 845
– GC–MS 489
– immunoassay 488
– medical assessment 491
– plasma half-life 492
– plasma/serum concentration 845
– serum concentration 492
glibenclamide 616
– blood/plasma/serum concentration 620
– elimination half-life 845
– GC–MS 109, 112, 116
– HPLC 109, 112, 116
– LC–MS 619
– plasma/serum concentration 845
glibornuride 616
– blood/plasma/serum concentration 620
– LC–MS 619
gliclazide 617
– blood/plasma/serum concentration 620
– LC–MS 619
glimepiride 617
– blood/plasma/serum concentration 620
– LC–MS 619
glipizide
– blood/plasma/serum concentration 620
– LC–MS 619
– metabolism 617
glucocorticoide
– anemia 751
glutethimide 872
– clinical interpretation 385
– creatine kinase activity 750
– elimination half-life 385, 845
– extractability 369
– GC 375
– GC–MS 109, 112, 116
– HPLC 109, 112, 116, 371
– medical assessment 385
– metabolism 385
– plasma/serum concentration 845
– serum concentration 385
glycerol
– nephrotoxicity 749
glyceryl trinitrate
– elimination half-life 845
– plasma/serum concentration 845
glycol ether 531
– BAT value 537
– clinical interpretation 539
– elimination half-life 537
– GC 534
– intoxication 539
– medical assessment 537
– metabolism 537
glycol monomethyl ether 537
glycols 531–539
– clinical interpretation 535
– GC 532
– medical assessment 535
– poisoning 531
glycosides 804
– cardiac, *see* cardiac glycosides
– cyanogenic 804, 806
– digitalis 806
– strophantidin 806
gonyautoxin 828

griseofulvin
– elimination half-life 845
– plasma/serum concentration 845
guanoxon
– TBPE test 179
Gymnopilus spectabilis 813
Gyromitra esculenta 812
– antidote 858
gyromitrin 812

h

halazepam 874
– metabolism 359
– nordazepam elimination half-life 845
– nordazepam plasma/serum concentration 845
halothane
– BAT value 854
haloperidol 434
– clinical interpretation 434
– elimination half-life 845
– GC 405
– GC–MS 109, 116, 409
– HPLC 109, 116, 397
– intoxication 434
– medical assessment 434
– metabolism 436
– plasma half-life 434
– plasma/serum concentration 845
– serum concentration 435
– TBPE test 179
haloxazolam 874
hallucinogen
– GC-MS 158
Hapalochlaena sp. 830
hashish 475
headspace-gas chromatography (headspace-GC) 40 – 43, 165
– calibration 42
– multiple headspace extraction 43
– practicability 43
– sample introduction 41
– sample preparation 41
– sampling 40
heavy metals
– AAS, ICP-MS, VM 159
Hedeoma pulegioides 799
Hedera helix 789, 794
hemorrhagic disorder, toxic 751
henbane (*Hyoscyamus niger* L.) 785–786
heparin
– antidote 858
hepatotoxic chemicals 749
heptabarbital
– elimination half-life 845
– plasma/serum concentration 845
heptachlor 578
– toxicokinetics 580
n-heptane
– headspace-GC 171, 173
1-heptanol
– headspace-GC 171, 174
heptenophos
– TLC 594
heroin 241
– antidote 858
– creatine kinase activity 750
– elimination half-life 247
– GC–MS 112
– HPLC 112
– pharmacokinetics 247
hexachlorobenzene 578, 582
– BAT value 854
– metabolite 582
– nephrotoxicity 749
hexachlorocyclohexane (HCH) 576–578
– ADI value 580
– MAK value 580
– toxicokinetics 580
γ- hexachlorocyclohexane (lindane) 576–578
– ADI value 580
– BAT value 854
– blood/plasma/serum concentration 581
– clinical interpretation 581–582
– drug-induced seizure 750
– MAK value 580
– medical assessment 580
– metabolite 582
– poisoning 581
– TLC 579
– toxicity 580
– toxicokinetics 580
hexachloroethane
– Fujiwara reaction 542–543
n-hexane
– BAT value 854
– headspace-GC 171, 173
1-hexanol
– headspace-GC 171, 174
2-hexanone
– BAT value 854
– headspace-GC 171, 173
hexobarbital
– elimination half-life 845
– immunoassay 346
– plasma/serum concentration 845
high – performance liquid chromatography, *see* HPLC

homofenazine
– TBPE test 179
hornet toxin
– creatine kinase activity 750
HPLC (high-performance liquid chromatography) 32–35
– background correction 128
– condition 120, 149
– dedicated system 400
– design of spectra library 129
– detection 33
– electrochemical detection 50
– gradient elution 125
– impact of the mobile phase 127
– impact of pH 127
– isocratic system 123
– library of absorption spectra 130
– limits of identification 128
– mechanized 148
– packing material 32
– photodiode array detector (DAD) 108
– spectra library 126
– Remedi 150
– sample preparation 108
– screening 190
– spectra library 149
– uniformity 128
HPLC-DAD 108
– metabolite profiles 134
HPLC-MS 430–432
HPTLC (high-performance thin-layer Chromatography 422–429
hydralazine
– anemia 751
– elimination half-life 845
– plasma/serum concentration 845
hydrazine
– antidote 858
– hepatotoxicity 749
hydrocarbons
– chlorinated, *see* chlorinated hydrocarbons
hydrochlorothiazide
– elimination half-life 845
– plasma/serum concentration 845
hydrocodone 241, 874
– elimination half-life 845
– plasma/serum concentration 845
– TBPE test 179
hydrofluoric acid
– antidote 858
hydrogen fluoride
– BAT value 854
hydromorphinol 869
hydromorphone 241, 874
– elimination half-life 845
– plasma/serum concentration 845
γ-hydroxybutyrate, *see* GHB
N_7-(2-hydroxyethyl thioethyl) guanine (HETEG) 723
hydroxypethidine 869
hydroxyzine
– GC–MS 409
– TBPE test 179
hyoscyamine 791
Hyoscyamus niger L. (henbane) 785–786
– antidote 858
hyperthermia
– antidote 858
hypnotics 339–350, 351–364, 367–390
– barbiturate 339–350
– benzodiazepine 351–364
– clinical interpretation 380
– GC 372
– GC-MS 158
– HPLC 368
– ion-selective electrode 376
– medical assessment 380
– photometry (color test) 377
hypo-osmolar infusion
– drug-induced seizure 750

i

ibotenic acid 813
ibuprofen
– clinical interpretation 201
– elimination half-life 201, 845
– extraction 190
– GC 198
– GC–MS 109, 112, 116
– HPLC 109, 112, 116, 193
– medical assessment 201
– metabolism 201
– plasma/serum concentration 845
– serum concentration 201
Ilex aquifolium 787–789, 795
imipramine 177, 418, 432, 440
– clinical interpretation 432, 440
– elimination half-life 845
– GC 405
– GC–MS 109, 112, 116, 409
– HPLC 109, 112, 116, 136, 397, 401
– HPLC-DAD 143
– immunoassay 781
– medical assessment 432, 440
– metabolism 441
– overdose 440
– plasma half-life 435, 440
– plasma/serum concentration 845

– TBPE test 179
imipramine metabolite
– HPLC-DAD 143
immunoassay 25–29
– detection 28
indicator tube method 647–650
indometacin
– anemia 751
– elimination half-life 845
– extraction 190
– GC–MS 110, 113, 116
– HPLC 110, 113, 116, 193
– plasma/serum concentration 845
INH
– elimination half-life 845
– plasma/serum concentration 845
inhalant
– abuse 553–555
– clinical interpretation 555
– GC 555
– medical assessment 555
insecticide carbamates, *see* carbamates
insects 832
– venom detection 832
insence 479
insulin 613
– clinical interpretation 620
– half-life 619
– immunoassay 613–614
– medical assessment 619
– poisoning 753
inverse voltammetry, *see* voltammetry
iodine (radioactive)
– antidote 858
1-iodobutane
– headspace-GC 171, 173
ion chromatography 694
Ipomoea violacea (morning glory) 806
Iris germanica 789, 795
iron
– antidote 858
– anemia 751
– nephrotoxicity 749
– toxic hemorrhagic disorder 751
isofenphos
– blood/plasma/serum concentration 597
– TLC 594
– toxicity 597
isolan
– TLC 569
isomethadone 872
isoniazid
– acid-base balance 752
– antidote 858
– drug-induced seizure 750
– elimination half-life 845
– plasma/serum concentration 845
isopropanol, *see* 2-propanol
isothipendyl
– TBPE test 179
itraconazole
– elimination half-life 845
– plasma/serum concentration 845

j

jellyfish 827
jequirity bean 804
jimsonweed, *see Datura stramonium L.*
judgment
– longitudinal 78
– transverse 79

k

kava 799
kebuzone
– extraction 190
– GC 198
– HPLC 193
ketamine 492–493
– clinical interpretation 493
– elimination half-life 493, 845
– GC–MS 492
– HPLC 492
– immunoassay 492
– medical assessment 493
– metabolism 493
– plasma/serum concentration 845
– serum concentration 493
ketanserin
– elimination half-life 845
– plasma/serum concentration 845
ketazolam 874
– elimination half-life 845
– metabolism 359
– plasma/serum concentration 845
ketobemidone
– elimination half-life 845
– plasma/serum concentration 845
– TBPE test 180
ketoconazole
– elimination half-life 845
– plasma/serum concentration 845
ketones
– clinical interpretation 522
– headspace-GC 517
– highly volatile 511–523
– medical assessment 520
ketoprofen

– elimination half-life 845
– HPLC 193
– plasma/serum concentration 845
khat (*Catha edulis*) 799
kraits 833

l

labetalol
– elimination half-life 845
– HPLC 457
– plasma/serum concentration 845
– TBPE test 180
lacrimator 733
LADME (liberation, absorption, distribution, metabolism, and excretion) 776
lamotrigine 296
– clinical interpretation 296
– elimination half-life 296, 845
– GC–MS 110, 113, 116, 292–294, 409
– HPLC 110, 113, 116, 289
– medical assessment 296
– plasma/serum concentration 845
– serum concentration 295
Lantana camara L. 805
lanthanum trifluoride electrode 48
Lathyrus sativus (chickling pea) 800
Latrodectus tredecimguttatus (black widow spider) 831
laxatives, anthraquinone-type
– GC-MS 161
laxatives, diphenol-type
– GC-MS 158
LD_{50} 560
lead
– acid-base balance 752
– antidote 858
– drug-induced seizure 750
– nephrotoxicity 749
lectins 797
lefetamine 869
Leonorus sibiricus 479
lepidella syndrome 812
Lepiota helveola 811
levacetylmethadol 874
levamfetamine 872
levodopa
– elimination half-life 845
– plasma/serum concentration 845
levomepromazine 418, 441
– acute intoxication 442
– antidote 858
– clinical interpretation 442
– elimination half-life 845, 846
– GC 405
– GC–MS 110, 113, 116, 409
– HPLC 110, 113, 116, 397
– medical assessment 442
– metabolism 442
– plasma half-life 441
– plasma/serum concentration 845, 846
– serum concentration 418, 435
levomethadone 874
– elimination half-life 846
– plasma/serum concentration 846
– TBPE test 179
levomethorphan 869
levomoramide 872
levorphanol 872
– elimination half-life 845
– plasma/serum concentration 845
levophenacylmorphan
– elimination half-life 846
– plasma/serum concentration 846
lewisite (L, 2-chlorovinylarsine dichloride) 703, 726
– hemoglobin adduct 729
– hydrolysis 726
– LD_{50} value 681
– metabolite 727
– toxicity 681
lidocaine 177
– clinical interpretation 279
– drug-induced seizure 750
– elimination half-life 278, 846
– GC 274
– GC–MS 110, 113, 116
– HPLC 110, 113, 116, 136, 273
– immunoassay 781
– medical assessment 278
– metabolism 279
– plasma/serum concentration 846
– serum concentration 278
linamarin 806
lindane, *see* γ-hexachlorocyclohexane
lionfish (*Pterois* spp.) 827
liquid chromatography–mass spectrometry (LC – MS) 43–46
– plasma 690–692
– urine 693–694
lisinopril
– elimination half-life 846
– plasma/serum concentration 846
lithium
– acid-base balance 752
– AAS, FAES, ISE 159
– drug-induced seizure 750
– elimination half-life 846
– plasma/serum concentration 846

– quality assurance 64
liver toxin 748
local anesthetics
– GC–MS 158
lofentanil 869
lofepramine
– elimination half-life 846
– GC–MS 409
– plasma/serum concentration 846
– TBPE test 180
Lophophora williamsii 498
loprazolam 874
– elimination half-life 846
– plasma/serum concentration 846
lorazepam 361, 874
– antidote 362
– clinical interpretation 362
– elimination half-life 361, 846
– GC–MS 110, 113, 116, 409
– HPLC 110, 113, 116
– medical assessment 362
– metabolism 359
– plasma/serum concentration 361, 846
lormetazepam 874
– antidote 363
– clinical interpretation 363
– elimination half-life 362, 846
– GC–MS 110, 113, 116
– HPLC 110, 113, 116, 136
– medical assessment 363
– plasma/serum concentration 363, 846
LSD (lysergic acid diethylamide) 463, 869
– clinical interpretation 498
– creatine kinase activity 750
– elimination half-life 846
– GC 498
– GC–MS 110, 113
– HPLC 110, 113, 495
– immunoassay 494
– medical assessment 498
– metabolism 494
– minimal concentration to be detected 75
– plasma concentration 498
– plasma half-life 498
– plasma/serum concentration 846
lung-damaging agent 729–732
β-lyase 715
– ESI-LC–MS/MS 708
– GC–MS 708, 715–717
– GC–MS/MS 708
– LC–MS/MS 717
– metabolite 708
lysergic acid diethylamide, *see* LSD
lysergide, *see* LSD

m

Ma huang 799
magnesium
– nephrotoxicity 749
MAK value 562
malathion
– ADI value 596
– blood/plasma/serum concentration 597
– LD_{50} value 596
– MAK value 596
– nephrotoxicity 749
– TLC 594
– toxicity 597
mamba 833
manganese
– hepatotoxicity 749
– nephrotoxicity 749
Manihot esculenta (cassava) 800, 806
maprotiline 418, 442
– clinical interpretation 442
– elimination half-life 846
– GC 405
– GC–MS 110, 113, 117, 409
– HPLC 110, 113, 117, 397, 400, 401
– intoxication 443
– medical assessment 442
– metabolism 443
– overdose 442
– plasma half-life 442
– plasma/serum concentration 846
– serum concentration 418, 435
– TBPE test 179
marihuanilla 479
marijuana 475
marine animal 826
materials for investigation 11
– containers 15
– evaluation 20
– identification
– judicial preconditions 11,23
– kind of materials
– request form 19
– sampling 11, 15
– storage 22
– transport 19
maximum allowable concentration (MAC) 562
maximum permissible concentration (MCP) 562
mazindol 872
MBDB (methylbenzodioxazolylbutanamine) 463–470
MDA (3,4-methylenedioxyamphetamine) 463–470

– elimination half-life 846
– GC–MS 110, 113, 117
– HPLC 110, 113, 117
– plasma/serum concentration 846
MDE
– GC–MS 110, 117
– HPLC 110, 117
MDEA (3,4-methylenedioxyethylamphetamine, Eve) 463–470
– metabolism 464
MDMA (3,4-methylenedioxymethamphetamine, Adam) 463–470
– elimination half-life 846
– GC–MS 110, 113, 117
– HPLC 110, 113, 117
– metabolism 465
– plasma/serum concentration 846
mebendazole
– elimination half-life 846
– plasma/serum concentration 846
mebhydrolin
– TBPE test 179
mecloqualone 869
mecloxamine
– TBPE test 179
meclozine
– TBPE test 180
medazepam 874
– elimination half-life 846
– GC–MS 408
– HPLC 136
– metabolism 360
– plasma/serum concentration 846
medical interpretation 89–91
mefenamic acid
– drug-induced seizure 750
mefenorex 872
melitracene
– elimination half-life 846
– plasma/serum concentration 846
– TBPE test 179
melperone 418
– elimination half-life 846
– GC 405
– GC–MS 408
– HPLC 397
– plasma/serum concentration 846
– serum concentration 419
– TBPE test 179
mematine
– GC–MS 408
Mentha pulegium 799
meperidine (pethidine) 228, 874
– antidote 858
– clinical interpretation 230
– drug-induced seizure 750
– elimination half-life 229, 846
– GC–MS 229
– HPLC 229
– immunoassay 228
– medical assessment 228
– metabolism 228
– plasma/serum concentration 846
– serum concentration 229
– TBPE test 179
mepindolol
– HPLC 457
mepivacaine
– elimination half-life 846
– plasma/serum concentration 846
– TBPE test 180
meprobamate 386, 872
– clinical interpretation 386
– elimination half-life 386, 846
– extractability 369
– GC 375
– GC-MS 110, 113, 158
– HPLC 110, 113, 371
– medical assessment 386
– plasma/serum concentration 846
– serum concentration 386
mequitazine
– TBPE test 179
mercaptodimethur
– TLC 572
6-mercaptopurine
– acid-base balance 752
– elimination half-life 842
– plasma/serum concentration 842, 846
mercury
– acid-base balance 752
– antidote 858
– BAT value 854
mesalazine
– elimination half-life 846
– plasma/serum concentration 846
mescaline 498
– metabolism 499
– plasma half-life 499
– serum concentration 499
mesitylene
– headspace-GC 171, 174
mesocarb 872
mesoridazine
– elimination half-life 850
– plasma/serum concentration 850
mesuximide
– elimination half-life 846

– plasma/serum concentration 846
metabolite
– identification 131
– shift of relative retention times 132
– UV spectra 132
metamfetamine 872
metamizole, *see also* dipyrone
– clinical interpretation 203
– elimination half-life 202, 846
– extraction 190
– GC 198
– GC–MS 110, 113, 117
– HPLC 110, 113, 117, 136, 193
– medical assessment 202
– metabolism 203
– plasma/serum concentration 846
– serum concentration 202
metazocine 869
metformin
– elimination half-life 846
– plasma/serum concentration 846
methadone 230, 874
– antidote 858
– clinical interpretation 239
– creatine kinase activity 750
– elimination half-life 846
– GC–MS 110, 113, 117, 236
– HPLC 110, 113, 117, 231
– immunoassay 231
– metabolism 240
– minimal concentration to be detected 75
– L- and D-Methadone 234
– medical assessment 238
– plasma concentration 238
– plasma/serum concentration 846
methamidophos
– TLC 594
methamphetamine
– elimination half-life 846
– GC–MS 110, 113, 117
– HPLC 110, 113, 117
– plasma/serum concentration 846
methanol
– acid-base balance 752
– antidote 523, 754, 858
– BAT value 520, 854
– blood concentration 521
– clinical interpretation 522
– elimination half-life 521
– headspace-GC 171, 173, 519
– hepatotoxicity 749
– intoxication 521–522
– marker of alcoholism 522
– medical assessment 521
– nephrotoxicity 749
– poisoning 753
– serum concentration 521
methanthelinium bromide
– TBPE test 179
methaqualone 387, 872
– clinical interpretation 387
– elimination half-life 386, 846
– extractability 369
– GC 198, 375
– GC–MS 113, 117
– HPLC 113, 117, 136, 371
– immunoassay 367
– medical assessment 387
– minimal concentration to be detected 75
– plasma/serum concentration 846
– serum concentration 386
methemoglobin 628
– clinical interpretation 630
– medical assessment 630
– oximetry 628
– spectrophotometry 628–630
methemoglobin-inducing agents 631
– acid-base balance 752
– methemoglobin measurement 159, 628–630
methemoglobinemia 631
– antidote 858
methidathion
– TLC 594
methimazole
– elimination half-life 846
– plasma/serum concentration 846
methiocarb
– ADI value 574
– MAK value 574
– medical assessment 573
– TLC 569
methods for toxicological analysis
– capillary electrophoresis 54
– electrochemical detection for HPLC 50
– electrochemical methods 46
– gas chromatography 35
– GC – mass spectrometry 38, 152
– headspace GC 40
– HPLC 32
– high-performance TLC 422
– immunoassay 25
– inverse voltammetry 49
– liquid chromatography – mass spectrometry 43
– potentiometry 46
– thin-layer chromatography 29
methohexital 340

– clinical assessment 345
– elimination half-life 345
– GC 343
– GC–MS 110, 113, 117
– HPLC 110, 113, 117, 340
– immunoassay 346
– medical assessment 345
methomyl 564
– ADI value 574
– blood concentration 575
– clinical interpretation 575
– LD_{50} value 573
– MAK value 574
– medical assessment 573
– poisoning 575
– TLC 568
methotrexate
– antidote 858
– immunoassay 781
methotrimeprazine, *see* levomepromazine
2-methoxyacetic acid
– elimation half-life 537
methoxychlor 576, 582
– ADI value 580
– blood/plasma/serum concentration 581
– MAK value 580
– TLC 579
methyl acetate
– headspace-GC 171, 173
2-methyl-1-butanol
– headspace-GC 171, 173
2-methyl-2-butanol
– headspace-GC 171, 173
N-methyl-*N*(*tert*-butyldimethylsilyl)-trifluoroacetamide (MTBSTFA) 695
methyl cellulose
– nephrotoxicity 749
methyl chloride
– hepatotoxicity 749
methylchloroform, see 1,1,1-trichloroethane
methyldesorphine 869
methyl diethanolamine (MDEAN) 724
methyldihydromorphine 869
methylenedioxyamphetamine
– elimination half-life 846
– plasma/serum concentration 846
methylenedioxymethylamphetamine
– elimination half-life 846
– plasma/serum concentration 846
β-methyldigoxin 336
methyldopa
– elimination half-life 846
– plasma/serum concentration 846
methyl ethyl ketone, *see* 2-butanone
methyl isobutyl ketone, *see* 4-methyl – pentane-2-on
4-methyl-pentane-2-on (methyl isobutyl ketone)
– BAT value 854
– headspace-GC 172, 173
2-methyl-2-pentanol
– headspace-GC 172, 173
4-methyl-2-pentanol
– headspace-GC 171, 173
methylphenidate 177, 419, 451, 872, 874
– clinical interpretation 451
– elimination half-life 846
– GC–MS 408
– medical assessment 451
– plasma half-life 451
– plasma/serum concentration 846
– serum concentration 435
– TBPE test 179
methylphenobarbital 874
methylphosphonic acid (MPA)
– derivative 689
– GC–FPD 698
– GC–MS 697
– ion chromatography 694
– LC–MS 694
2-methyl-1-propanol
– headspace-GC 171, 173
methyprylone 872
– elimination half-life 846
– extractability 369
– plasma/serum concentration 846
4-methylpyrazole (fomepizole) 523, 536, 539
methylxanthines
– drug-induced seizure 750
– GC-MS 158
metildigoxin
– elimination half-life 847
– plasma/serum concentration 847
metipranolol
– elimination half-life 847
– HPLC 457
– plasma/serum concentration 847
– TBPE test 180
metixene
– TBPE test 179
metoclopramide
– antidote 858
– elimination half-life 847
– plasma/serum concentration 847
– TBPE test 179
metofenazate
– HPLC 136

metopon 870
metoprolol
– elimination half-life 847
– GC–MS 110, 113, 117
– HPLC 110, 113, 117, 457
– metabolism 459
– plasma half-life 458
– plasma/serum concentration 459, 847
– TBPE test 180
Metrology of Qualitative Chemical Analysis (MEQUALAN) 68
metronidazole
– drug-induced seizure 750
– elimination half-life 847
– plasma/serum concentration 847
mevinphos/phosdrin
– ADI value 596
– LD_{50} value 596
– MAK value 596
– TLC 594
mexiletine
– clinical interpretation 280
– elimination half-life 280, 847
– GC 274
– GC–MS 110, 113, 117
– HPLC 110, 113, 117, 273
– medical assessment 280
– metabolism 280
– plasma/serum concentration 847
– serum concentration 280
mianserin
– elimination half-life 847
– GC–MS 408
– plasma/serum concentration 847
– TBPE test 179
midazolam 874
– antidote 364
– clinical interpretation 364
– elimination half-life 363, 847
– GC–MS 110, 113, 117
– HPLC 110, 113, 117, 136
– medical assessment 364
– metabolism 360, 364
– plasma/serum concentration 363, 847
minacide
– TLC 569
mirtazapine 419, 443
– clinical interpretation 443
– elimination half-life 443, 847
– GC–MS 408
– intoxication 444
– medical assessment 443
– plasma/serum concentration 847
– serum concentration 419, 435
moclobemide 419
– elimination half-life 847
– GC–MS 408
– plasma/serum concentration 847
modafinil 452
– clinical interpretation 452
– elimination half-life 452, 847
– GC–MS 408
– medical assessment 452
– plasma/serum concentration 435, 847
mollusca 830
molsidomine
– elimination half-life 847
– plasma/serum concentration 847
monitoring
– therapeutic drug, *see* TDM
monkshood (*Aconitum napellus* L.) 785–786
monoamine oxidase inhibitor
– toxic hemorrhagic disorder 751
monocrotophos
– blood/plasma/serum concentration 597
– toxicity 597
morazone
– TBPE test 179
morning glory 806
morpheridine 870
morphine 240
– antidote 858
– clinical interpretation 248
– elimination half-life 247, 847
– GC–MS 110, 113, 117, 244, 247
– HPLC 110, 113, 117, 244
– immunoassay 243
– medical assessment 247
– metabolism 242
– pharmacokinetics 247
– plasma/serum concentration 847
D_3-morphine
– GC–MS 247
muscarine 812
muscazon 813
muscimol 813
muscle relaxants
– acid-base balance 752
– central, GC-MS 158
– peripheral, LC–MS 159
mushroom
– classification 810
– detection of toxin 816
– identification 815
– poisonous 809–822
– toxic 748–749
– toxic hemorrhagic disorder 751
myrophine 870

n

nabilone 874
nadolol
– HPLC 457
naftidrofuryl
– TBPE test 179
nalidixic acid
– drug-induced seizure 750
naloxone
– elimination half-life 847
– plasma/serum concentration 847
naphthalene
– chlorinated 749
– hepatotoxicity 749
Narcissus pseudonarcissus 789, 796
neostigmine
– antidote 858
– elimination half-life 847
– plasma/serum concentration 847
nephrotoxic compound 749
Nerium oleander (oleander) 806
nerve agent 682–703
– free 684–686
– GC–MS 685, 695–699
– LC–MS 690–694
– metabolism 684
– metabolite 686
netilmicin
– elimination half-life 847
– plasma/serum concentration 847
neuroleptics 393–452
– clinical interpretation 432
– color test 394
– GC 404
– GC–MS 112, 407
– group assay 393
– HPLC 112, 396
– HPLC-MS 430–432
– HPTLC 422–429
– immunoassay 393
– medical assessment 432
nicocodine 872
nicodicodine 872
nicomorphine 870
Nicotiana tabacum L. 639, 806
nicotine 639
– clinical interpretation 641
– elimination half-life 847
– GC–MS 195, 640
– HPLC 190–194, 640
– immunoassay 640
– medical assessment 641
– metabolism 641
– plasma elimination half-life 641
– plasma/serum concentration 847
– poisoning 641, 806
– serum concentration 641
– TBPE test 180
– toxicity 641
nicotinic acid
– elimination half-life 847
– plasma/serum concentration 847
nifedipine 317
– clinical interpretation 318
– elimination half-life 847
– GC–MS 110, 113, 117
– HPLC 110, 113, 117, 318
– medical assessment 318
– metabolism 319
– plasma/serum concentration 847
– serum concentration 318
niflumic acid
– elimination half-life 847
– plasma/serum concentration 847
nimetazepam 874
nitrazepam 874
– elimination half-life 847
– GC–MS 110, 113, 117
– HPLC 110, 113, 117, 136
– metabolism 360
– plasma/serum concentration 847
nitriles
– antidote 858
nitrobenzene
– anemia 751
– BAT value 854
– headspace-GC 172, 174
– hepatotoxicity 749
nitrofurantoin
– anemia 751
– elimination half-life 847
– plasma/serum concentration 847
nitrogen mustard 723
– antidote 858
– DNA-adduct 725
– guanine alkylated with HN-2 726
– HN-1 (*N*-ethyl-2,2′-dichloroethyl amine) 703, 724
– HN-2 (*N*-methyl-2,2′-dichloroethyl amine 703, 724
– HN-3 (tris-(2-chloroethyl amine) 703, 724
– hydrolysis 723–724
– LC–MS 725
– LC–MS/MS 724
– LD_{50} value 681
– metabolite 723
– toxicity 681
nitroglycerin

– elimination half-life 847
– plasma/serum concentration 847
1-nitropropane
– headspace-GC 172, 173
nitroprusside
– elimination half-life 847
– plasma/serum concentration 847
nomifensine
– elimination half-life 847
– plasma/serum concentration 847
– TBPE test 180
nonachlor 582
n-nonane
– headspace-GC 172, 173
noracymethadol 870
norclomipramine
– HPLC 401
norclozapine
– HPLC 401
– HPLC – MS 432
– plasma/serum concentration 843
norcodeine 870
nordazepam, 360, 874
– elimination half-life 847
– GC–MS 408
– HPLC 143, 355
– plasma/serum concentration 847
nordiazepam, *see* nordazepam
nordoxepin
– HPLC 401
– immunoassay 781
norflunitrazepam
– HPLC 355
norfluoxetine
– elimination half-life 844
– HPLC-DAD 146
– plasma/serum concentration 844
norlevorphanol 870
normethadone 874
normorphine 870
norpipanone 870
norpropoxyphene
– plasma/serum concentration 843
nortilidine
– GC–MS 261
– TBPE test 179
nortriptyline, *see also* amitriptyline 414, 432–434
– clinical interpretation 432–434
– elimination half-life 841, 847
– GC 405
– GC–MS 110, 113, 117, 408
– HPLC 110, 113, 117, 397, 401
– immunoassay 781
– medical assessment 432–434
– metabolism 433
– overdose 434
– plasma half-life 434
– plasma/serum concentration 841, 847
– serum concentration 414, 435
– TBPE test 179
nose and throat irritant 734
noxiptiline (noxiptyline)
– HPLC 136
– TBPE test 179
nutmeg ingredients
– GC-MS 158

o

1-octanol
– headspace-GC 172, 174
ofloxacin
– TBPE test 180
olanzapine 419, 444
– clinical interpretation 444
– elimination half-life 444, 847
– GC–MS 408
– intoxication 444
– medical assessment 444
– plasma/serum concentration 847
– serum concentration 419, 435
oleander 806
omethoate
– LD_{50} value 596
– TLC 594
opiates 215–265
– acid-base balance 752
– antidote 858
– clinical interpretation 248
– GC-MS 158
– group immunoassay 216
– medical assessment 247
– minimal concentration to be detected 75
– pharmacokinetics 247
opioids 215–265
– antidote 858
– drug-induced seizure 750
– GC-MS 158
– group immunoassay 216
– poisoning 753
opipramol 419, 432, 445
– clinical interpretation 445
– elimination half-life 847
– GC 405
– GC–MS 110, 113, 117, 408
– HPLC 110, 113, 117, 136, 397
– immunoassay 781
– intoxication 445

– medical assessment 445
– plasma half-life 445
– plasma/serum concentration 847
– serum concentration 420, 435
– TBPE test 180
orellanine 812
orelline 812
organophosphorus compounds (OPC, pesticides) 591–599
– ADI value 596
– antidote 858
– clinical interpretation 598
– drug-induced seizure 750
– enzyme complex 700
– GC–MS 113
– general unknown analysis 593
– HPLC 113
– intoxication 598
– LD_{50} value 596
– MAK value 596
– spectrophotometry 595
– TLC 593–595
– toxicity 595, 597
osmolal gap 747
oxalates 790–792
– antidote 858
oxalic acid 791–797
– nephrotoxicity 749
Oxalis acetosella L. (wood sorrel) 804
oxamyl
– ADI value 574
– LD_{50} value 573
– MAK value 574
oxazepam 874
– elimination half-life 847
– GC–MS 408
– HPLC 355
– metabolism 360
– plasma/serum concentration 847
oxazolam 874
– metabolism 360
oxcarbazepine 297
– elimination half-life 297, 847
– GC–MS 110, 113, 117
– HPLC 110, 113, 117
– metabolism 297
– plasma/serum concentration 847
– serum concentration 295
oxprenolol 459
– elimination half-life 847
– GC–MS 110, 113, 117
– HPLC 110, 113, 117, 457
– plasma half-life 457
– plasma/serum concentration 460, 847
– TBPE test 180
oxychlordane 582
oxycodone 241, 249, 874
– clinical interpretation 252
– elimination half-life 251, 847
– GC–MS 250
– HPLC 250
– immunoassay 250
– medical assessment 251
– metabolism 251
– plasma/serum concentration 847
– serum concentration 251
oxydemeton-methyl
– TLC 594
oxytocin
– drug-induced seizure 750
oximetry 624
oxymorphone 241, 872
oxyphenonium bromide
– TBPE test 180

p

Panaeolus subalteatus 503
pancuronium
– elimination half-life 847
– plasma/serum concentration 847
pancuronium bromide
– TBPE test 180
Paneolina foensecii 813
Paneolina spp. 813
pantherine syndrome 813
Papaver somniferum 800
papaverine
– elimination half-life 847
– plasma/serum concentration 847
– TBPE test 180
paper strip 650–652
paracetamol, *see also* acetaminophen
– antidote 858
– clinical interpretation 205
– elimination half-life 204, 847
– extraction 190
– GC 110, 113, 117, 198
– HPLC 110, 113, 117, 193
– immunoassay 203
– liver toxin 748
– medical assessment 204
– metabolism 206
– plasma/serum concentration 847
– poisoning 753
– serum concentration 204
paraldehyde
– acid-base balance 752
– elimination half-life 848

– plasma/serum concentration 848
paralytic shellfish poisoning (PSP) 828
paraoxon
– elimination half-life 848
– plasma/serum concentration 848
– toxicokinetics 597
paraquat 582–591
– clinical interpretation 591
– color test 159, 583
– GC–MS 113
– general unknown analysis 583
– half-life 591
– HPLC 113
– intoxication 591
– LD_{50} value 590
– liver toxin 748
– MAK value 590
– medical assessment 590
– nephrotoxicity 749
– poisoning 588, 590
– spectrophotometry 585–590
– toxicity 591
parasorbic acid 804
parathion
– ADI value 596
– BAT value 854
– blood/plasma/serum concentration 597
– clinical interpretation 598
– elimination half-life 848
– intoxication 598
– LD_{50} value 596
– MAK value 596
– plasma/serum concentration 848
– toxicity 597
– toxicokinetics 597
parathion-ethyl 592
– TLC 594
parathion-methyl
– ADI value 596
– LD_{50} value 596
– MAK value 596
parkinsonian symptoms 858
paroxetine
– elimination half-life 848
– GC–MS 110, 113, 117, 408
– HPLC 110, 113, 117
– plasma/serum concentration 848
Paxillus involutus (poison paxillus) 813
PBB *and* PCB, *see* biphenyls
PCP, *see* phencyclidine
peace lily 789, 797
Pelagia noctiluca 826
pemoline 874
– GC–MS 408
penbutolol
– HPLC 457
– TBPE test 179
pennyroyal oil 799
pentachloroethane
– Fujiwara reaction 542
pentafluorobenzyl bromide 695
1-pentanol
– headspace GC 172, 173
2-pentanone
– headspace GC 172, 173
pentazocine 177, 252, 874
– antidote 858
– clinical interpretation 257
– drug-induced seizure 750
– elimination half-life 257, 848
– GC–MS 110, 113, 117, 255
– HPLC 110, 113, 117, 253
– medical assessment 257
– metabolism 253
– plasma/serum concentration 848
– serum concentration 257
– TBPE test 179
pentobarbital 874
– elimination half-life 848
– GC–MS 110, 113, 117, 346
– HPLC 110, 113, 117, 345
– immunoassay 345–346
– plasma/serum concentration 848
pentoxifylline
– elimination half-life 848
– plasma/serum concentration 848
pentyl acetate
– headspace-GC 172, 174
iso-pentyl acetate
– headspace-GC 172, 174
perazine 420, 446
– clinical interpretation 446
– elimination half-life 848
– GC 405
– GC–MS 110, 113, 117, 408
– HPLC 110, 113, 117, 136, 397
– intoxication 446
– medical assessment 446
– plasma half-life 446
– plasma/serum concentration 848
– serum concentration 420, 435
– TBPE test 179
perchloroethylene, *see* 1,1,2,2-tetrachloroethene
permethrin 600
– ADI value 602
– LD_{50} value 602
permissible exposure limit (PEL) 562

perphenazine 177
– GC–MS 408
– TBPE test 179
pesticides 559–603
– ADI value 561
– BAT value 563
– BEI value 563
– biological monitoring 562
– classification 559–560
– GC–MS 112, 113
– HPLC 112, 113
– poisoning 563
– TLV 562
– toxicity 560–561
pethidine, *see* meperidine
phallotoxin 811
phenacetin
– anemia 751
– elimination half-life 848
– extraction 190
– GC 198
– HPLC 193
– plasma/serum concentration 848
phenadozone 870
phenampromide 870
phenarsazine chloride (Adamsite) 734
phenazocine 870
phenazone, *see also* antipyrine
– clinical interpretation 208
– elimination half-life 207, 848
– extraction 190
– GC 198
– GC–MS 110, 113, 117
– HPLC 110, 113, 117, 193
– medical assessment 207
– metabolism 208
– elimination half-life 848
– plasma/serum concentration 848
– serum concentration 207
phencyclidine (1-phenylcyclohexylpiperidine, PCP) 177, 499–502, 870
– clinical interpretation 502
– creatine kinase activity 750
– drug-induced seizure 750
– elimination half-life 848
– GC–MS 110, 113, 117, 500
– HPLC 110, 113, 117, 500
– immunoassay 499
– intoxication 502
– medical assessment 502
– metabolism 500
– minimal concentration to be detected 75
– plasma/serum concentration 502, 848
– TBPE test 179
– toxic hemorrhagic disorder 751
phendimetrazine 872
phenformin
– elimination half-life 848
– plasma/serum concentration 848
pheniramine
– elimination half-life 848
– plasma/serum concentration 848
– TBPE test 179
phenmedipham
– TLC 568–572
phenmetrazine 872
phenobarbital 345, 874
– antidote 754, 859
– clinical interpretation 297, 347
– drug-induced seizure 750
– elimination half-life 347, 848
– GC–MS 110, 113, 117, 292–294
– HPLC 110, 113, 117, 289
– immunoassay 288, 345–346
– medical assessment 297, 347
– metabolism 297, 346
– pharmacokinetics 347
– plasma/serum concentration 848
– quality assurance 64
– serum concentration 347
phenol
– drug-induced seizure 750
– hepatotoxicity 749
phenomorphan 870
phenoperidine 870
phenothiazine neuroleptics
– antidote 858
– GC-MS 158
phenprocoumon 303
– elimination half-life 848
– GC–MS 113, 117
– half-life 306
– HPLC 113, 117, 305
– pharmacodynamics 307
– pharmacokinetics 306
– plasma/serum concentration 848
– serum concentration 307
phenpromethamine 870
phentermine 874
– elimination half-life 848
– plasma/serum concentration 848
phenylbutazone
– anemia 751
– elimination half-life 848
– extraction 190
– GC 198
– HPLC 136, 193
– plasma/serum concentration 848

p-phenylenediamine
– creatine kinase activity 750
phenylephrine
– elimination half-life 848
– plasma/serum concentration 848
phenylhydrazine
– hepatotoxicity 749
phenylphosphonic acid (PhPA) 689
phenylpropylamine
– creatine kinase activity 750
phenytoin 298
– clinical interpretation 298
– drug-induced seizure 750
– elimination half-life 848
– GC–MS 110, 113, 117, 292–294
– HPLC 110, 113, 117, 289
– immunoassay 288
– medical assessment 298
– metabolism 298
– plasma/serum concentration 848
– quality assurance 64
– serum concentration 295
Philodendron spp. 786–789, 796
pholcodine 872
Pholiotina (Conocybe) cyanopus 813
Pholiotina (Conocybe) filaris 811
phosalone
– blood/plasma/serum concentration 597
– TLC 594
– toxicity 597
phosdrin, *see* mevinphos
phosgene 730
– albumin adduct 731
– globin adduct 730
– LC–MS 731–732
– LC–MS/MS 730–731
– metabolite 730
phosphamidon
– TLC 594
phosphonic acid ester 690
– GC–MS 695–699
– LC–MS 690–694
phosphorus
– hepatotoxicity 749
– liver toxin 748
phoxim
– TLC 594
Physalia physalis 826
physostigmine
– antidote 859
– drug-induced seizure 750
– elimination half-life 848
– plasma/serum concentration 848
Phytolacca americana (Virginian poke) 789, 797
piminodine 870
pimozide
– GC–MS 408
pinacolyl methylphosphonic acid (PMPA) 689
– CE-UV 694
– GC–FPD 697
– GC–MS 697
– GC–MS/MS 698
– GC–NICI-MS 697
– GC–NICI-MS/MS 697
– ion chromatography 694
– LC–MS 694
– LC–MS/MS 694
pinazepam 874
– metabolism 360
pindolol
– elimination half-life 848
– HPLC 457
– plasma/serum concentration 848
pindone
– HPLC 305
pipamperone 420, 446
– acute poisoning 447
– clinical interpretation 447
– elimination half-life 848
– GC 405
– GC–MS 110, 113, 117, 408
– HPLC 110, 113, 117, 397
– medical assessment 446
– plasma/serum concentration 848
– serum concentration 420, 435
– TBPE test 179
Piper methysticum (kava) 799
Piper nigrum 797
piperazine
– elimination half-life 848
– plasma/serum concentration 848
pipradol 874
pirenzepine
– elimination half-life 848
– plasma/serum concentration 848
pirimicarb
– ADI value 574
– MAK value 574
– medical assessment 573
– TLC 568–572
pirimiphos-methyl
– ADI value 596
– MAK value 596
– TLC 594
piritramide 875

piroxicam
– elimination half-life 848
– plasma/serum concentration 848
plant, poisonous 785–806
plant alkaloids (volatile)
– GC-MS 158
Pluteus salicinus 813
plutonium
– antidote 859
poison
– detection 826
– diagnosis 745
poison hemlock (*Conium maculatum L.*) 785–786
Poison Information Center (PIC) 90
– telephone numbers 861–865
poisoning
– acid-base balance 752
– biochemical investigation 745–769
– creatine kinase activity 750
– diagnosis 96, 745
– diagnosis of poisoning caused by animals 825
– epidemiology 95
– fish 829
– frequency 1
– monitoring intoxicated patients 767
– plant 788–806
  – first diagnostic step 800
  – identification of plant 801
  – macroscopic identification 802
  – microscopic identification 803
– symptoms 97, 788
– toxic agent 97
– treatment 752
– venomous and poisonous animal 825–834
polonium
– antidote 859
polyene 804
potentiometry 46, 656–659, 661–663
practicability of toxicological analyses 59
prajmaline 177
– elimination half-life 848
– GC 274
– HPLC 273
– plasma/serum concentration 848
prajmalium bitartrate
– TBPE test 179
prazepam 875
– elimination half-life 848
– GC–MS 408
– metabolism 360
– plasma/serum concentration 848
prazosin
– elimination half-life 848
– plasma/serum concentration 848
prednisolone
– elimination half-life 848
– plasma/serum concentration 848
primaquine
– elimination half-life 848
– plasma/serum concentration 848
primidone 299
– elimination half-life 299, 848
– GC–MS 110, 113, 117, 292–294
– HPLC 110, 113, 117, 289
– immunoassay 288
– medical assessment 299
– metabolism 299
– plasma/serum concentration 848
– quality assurance 64
– serum concentration 295
probenecid
– elimination half-life 848
– plasma/serum concentration 848
procainamide
– clinical interpretation 281
– elimination half-life 848
– GC 110, 113, 117, 274
– HPLC 110, 113, 117, 273
– immunoassay 781
– medical assessment 281
– metabolism 281
– plasma/serum concentration 848
– serum concentration 281
– TBPE test 180
procaine
– elimination half-life 848
– plasma/serum concentration 848
– TBPE test 180
profenofos
– blood/plasma/serum concentration 597
– toxicity 597
proheptazine 870
proinsulin
– immunoassay 613–614
promazine
– GC–MS 408
– HPLC 136
promecarb
– TLC 572
promethazine 420, 447
– acute poisoning 447
– clinical interpretation 447
– elimination half-life 848
– GC 405
– GC–MS 110, 113, 117, 408
– HPLC 110, 113, 117, 136, 397

– medical assessment 447
– plasma half-life 447
– plasma/serum concentration 848
– serum concentration 420, 435
– TBPE test 179
propafenone
– clinical interpretation 282
– elimination half-life 282, 848
– GC 274
– GC–MS 110, 113, 117
– HPLC 110, 113, 117, 136, 273
– medical assessment 282
– metabolism 282
– plasma/serum concentration 282, 848
propallylonal
– elimination half-life 848
– plasma/serum concentration 848
1-propanol
– headspace-GC 172, 173
2-propanol (*iso*-propanol, propan-2-ol) 521, 700
– BAT value 520, 854
– blood concentration 521
– clinical interpretation 522
– creatine kinase activity 750
– GC–FID 700
– headspace-GC 173, 519
– medical assessment 521
– toxicity 522
properidine 870
propham
– TLC 568–572
propiram 872
propofol 387
– clinical interpretation 387
– elimination half-life 387, 849
– GC 375
– HPLC 371
– medical assessment 387
– plasma/serum concentration 849
– serum concentration 387
propoxur 564
– ADI value 574
– blood concentration 575
– clinical interpretation 575
– LD_{50} value 573
– MAK value 574
– medical assessment 573
– poisoning 574–575
– TLC 568–572
propoxyphene, *see also* dextropropoxyphene 218
– drug-induced seizure 750
– elimination half-life 849
– metabolism 218
– minimal concentration to be detected 75
– plasma/serum concentration 849
propranolol 460
– drug-induced seizure 750
– elimination half-life 849
– GC–MS 110, 113, 117
– HPLC 110, 113, 117, 136, 457
– metabolism 460
– plasma half-life 460
– plasma/serum concentration 461, 849
– TBPE test 179
propylene carbonate
– headspace-GC 172, 174
1,2-propylene glycol
– GC 534
i-propyl methylposphonic acid (IMPA) 689
– CE-UV 694
– GC–FPD 697–698
– GC–MS 697
– GC–MS/MS 698
– GC–NICI-MS 697
– GC–NICI-MS/MS 697
– ion chromatography 694
– LC–MS 694
– LC–MS/MS 694
propyphenazone
– elimination half-life 849
– extraction 190
– GC 198
– HPLC 136, 193
– plasma/serum concentration 849
protein precipitation 119, 138
protein–organophosphate complex 702
– LC–MS/MS 702
prothipendyl
– elimination half-life 849
– GC–MS 408
– plasma/serum concentration 849
– TBPE test 179
protriptyline
– elimination half-life 849
– plasma/serum concentration 849
– serum concentration 435
Prunus spp. 804, 806
pseudoephedrine
– elimination half-life 849
– plasma/serum concentration 849
psilocin 503, 813
– clinical interpretation 506
– GC–MS 504
– HPLC 503
– medical assessment 506
Psilocybe mexicana 503

Psilocybe spp. *813*
psilocybin 503, 870
– clinical interpretation 506
– GC–MS 504
– HPLC 503
– medical assessment 506
– metabolism 507
– syndrome 813
psychopharmaceuticals
– intoxication 142
psychotomimetic agent 732–733
Pyracantha spp. 787
pyramat
– TLC 569
pyranocoumarin
– HPLC 305
pyrazophos
– TLC 594
pyrethroids 599–603
– ADI value 602
– blood/plasma half-life 602
– clinical interpretation 603
– GC–MS 114, 599, 601
– HPLC 114
– intoxication 603
– LD_{50} value 602
– medical assessment 601
– metabolism 602
– metabolite 601
– toxicity 601, 603
– toxicodynamics 603
– toxicokinetics 602
pyridine
– headspace-GC 172, 173
– hepatotoxicity 749
pyridostigmine
– antidote 859
– elimination half-life 849
– plasma/serum concentration 849
pyrithyldione
– elimination half-life 849
– extractability 369
– GC 375
– HPLC 371
– plasma/serum concentration 849
pyrovaleron 873
pyrrolizidine alkaloids
– liver toxin 748

q
qualitative analysis 66–73
– Metrology of Qualitative Chemical Analysis (MEQUALAN) 68
– report 85
– standard 66f.
– validation 69
– uncertainty and unreliability 71–73
quality assessment
– external 65–73
– internal 63
quality assurance (QA) 63–75
– qualitative examination 66
– quantitative measurement 63
quantitative analysis
– report 85
quetiapine 421, 448
– clinical interpretation 448
– elimination half-life 448, 849
– GC–MS 408
– HPLC–MS 432
– medical assessment 448
– overdose 448
– plasma/serum concentration 849
– serum concentration 421, 435
quinidine 177
– clinical interpretation 283
– elimination half-life 283
– GC 274
– GC–MS 110, 114, 117
– HPLC 110, 114, 117, 273
– immunoassay 781
– medical assessment 283
– metabolism 284
– serum concentration 283
quinine
– TBPE test 179
3-quinuclidinyl benzilate (BZ) 732–733
– hydrolysis 733

r
racemethorphan 871
racemoramide 873
racemorphan 873
ranitidine
– elimination half-life 849
– plasma/serum concentration 849
rare elements
– AAS, ICP-MS, VM 159
rattlesnake 833
reboxetine
– GC–MS 408
β-receptor blocking agents, *see* beta-receptor blocking agents
red blood cell acetylcholinesterase (RBC-AChE)
– variability 763
remifentanil 875
report, analytical toxicological 81
– delivery 86

– identification 81
– qualitative results 85
– quantitative results 85
request 100
reserpine
– anemia 751
Rhododendron ponticum 786–787
Ricinus communis L. (castor oil plant) 785, 804
risperidone
– elimination half-life 849
– GC–MS 111, 117, 409
– HPLC 111, 117
– plasma/serum concentration 849
rodenticide coumarin derivatives 309
– HPLC 309
– pharmacokinetics 310
– screening method 309
– serum concentration 311
rodenticides
– antidote 859
rolicyclidine 871
Russian VX (VR) 683
– GC–MS 687
Russula species 814, 819

S
salbutamol
– poisoning 753
salicylamide 212
– extraction 190
– GC 198
– HPLC 193
salicylate 212
– acid-base balance 752
– anemia 751
– antidote 754, 859
– clinical interpretation 213
– creatine kinase activity 750
– elimination half-life 212
– GC-MS 111, 114, 117, 161
– HPLC 111, 114, 117
– immunoassay 211
– medical assessment 212
– metabolism 209
– photometry 209
– poisoning 753
– serum concentration 212
– toxic hemorrhagic disorder 751
salicylic acid 209
– elimination half-life 849
– GC 198
– HPLC 193
– metabolism 209
– plasma/serum concentration 849
Sambucus nigra (elderberry) 806
saponins 794–797, 804
sarin (GB) 683
– 2D-GC 688
– antidote 859
– GC–MS 687–688
– GC–NPD 688
– HPLC 688
– LD_{50} value 681
– radiometric detection 688
– toxicity 681
saxitoxin 828
Schefflera spp. 787
Schrader formula 592
scombrotoxism 830
– toxin detection 830
scopolamine 791
– elimination half-life 849
– plasma/serum concentration 849
scorpionfish (*Scorpaena* spp.) 827
scorpions 831
– venom detection 831
sea snake 833
sea urchin 827
seafood poisoning 828
secbutabarbital 873
– elimination half-life 849
– plasma/serum concentration 849
secobarbital 875
– elimination half-life 849
– immunoassay 346
– plasma/serum concentration 849
sedatives 351–364, 367–391
– acid-base balance 752
– clinical interpretation 380
– GC 372
– HPLC 368
– ion-selective electrode 376
– medical assessment 380
– photometry (color test) 377
seizure
– drug-induced 750
selenium
– hepatotoxicity 749
sence 479
sertindole
– GC–MS 111, 114, 117
– HPLC 111, 114, 117
sertraline 421, 449
– clinical interpretation 449
– elimination half-life 449, 849
– GC–MS 111, 114, 117, 409
– HPLC 111, 114, 117
– medical assessment 449

– overdosage 449
– plasma/serum concentration 849
– serum concentration 421, 435
shellfish 828
– toxin detection 828
sildenafil
– GC–MS 117
– HPLC 117
silver
– nephrotoxicity 749
silver sulfide electrode 47
smok(e) 479
snake 832
snake toxin
– toxic hemorrhagic disorder 751
sodium warfarin 303
Solanum spp. 787
solid-phase micro extraction (SPME) 163
solvent 511–555
soman (GD) 683
– acetylcholinesterase (AChE) activity 688
– antidote 859
– enzymatic assay 688
– GC–MS 687–688
– GC–MS/MS 688
– GC–NPD 688
– immunoassay 688
– LD_{50} value 681
– TLC 688
– toxicity 681
Sorbus aucuparia L. 804
Sorghum spp. 806
sotalol 461
– elimination half-life 849
– GC–MS 111, 114, 117
– HPLC 111, 114, 117, 457
– plasma half-life 461
– plasma/serum concentration 461, 849
space 479
sparteine 177
L-sparteine
– TBPE test 179
spasmolytics
– antidote 859
Spathiphyllum floribundum (peace lily) 789, 797
spectrophotometry 625, 628
spice 479
spider 831
spider toxin
– creatine kinase activity 750
spironolactone
– elimination half-life 849
– plasma/serum concentration 849
stibine
– antidote 859
stilbine
– anemia 751
stimulants/hallucinogen
– GC-MS 158
stonefish (*Synanceja* spp.) 827
strategy of investigations
– brain death 104
– differential diagnosis 103
– drug abuse 104
– materials 96
– monitoring 103
– suspected poisoning
streptomycin
– elimination half-life 849
– plasma/serum concentration 849
Stropharia coronilla 503, 813
strychnine 642, 806
– antidote 646
– clinical interpretation 646
– creatine kinase activity 750
– drug-induced seizure 750
– elimination half-life 849
– GC–MS 114
– HPLC 114
– medical assessment 645
– plasma elimination half-life 645
– plasma/serum concentration 849
– poisoning 643
– serum concentration 645
– TBPE test 179
– TLC 643
– toxicity 645
Strychnox nux vomica 806
styrene
– BAT value 854
– headspace-GC 172, 174
– hepatotoxicity 749
succinylcholine
– creatine kinase activity 750
sufentanil 875
– elimination half-life 849
– plasma/serum concentration 849
sulfadiazine
– HPLC-DAD 146
sulfamethoxazole
– HPLC 136
sulfapyridine
– HPLC 136
sulfhemoglobin 630, 632
– spectrometry 632
1,1′-sulfonyl-bis[2-*S*-(*N*-acetylcysteinyl)ethane]
– ESI-LC–MS 708

– LC–MS 715
sulfonylurea 615
– blood/plasma/serum concentration 620
– clinical interpretation 621
– medical assessment 620
sulforidazine
– elimination half-life 850
– plasma/serum concentration 850
sulfotepp 592
– TLC 594
sulfur mustard (HD, 2,2′-bis-chloroethyl sulfide) 703
– albumin adduct 721
– antidote 859
– DNA adduct 722
– GC 705
– GC–MS 706, 719
– hemoglobin adduct 719
– LC–MS 722
– LC–MS/MS 721
– LD_{50} value 681
– metabolism 707, 716
– metabolite 706
– protein adduct 718–722
– toxicity 681
sulindac
– elimination half-life 849
– plasma/serum concentration 849
sulpiride
– elimination half-life 849
– GC–MS 409
– plasma/serum concentration 849
– TBPE test 179
sultiame
– elimination half-life 849
– plasma/serum concentration 849
suramin
– elimination half-life 849
– plasma/serum concentration 849
surfactant
– antidote 859
sympathomimetic drug
– drug-induced seizure 750
systematic toxicological analysis (STA) 103

t

tabun (GA) 683
– antidote 859
– GC–MS 687
– hydrolysis 691
– LD_{50} value 681
– toxicity 681
tacrolimus
– elimination half-life 849
– plasma/serum concentration 849
talinolol
– elimination half-life 850
– GC–MS 111, 114, 117
– HPLC 111, 114, 117, 457
– metabolism 462
– plasma half-life 461
– plasma/serum concentration 461, 850
tannic acid 794
tartrate
– nephrotoxicity 749
Taxus spp. 787
TBPE 175, 177
– deprotonated 177
– protonated 177
TDM (therapeutic drug monitoring) 775–782
– analytical method 780
– blood concentration 778
– interpretation 777
– pharmacokinetics 776
teicoplanin
– elimination half-life 850
– plasma/serum concentration 850
temazepam 875
– elimination half-life 850
– HPLC 355
– HPLC-DAD 143
– metabolism 360
– plasma/serum concentration 850
temephos
– TLC 594
tenamfetamine 871
tenocyclidine 871
tenoxicam
– elimination half-life 850
– plasma/serum concentration 850
terbufos
– TLC 594
terbutaline
– elimination half-life 850
– plasma/serum concentration 850
terfenadine
– elimination half-life 850
– plasma/serum concentration 850
terpene 804
1,1,2,2-tetrabromoethane
– Fujiwara reaction 543
– GC artifact in headspace-GC 172, 174
– headspace-GC 172, 174
tetrabromomethane (carbon tetrabromide)
– Fujiwara reaction 542
– hepatotoxicity 749
tetrabromophenolphthalein ethyl ester, *see* TBPE

tetracaine
– TBPE test 179
1,1,2,2-tetrachloroethane
– Fujiwara reaction 542–543
– hepatotoxicity 749
– liver toxin 748
– nephrotoxicity 749
1,1,2,2-tetrachloroethene (tetrachloroethylene, perchloroethylene)
– BAT value 549
– clinical interpretation 551–552
– Fujiwara reaction 542–543
– headspace-GC 548
– hepatotoxicity 749
– medical assessment 548, 550
– nephrotoxicity 749
tetrachloromethane (carbon tetrachloride)
– BAT value 549, 551, 853, 854
– clinical interpretation 551, 553
– headspace-GC 172, 173, 548
– hepatotoxicity 749
– intoxication 553
– liver toxin 748
– MAK value 551
– medical assessment 548, 551
– nephrotoxicity 749
– toxic hemorrhagic disorder 751
tetrachlovinphos
– TLC 594
tetracycline
– elimination half-life 850
– plasma/serum concentration 850
tetraethyl lead
– BAT value 854
Δ^9-tetrahydrocannabinol (THC) 472
– elimination half-life 850
– plasma/serum concentration 850
tetrahydrofuran
– BAT value 854
– headspace-GC 172, 173
tetramethyl lead
– BAT value 854
tetrazepam 875
– elimination half-life 850
– plasma/serum concentration 850
thalidomide
– elimination half-life 850
– plasma/serum concentration 850
thallium 664
– absorption spectrophotometry 668–671
– antidote 672, 859
– atomic absorption spectrometry 664
– elimination half-life 672
– inverse voltammetry 665–668
– liver toxin 748
– medical assessment 671
– poisoning 672
– toxicity 671
thebacon 873
thebaine
– TBPE test 179
theobromine
– elimination half-life 850
– plasma/serum concentration 850
theophylline
– acid-base balance 752
– clinical interpretation 315
– elimination half-life 850
– GC–MS 111, 114, 117, 315
– half-life 315
– HPLC 111, 114, 117, 315
– immunoassay 315, 782
– medical assessment 315
– pharmacokinetics 315
– plasma/serum concentration 850
– poisoning 753
– quality assurance 64
– serum concentration 315
thiazide diuretics
– acid-base balance 752
thin layer chromatography (TLC) 29–32
– detection 30
– limitation 31
– procedure 565–573
– toxic plant 805
thiobencarb
– TLC 572
thiocyanate
– elimination half-life 850
– plasma/serum concentration 850
thiodiglycol (TDG) 706
– blood plasma 709
– cleavage from blood protein 711
– GC–MS 708, 710–712
– GC–MS/MS 708
– urine 710
thiodiglycol-sulfoxide
– GC–MS 708, 714
– reduction to TDG 714
– urine 713
thiofanox
– TLC 568
thiometon
– ADI value 596
– LD_{50} value 596
– MAK value 596
thiopental
– clinical interpretation 350

– elimination half-life 350, 850
– GC 348
– GC–MS 111, 114, 118
– HPLC 111, 114, 118, 136, 347
– immunoassay 346
– medical assessment 350
– pharmacokinetics 350
– plasma/serum concentration 850
– serum concentration 350
thioridazine 421, 435, 449
– clinical interpretation 449
– elimination half-life 850
– GC 405
– GC–MS 111, 114, 118, 409
– HPLC 111, 114, 118, 136, 397
– intoxication 449
– medical assessment 449
– metabolism 450
– plasma half-life 449
– plasma/serum concentration 850
– serum concentration 421, 435
– TBPE test 179
thorn apple, *see Datura stramonium L.*
thorotrat (thorium dioxide)
– liver toxin 748
threshold limit value (TLV) 562
tiapride
– elimination half-life 850
– plasma/serum concentration 850
tilidine 257, 875
– antidote 858
– clinical interpretation 262
– elimination half-life 262, 850
– GC–MS 111, 114, 118, 259
– HPLC 111, 114, 118, 258
– medical assessment 261
– metabolism 258
– plasma/serum concentration 850
– serum concentration 261
– TBPE test 179
cis-tilidine 873
timolol
– elimination half-life 850
– HPLC 457
– plasma/serum concentration 850
tizanidine
– TBPE test 180
tobacco plant 639
tobramycin
– elimination half-life 850
– immunoassay 781
– plasma/serum concentration 850
tocainide
– elimination half-life 850
– GC 274
– HPLC 273
– plasma/serum concentration 850
tolbutamide 617
– blood/plasma/serum concentration 620
– elimination half-life 850
– LC–MS 619
– metabolism 617
– plasma/serum concentration 850
tolclofos-methyl
– TLC 594
toliprolol
– HPLC 457
– TBPE test 180
toluene
– acid-base balance 752
– BAT value 527, 854
– clinical interpretation 530
– creatine kinase activity 750
– headspace-GC 172, 173, 526
– hepatotoxicity 749
– intoxication 529
– medical assessment 529
– metabolism 529
toxalbumin 804
toxaphene 578, 582
– toxicokinetics 580
toxic plant 785–806
– investigation of biological material 805
– methods for detection 805
– substances 803
– TLC 805
Toxicodendron spp. 787
toxicology
– biochemical investigation 745–769
tramadol 262
– antidote 858
– clinical interpretation 263
– elimination half-life 263, 850
– GC–MS 111, 114, 118, 262
– HPLC 111, 114, 118
– immunoassay 262
– medical assessment 263
– metabolism 264
– plasma/serum concentration 850
– serum concentration 263
– TBPE test 180
tranylcypromine
– GC–MS 409
trazodone
– GC–MS 409
– HPLC 401
triamterene
– elimination half-life 850

– plasma/serum concentration 850
triazolam 875
– elimination half-life 850
– plasma/serum concentration 850
triazophos
– TLC 594
tribromoethene
– Fujiwara reaction 542–543
tribromomethane (bromoform)
– Fujiwara reaction 542–543
– headspace-GC 171, 174, 548
– hepatotoxicity 749
trichlorfon (trichlorphos)
– TLC 594
trichloroacetic acid
– Fujiwara reaction 542–543
trichloro-*tert*-butanol
– Fujiwara reaction 543
1,1,1-trichloroethane (methylchloroform)
– BAT value 549, 854
– clinical interpretation 551–552
– Fujiwara reaction 542–543
– headspace-GC 172, 173, 548
– MAK value 550
– medical assessment 548, 550
– metabolism 550
– toxicity 550, 552
1,1,2-trichloroethane
– clinical interpretation 551, 552
– MAK value 550
– medical assessment 548, 550
– toxicity 552
2,2,2-trichloroethanol
– headspace - GC 172, 174
trichloroethanol
– Fujiwara reaction 543
trichloroethene (trichloroethylene)
– clinical interpretation 551–552
– Fujiwara reaction 542–543
– half-life 549
– headspace-GC 548
– hepatotoxicity 749
– medical assessment 548–549
– metabolism 550
– poisoning 552
1,1,1-trichlorohydrocarbon
– Fujiwara reaction 540
trichloromethane (chloroform)
– BAT value 549, 551
– clinical interpretation 551–552
– Fujiwara reaction 542–543
– headspace-GC 548
– hepatotoxicity 749
– MAK value 551
– medical assessment 548, 551
– metabolism 551
– poisoning 552
α,α,α-trichlorotoluene
– Fujiwara reaction 542
trichlorphos (trichlorfon)
– TLC 594
Tricholoma flavovirens 813
triethanolamine (TEA) 724
triethylene glycol
– GC 534
trifluoperazine
– elimination half-life 850
– plasma/serum concentration 850
– TBPE test 179
trifluperidol
– GC 405
– HPLC 397
triflupromazine
– elimination half-life 850
– plasma/serum concentration 850
triiodomethane (iodoform)
– Fujiwara reaction 542–543
trimeperidine 871
trimethoprim
– TBPE test 179
trimethylchlorosilane (TMCS) 695
2,4,6-trimethylpyridine
– headspace-GC 172, 174
trimipramine 422, 432, 450
– clinical interpretation 450
– elimination half-life 850
– GC 405
– GC–MS 111, 114, 118, 409
– HPLC 111, 114, 118, 136, 397, 401
– immunoassay 781
– intoxication 451
– medical assessment 450
– overdosage 451
– plasma half-life 450
– plasma/serum concentration 850
– serum concentration 422, 435
– TBPE test 179
trinitrotoluene
– liver toxin 748
triterpene 797
tryptophan
– GC–MS 409

u

n-undecane
– headspace-GC 172, 174
uranium
– nephrotoxicity 749

v
valproate
– GC–MS 111, 114
– HPLC 111, 114
– immunoassay 288
valproic acid 300
– clinical interpretation 300
– elimination half-life 300, 850
– immunoassay 781
– medical assessment 300
– metabolism 299
– plasma/serum concentration 850
– quality assurance 64
vancomycin
– elimination half-life 850
– immunoassay 781
– plasma/serum concentration 850
venlafaxine 422
– elimination half-life 851
– GC–MS 111, 114, 118, 409
– HPLC 111, 114, 118
– plasma/serum concentration 851
– serum concentration 422
venom
– animal 826
verapamil 320
– clinical interpretation 324
– elimination half-life 851
– GC 322
– GC–MS 111, 114, 118, 320
– HPLC 111, 114, 118, 136, 320
– medical assessment 324
– plasma/serum concentration 851
– serum concentration 324
– TBPE test 179
vesicants 703–729
Vicia faba (broadbean) 800
vigabatrin
– elimination half-life 851
– plasma/serum concentration 851
viloxazine
– GC–MS 409
– TBPE test 180
vinylbital 873
– elimination half-life 851
– plasma/serum concentration 851
vinyl chloride
– liver toxin 748
Vipera spp. (viper) 832–833
Virginian poke 789, 797
vitamin A
– liver toxin 748
volatile halogenated hydrocarbons (VHHC) 539 – 553
– clinical interpretation 551
– color test 540
– GC 544–548
– medical assessment 548
– poisoning 551
voltammetry
– inverse 49, 665
VX 683
– antidote 859
– CIEIA (competitive inhibition enzyme immunoassay) 688
– GC–MS 687
– HPLC–ECD 688
– hydrolysis 690
– LC–MS (APCI, atmospheric pressure chemical ionization) 688
– LD_{50} value 681
– toxicity 681

w
warfarin
– antidote 859
– clinical interpretation 306
– elimination half-life 851
– GC–MS 111, 114, 118
– half-life 306
– HPLC 111, 114, 118, 305
– medical assessment 306
– pharmacodynamics 307
– pharmacokinetics 306
– plasma/serum concentration 851
– serum concentration 307
– toxicity 310–311
wasp toxin
– creatine kinase activity 750
weeping fig (*Ficus benjamina*) 794
weeverfish (*Echiichthys* spp.) 827
wood sorrel 804

x
m-xylene
– BAT value 527, 854
– clinical interpretation 531
– elimation half-life 529
– headspace-GC 172, 173, 526
– medical assessment 529
– metabolism 530
o-xylene
– BAT value 527, 854
– clinical interpretation 531
– elimation half-life 529
– headspace-GC 172, 174, 526
– medical assessment 529
– metabolism 530

p-xylene
– BAT value 527, 854
– clinical interpretation 531
– elimination half-life 529
– headspace-GC 172, 173, 526
– medical assessment 529
– metabolism 530
xylite
– nephrotoxicity 749

y

yohimbine
– elimination half-life 851
– HPLC-DAD 146
– intoxication 145
– plasma/serum concentration 851
– TBPE test 180
yucatan fire 479

z

zaleplon 388
– antidote 389
– clinical interpretation 388
– elimination half-life 388
– GC 375
– HPLC 371
– medical assessment 388
– metabolism 389
– serum concentration 388
Zantedeschia aethiopica 789, 798
zectram
– TLC 569
zidovudine
– elimination half-life 851
– plasma/serum concentration 851
zinc
– antidote 859
zipeprol 873
ziprasidone
– GC–MS 409
zolpidem 389, 875
– clinical interpretation 389
– elimination half-life 389, 851
– extractability 369
– GC 375
– GC–MS 111, 114, 118
– HPLC 111, 114, 118, 371
– medical assessment 389
– metabolism 390
– plasma/serum concentration 851
– serum concentration 389
zopiclone 390
– clinical interpretation 390
– elimination half-life 390, 851
– extractability 369
– GC 375
– GC–MS 111, 114, 118
– HPLC 111, 114, 118, 371
– HPLC-DAD 146
– intoxication 145
– medical assessment 390
– plasma/serum concentration 851
– serum concentration 390
zotepine
– elimination half-life 851
– GC–MS 409
– plasma/serum concentration 851
zuclopenthixol
– GC–MS 409